Smart Water Quality Monitoring

Daniel Gutiérrez • Pablo Millán •
Julian Blasco
Editors

Smart Water Quality Monitoring

Artificial Intelligence, Automation and Analytical Chemistry

Springer

Editors
Daniel Gutiérrez
Electronic Engineering Department
University of Seville
Sevilla, Sevilla, Spain

Pablo Millán
Engineering Department
Universidad Loyola Andalucía
Sevilla, Sevilla, Spain

Julian Blasco
Department of Ecology and Coastal Management
Institute of Marine Sciences of Andalusia (CSIC)
Puerto Real (Cádiz), Spain

ISBN 978-3-032-03048-1 ISBN 978-3-032-03049-8 (eBook)
https://doi.org/10.1007/978-3-032-03049-8

This Springer imprint is published by the registered company Springer Nature Switzerland AG
The registered company address is: Gewerbestrasse 11, 6330 Cham, Switzerland

Preface

This book presents innovative approaches to monitoring water ecosystems using state-of-the-art techniques, methodologies and technologies, including data analytics, artificial intelligence, the Internet of Things (IoT), sensor systems and robotics, to name a few.

Water is fundamental to human beings and to the planet's ecosystems. Unfortunately, water ecosystems are under threat from various natural and human-caused factors. For example, many bodies of water around the world are currently suffering from eutrophication-related contamination problems. Furthermore, water acidification is a current issue associated with the large quantity of CO_2 in the atmosphere, which has a negative impact on water ecosystems. Plastics, microplastics, and emerging pollutants are also affecting the quality of freshwater sources. Furthermore, most climate change projections predict a generalised reduction in freshwater resources. Therefore, to ensure an adequate supply of clean water, it is essential to protect and monitor water resources effectively. It is also vital to have monitoring systems in place to supervise water ecosystems, improve their quality and avoid accidents that could damage them.

This book is intended for researchers and practitioners in the field of water resources monitoring, as well as PhD students and anyone else interested in the subject. Adopting a multidisciplinary and interdisciplinary approach, the book addresses water body monitoring problems through various analytical and data-driven techniques, including artificial intelligence and deep and machine learning.

The book is divided into 15 chapters that explore novel approaches to water quality monitoring. Chapters "Data Science and Public Policies: Towards Water Security", "Sources and Effects of Water Contamination: Characteristics and Ecological Implications", and "Use of Omics Techniques for Assessing Water Quality: Omics to Evaluate Water Quality" are dedicated to the fundamentals of water quality monitoring. Chapters "Hyperspectral Technology to Monitor Marine Pollution", "Machine and Deep Learning Approaches for Water Pollution Detection Using Hyperspectral Imaging", "Intelligent Real-Time Anomaly Detection for Optimization of Water Monitoring Systems", "Smart Sensors for Water Quality Monitoring in Aquaculture Systems", "Drought Impacts on Hydrological Ecosystem Services: Indicators and Methodological Processes", and "Miniaturised (Bio)sensors in Aquatic Environmental Monitoring" then contain novel sensing technology and water management approaches. The final few chapters (Chapters "Autonomous

Surface Vehicle for Water Monitoring Using Artificial Intelligence Methodologies", "From Concept to Control: Development of an Advanced ASV Platform for Testing", "Model-Based Online Planning Strategy for Environmental Disaster Scenarios with Autonomous Vehicles", "Unmanned Underactuated Surface Vehicle Formation Control Using Deep Reinforcement Learning", "Multitask Multiagent Deep Reinforcement Learning for a Fleet of Autonomous Surface Vehicles in Environmental Cleanup Missions", and "Deep Reinforcement Learning and Informative Path Planning: Diving into the Cooperation of Heterogeneous Aquatic Surface Vehicles") cover AI-based applications of autonomous systems for water monitoring. A brief summary of each chapter can be found below.

Chapter "Data Science and Public Policies: Towards Water Security" discusses the application of data science and AI to challenges relating to water security and environmental monitoring in smart cities. It emphasises the importance of processing heterogeneous data and the role of data collection and analysis in achieving sustainable development goals and water resource policies.

Chapter "Sources and Effects of Water Contamination: Characteristics and Ecological Implications" examines the origins and consequences of water contamination, including its ecological, social and economic impacts. It discusses sources such as heavy metals, POPs (persistent organic pollutants), nutrient runoff and microplastics, emphasising their detrimental effects on biodiversity and ecosystem services, and the importance of monitoring and data-driven decisions for conservation.

Chapter "Use of Omics Techniques for Assessing Water Quality: Omics to Evaluate Water Quality" reviews the application of omics techniques (genomics, transcriptomics, proteomics and metabolomics) to assess water quality and detect the effects of pollution by analysing biomolecules, particularly those from microbial communities. It discusses metagenomic analysis of environmental DNA (eDNA) and highlights its potential and the associated challenges, including the role of AI, machine learning and deep learning (AI/ML/DL) in overcoming database limitations.

Chapter "Hyperspectral Technology to Monitor Marine Pollution" discusses hyperspectral technology for monitoring marine pollution. This chapter explores hyperspectral imaging (HSI) for monitoring marine pollutants such as oil spills and plastics. It contrasts HSI with other sensing methods and discusses spectral indices and AI techniques for detection. It also addresses the challenges posed by large volumes of data and highlights the potential of miniaturised sensors and cloud computing.

Chapter "Machine and Deep Learning Approaches for Water Pollution Detection Using Hyperspectral Imaging" explores machine learning and deep learning for hyperspectral image classification, particularly for oil spill detection. The chapter covers data acquisition and pre-processing, as well as dimensionality reduction techniques such as principal component analysis (PCA) and autoencoders. It also evaluates the performance of deep neural network approaches for classifying water and hydrocarbon samples.

Chapter "Intelligent Real-Time Anomaly Detection for Optimization of Water Monitoring Systems" presents a novel CEP-based software architecture for real-time anomaly detection in water monitoring systems that leverages IoT data. It describes the components of the architecture, the event processing patterns and demonstrates the efficiency and stability of the system through performance tests.

Chapter "Smart Sensors for Water Quality Monitoring in Aquaculture Systems" discusses smart sensors for water quality monitoring (WQM) in aquaculture systems that align with sustainability goals. It describes different types of aquaculture, how technology such as machine vision and sonar is used for monitoring, and the key WQM parameters. It also covers the hardware and communication systems (e.g. sensors, microcontrollers, WSNs, 5G and LoRaWAN) used for real-time data collection, and highlights the benefits and challenges of pollutant sensing.

Chapter "Drought Impacts on Hydrological Ecosystem Services: Indicators and Methodological Processes" focuses on understanding the impact of drought on hydrological ecosystem services (HES) and proposes an integrated management approach using physical and socio-economic indicators. It classifies drought types, discusses the limitations of traditional indicators and advocates multidimensional indicators and multicriteria decision-making (MCDM) techniques for a more comprehensive assessment and adaptive strategies.

Chapter "Miniaturised (Bio)sensors in Aquatic Environmental Monitoring" reviews miniaturised sensors, including microfluidic paper-based analytical devices (μPADs) and electrochemical biosensors, for detecting heavy metals and pesticides in water. The chapter discusses their advantages and the different detection techniques involved, as well as the potential integration of these sensors with the Internet of Things (IoT) and machine learning for portable, real-time monitoring.

Chapter "Autonomous Surface Vehicle for Water Monitoring using Artificial Intelligence Methodologies" introduces a real-world ASV prototype equipped with advanced sensors and AI capabilities for continuous water monitoring. It details the ASV's hardware and software components, as well as the AI techniques employed for tasks such as macroplastic detection and adaptive path planning. The chapter presents experimental validation of the ASV's capabilities.

Chapter "From Concept to Control: Development of an Advanced ASV Platform for Testing" details the development of control algorithms for ASVs, with a focus on the Yellowfish ASV platform and its hardware. It presents the implementation of a path-following control strategy using the open-source Robot Operating System 2 (ROS 2), and describes the design of waste collection tools and experimental validation in real-world conditions.

Chapter "Model-Based Online Planning Strategy for Environmental Disaster Scenarios with Autonomous Vehicles" proposes the use of Gaussian processes (GPs) to model physicochemical variables, as well as the CMA-ES algorithm, to plan optimal ASV sampling trajectories in environmental disaster scenarios. The chapter explains how GPs quantify uncertainty for real-time decision-making and how CMA-ES adapts to optimise information gain through receding horizon control, as validated in simulations.

Chapter "Unmanned Underactuated Surface Vehicle Formation Control Using Deep Reinforcement Learning" focuses on achieving autonomous operation and formation control for multiple USVs using deep reinforcement learning (DRL), specifically proximal policy optimization (PPO). The chapter defines the DRL framework, state/action spaces and reward functions, enabling coordinated movement without the need for precise hydrodynamic models. It also presents simulation results.

Chapter "Multitask Multiagent Deep Reinforcement Learning for a Fleet of Autonomous Surface Vehicles in Environmental Cleanup Missions" presents a multi-task, multi-agent deep reinforcement learning framework for a fleet of ASVs cleaning environmental waste. The mission is decoupled into exploration and cleaning phases. It uses a parameter-sharing multi-agent multi-task deep Q-network (PSMA-MDQN) and a multi-objective approach to optimise coordination and explore the trade-offs between the two phases in different environments.

Chapter "Deep Reinforcement Learning and Informative Path Planning: Diving into the Cooperation of Heterogeneous Aquatic Surface Vehicles" analyses the use of deep reinforcement learning (DRL) and informative path planning (IPP) for a heterogeneous fleet of ASVs (scouts and cleaners) cooperating in plastic debris clean-up. It details the DRL methodology and network architecture and examines how configurations of the reward and observation functions impact inter-team cooperation and mission success.

Sevilla, Spain — Daniel Gutiérrez
Sevilla, Spain — Pablo Millán
Puerto Real (Cádiz), Spain — Julian Blasco

Acknowledgements The REINFORCED research network, Research Network on Autonomous Vehicles and Artificial Intelligence for Water Resources Monitoring, has motivated this book. The network was funded by the University of Seville's Research Plan, through the I.10 call for grants to support the formation and consolidation of thematic networks.

Contents

Data Science and Public Policies: Towards Water Security

Regina León-Ovelar

Contents

Abstract

A key pillar of sustainable development is water security as it has socio-economic and environmental impacts. This chapter presents the fundamental role of data science and public policy in ensuring equitable access to water with an emphasis on the connection to the United Nations Sustainable Development Goal 6 (SDG 6). A multidisciplinary view highlights how technological advances, including big data, artificial intelligence (AI) and the Internet of Things (IoT), are transforming water governance. UNESCO (2022) IHP-IX: strategic plan of the intergovernmental hydrological programme: science for a water secure world in a changing environment, ninth phase 2022-2029 is seen as a strategic framework for scientific collaboration and policy development to address global

R. León-Ovelar (✉)
Departamento de Ingeniería Industrial, Facultad de Ingeniería de la Universidad Nacional de Asunción, San Lorenzo, Paraguay
e-mail: rleon@ing.una.py

D. Gutiérrez et al. (eds.), *Smart Water Quality Monitoring*,
https://doi.org/10.1007/978-3-032-03049-8_1

water challenges. To mitigate climate-related risks such as floods, droughts and water pollution, the integration of predictive models driven by AI, remote sensing and digital water management systems is crucial. In addition, the chapter stresses the importance of addressing data gaps, fostering international cooperation and strengthening science-based water governance. Data transparency, open-access platforms and digital monitoring tools have been found to improve decision-making processes, contributing to sustainable water security. By harnessing new technologies and interdisciplinary research, policymakers can develop resilient strategies to address the growing complexities of integrated water resources management.

Keywords

Water security · Data science in water management · Artificial intelligence (AI) and hydrology · Water governance

Introduction

Water as a Human Right

The 17 Sustainable Development Goals (SDGs) each have their own targets and implementation pathways, but which of them is the most crucial? Are some SDGs more important than others? The answer to this question can be complex when considering each goal in isolation. However, it can be asserted that without SDG 6: Clean Water and Sanitation, life itself is at such significant risk that the achievement of other goals becomes secondary.

According to Resolution 64/292 of the United Nations General Assembly (2010), access to safe drinking water and sanitation is recognised as a fundamental human right, serving as the foundation for the realisation of other human rights. Furthermore, the resolution urges states and institutions to provide financial resources, strengthen technical capacity and facilitate the transfer of technology and knowledge to ensure equitable access to these essential services, particularly in developing countries. Figure 1 shows globally how SDG 6 has an impact on the other SDGs and thus ends up affecting social, economic, environmental and institutional aspects of a society.

Figure 1 was developed using a qualitative literature-based approach, based on a review of global policy documents and scientific literature. Sources include mostly UN reports, UN-Water, and the World Health Organisation (WHO). The graphic is a conceptual synthesis based on data that systematically highlights SDG 6 and its impact on health, education, gender equality, economic growth, climate resilience and institutional capacity. The interrelationships represented are not merely thematic; they reflect functional dependencies, where progress in water and sanitation contributes directly or indirectly to outcomes in other areas. They

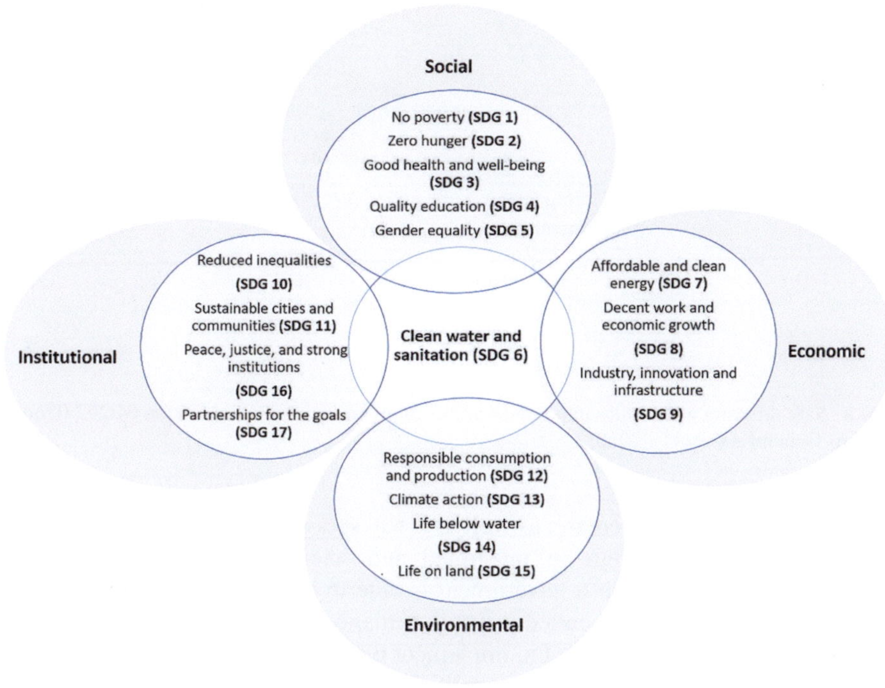

Fig. 1 Impact of SDG 6 on the other SDGs

seek to show the strategic influence of SDG 6, reinforcing its position as a catalyst for sustainable development across the 2030 Agenda.

MacAlister et al. (2023), in the report Global Water Security Assessment 2023, mention that, worldwide, more people die from lack of safe drinking water and sanitation than from the consequences of disasters, which is consistent with the global increase in mortality rates attributed to lack of water, sanitation and thus minimum hygiene conditions.

After air, water is the most vital element for life. Ensuring equitable, safe and high-quality access to water for all has been established as a fundamental human right. However, this remains largely a utopian ideal, as it is not fully realised even in developed countries. This situation is further exacerbated in developing and underdeveloped nations, where disparities in water availability and quality continue to grow. The World Health Organisation (WHO) estimates that approximately two million deaths per year are attributable to unsafe water conditions (Kekes et al. 2023). Moreover, by 2025, an estimated 3 billion people will lack access to potable water, underscoring the urgent need for immediate intervention (Ulusoy et al. 2024).

It is important to consider a socio-economic balance since water is not only the vital liquid for life but also the raw material for economic development of a country or region, and inequity in access to water will be a growing problem as population

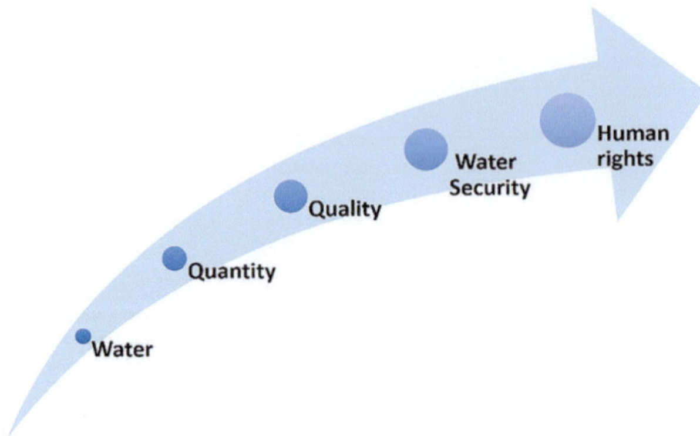

Fig. 2 Safe drinking water and sanitation as pillars of all human rights Resolution 64/292 (United Nations General Assembly, 2010)

and industrial activities increase, as well as urban sprawl. Countries need to address water management through an integrated and science-based policy approach. Mishra et al. (2021) add: if a government is able to manage water efficiently and equitably, it will achieve so-called water governance. This governance depends on how long it takes the government to implement measures that ensure the availability of water in adequate quantity and quality, providing the population with water security and guaranteeing such a basic human right as access to water (Fig. 2).

Kekes et al. (2023) examined the differences between surface and groundwater, emphasising how water sources impact availability and treatment processes. Their findings highlight the critical importance of water resource management, as nearly 1 billion people worldwide currently lack access to clean drinking water.

The pressure on water resources has become critical, driven by various sectors, including domestic consumption, industrial use, and horticultural production. This has led to an increasing competition for water, necessitating the sustainable management of this resource, particularly in regions already experiencing socio-environmental vulnerability (Ferreira et al. 2024). The inequity in access to safe water is particularly pronounced in low-income countries and rural areas, significantly increasing health risks. While the global rate of access to water supply services has improved from 70% to 74% in recent years, this progress remains insufficient. To meet global demand, the expansion rate must quadruple, while in developing countries, it must increase tenfold, and in highly water-vulnerable contexts, it must multiply by 23 (Lopes et al. 2022).

In the agricultural sector, soil properties have traditionally been the primary determinant for selecting irrigation methodologies. However, in the near future, decision-making will be based primarily on water accessibility, which will, in turn, be closely linked to crop production costs (Ferreira et al. 2024). Fruits, vegetables, and ornamental plants play a significant role in the global economy, with their production on an upward trend. However, this sector is highly water-intensive and faces growing challenges due to the effects of climate change and water scarcity. These factors highlight unequal access to water, forcing greater reliance on irrigation, particularly in semiarid and arid regions (Ferreira et al. 2024).

The demand for water has increased significantly due to population growth and industrialisation. This overexploitation of water resources represents a severe threat to this finite and vital natural resource (Ulusoy et al. 2024). Population growth has also driven the expansion of horticultural activities, which not only intensifies water extraction but also increases contamination from agrochemical residues. Therefore, effective water management tools and eco-friendly food production methods must be implemented (Ferreira et al. 2024).

The adoption of sustainable water treatment technologies plays a crucial role in ensuring water sustainability, as the problem is directly linked to anthropogenic pressures. Under traditional treatment methods, wastewater treatment alone accounts for 3–4% of global energy consumption and contributes 300 million tons of carbon emissions annually (Ulusoy et al. 2024).

A wastewater treatment plant is a complex plant typically used to remove organic matter, nitrogen and phosphorus from polluted water from domestic, industrial and agricultural activities. Treatment processes are usually organised in three stages: primary (or mechanical) treatment, secondary (or biological) treatment and tertiary (chemical) treatment (Selișteanu et al. 2021).

In the Global Water Security 2023 Assessment report, MacAlister et al. (2023) assessed indicators concerning water treatment for 186 countries finding that 26 countries (14%) have zero value for their water treatment, and the other countries have values in various ranges of wastewater treatment percentage. Figure 3 shows these results with respect to the level of water treatment in the countries assessed.

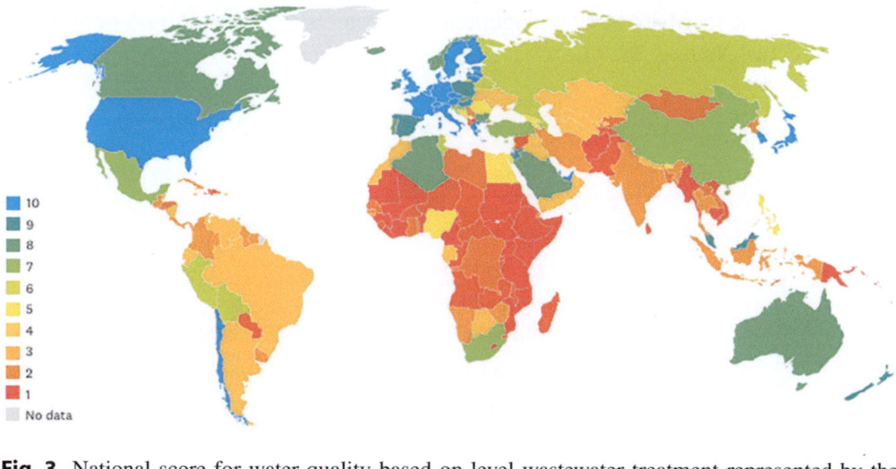

Fig. 3 National score for water quality based on level wastewater treatment represented by the proportion of safely managed household wastewater (Global Water Security 2023 Assessment report)

MacAlister et al. (2023) consider that the data reported as the 'domestic' part of inDicator 6.3. 1 (treated wastewater) in 2020 are estimates of flows generated and safely treated using a combination of nationally reported data, and, in their absence, the analysis of domestic wastewater is based on data from international organisations such as the Organisation for Economic Co-operation and Development (OECD), the United Nations Environment Programme (UNEP) or Eurostat, the United Nations Environment Programme (UNEP) or Eurostat, as well as data collected directly from national statistical agencies, regulators, line ministries, utilities and the JMP (UN-Habitat and WHO 2021). Estimates of domestic wastewater treatment refer to 2020; however, the components of the estimate are derived from data from several years, including the most recent data available.

In developed countries, domestic wastewater treatment is assumed to be already implemented, research is currently focused on the methodologies to be used and optimisation is sought as well as environmentally friendly treatments; however, as can be seen in the figure above, there are still countries with serious problems associated with the treatment of their water, which represents a risk to soil, water and human health due to the presence of multiple types of pollutants. In this context, it is possible to foresee that the sustainable development goals for 2030 as well as the longed-for water justice will not be achieved and will continue to be a pending issue with its foreseeable socio-economic consequences.

Understanding Water Security

The concept of water security encompasses both the quantity and quality of water while considering human health, ecosystems, and economic productivity, Fig. 4. Additionally, it involves risk management to protect both people and economies (Van Beek and Arriens 2014).

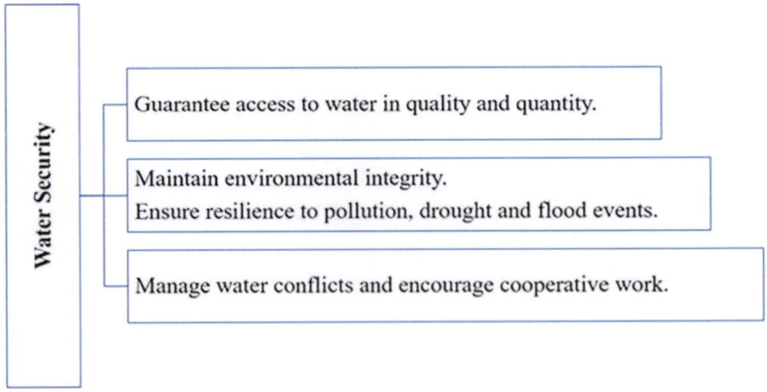

Fig. 4 Water security goals (Adapted from Van Beek and Arriens 2014)

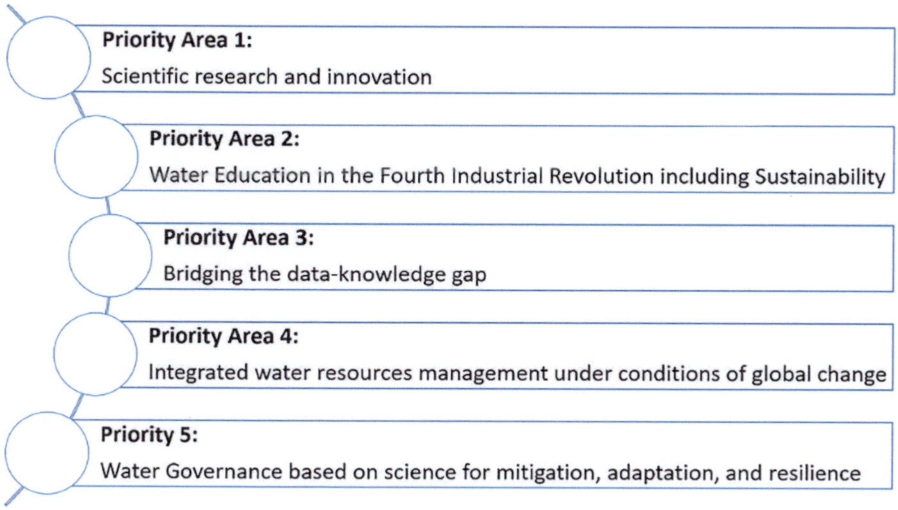

Fig. 5 Priority areas set out in the IHP-IX Operational Implementation 2022–2029

The term 'water security' can be cross-cutting or complex because it links multiple areas or sectors. UN-Water refers to water security as 'The ability of people to protect sustainable access to water in adequate quantity but also in quality to sustain livelihoods while considering human well-being and socio-economic development; to ensure protection against water pollution and water-related disasters; and to preserve ecosystems in a climate of peace and political stability' (MacAlister et al. 2023).

UNESCO has been developing programmes and plans aimed at achieving Water Security for several years. Currently, the Intergovernmental Hydrological Programme (IHP-IX) 2022–2029 is in effect, with the objective of promoting water security in a changing environment. This programme strongly emphasises scientific research and education as essential supports for governance strategies. The IHP-IX aligns with the United Nations Sustainable Development Goals (SDGs), fostering international collaboration, capacity building, and the development of science-based policies for sustainable water management. The initiative defines five interconnected priority areas (Fig. 5), each with its own performance indicators and expected outcomes, ensuring a structured approach towards addressing global water challenges.

Water Insecurity: Extreme Weather Events and Pollution

Extreme weather events are a direct consequence of climate change and affect both the quantity and quality of available water and therefore pose a significant risk to populations. Díaz-Cruces et al. (2024) address this issue by introducing the concept

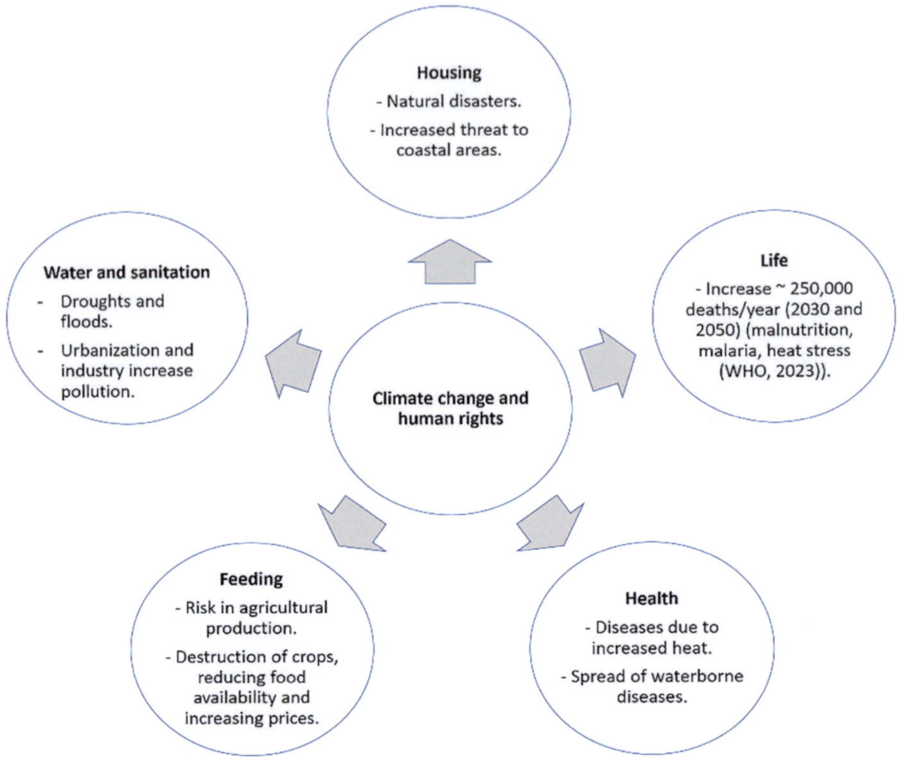

Fig. 6 Climate change and human rights (adapted from Díaz-Cruces et al. 2024)

of the 'right to a stable climate', arguing that recognising this right as a fundamental human right is essential for ensuring justice and sustainability amid increasingly unpredictable climate changes. It is possible to estimate that these climatic events are of high risk to the quality of life and would affect society in multiple aspects. Figure 6 aims to exemplify some of these risks in a simple way.

In this context, an emerging concept known as the 'right to climate' can be introduced. This concept is crucial in explaining the link between climate events and human health, quality of life and even survival (Díaz-Cruces et al. 2024). The right to climate not only concerns extreme weather events but also biodiversity loss. Its recognition seeks to establish commitments towards both climate change mitigation and resilience, as well as ensure the transfer of technologies to countries most vulnerable to these changes (Díaz-Cruces et al. 2024).

Access to water is increasingly affected by extreme events such as floods and droughts, directly threatening the human right to water. This situation is particularly severe in vulnerable nations. Recognising the 'right to climate' as a human right could serve as a tool to strengthen regulations and laws, ensuring climate justice (Díaz-Cruces et al. 2024).

Urban expansion linked to population growth presents challenges such as ecosystem degradation and extreme climatic events, including floods. This highlights the importance of integrated water management, as urbanised areas exhibit unique dynamics. Urban water security has emerged as a more specific concept tailored to these particular scenarios (Aboelnga et al. 2019). Lack of safe drinking water due to poor water quality has multiple consequences on health and the environment, aggravating availability. It is estimated that more than half of the world's population is settled in urban areas, which is a growing threat in poor countries, making it necessary to consider urban sprawl and its pattern for water supply and wastewater treatment (Mishra et al. 2021).

Water has become the most extensively extracted natural resource worldwide, with reservoirs even present in arid regions. In the case of groundwater, its properties generally remain stable over the years. However, it can be more vulnerable than surface water, as once contaminated, decontamination becomes significantly more challenging. Another associated issue is saline water intrusion into wells located in coastal areas—this salinisation is difficult to predict, and desalination is an expensive treatment process (Kekes et al. 2023). Additionally, poor soil drainage in agricultural areas, coupled with intensified irrigation, contributes to soil salinisation. In fact, salinisation caused by horticultural activities has become one of the most pressing issues in countries such as Australia and India (Ferreira et al. 2024). Understanding the various local contexts and scenarios is essential for developing effective strategies and action plans to achieve water security.

Both surface and groundwater sources are subjected to over-extraction for agricultural use, exacerbating soil degradation. Intensive farming practices increase soil erosion, leading to runoff of agrochemical residues and sediments into water bodies, which affects water turbidity and disrupts aquatic ecosystems (Ferreira et al. 2024). Globally, 20–30% of water consumption in the horticultural sector is attributed to crop irrigation, while intensive agriculture is also associated with water contamination due to agrochemical use and the presence of heavy metal pollutants such as cadmium and zinc (Ferreira et al. 2024).

Water pollution has always been a problem in the world; however, with the increase of agricultural and industrial activities and urban growth, this problem has been increasing and adding new pollutants such as emerging pollutants or the increase of agrochemicals due to the generation of microorganism resistance. The IHP-IX Operational Implementation Plan 2022–2029 proposes a series of tasks and activities linked to water pollution in the framework of its priority areas, including knowledge of water quality and monitoring through scientific research on emerging pollutants, pollutant transport and health risks. Kekes et al. (2023) made a classification of pollutants based on World Health Organisation standards according to their potential health impact as well as the properties of the pollutants, Table 1. In addition to mitigation strategies or circular economy, the importance of open databases and increased transparency is also considered.

For pollutant mitigation, data science can play an important role in enabling real-time detection, classification and prediction of pollutants with monitoring and analysis tools. Multispectral imaging, IoT sensors and machine learning are useful

Table 1 Pollution of water according to the World Health Organisation (WHO) (adapted from Kekes et al. 2023)

Classification	Subclassification	Examples
Chemical	Organic pollutants	Food processing waste, petroleum hydrocarbons, disinfection by-products (DBPs), volatile organic compounds (VOCs), insecticides, herbicides, tree and bush debris from logging operations
	POPs, PBTs, TOMPs	Persistent organic pollutants (POPs) also called persistent, bioaccumulative, and toxic (PBTs) or toxic organic micro pollutants (TOMPs) like carcinogenic polycyclic aromatic hydrocarbons (PAHs), brominated flame retardants, tributyltin (TBT)
	Inorganic pollutants	Arsenic (As), cadmium (Cd), lead (Pb), zinc (Zn), mercury (Hg), nitrate, phosphate
	Emerging contaminants	Endocrine-disrupting chemicals, pharmaceuticals, personal care products, microplastics, pesticides, herbicides
Microbial	Pathogens	Bacteria: Campylobacter, *Escherichia coli*, *salmonella* Protozoa: *Cryptosporidium*, giardia
	Sources	Animal excreta, plants, human activity (e.g. dust, saliva, animal dander)
Radiological	Naturally occurring	Radionuclides from geological settings
	Human-made	Controlled radioactive substances from industrial processes
Physical	Organic material	Leaves, sediments, heat
	Inorganic	Iron and other metals affecting sensory characteristics like colour and odour
	Treatment by-products	Chlorination by-products (e.g. bromophenols), ozonation intermediates
Sensory characteristics	Biological sources	Decaying vegetation, microbiota metabolites
	Chemical sources	Agricultural runoff, industrial discharges, wastewater and metal manufacturing

tools for continuous monitoring of physicochemical parameters such as turbidity, pH, nutrients, detecting anomalies and pollution episodes with high spatial and temporal resolution. The results of these controls can support immediate response to situations but can also be the basis for long-term mitigation strategies and public policies, modelling risk scenarios.

The 'national level of water security' has been estimated according to the ten indicators, including all countries with data to assess the indicators, regardless of geography, size or population. The metrics are based on SDG indicators, although in some cases data were only available up to 2020 and more than half of the water indicators had data limitations, so sub-indicators and proxy values from datasets were used. The ten components of water security assessed are (1) drinking

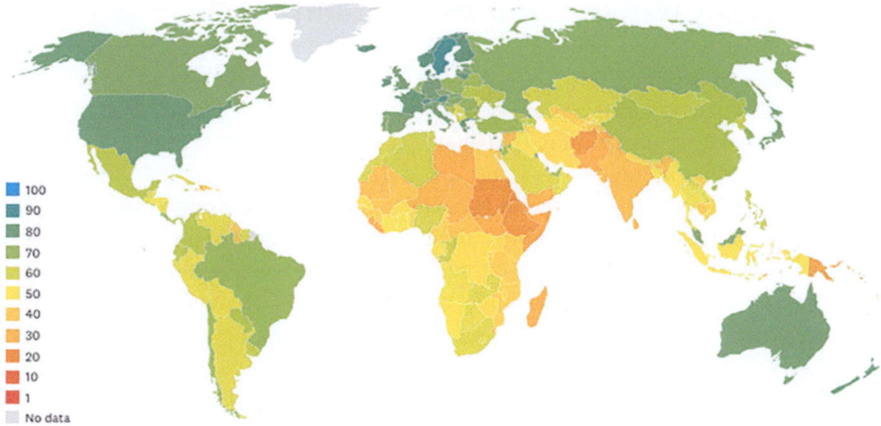

Fig. 7 National water security mapped globally, based on a score of 1–100 (Global Water Security 2023 Assessment report)

water, (2) sanitation, (3) good health, (4) water quality, (5) water availability, (6) water value, (7) water governance, (8) human safety, (9) economic safety, and (10) Water resource stability (MacAlister et al. 2023). Figure 7 differentiates countries according to the percentage of implementation or development of national water security measures.

Latin American countries have a relative abundance of renewable water resources, which together account for 33% of the world's water capital; however, Arroyo et al. (2015) in the report 'Economic Water Insecurity in Latin America: from abundance to insecurity' resulting from the VII World Water Forum (Korea, 2015) describe Latin America as a region with high water availability and growing scarcity.

The VII World Water Forum introduces the concept of economic water insecurity, which is really important for the region as it considers that it is not enough to estimate the amount of water per capita, but rather it adopts a more comprehensive vision for decision-making in terms of water demand and sanitation, considering infrastructure for service management, water deficit due to pollution and floods, vital points in the search for poverty reduction and economic development.

The concept of Economic Water Insecurity (EWI) was introduced in 2007 by the International Water Management Institute (IWMI) in the Comprehensive Assessment of Water Management in Agriculture. The aim is to distinguish between physical and economic constraints on access to water resources. While water stress is based on water availability, water scarcity also depends on access. EWI occurs when factors such as human, institutional and financial capital limit access to water, even when there is sufficient resource available in nature. It can be a consequence of poor water resource management, with symptoms such as insufficient infrastructure or poor water quality control (Arroyo et al. 2015).

Transboundary Waters: Global Water Security

When surface or groundwater resources are shared between two or more countries, they are classified as 'transboundary waters'. These waters play a crucial role in global water security and require transnational management, a necessity recognised by UNESCO. The increasing pressure on water resources, coupled with unequal distribution, presents a major threat to water justice. In this context, it is essential to consider that transboundary rivers alone account for approximately 60% of the world's freshwater flows, directly affecting the environmental conditions, economic activities and livelihoods of over three billion people. The management and flow of these waters represent an additional challenge in the pursuit of water security (UNECE, UNESCO and UN-Water 2024).

In Latin America, despite ongoing efforts, an effective management system for transboundary aquifers has yet to be established. One notable example is the Guaraní Aquifer, for which an international cooperation agreement was signed in 2020. While this represents progress in international water governance, it remains only partially operational (UNECE, UNESCO and UN-Water 2024).

In the region, Paraguay plays a key role in the management of transboundary waters, as it shares major water resources, such as the La Plata River Basin, with neighbouring Brazil, Bolivia and Argentina. Additionally, the Guaraní Aquifer System itself is shared among Argentina, Brazil, Paraguay and Uruguay. This underscores the urgent need to strengthen water governance in Paraguay and across the region, ensuring a more coordinated, sustainable and equitable management of shared water resource. Figure 8 shows the transboundary river and lake basins, transboundary aquifers and international borders.

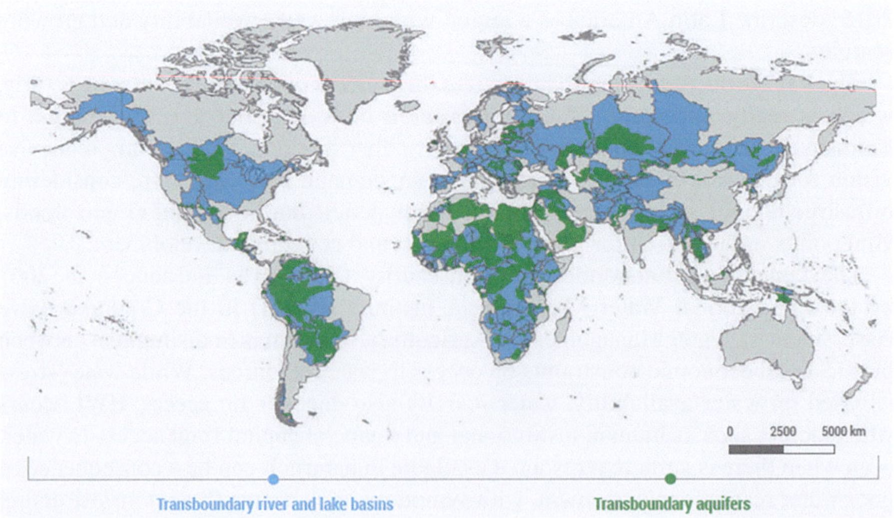

Fig. 8 Transboundary river and lake basins, transboundary aquifers and international borders (UNECE, UNESCO and UN-Water 2024)

In terms of operational agreements for transboundary water cooperation, Europe and North America represent the most advanced regions, where 42–45 countries share surface and groundwater resources. This cooperation is reflected in well-established regulatory frameworks and policies. In Europe, the EU Water Framework Directive (2000/60/EC) and the 1992 Water Convention set standards for the management of shared water resources. In North America, cooperation is supported by the 1909 Boundary Waters Treaty between the United States and Canada, as well as the 1944 Treaty on the Colorado, Tijuana and Bravo Rivers between the United States and Mexico. Despite this extensive regulatory coverage, 18 countries have yet to secure effective agreements covering 90% of their transboundary basins, highlighting the need for further expansion of cooperation.

Kim et al. (2020) conducted a study on surface water variations using remote sensing in the Imjin River basin, a transboundary watercourse between North and South Korea that has been a source of conflict between the two nations. The study focused on surface water storage, identifying distinct patterns of surface variability in the Hwanggang Reservoir before and after 2008. These changes were attributed to the construction of the Hwanggang Dam by North Korea in December 2007.

Transboundary water resources can be leveraged as a political tool, as shared waters often become a source of international tensions between upstream and downstream countries. Infrastructure projects such as dams can significantly impact downstream communities. Conflicts are frequently linked to water scarcity or the water-energy-food nexus. For instance, clashes along the Kenya-Ethiopia border due to a lack of grazing land, livestock, and water scarcity, as well as the humanitarian crisis in the Gaza Strip following Israel's sanctions in 2007—which restricted access to fuel for water supply—illustrate how the exploitation of water resources can serve as a geopolitical instrument (Kim et al. 2020).

Large volumes of data, for example, from environmental monitoring can be processed by Data Science and converted into inputs for expert studies in various areas of science to support water quality control, a clear case for decision-making with accurate information is the use of machine learning for the assessment of chemical contaminants in groundwater and the generation of pseudo-contour maps of the contaminants found (Wu and Polya 2024).

Transboundary water management can be strengthened by emerging technologies such as geospatial analysis, remote sensing or cloud-based platforms. Escobar et al. (2020) conducted a local study in Valencia (Spain) whose methodology can be extrapolated due to standardisation and shared semantics to a shared water situation, in this case demonstrating how ontology-based frameworks and linked open data contribute to semantic integration and decision-making in complex water distribution networks. For the case study of Arroyo-Ortega et al. (2024) on transitional waters, water-land transition waters including watersheds, deltas, among others, they have employed interpolation and spatial analysis methodologies for pollutant monitoring. These tools generate precise maps and diagnostics that can be scaled to shared watersheds. Polpanich et al. (2022) showed how a water monitoring system based on spectral satellite imagery, cloud computing and data science applied in the Lower Mekong Region (LMR) (Thailand) improves early warning capacity and coordinated response to extreme events.

In Central Asia, transboundary water management is complex, involving multiple countries, Kazakhstan, Kyrgyzstan, Tajikistan, Turkmenistan and Uzbekistan with uneven water distribution and limited water information. In this case, De Keyser et al. (2023) offer a solution to the problem through open-source geodata by grouping data into categories such as climate, hydrology, geography and topography, among others, which facilitates transboundary cooperation.

The countries that have the capacity to manage shared waters and also generate data on water level monitoring, water quality, loading and discharge zones, among others, will be the countries that will finally have the possibility of making decisions by processing this data for modelling or projections, forecasting vulnerable geographical areas, etc. These are some of the reasons why it is important for each government to take care of the interests of its country and its population. This may also represent a problem that increases inequality in access to water in quantity and quality in less developed countries as they will not have adequate technologies or qualified human resources; therefore it is essential that international organisations promote international cooperation and open databases.

Data Science as a Tool for Water Governance

Data science enables real-time data collection, analysis, and modelling to improve water management and prevent water crises. Defining *Data Science* (DS) can be complex, yet it can be broadly understood as the integration of fields such as data engineering, machine learning and advanced data processing. Its applications are extensive, enabling the identification of patterns and the prediction of scenarios (Jahani et al. 2023; Pessanha Santos 2023). Therefore, DS is not merely a statistical tool but an effective data-driven management solution (Jahani et al. 2023).

Advancements in computational power and data storage have made real-time information accessible. Data Science facilitates the extraction and analysis of vast amounts of data through algorithms. However, it still faces challenges regarding interoperability, Fig. 9, when accessing multiple data sources and software platforms, underscoring the need for standardisation. In this regard, the availability of open data and its preprocessing are critical for ensuring reliable analysis (Pessanha Santos 2023).

In domains where large volumes of data are generated, such as supply chains that receive input from multiple sources—including sales records, sensors, and customer purchasing patterns—the decision-making process can be enhanced through big data analytics (BDA). This approach must always consider the nature of the data, which may be either structured or unstructured (Jahani et al. 2023).

To maximise the utility of data and its applications, advanced tools and methodologies that enable large-scale data processing (Fig. 10) are rapidly emerging. Deep neural networks, classical machine learning and scalable graphics processing units

Fig. 9 Dataset standardisation challenges (adapted from Pessanha Santos 2023)

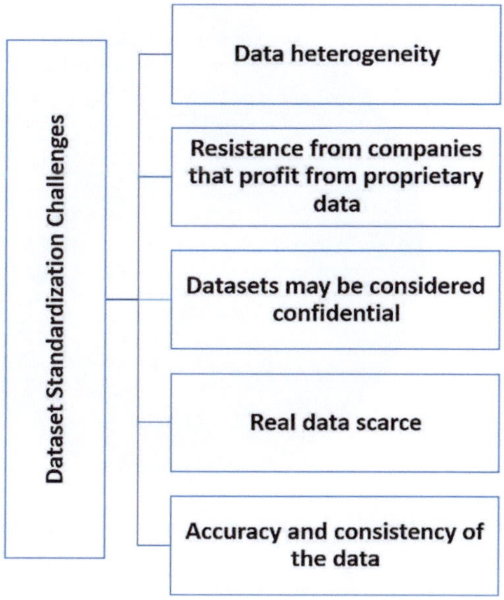

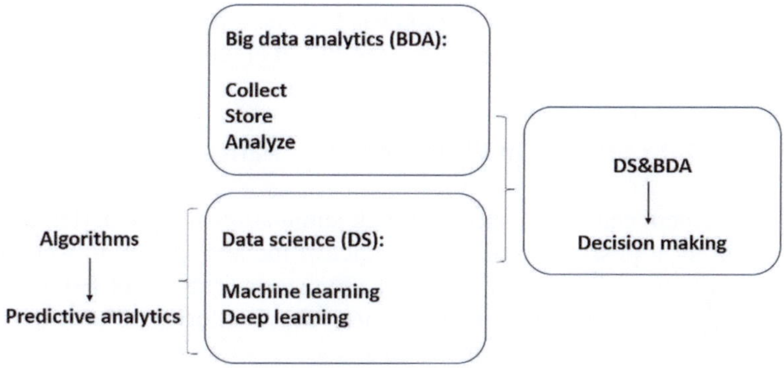

Fig. 10 Big data and data science tools (adapted from Jahani et al. 2023)

(GPUs) have become fundamental pillars of artificial intelligence (AI) (Raschka et al. 2020).

In the field of computer science, artificial intelligence (AI) is introduced with the objective of performing human-like tasks such as natural language understanding and image recognition through programming. As a subset of AI, machine learning (ML) takes this concept a step further by drawing inspiration from the human brain and cognitive processes (Raschka et al. 2020; Sharma et al. 2021). ML is a powerful tool for scientific applications, as it can develop models and algorithms capable of recognising patterns without the need for explicit programming (Raschka et al. 2020). Deep learning is a specialised subset of ML, further enhancing its capabilities through complex multi-layered neural networks (Fig. 11).

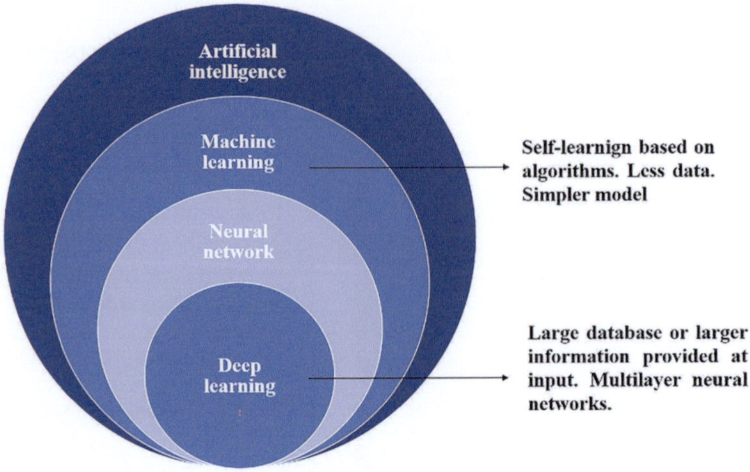

Fig. 11 Machine learning and deep learning (adapted from Raschka et al. 2020)

From labelled datasets, ML enables complex decision-making supported by predictive rules and data patterns. Python has become the most popular and widely used programming language in sciences (Raschka et al. 2020).

The Rise of Data Science in Water Management

Water management can be enhanced through automation and prediction using ML and DL, making it possible to develop predictive models applied to water quality and even anticipate risks or issue contamination alerts. The use of IoT sensors for real-time data collection would greatly facilitate the monitoring of basic parameters such as pH, salinity, or certain contaminants.

Currently, artificial intelligence is already fundamental in certain areas like autonomous navigation and is even referred to as universal learning (Sharma et al. 2021). Its utilities extend to all areas, including environmental monitoring and specifically for water resource management. Industry 4.0 and IoTs require analysis techniques based on DS and BDA, which include social and environmental topics in line with the advancement of these analysis techniques (Jahani et al. 2023). Lopes et al. (2022) explain how it is possible to prevent diseases and make policy decisions through effective health surveillance systems where data science is clearly linked to public health. Concrete examples of the application of AI or data science and the changes it has produced can be found in the health sector, from the data-driven approach, the possibility of image-based diagnostics and even drug discovery, representing a profound transformation (Costa and Aparicio 2023). Through machine learning algorithms, it is possible to evaluate personal medical history and genetic backgrounds to provide treatment recommendations.

In the business area, it is possible to predict risks, price variations and even fraud, and for customer service, virtual assistants or chatbots improve the experience and streamline procedures (Costa and Aparicio 2023). These applications extend to marketing, transportation and logistics, industrial processes and even political aspects.

Challenges still exist, such as the use of deep learning for big data analysis, where both the quantity of data and its veracity affect its effectiveness. Therefore, it is necessary to ensure scalability with large amounts of data or adequately consider situations where there is not much data. In these cases, methods like machine learning may be more appropriate; however, situations may arise where, beyond a certain threshold, no further progress is seen, unlike the performance obtained with deep learning, which improves as the volume of data increases (Sharma et al. 2021).

Typically, deep learning learns with a large amount of data, but it is necessary to perfect models with smaller amounts of data, as this is fundamental in the water sector, where big data is mostly unavailable for urban water infrastructures. Although data collection in the sector has advanced at a rapid pace, it is still not possible to compare big data with sectors like finance, for example (Fu et al. 2022).

Challenges related to water are already being addressed through machine learning for the analysis and prediction of droughts, floods, water pollution and even availability. This will undoubtedly contribute to achieving the targets within the framework of SDG 6 (Fu et al. 2022). Therefore, it is foreseeable that AI tools applied to the monitoring, modelling and prediction of water resources behaviour would lead to attaining water security (Fig. 12) and making it sustainable.

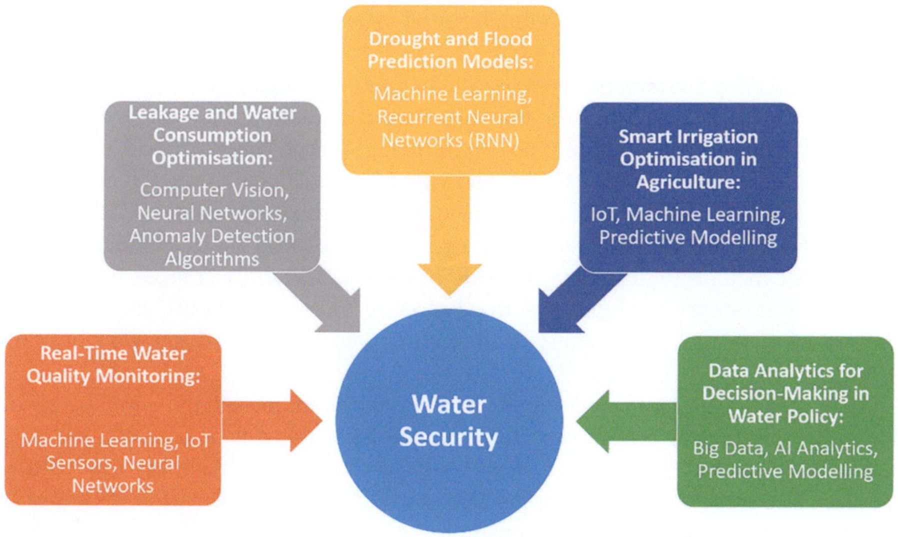

Fig. 12 AI tools for water security

Deep learning has been employed in urban water management for system planning or anomaly detection such as leaks or pollution; DL can be used at all stages of the urban water management system (Fu et al. 2022). It is possible to predict daily water demand within 1 h in an online environment and with a few days of learning, thus optimising the operation of water providers' pumps (Fu et al. 2022). Ma et al. (2024) state that the combination of remote sensing and machine learning (ML) has been applied in the estimation of chlorophyll-a, nutrients, dissolved organic carbon and water temperature. Nevertheless, field data collection remains a challenge due to the need for in situ sampling and laboratory analysis, as well as the variability of water resources according to climate and seasons. To address these issues, transfer learning (TL) has been implemented to improve the temporal and spatial generalisation capacity of ML models in water management. Pires de Lima and Marfurt (2019) present transfer learning as a technique that uses information learned in a primary machine learning task to perform a secondary machine learning task and confirm the idea that models trained on large datasets of natural images offer better results than those trained with smaller remote sensing datasets. These findings reinforce the value of transfer learning for satellite image classification and water resource monitoring.

Huang et al. (2020) mention that to improve the accuracy of hydrological prediction, it is necessary to integrate optimisation algorithms with satellite images and spatial information models. This integration can facilitate water management and also be useful for preventing extreme climate-related events such as flooding.

Various areas of science, including remote sensing, have been able to exploit the success of natural image classification, a set that can contain millions of samples enabling their study with deep learning techniques; this classification can be performed with convolutional neural network models using a technique commonly called transfer learning (Pires de Lima and Marfurt 2019). Machine learning and deep learning have become techniques to evaluate and quantify changes in the environment resulting from human activities or natural disasters, with models capable of studying cartography or even predicting crop yields. Based on surface temperature, land elevation or slopes, and with ML and DL techniques accompanied by remote sensing, it is possible to estimate soil properties (Ma et al. 2024). Fu et al. (2022) explain how through deep learning it would be possible to conduct big data analysis of meteorological data, LiDAR images and satellite images to generate multispectral images for input into flood models. There are already precedents for generating flood risk maps using deep learning and water flood models (Fu et al. 2022).

Both satellite sensors and remote sensing have improved the monitoring of hydrological data; several models facilitate the measurement of precipitation with resolutions between 4 and 25 km, although accuracy may decrease due to low resolution. Currently, remote sensing optimises the design of monitoring networks through artificial intelligence; for example, to estimate precipitation from satellite images, machine learning models such as artificial neural networks can be used (Huang et al. 2020). Data science and other emerging technologies and the consideration of societal and economic aspects are important in water management with a holistic view. Table 2 shows a summary of some methodologies or tools applied to water management and their possible impact in three areas, public policy, technical management and risk prevention.

Transforming Water Governance: The Power of Scientific Research and Data

Just as there are multiple concepts for water security, there are also various perspectives or application strategies. According to Aboelnga et al. (2019), while some policies or strategies focus on the risks or problems arising from inadequate water management, others have adopted an interpretation primarily based on water resources and the benefits for people. These differences in concepts and the lack of standardised strategies can be detrimental to the sustainable development of water resources and water security itself, resulting in unclear regulatory frameworks or laws that are not adapted to local needs or scenarios. These are some of the reasons that highlight the importance of data monitoring and processing as pillars for decision-making.

Water governance becomes effective when there are participatory or inclusive monitoring programmes for stakeholders, with an emphasis on disseminating information related to water in all its aspects. Thus, monitoring measures carried out by governmental entities can serve as models, but water governance is complex and must always adapt to the local context due to disparities between regulations, quality indices, and local accessibility (Lopes et al. 2022). 'Collaborative governance' is also presented as an approach to water stewardship, seeking greater democracy and better environmental outcomes. It is presumed to have a more equitable vision than classical or traditional governance as it employs a participatory methodology to involve diverse sectors in decision-making, although if power dynamics are ignored in the process, inequalities are likely to widen and transparency and communication are necessary for equity in decision-making and benefit-sharing (Koebele et al. 2024).

The role of the private sector is fundamental to achieving the Sustainable Development Goals (SDGs), including SDG 6. Corporate sustainability encompasses social and environmental aspects in a company's economy and can allocate resources towards achieving the SDGs. Additionally, it can become a competitive advantage by offering business models based on innovation and sustainability. Companies can see increased profits by reducing costs and risks while enhancing brand value and improving performance (Ficco et al. 2023).

Some companies already place a socio-environmental purpose at the core of their business model, known as 'B Corporations', which are for-profit but socially committed (Ficco et al. 2023). As an example, the case study by Ficco et al. (2023) on the importance of monitoring and data collection is the study on the contribution of the Argentine B corporations and their progress towards achieving the SDGs. An analysis through social networks facilitated the understanding of the relationship between B Corporations and their connection to the SDGs. It was found that SDG 8

Table 2 Some methodologies or tools applied to water management and their possible impact

Legend: High (green) | Medium (yellow) | Low (red)

Application / Case	Tool/ Method Evaluated	Public Policy Impact	Impact on Technical Management	Risk Prevention Impact	Ref.
Weather pattern analysis	Machine Learning	High	Medium	High	Fu et al., 2022
Flood risk mapping	Deep Learning + Satellite imagery	Medium	High	High	Fu et al., 2022
Groundwater contamination	Machine Learning	Medium	High	Medium	Wu & Polya, 2024
Estimation of soil properties	ML + Remote Sensing	Low	Medium	Medium	Ma et al., 2024
Environmental image classification	Transfer Learning	Low	Medium	Medium	Pires de Lima & Marfurt, 2019
Crop diseases (impact on water)	Data Science	Low	Low	Low	Rodríguez-Almonacid et al., 2023
Sanitary surveillance	Data Science	High	Medium	Medium	Lopes et al., 2023
Water market in Chile	Multivariate Analysis	High	Medium	Low	Araos & Roco, 2025
Science-governance integration (IHP-IX)	Data Science + SDG Indicators	High	High	High	UNESCO IHP-IX
Global water security assessment	Open Data + SDG Indicators	High	Medium	Medium	MacAlister et al., 2023
Improving drinking water policies linked to agriculture in the EU	Applied science + real time monitoring + dissemination strategies	High	Medium	Medium	Glavan et al., 2019
Socio-economic vision for water management in Brazil and Portugal	Document analysis + SWOT + reference to digital transformation	High	Medium	Medium	Santos et al., 2023

(continued)

Table 2 (continued)

Intelligent monitoring and control in agricul-ture (IoT)	IoT (sensors, WSNs, drones, cloud, ML)				Rehman et al., 2022
IA applied to the Wa-ter-Energy-Food nexus	Artificial In-telligence + SLR review				D'Amore et al., 2022
Integrated approaches to water management	Systematic review + governance approach				Bachmann et al., 2022

('Decent Work and Economic Growth') and SDG 12 ('Responsible Consumption and Production') are of greatest importance, followed by a higher commitment to SDG 13 ('Climate Action'), 9 ('Industry, Innovation, and Infrastructure'), 17 ('Partnerships for the Goals') and SDG 5 ('Gender Equality'). To a lesser extent, SDGs 4 ('Quality Education') and 11 ('Sustainable Cities and Communities') were noted.

Studies like this can serve as the basis for developing public policies for corporate sustainability and water management itself. There is an interconnection between the 17 SDGs, so it is necessary to discuss how to support SDG 6 transversally.

Kekes et al. (2023) conducted a study on methods for water treatment and Hazard Analysis and Critical Control Points (HACCP), highlighting how the synergy between technological advancements, regulatory frameworks and a multidisciplinary work vision could become a tool towards water security. In particular, science and technology (Fig. 13) are the pillars on which fair and sustainable public policies must be based.

The sustainable management of water is currently advancing through the application of innovative technologies such as the Internet of Things (IoT) and artificial intelligence (AI), which enable effective and precise irrigation management in agriculture. Technology-based decision support systems are utilised to make decisions based on soil and irrigation control data in certain horticultural crops like tomatoes and lettuces (Ferreira et al. 2024). Technological control represents the future of sustainable agriculture; decision-making based on the collection and processing of real-time data is made possible by IoT and AI tools (Ferreira et al. 2024). Therefore, the application of sensors and IoT can be fundamental for policy decisions and encourage the adoption of smart technologies in agriculture.

In urban areas, water security outcomes are influenced by local characteristics. Multiple countries adopt their public policy strategies based on Integrated Water Resources Management (IWRM); however, challenges in water management persist. It is essential to raise awareness of the importance of water in urban planning

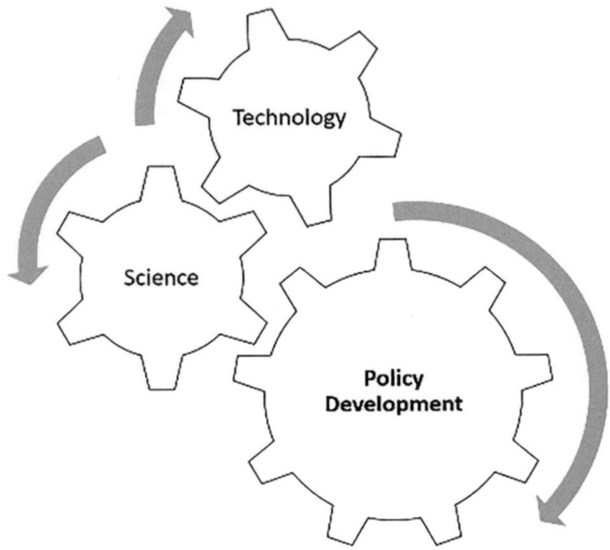

Fig. 13 Science and technology in decision-making

and sustainable development to establish local policies that ensure urban water security, thereby enhancing urban resilience and adaptability (Aboelnga et al. 2019).

According to the definition of the 'right to climate protection', the state has the obligation or responsibility to adopt measures for mitigation and/or resilience to climate change. For this, it is important to have data demonstrating the changes that have occurred over the years concerning human rights and climate change. In this regard, science, through academia, will be the tool for decision-makers to have information for policy development (Díaz-Cruces et al. 2024).

Both data science and artificial intelligence have become fundamental disciplines, not only due to the speed at which they process data but also for their capacity to identify patterns and project outcomes or models. These fields are applicable across various sectors and have significantly transformed how data is collected and analysed (Costa and Aparicio 2023).

Sustainable development can benefit in multiple ways from AI and data science tools. In urban environments, these technologies facilitate traffic analysis, optimise energy consumption and improve public services (Costa and Aparicio 2023). Data-driven management has also strengthened vital fields such as climate science and agriculture. Historically, agricultural data have been vast and diverse depending on the region or type of crops, which complicates data processing. Rodríguez-Almonacid et al. (2023) analysed 8 years of data on climate variations and rice crop diseases by integrating data science tools. This methodology enables the understanding of multitemporal datasets and is applicable to water management, allowing for the analysis of water availability and climate changes that may affect this vital resource.

In the case of water management in Chile, it should be noted that the legislation declares water to be a national asset in the public domain, although it grants appropriation rights to individuals, companies and organisations. To acquire this 'right', an application is made to the state, and once granted, they can be bought, sold and leased in a free market. To estimate the cost of water resources in Chile, Araos and Roco (2025) have evaluated certain variables such as annual rainfall and maximum temperatures, regional GDP per capita and factors such as rurality index, among others. The results show a heterogeneous market where rainfall, industrialisation or geopolitical and demographic characteristics significantly influence the price of water rights. By identifying these variables and their behaviour, the aim is to provide data for the formulation of sound regulations and to contribute to the equitable distribution of water in Chile.

Santos et al. (2023) mention that data integration in the framework of digital transformation can make an important contribution to understanding the impact of complex factors such as socio-economic and cultural factors on water governance. Considering economic or cultural differences, as well as low citizen participation, is key to ensuring access to water and its management; not assessing these parameters will undoubtedly affect more vulnerable populations or those already excluded from society.

Regardless of the technology or data science that a country or region is able to use, the local scenario directly affects the governance model to be considered, so decision-making based on local data and with a holistic vision is crucial. It is common to associate the quantity or availability of water in certain countries or regions such as Latin America; however, the quality of water is affected and equitable access to drinking water is not fulfilled with the aggravating factor of a great socio-economic difference even within the same country.

A comparison of water governance or management between Latin American and European countries is interesting. Santos et al. (2023) conducted a cross-sectional analysis of water governance between Portugal and Brazil, finding how their socio-economic and institutional contexts are determinant in assessing both common challenges and divergent strategies between these countries. Although the study is specific to two countries, their models are similar or applicable to other countries in their respective regions. EU countries are governed by stronger sustainable policies than Latin American countries where the impact of lack of infrastructure or lack of participatory management is greater. Figure 3 in this chapter shows a clear disparity between the wastewater treatment situation between the EU and Latin America. Table 3 contains a summary of the governance characteristics between the two countries.

In the case of Portugal, it can be highlighted for its integration into the European Union regulatory framework, in particular the EU Water Framework Directive (WFD), which promotes basin planning, stakeholder participation and ecological protection. Portugal also shows progress in digital transformation initiatives, such as the use of Geographic Information Systems (GIS) to optimise water distribution or the implementation of smart water meters in order to increase transparency. The case of Brazil is presented as operating under the National Water Resources Policy

Table 3 Water governance features in Portugal and Brazil (Adapted from Santos et al. 2023)

	Portugal	Brazil
Legal framework	Water framework directive (EU)	Law 9433/1997—National Policy on Water Resources
Level of institutionalisation	High, with integration with European policies	Medium, with challenges in inter-institutional articulation
Basin management	Implemented with integrated ecological planning	Existence of river basin committees but limited effectiveness
Citizen participation	Formalised, with public consultation on decisions	Present, but with unequal access depending on the region
Water reuse	High, with installed technologies and active regulations	Low, with technical and regulatory barriers
Digital transformation	Advanced: use of smart meters, GIS, automated monitoring	Developing, with disparity between regions
Key challenges	Seasonal water scarcity, adaptation to climate change	Regional inequity, poor infrastructure and pollution

(Law No. 9433/1997), which establishes river basin committees and decentralised governance structures, this participatory model represents a step forward; however, its effectiveness is limited by institutional fragmentation, regional inequality and low technological capacity.

Both Portugal and Brazil share concerns about climate change, pollution and water scarcity, but Portugal shows greater institutional stability and infrastructure. Reuse of treated wastewater is more widespread in Portugal, as in the EU, and in Brazil there are limitations to implementing these practices at the national level. This comparison can be useful to illustrate how policy coherence, technological adoption and social equity influence the success of water governance models, offering valuable insights for other countries seeking improvements in sustainability and resilience in the management of their water resources (Santos et al. 2023).

Towards Water Security

Climatic catastrophes such as floods or droughts are the most devastating and are often influenced by human activities as they are not only due to climate but also to urbanisation, intensive agriculture and poor land use management, which allows the use of vulnerable or high-risk areas. These catastrophes have consequences in various areas, e.g. the economy because they significantly affect crops and urban infrastructure, as well as energy production. Competition between human and industrial water use has also been exacerbated, as well as pressure on groundwater (Mishra et al. 2021). These are some of the reasons that highlight the need for data and studies to enable data-driven decision-making as well as the expertise of skilled people, the need to strengthen the legal framework of countries with respect to water and its uses to ensure water availability and avoid future conflicts and the need for a more effective water management system.

The issue of emerging contaminants remains widely debated, with the absence of regulations or laws to address this challenge, despite its potential risks to human and environmental health. For instance, the increasing identification of microplastics and chemical residues affecting the endocrine system has been noted. The presence and rise of emerging contaminants pose a significant challenge to global water security and highlight the urgent need for innovative solutions in water protection and management (Kekes et al. 2023). This point highlights the importance of scientific research for the identification and monitoring of emerging pollutants, such as drug residues or new synthetic agrochemicals due to the urgent need to update water quality legislation in many countries; technology and innovation in integrated water resources management would shorten the time to understand the problem, to develop international collaboration plans and to prevent long-term health problems, among others.

UNESCO has launched a set of priorities aimed at ensuring sustainable and inclusive water security on a global scale. The Intergovernmental Hydrological Programme can be considered a roadmap that opens the door to the integration of science in water management and governance by facilitating decision-making based on scientific data and studies for the development of public policies. The first three priority areas are directly related to education, science and the strengthening of human capital. The IHP-IX Implementation Plan outlines five priority areas, which are briefly described below:

Scientific Research and Innovation: This priority focuses on hydrological research, technological advancements and international cooperation to enhance water quality monitoring, disaster risk management for extreme climatic events and climate change adaptation.

Water education for the fourth industrial revolution: This area seeks to raise public awareness, reinforce multidisciplinary collaboration and develop educational resources to bridge the gap between science and policymaking.

Bridging the data and knowledge gaps: This priority aims to improve data accessibility, transparency and integration through advanced hydrological research methods, open-access platforms and capacity-building initiatives.

Integrated and inclusive water resources management: This focuses on resilient water governance, addressing water conflicts, exploring nonconventional water sources, mitigating pollution and supporting transboundary water cooperation.

Science-based water governance for mitigation, adaptation and resilience: This priority reinforces science-driven policies, raising awareness among decision-makers, integrating science into governance frameworks and developing adaptive water management strategies.

In order to mitigate the impact of the agricultural sector on drinking water quality, the European Union can be an important example of science-based water governance. Glavan et al. (2019) mentions how the EU has implemented a framework of directives or policies such as the Water Framework Directive (WFD), the Nitrates Directive (ND) and the Sustainable Use of Pesticides Directive (SUPD) as a basis for the integration of data science into public policy. It is also mentioned that it is ultimately the political will and capacity of states that determines the effectiveness of these policies or directives in moving from science to practical application. Success stories include Scotland's specific advisory model and Ireland's Agricultural Catchment Programme, both of which show how a collaborative stakeholder strategy and scientific monitoring can achieve improvements in water quality governance. Both cases demonstrate how integrated collaboration between science, policy and real-time monitoring and the active participation of scientists, decision-makers and local stakeholders is necessary to ensure policy implementation and effectiveness (Glavan et al. 2019).

Advanced monitoring tools such as AI-powered satellite imagery, IoT-based sensor networks and open-access big data platforms are required to mitigate data gaps. Water security will be enhanced and strengthened with these technologies by providing real-time information for decision-makers, politicians or civil organisations to proactively engage or act on issues such as pollution, droughts and resource allocation.

A key factor in the future of water and people's quality of life is data transparency, which is the only way to strengthen water governance. Anticipating events, predicting or modelling, through machine learning and AI modelling, will be the tools for decision-makers to anticipate water-related risks and design policies based on scientific data. In addition, open data platforms foster accountability in water management and greater public participation, making society a user and protector of natural resources.

References

Aboelnga HT, Ribbe L, Frechen FB, Saghir J (2019) Urban water security: definition and assessment framework. Resources 8(4):178

Araos A, Roco L (2025) Identifying the determinants of water rights price: the Chilean case. Water 17(3):395

Arroyo V, Ballestero M, Mejía A (2015) Inseguridad Económica del Agua en Latinoamérica: de la abundancia a la inseguridad

Arroyo-Ortega I, Chavarin-Pineda Y, Torres E (2024) Assessing contamination in transitional waters using geospatial technologies: a review. ISPRS Int J Geo Inf 13(6):196

Costa CJ, Aparicio M (2023) Applications of data science and artificial intelligence. Appl Sci 13(15):9015

De Keyser J, Hayes DS, Marti B, Siegfried T, Seliger C, Schwedhelm H, Habersack H (2023) Integrating open-source datasets to analyze the transboundary water–food–energy–climate nexus in Central Asia. Water 15(19):3482

Díaz-Cruces E, Méndez Rocasolano M, Zamora-Ledezma C (2024) Human rights at the climate crossroads: analysis of the interconnection between human rights, right to climate, and intensification of extreme climate events. Laws 13(5):63

Escobar P, Roldán-García MDM, Peral J, Candela G, Garcia-Nieto J (2020) An ontology-based framework for publishing and exploiting linked open data: a use case on water resources management. Appl Sci 10(3):779

Ferreira CS, Soares PR, Guilherme R, Vitali G, Boulet A, Harrison MT, Ferreira AJ (2024) Sustainable water management in horticulture: problems, premises, and promises. Horticulturae 10(9):951

Ficco C, Bersía P, Bressan C, Herrera-Rodríguez EE (2023) Contribution of Argentinian B corporations to sustainable development goals: empirical analysis based on their practices and solutions to socioenvironmental problems. Sustainability 16(1):6

Fu G, Jin Y, Sun S, Yuan Z, Butler D (2022) The role of deep learning in urban water management: a critical review. Water Res 223:118973

Glavan M, Železnikar Š, Velthof G, Boekhold S, Langaas S, Pintar M (2019) How to enhance the role of science in European Union policy making and implementation: the case of agricultural impacts on drinking water quality. Water 11(3):492

Huang Y, Zhao H, Jiang Y, Lu X (2020) A method for the optimized design of a rain gauge network combined with satellite remote sensing data. Remote Sens 12:194

Jahani H, Jain R, Ivanov D (2023) Data science and big data analytics: a systematic review of methodologies used in the supply chain and logistics research. Ann Oper Res. https://doi.org/10.1007/s10479-023-05390-7

Kekes T, Tzia C, Kolliopoulos G (2023) Drinking and natural mineral water: treatment and quality–safety assurance. Water 15(13):2325

Kim D, Lee H, Jung HC, Hwang E, Hossain F, Bonnema M, Getirana A (2020) Monitoring river basin development and variation in water resources in transboundary Imjin River in north and South Korea using remote sensing. Remote Sens 12(1):195

Koebele EA, Méndez-Barrientos LE, Nadeau N, Gerlak AK (2024) Beyond engagement: enhancing equity in collaborative water governance. Wiley Interdiscip Rev Water 11(2):e1687

Lopes RH, Silvian CRDV, Salvador PTCDO, Silva IDS, Heller L, Uchôa SADC (2022) Surveillance of drinking water quality worldwide: scoping review protocol. Int J Environ Res Public Health 19(15):8989

Ma Y, Chen S, Ermon S, Lobell DB (2024) Transfer learning in environmental remote sensing. Remote Sens Environ 301:113924

MacAlister C, Baggio G, Perera D, Qadir M, Taing L, Smakhtin V (2023) Global water security 2023 assessment; United Nations. University Institute for Water, Environment and Health, Hamilton

Mishra BK, Kumar P, Saraswat C, Chakraborty S, Gautam A (2021) Water security in a changing environment: concept, challenges and solutions. Water 13(4):490

Pessanha Santos N (2023) The expansion of data science: dataset standardization. Standards 3(4):400–410

Pires de Lima R, Marfurt K (2019) Convolutional neural network for remote-sensing scene classification: transfer learning analysis. Remote Sens 12(1):86

Polpanich OU, Bhatpuria D, Santos Santos TF, Krittasudthacheewa C (2022) Leveraging multi-source data and digital technology to support the monitoring of localized water changes in the Mekong region. Sustainability 14(3):1739

Raschka S, Patterson J, Nolet C (2020) Machine learning in python: main developments and technology trends in data science, machine learning, and artificial intelligence. Information 11(4):193

Rehman A, Saba T, Kashif M, Fati SM, Bahaj SA, Chaudhry H (2022) A revisit of internet of things technologies for monitoring and control strategies in smart agriculture. Agronomy 12(1):127

Rodríguez-Almonacid DV, Ramírez-Gil JG, Higuera OL, Hernández F, Díaz-Almanza E (2023) A comprehensive step-by-step guide to using data science tools in the gestion of epidemiological and climatological data in rice production systems. Agronomy 13(11):2844

Santos E, Carvalho M, Martins S (2023) Sustainable water management: understanding the socioeconomic and cultural dimensions. Sustainability 15(17):13074

Selișteanu D, Popescu IM, Roman M, Sulea-Iorgulescu C, Mehedinteanu S (2021) A software emulator for the modelling and control of an activated sludge process in a wastewater treatment plant. Processes 9(11):2054

Sharma N, Sharma R, Jindal N (2021) Machine learning and deep learning applications-a vision. Glob Trans Proc 2(1):24–28

Ulusoy A, Atılgan A, Rolbiecki R, Jagosz B, Rolbiecki S (2024) Innovative approaches for sustainable wastewater resource management. Agriculture 14(12):2111

UNECE, UNESCO and UN-Water (2024) Progress on transboundary water cooperation: mid-term status of SDG indicator 6.5.2, with a special focus on Climate Change

UNESCO (2022) IHP-IX: strategic plan of the intergovernmental hydrological programme: science for a water secure world in a changing environment, ninth phase 2022-2029

United Nations General Assembly (2010) Resolution A/RES/64/292: The human right to water and sanitation (adopted 28 July 2010). https://digitallibrary.un.org/record/687002?v=pdf

UN Habitat and WHO (2021) Progress on wastewater treatment–Global status and acceleration needs for SDG indicator 6.3.1. United Nations Human Settlements Programme (UN-Habitat) and World Health Organization (WHO), Geneva. https://unhabitat.org/progress-on-wastewater-treatment-%E2%80%93-2021-update

Van Beek E, Arriens WL (2014) Water security: putting the concept into practice. Global Water Partnership, Stockholm, p 55

Wu R, Polya D (2024) Data science and environmental analytics for water for better health. In: Water matters. Elsevier, Amsterdam, pp 29–41

Sources and Effects of Water Contamination: Characteristics and Ecological Implications

Carlos Morales-Polo and María del Mar Cledera-Castro

Contents

C. Morales-Polo (✉) · M. del Mar Cledera-Castro
Research Group on Emerging Pollutants and Resource Recovery (gCEV), ICAI School of Engineering, Comillas Pontifical University, Madrid, Spain
e-mail: cmorales@comillas.edu

D. Gutiérrez et al. (eds.), *Smart Water Quality Monitoring*,
https://doi.org/10.1007/978-3-032-03049-8_2

Abstract

Water contamination is a global issue with significant ecological, social, and economic consequences. This chapter examines the origins of water pollution, differentiating between natural and anthropogenic sources, and categorizes key pollutants, including organic, inorganic, microbiological, and emerging pollutants. It also explores their environmental pathways, persistence, and interactions within aquatic ecosystems.

The ecological impacts of contamination are analysed, with emphasis on eutrophication, bioaccumulation, and biomagnification. These processes not only disrupt biodiversity and ecosystem services but also pose risks to human health through waterborne exposure and trophic transfer. Case studies illustrate how pollution sources and contaminant dynamics vary across different geographical and climatic contexts.

Furthermore, the chapter discusses current challenges in water quality management, highlighting the role of monitoring strategies in pollution assessment and mitigation. Advances in real-time monitoring and predictive models offer new tools for tracking contamination trends, enabling more effective regulatory and remediation efforts. Ensuring the sustainability of water resources requires an integrated approach that combines scientific research, policy development, and technological innovation to address both existing and emerging threats to aquatic ecosystems.

Keywords

Water quality · Pollution sources · Contaminant characteristics · Ecological impact · Monitoring technologies

Abbreviations

AI	Artificial intelligence
BOD	Biochemical oxygen demand
COD	Chemical oxygen demand
GIS	Geographic information systems
IoT	Internet of Things
PCB	Polychlorinated biphenyls
POP	Persistent organic pollutant

Introduction

Historical Evolution of Water Mass Contamination

Water mass contamination has its roots in human activities, though its forms and scales have evolved significantly throughout history. Early civilizations impacted water bodies through the disposal of organic waste, but the Industrial Revolution marked a dramatic shift in the composition and scale of pollutants. The introduction of chemicals, heavy metals, and synthetic compounds such as polychlorinated biphenyls (PCB) in the twentieth century led to the accumulation of toxic substances in rivers and lakes near industrial centres (Chowdhary et al. 2020). In this context, environmental movements began to highlight the harmful effects of these substances on ecosystems and human health.

Recent studies confirm that pollutants introduced during the Industrial Revolution persist in the sediments of rivers and lakes, posing a constant threat to aquatic ecosystems and human health. Moreover, according to Vasilachi et al. (2021), historical pollution has laid the groundwork for the accumulation of emerging pollutants, such as pharmaceuticals and microplastics, which exacerbate the existing problem. Altenburger et al. (2019) emphasize that the chemical mixtures in aquatic environments create synergistic effects that are not always detected by traditional monitoring methods, underscoring the need for integrated approaches.

From the 1970s onwards, the tightening of environmental policies in many regions of the world, such as the Clean Water Act in the USA (1972) and the European Union's Water Framework Directive (2000), sought to mitigate the effects of pollution through regulations mandating waste discharge reductions and wastewater treatment. However, while these regulatory frameworks have helped improve water quality in many areas, challenges persist due to the emergence of new pollutants and the difficulty of controlling diffuse pollution sources such as agricultural runoff (Morin-Crini et al. 2022).

According to Carvalho et al. (2019), the European regulatory framework has been effective in improving the chemical quality of water; however, it faces significant limitations in detecting emerging pollutants, highlighting the need for adaptive approaches and advanced monitoring technologies. Meanwhile, Cho et al. (2020) note that managing diffuse pollution sources remains one of the greatest global challenges, particularly in regions with limited monitoring and control systems.

Advances in predictive and modelling technologies have also allowed for more accurate analysis of water quality changes in response to regulatory policies. The use of machine learning models, such as those discussed by Yaseen et al. (2019), has proven crucial in predicting long-term variations in contamination levels, providing key information for adjusting existing regulations.

Current Challenges in Water Quality

Water quality faces multiple challenges, ranging from historical issues, such as eutrophication, to more recent concerns, such as microplastics and pharmaceuticals. Eutrophication, caused by excess nutrients like nitrogen and phosphorus, leads to uncontrolled algal growth that reduces oxygen levels in water, creating "dead zones" that threaten aquatic biodiversity (Khan and Mohammad 2014). This issue is particularly persistent in areas of intensive agriculture. According to Grizzetti et al. (2021), agricultural activity is one of the primary sources of nutrients responsible for eutrophication, highlighting the need for real-time monitoring strategies to mitigate impacts before they become irreversible.

Increasing urbanization and inefficient water infrastructure management have heightened the pressure on freshwater resources, exacerbated by climate change. These factors complicate the management of both traditional and emerging pollutants (Alshami et al. 2024). Among the newer challenges is the impact of microplastics, small fragments derived from the degradation of larger plastics or specific sources, such as microbeads in personal care products. These fragments can enter the food chain, accumulating in aquatic organisms and potentially in humans, with health consequences that remain uncertain (Lambert et al. 2013). Additionally, microplastics can adsorb chemical pollutants present in water, increasing their toxicity (Vasilachi et al. 2021). However, their detection remains limited, driving the adoption of tools based on artificial intelligence (AI) to improve monitoring (Tiyasha et al. 2020).

Pharmaceuticals, such as antibiotics, hormones, and antidepressants, pose another significant challenge. They persist in surface and groundwater because wastewater treatment plants are not designed to remove them completely (Tran et al. 2018). These compounds affect the reproduction and behaviour of aquatic organisms, altering ecosystem structures. Altenburger et al. (2019) emphasize the importance of studying the cumulative effects of these pollutants, which often interact with other compounds, amplifying their toxicity.

Climate change intensifies these problems. Higher temperatures accelerate chemical reactions and alter contaminant behaviour, while extreme precipitation events increase the runoff of agricultural and industrial pollutants into water bodies (Shukla et al. 2023). Cho et al. (2020) highlight that these extreme events elevate concentrations of both emerging and traditional pollutants, challenging treatment infrastructure. Meanwhile, Yaseen et al. (2019) stress the value of machine learning models for predicting the impacts of climate change on water quality, enabling proactive planning and more effective management strategies.

Sources, Characteristics, and Implications of Pollutants

Understanding the sources and characteristics of pollutants is fundamental for developing effective monitoring and management strategies. Contamination sources can be classified as point sources, such as industrial discharges, and diffuse sources, such as agricultural runoff. Point sources are easier to regulate through specific policies, whereas diffuse sources require more complex and adaptive approaches due to their widespread and variable nature (Schwabe 2001).

According to Altenburger et al. (2019), effective monitoring of diffuse sources requires a combination of detailed chemical analyses and biological tools to evaluate cumulative effects on ecosystems. This integrated approach provides a more accurate view of how pollutants interact in complex aquatic environments. Similarly, Kay et al. (2018) identify that urban areas significantly contribute to diffuse sources of contamination through runoff that transports oils, heavy metals, and microplastics from impervious surfaces. Both studies emphasize the need for simultaneous monitoring of multiple types of pollutants to enhance management efficiency.

The properties of pollutants, such as solubility, persistence, and toxicity, determine their ability to accumulate in ecosystems and affect wildlife. Processes like bioaccumulation and biomagnification concentrate pollutants in living organisms and transfer them through the food chain, impacting top predators such as birds, large fish, and even marine mammals. For example, Alexander (1999) demonstrated the biomagnification of persistent organic pollutants (POPs) in marine food webs, with higher concentrations observed at upper trophic levels. Furthermore, Vasilachi et al. (2021) highlight that microplastics, besides bioaccumulating, act as vectors for transporting other persistent pollutants, such as pesticides and heavy metals, amplifying their effects along the food chain.

The presence of toxic pollutants reduces aquatic biodiversity, disrupts ecological relationships, and promotes antimicrobial resistance—a phenomenon driven by the discharge of antibiotics into water bodies (Singh et al. 2024a). Cho et al. (2020) underline that antimicrobial resistance spreads not only through wastewater but also via contact with contaminated sediments, where resistance genes can transfer between microorganisms. Meanwhile, Tiyasha et al. (2020) emphasize that predictive models based on machine learning help evaluate how antimicrobial resistance evolves under different contamination scenarios, enabling the implementation of more effective mitigation strategies.

Water contamination also affects ecosystem services provided by these bodies, such as potable water supply, climate regulation, biodiversity support, and recreational activities like fishing and tourism. Its degradation has long-term socioeconomic implications, impacting both human well-being and environmental health (Amatucci et al. 2024). Altenburger et al. (2019) highlight that the loss of water-related ecosystem services can have significant economic impacts, particularly in sectors like tourism and fishing, which depend directly on water quality. According to Yaseen et al. (2019), deep learning-based approaches are valuable tools for modelling how aquatic ecosystem services are affected by specific pollutants and climatic fluctuations, helping prioritize interventions in critical areas.

Finally, Hemdan et al. (2023) recommend integrating smart sensors and real-time data with predictive models to overcome the limitations of traditional monitoring approaches and enhance responsiveness to emerging threats.

Measurement and Continuous Monitoring

Measurement and continuous monitoring are essential for the effective management of water quality. Real-time monitoring enables early problem detection, facilitating data-driven decision-making and rapid responses to pollution events. Traditionally, water quality monitoring has relied on sample collection and laboratory analysis—a process that, while accurate, is costly and lacks immediacy (Cho et al. 2020).

Recent studies, such as that by Bhardwaj et al. (2022), highlight that the use of advanced sensors integrated with geographic information systems (GIS) has significantly improved the accuracy of continuous monitoring. These sensors capture not only physicochemical parameters but also provide real-time information on contaminant dynamics.

Technological advancements have driven the development of automated monitoring systems capable of collecting data on critical parameters, such as pH, turbidity, dissolved oxygen, and concentrations of nutrients and specific pollutants. Combined with technologies like the Internet of Things (IoT) and GIS, these systems offer broader and more detailed coverage of environmental conditions, optimizing water resource management (Yaroshenko et al. 2020). Continuous monitoring is particularly critical in mitigating the effects of climate change, which alters both the distribution and concentration of pollutants. This constant monitoring provides water managers with a better understanding of contaminant dynamics, enabling swift adaptation to changing conditions.

A recent approach based on machine learning and hybrid modelling, as described by Yaseen et al. (2019), has proven effective in predicting pollution events and reducing uncertainties in water quality monitoring. These models integrate large volumes of data, optimizing planning and adaptive responses.

However, implementing these systems requires investments and training to ensure that the data collected is accurate and actionable. Tiyasha et al. (2020) emphasize that combining predictive models with automated monitoring systems can help prioritize actions and allocate resources in critical areas, improving operational efficiency.

Drones and Artificial Intelligence as Emerging Monitoring Tools

Currently, aquatic and aerial drones, alongside AI technologies, are emerging as powerful tools for water quality monitoring. Drones equipped with specialized sensors enable remote monitoring of extensive and hard-to-reach areas, facilitating data collection in water bodies that would otherwise be inaccessible (Nishan et al. 2024). These devices can measure parameters such as temperature, turbidity, and dissolved oxygen levels, offering detailed and real-time mapping of water quality (Alshami et al. 2024).

According to Cho et al. (2020), drones equipped with multispectral cameras and advanced sensors have demonstrated remarkable efficiency in identifying both point and diffuse pollution sources in real-time. Their ability to cover large areas in a short period makes them ideal for monitoring complex watershed systems.

AI, when applied to the analysis of data collected by drones and other sensors, enables the development of predictive models that forecast changes in water quality. Machine learning algorithms identify complex patterns and correlations within large datasets, aiding water managers in making evidence-based decisions and predicting potential pollution events (Talukdar et al. 2023). The study by Altenburger et al. (2019) highlights that combining sensor data with AI algorithms significantly enhances the detection of emerging pollutants and their synergistic interactions in aquatic environments.

In conjunction with drones, AI enables rapid and adaptive responses to environmental changes, fostering more effective and sustainable water resource management. Integrated drone and AI systems, as discussed by Tiyasha et al. (2020), have shown potential in identifying long-term trends in water quality, which is critical for strategic planning and policy evaluation.

The use of drones and AI in water quality monitoring is especially relevant in the context of climate change, where environmental variations significantly alter contaminant dynamics. These technological advances present a unique opportunity to improve monitoring and response efforts in water quality management, enabling better protection of aquatic ecosystems and the ecosystem services they provide.

Origins of Water Pollution

Natural and Anthropogenic Pollution

Water pollution can originate from both natural processes and human activities. Natural pollution arises from geological and biological phenomena that introduce compounds and elements into water without human intervention. Examples include soil erosion, which deposits sediments rich in metals and other elements into water bodies, and volcanic activity, which can release heavy metals and minerals. In coastal systems such as the aquifers of Recife, Brazil, processes like rock-water interaction and natural redox gradients increase salinity and mobilize compounds such as sulphate and nitrate (Bertrand et al. 2016). Similarly, phenomena like landslides or the leaching of organic and inorganic compounds from soil contribute to groundwater contamination, particularly in complex geological systems (Akhtar et al. 2021). Moreover, the decomposition of organic matter in ecosystems such as wetlands can naturally elevate nutrient and organic compound levels in water, altering its quality without human intervention. Although natural, this contamination can significantly affect ecosystems if the concentrations of these compounds exceed the adaptive capacity of organisms. For instance, excessive levels of minerals like iron and manganese can disrupt the balance of aquatic ecosystems (Weng et al. 2007). In the volcanic basin of Campi Flegrei (Italy), hydrothermal activity increases the natural concentrations of heavy metals in groundwater (Tarzia et al. 2002).

Anthropogenic pollution, caused by human activities, represents a more substantial load and greater variety of pollutants in rivers, lakes, and seas. Recent studies show that urban and industrial systems are major sources of heavy metals and persistent pollutants, driving research into their cumulative effects on aquatic ecosystems (de Almeida Ribeiro Carvalho et al. 2022). Examples include nitrate and phosphate discharges from fertilizers and pesticides, which contaminate both surface and groundwater through aquifer infiltration (Akhtar et al. 2021).

Anthropogenic pollution also includes the introduction of toxic substances such as pesticides, fertilizers, heavy metals, and emerging pollutants like microplastics (Fig. 1) and pharmaceuticals (Schwarzenbach et al. 2010). In addition to microplastics, other micro- and nanoparticles play a significant role in aquatic contamination. Metal-based nanomaterials, such as titanium dioxide and zinc oxide, are widely used in sunscreens and cosmetics, where they eventually reach water bodies through wastewater discharge. Carbon-based nanomaterials, including carbon nanotubes and graphene derivatives, are used in electronics and coatings, raising concerns about their persistence and toxicity in aquatic ecosystems. Synthetic fibres, often released during washing processes, accumulate in sediments, affecting filter-feeding organisms. Nanoplastics, with their ability to penetrate biological membranes, present an additional risk, with potential long-term consequences for aquatic food webs and human exposure (López et al. 2022). Additionally, industrial activities release pollutants via solid and liquid waste, some of which persist in the environment for decades (Tarzia et al. 2002). With the growth of human activity and global population, the quantity and diversity of pollutants discharged into the aquatic environment have increased, negatively impacting biodiversity and limiting the availability of clean water. Current research focuses on the cumulative effects of these pollutants, many of which persist for decades, causing long-term damage.

Point and Diffuse Sources

The study of water pollution focuses on identifying point and diffuse sources, as each type poses distinct management challenges. Point sources are localized emissions that discharge pollutants from a specific location, such as industrial plants or wastewater treatment facilities. These sources are easier to regulate and control through laws and emission reduction policies. In highly industrialized areas, such as urban zones or industrial parks, point sources can contribute significant quantities of pollutants, including heavy metals, persistent organic chemicals, and nutrients (Khanam et al. 2022). For instance, in the Tha Chin River basin in Thailand, activities like aquaculture and pig farming are significant point sources of nitrogen and phosphorus, contributing to eutrophication and oxygen depletion (Schaffner et al. 2009). In rural-industrial areas of China, the regulation of industrial discharges has shown limited progress due to challenges in enforcement and monitoring, despite stringent regulations (Ongley et al. 2010). Currently, many environmental regulations require industries to monitor and limit their discharges to reduce pollution.

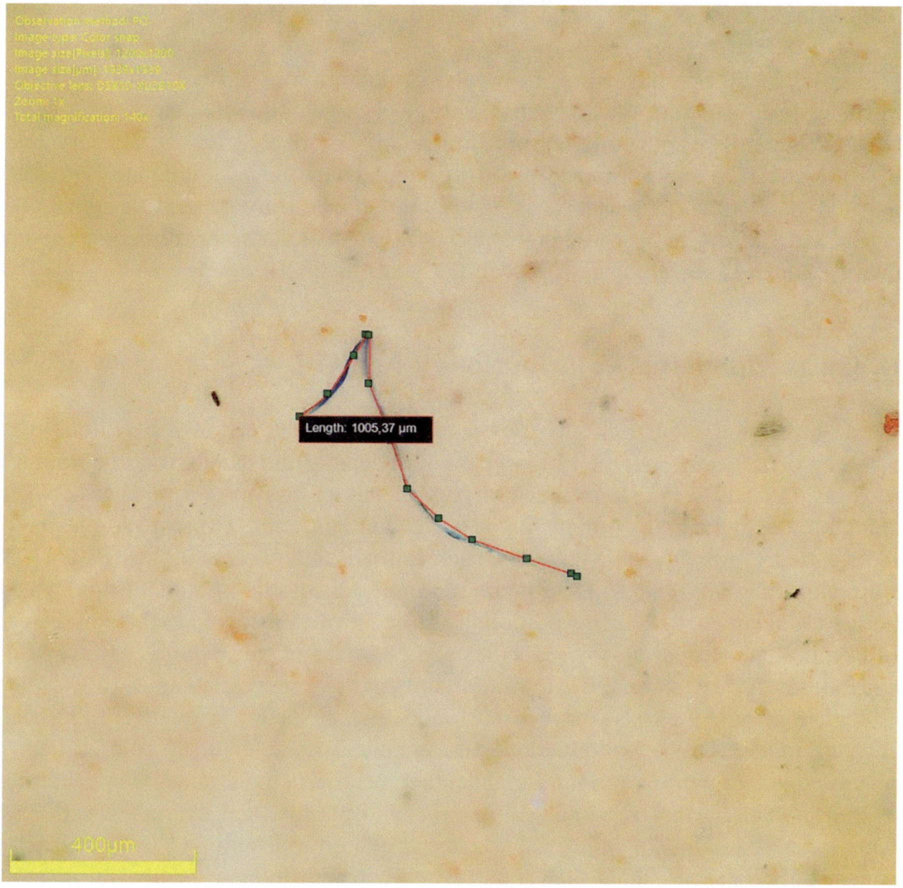

Fig. 1 An example of microplastic fibre from an aquatic lagoon

On the other hand, diffuse sources are more dispersed and harder to control. Examples include agricultural runoff, which transports pesticides, fertilizers, and sediments from fields to water bodies, particularly during heavy rains. In the Taihu Lake basin (China), 52% of nitrogen and 54% of phosphorus originate from diffuse sources, exacerbating eutrophication (Xia et al. 2020). Studies in China estimate that diffuse agricultural pollution accounts for up to 81% of nitrogen and 93% of phosphorus in certain basins, highlighting its importance in water quality management (Ongley et al. 2010). This type of pollution is highly variable, depending on factors such as climate, soil type, topography, and regional agricultural practices (Novotny 2002). Additionally, urban runoff, where rainwater collects pollutants from streets and impermeable surfaces, represents a significant diffuse source in urban areas. Mathematical models, such as material flow analysis, have quantified these sources in complex watersheds and proposed targeted mitigation measures (Schaffner et al. 2009).

Studies in Europe have shown how the combination of agricultural runoff and intensive fertilization in the Po Valley (Italy) has caused eutrophication in the Adriatic Sea, negatively impacting marine biodiversity (Giani et al. 2012). Conservation tillage techniques and optimized fertilizer management have significantly reduced nutrient losses to surface water, easing the burden of diffuse sources in vulnerable regions. In Taihu (China), these practices have demonstrated improvements in water quality in areas affected by eutrophication (Xia et al. 2020).

Urban and Industrial Contributions

Urban areas and industrial activities significantly contribute to the pollution of water bodies. Cities, with their high population density and extensive infrastructure for transport and services, produce large volumes of stormwater and wastewater containing heavy metals, organic compounds, and nutrients. Urbanization leads to the creation of impermeable surfaces, such as roads and pavements, which prevent water infiltration into the soil and increase runoff, transporting pollutants into water bodies (Khanam et al. 2022). Additionally, urban growth is linked to the emergence of new pollutants, such as microplastics and pharmaceuticals, which challenge the capabilities of conventional wastewater treatment systems (McGrane 2016).

Industrial activities also have a profound impact on water quality. Industries such as mining, energy production, and chemical manufacturing release heavy metals, toxic organic compounds, and other persistent pollutants. For example, in the Qiantang River in China, ammonia, phosphorus, and cyanide discharges were identified as major industrial sources of pollution, stemming from chemical and metallurgical activities that adversely affect both human health and aquatic biodiversity (Huang et al. 2010). In many countries, industrial processes contribute substantially to water pollution, and the effects of these pollutants often extend beyond production sites, as river systems carry pollutants to downstream water bodies (Weerasooriya et al. 2021).

In China, the implementation of green progress technologies, such as advanced treatments for reducing chemical oxygen demand (COD) and ammonia, has shown promising results. However, the investment required to adopt these technologies remains a critical challenge (Li et al. 2019). Embracing advanced treatment technologies and sustainable practices can significantly mitigate industrial impacts on water quality, but these changes demand substantial investments and sustained commitment from stakeholders.

Climate Change Impact on Pollutants

Climate change significantly influences the dynamics and distribution of pollutants in water bodies (Shukla et al. 2023). Changes in precipitation patterns and rising global temperatures directly affect the mobility and accumulation of these substances in aquatic ecosystems. Intense rainfall, increasingly common due to climate change, exacerbates surface runoff, transporting pesticides, fertilizers, and other pollutants into rivers and lakes. This phenomenon is particularly concerning in agricultural regions, where chemicals used in crop production drive eutrophication in water bodies (Kundzewicz et al. 2007). Additionally, torrential rains combined with strong winds increase the dispersion of microplastics, which often adsorb pollutants like pesticides and heavy metals, intensifying their accumulation and compounding ecological impacts (An et al. 2024).

Prolonged droughts, another consequence of climate change, reduce river flow, concentrating pollutants in specific areas and worsening their effects on aquatic organisms. Simultaneously, rising water temperatures alter the solubility and toxicity of pollutants such as heavy metals, making them more hazardous to ecosystems. Moreover, global warming enhances the volatilization of POPs from soils and sediments into the atmosphere, redistributing these substances on a global scale. This process is especially relevant in polar regions, where melting ice releases long-trapped pollutants into the food chain (Ma et al. 2016).

The thawing of permafrost releases previously sequestered pollutants, including PCBs and heavy metals, disrupting biogeochemical cycles in sensitive ecosystems such as Antarctica (Potapowicz et al. 2019). In some cases, higher temperatures accelerate contaminant degradation, while in others, they extend persistence due to alterations in organic carbon cycles in aquatic environments (Noyes et al. 2009; Ma et al. 2016). Lastly, meteorological conditions and atmospheric circulation influence contaminant deposition in coastal zones, as observed in the Baltic Sea (Störmer 2011). In polar regions, POPs are mobilized from ice reservoirs into aquatic ecosystems, significantly impacting bioaccumulation processes (Ma et al. 2016).

Representative Cases of Water Pollution

Water pollution is a global challenge affecting both aquatic ecosystems and human health, stemming from diverse sources, including natural and anthropogenic origins.

In the Subarnarekha River, India, mining and industrial activities have caused severe heavy metal pollution, including arsenic, posing significant carcinogenic and noncarcinogenic risks, particularly to children. This highlights the impact of anthropogenic point sources on water quality and public health (Giri and Singh 2015). Similarly, in Nirmal Province, India, excessive fertilizer use has resulted in diffuse nitrate contamination in groundwater, primarily affecting children (Adimalla et al. 2019).

In Europe, the Ebro River Basin in Spain suffers from diffuse pesticide contamination stemming from intensive agriculture. This pollution disrupts aquatic ecosystems, causing biodiversity loss and chronic effects on key organisms (Ccanccapa et al. 2016). A comparable issue is observed in Lake Taihu, China, where agricultural runoff rich in nitrogen and phosphorus has led to massive cyanobacterial blooms, impacting the lake's potable water supply and ecosystem services (Xia et al. 2020). Similarly, in the northern Adriatic Sea, nutrient discharges exacerbated by climate change have intensified hypoxia and anoxia in deep waters, resulting in significant biodiversity loss (Giani et al. 2012).

Examples such as the Jiulong River and the Chang Jiang River Estuary in China illustrate how point and diffuse sources interact with climatic factors to amplify ecological impacts (Zheng et al. 2016; Xu et al. 2018). Moreover, the Thames River in the UK underscores the growing issue of microplastics from urban sources such as thermoplastic paints and wastewater. These microplastics accumulate in river sediments, posing ecological and human health risks (Horton et al. 2017).

The review of these cases, summarized in Table 1, underscores the diversity of pollution problems and their impacts. Nutrient, heavy metal, and microplastic pollution exhibit varied patterns depending on sources and geographic contexts. This diversity highlights the need for integrated approaches, including the regulation of anthropogenic sources, contaminant load management, and climate-resilient solutions to sustainably address these issues.

The review of the presented cases confirms that water pollution is a complex global issue with significant repercussions for both aquatic ecosystems and human health. The diversity of analysed scenarios highlights the need for implementing comprehensive approaches tailored to the specificities of each context. Mitigation strategies should include the strict regulation of anthropogenic sources, control of nutrient and metal loads, improvements in wastewater treatment, and the sustainable management of agricultural and industrial activities. Furthermore, it is essential to address the interaction between pollution sources and natural factors, such as climate change, to develop resilient and effective long-term solutions.

Precise Monitoring for Adaptive Management

Given the significant impact of various pollution sources on water quality, it is crucial to implement monitoring systems capable of accurately tracking key parameters in real time (Talukdar et al. 2023). Continuous monitoring systems, such as real-time sensors and distributed station networks, enable the collection of detailed data on the presence and concentration of pollutants in different water bodies. This capability not only facilitates the identification of trends and patterns in water quality but also allows for rapid responses to acute pollution events (Hemdan et al. 2023).

The adaptive monitoring approach is essential to adjust management strategies based on changing environmental conditions, especially in climate change and increasing human activities in sensitive areas. Moreover, emerging technologies such as IoT, drones, and AI offer innovative tools that enhance the scope and precision of real-time monitoring. These technological solutions optimize informed decision-making and significantly reduce response times to critical pollution events (Chen et al. 2023).

Table 1 Representative cases of water pollution

Water body	Type of contamination	Source	Origin	Spread	Effects on ecosystem and humans	References
Global ecosystems	Heavy metals	Industry, agriculture, waste management	Anthropogenic	Diffuse	Bioaccumulation, trophic transfer, health risks for humans	Singh et al. (2024a)
Chang Jiang estuary, China	Microplastics	Fisheries and hydrodynamics	Anthropogenic	Diffuse	Ecological risks; ingestion of microplastics	Xu et al. (2018)
Lake Taihu, China	Nitrogen and phosphorus	Agricultural runoff	Anthropogenic	Diffuse	Algal blooms; potable water risks	Xia et al. (2020)
North Adriatic Sea	Nitrogen and phosphorus	Agriculture, wastewater, climate change	Anthropogenic/natural	Diffuse	Hypoxia, anoxia, biodiversity loss	Giani et al. (2012)
Nirmal Province, India	Nitrates	Fertilizers and septic tanks	Anthropogenic	Diffuse	Significant noncarcinogenic risks, especially in children	Adimalla et al. (2019)
Ebro River, Spain	Pesticides	Intensive agriculture	Anthropogenic	Diffuse	Biodiversity loss; ecosystem alterations	Ccanccapa et al. (2016)
Rivers of Zhengzhou, China	Urban waste	Urbanization	Anthropogenic	Diffuse	Biodiversity reduction; loss of ecological functionality	Li et al. (2023)
Cikakembang River, Indonesia	BOD, COD, ammonia	Industry and wastewater	Anthropogenic	Mixed	High BOD and COD levels; water quality degradation	Fitriana et al. (2024)
Qiantang River, China	Nutrients and heavy metals	Industry, urban, and agricultural runoff	Anthropogenic	Mixed	Eutrophication, biodiversity loss, anaerobic conditions	Huang et al. (2010)
Tha Chin River, Thailand	Nitrogen and phosphorus	Aquaculture, rice paddies, livestock	Anthropogenic	Mixed	Eutrophication, anoxia, massive fish die-offs	Schaffner et al. (2009)
Thames River, UK	Microplastics	Urbanization, road paints, wastewater	Anthropogenic	Mixed	Sediment accumulation; potential human health impacts	Horton et al. (2017)
Jiulong River, China	Pesticides	Intensive agriculture	Anthropogenic	Point	Endocrine disruption; biodiversity loss	Zheng et al. (2016)
Subarnarekha River, India	Heavy metals	Mining and industry	Anthropogenic	Point	Cancer risk; high noncarcinogenic toxicity	Giri and Singh (2015)
Bagnoli site, Italy	Heavy metals	Industry and hydrothermal activity	Anthropogenic/natural	Point	Non-bioaccessible metals; reduced risk for future developments	Tarzia et al. (2002)

Characteristics of Pollutants

Types of Pollutants: Organic, Inorganic, Microbiological, and Emerging

Water pollution encompasses a wide variety of pollutants classified by their origin and chemical nature. This categorization is essential for assessing associated risks and defining appropriate control strategies.

- Organic pollutants: These include carbon-containing compounds such as pesticides, hydrocarbons, and POPs, primarily derived from agriculture and industry. These pollutants are highly toxic and persist in the environment for years due to their long half-life (Schwarzenbach et al. 2010). POPs like PCBs are banned in many countries because of their toxicity and bioaccumulation potential (Arnold et al. 2013). Recently, pharmaceutical compounds have emerged as a concerning subgroup due to their role in inducing antibiotic resistance and affecting microbial ecology (Arnold et al. 2013). For instance, in the Ebro River basin, Spain, high levels of pesticides and POPs have been identified, mainly due to intensive agricultural practices. Studies show that these pollutants accumulate in river sediments, directly affecting local biodiversity (Navarro-Ortega and Barceló 2011). Additionally, biochar has been highlighted as an effective tool for enhancing the retention of these pollutants in soils and reducing their mobility toward water bodies (Qiu et al. 2022).
- Inorganic pollutants: This group includes heavy metals like lead, mercury, and cadmium, which are toxic and resistant to degradation. These originate from industrial activities such as mining and metallurgy, with prolonged exposure leading to neurological and hepatic damage in humans (Singh and Sharma 2024). For example, in Lake Baikal, one of the world's deepest lakes, elevated concentrations of heavy metals such as mercury and lead have been reported, largely stemming from industrial activities in Russia (Ciesielski et al. 2016). These pollutants have polluted the lake's sediments, affecting biodiversity and causing mercury bioaccumulation in local fish species (Singh and Sharma 2024). This phenomenon has serious implications for human health, as these pollutants enter the food chain, and for local fauna relying on the lake for sustenance (Gogoi et al. 2018).
- Microbiological pollutants: These include bacteria, viruses, and parasites introduced through untreated wastewater. Agents like *Escherichia coli* and *Salmonella* are indicators of faecal contamination and are linked to waterborne diseases (Ashbolt 2004). Additionally, antibiotic residues exacerbate microbial resistance, creating complex challenges for public health management (Witte 1998). For example, in Latin America, both rural and urban areas face significant water contamination due to untreated wastewater discharge (Vogelmann et al. 2022), compounded by agricultural and urbanization growth. This issue impacts water quality for human consumption and economic activities reliant on water resources (Kristanti et al. 2022).

- Emerging pollutants: This category includes substances like microplastics, pharmaceuticals, and other endocrine disruptors. These pollutants are not fully removed by conventional wastewater treatment processes and persist in the environment (Verlicchi et al. 2012). Microplastics, for instance, are ingested by aquatic organisms and bioaccumulate in the food chain (Koelmans et al. 2017), while endocrine disruptors like bisphenol A affect reproductive systems even at low concentrations (Gogoi et al. 2018). In the Thames River, UK, recent studies have identified elevated concentrations of microplastics originating from urban waste and road paint (Horton et al. 2017). These pollutants accumulate in river sediments, impacting biodiversity (Morin-Crini et al. 2022) and posing potential risks to aquatic organisms and human health (Horton et al. 2017).

Physicochemical Properties and their Ecological Impact

The physicochemical properties of pollutants determine their behaviour in aquatic ecosystems and their capacity to generate adverse impacts. Understanding these properties is crucial for assessing pollutant dynamics and designing effective management strategies.

- Solubility: Solubility directly affects the mobility and bioavailability of pollutants in water bodies. Heavy metals, for example, exhibit higher solubility in low pH environments, such as acidic waters, increasing their bioavailability and toxicity to aquatic organisms (Zhao et al. 2024). This phenomenon can disrupt entire communities of microorganisms and sensitive aquatic species, impacting local food chains. For instance, the solubility of pesticides in water facilitates their dispersion, potentially contaminating distant areas and affecting non-target organisms such as fish or amphibians dependent on these habitats (Walker et al. 2005). Additionally, the accumulation of low-solubility pollutants, such as hydrocarbons, in marine sediments creates significant ecological impacts, directly affecting benthic communities and reducing seafloor biodiversity.
- Persistence: The persistence of a pollutant determines how long it remains in the environment and its potential to generate long-term effects. Persistent compounds, such as POPs and certain antibiotics, can remain active for decades, accumulating in food webs and affecting the health of organisms across all ecosystem levels (Najam and Alam 2023). For instance, penicillin's slow degradation in aquatic environments negatively impacts microbiota, disrupting biogeochemical cycles and reducing ecosystems' capacity to recover functionality (Mohanty et al. 2024). Furthermore, interactions between persistent pollutants and other chemical compounds can amplify adverse effects, impacting both individual organisms and entire populations. This behaviour underscores the need for advanced remediation technologies capable of efficiently and sustainably degrading these compounds (Goodarzi et al. 2024).
- Toxicity: Toxicity measures a pollutant's ability to harm living organisms by disrupting key biological processes. Heavy metals such as mercury and lead

bioaccumulate throughout the food chain, affecting top predators like fish and aquatic birds (Giroux et al. 2024). This not only diminishes biodiversity in impacted ecosystems but also compromises essential ecosystem services, such as food provision and biogeochemical cycle regulation. Emerging pollutants, including microplastics and pharmaceutical residues, exert toxic effects even at minimal concentrations, altering the physiology and reproduction of aquatic organisms (Tang 2024). Synergistic interactions between different pollutants, such as POPs and heavy metals, exacerbate these effects, increasing pressures on vulnerable ecosystems and hindering their natural recovery.

Mechanisms of Transport and Dispersion in Water

Pollutants in water bodies move and disperse through various mechanisms that determine their concentration and distribution in aquatic environments. Understanding these processes is essential to assess how pollutants impact ecosystems and develop effective mitigation strategies.

- Currents and water flow: Currents and flows in rivers, lakes, and oceans are key mechanisms for transporting pollutants. In rivers, pollutants travel downstream, potentially affecting multiple ecosystems along their course (Sanchez-Vila et al. 2016). This process can facilitate contamination of downstream areas from the initial source, including sensitive zones like wetlands or estuaries. For instance, in large river systems, currents can distribute pollutants over kilometres, increasing the affected area and complicating remediation efforts.
- Molecular diffusion: Molecular diffusion is a natural process where pollutants move from areas of higher concentration to areas of lower concentration. This mechanism is particularly significant in static water bodies such as lakes or reservoirs, where the absence of strong currents promotes slow and uniform dispersion of pollutants (Berkowitz et al. 2014). This prolonged presence can intensify impacts on local aquatic flora and fauna. For example, molecular diffusion plays a critical role in spreading nutrients and chemical compounds in eutrophic lakes, exacerbating ecological imbalances.
- Adsorption in sediments: Adsorption is the process by which pollutants, such as heavy metals or persistent organic compounds, adhere to sediment particles at the bottom of water bodies (John and Gopinath 2024). While this mechanism can reduce pollutant concentrations in surface water, it poses long-term risks to benthic organisms. Furthermore, adsorbed pollutants can be re-released into the water under changing environmental conditions, such as increased pH levels or enhanced biological activity in sediments. This underscores the complexity of managing pollutants in aquatic ecosystems, where sediments act as dynamic reservoirs that influence pollutant behaviour and persistence.

Trophic Impacts: Bioconcentration, Bioaccumulation, and Biomagnification

Bioconcentration, bioaccumulation, and biomagnification are key processes that explain how pollutants enter, concentrate, and transfer within living organisms and through food chains. These mechanisms generate persistent impacts on ecosystems and human health.

- Bioconcentration: Bioconcentration refers to the accumulation of pollutants in an organism directly from the environment, typically through exposure to water or air, without the need to consume contaminated food. This process is particularly relevant for pollutants such as lipophilic pesticides and heavy metals, which can enter organisms through their skin, gills, or cell membranes (Singh and Sharma 2024). For example, fish in polluted aquatic environments can absorb pollutants directly from the water, resulting in higher concentrations in their tissues compared to the surrounding medium. This phenomenon is particularly problematic in areas with high diffuse pollution, such as eutrophic lakes and rivers, where aquatic organisms are continuously exposed to persistent pollutants (Gagneten 2011).
- Bioaccumulation: Bioaccumulation occurs when an organism absorbs pollutants at a rate faster than it can eliminate them. This process is especially common with lipophilic compounds, such as PCBs and DDT, which accumulate in the fatty tissues of organisms (Walker et al. 2005). These chemicals can cause toxic effects, such as reproductive alterations and developmental problems in affected organisms. For instance, endocrine disruptors, a group of emerging pollutants, exhibit high bioaccumulation rates due to their resistance to degradation, increasing risks to the health of exposed organisms and their future generations (Ruhí et al. 2016). This phenomenon directly impacts biodiversity and the stability of aquatic ecosystems, as lower trophic organisms, such as aquatic invertebrates, accumulate pollutants that are then transferred to their predators.
- Biomagnification: Biomagnification describes the progressive increase in the concentration of a pollutant as it moves up the food chain. When a predator consumes contaminated prey, the pollutants accumulate in its tissues, and this process intensifies at each trophic level. For instance, organochlorine compounds, such as PCBs and DDT, exhibit trophic magnification factors exceeding 5, particularly in predatory species like seabirds (Skarphedinsdottir et al. 2010). The trophic magnification factor quantifies the extent to which a contaminant increases in concentration from one trophic level to the next. A TMF greater than 1 indicates biomagnification, with values above 5 suggesting a significant accumulation of pollutants at higher trophic levels. Additionally, heavy metals like mercury and arsenic also biomagnify, reaching hazardous concentrations in top-level organisms, including large fish, birds, and marine mammals (Ali and Khan 2019). This poses a significant risk to humans, who may be exposed through the consumption of contaminated fish and seafood (Alexander 1999). This process not only affects higher organisms but also triggers cascading effects that reduce the functionality of ecosystems.

Smart Monitoring for Early Detection of Pollutants

Smart water quality monitoring employs advanced technologies such as real-time sensors, IoT, and AI to detect and analyse pollutants in water bodies. These tools enable continuous tracking of key parameters, delivering precise and real-time data on the presence of pollutants. This facilitates the early identification of hazardous substances, including emerging pollutants, and allows for corrective measures to be implemented before concentrations reach critical levels (Nishan et al. 2024).

IoT-based systems are particularly effective in detecting pollutants at low levels, enhancing the sensitivity and efficiency of early warning systems (Bhardwaj et al. 2022). Integration with AI enables the analysis of large volumes of real-time data, identifying patterns and pollutant behaviours that might otherwise go unnoticed. For example, machine learning algorithms can predict the spread of pollutants in rivers or lakes, enabling authorities to make proactive decisions to safeguard water quality and associated ecosystems.

Ecological Implications of Water Mass Contamination

Effects on Aquatic Biodiversity

Water body pollution significantly impacts aquatic biodiversity, resulting in species mortality, habitat degradation, and altered ecological interactions. Pollutants such as heavy metals, POPs, and emerging pollutants impair the reproductive capacity and health of aquatic organisms, ultimately increasing mortality rates (Singh et al. 2024b). For example, anthropogenic reactive nitrogen inputs foster eutrophication processes, shifting aquatic community compositions and displacing species adapted to nutrient-poor (oligotrophic) environments (Porter et al. 2013).

Habitat loss is often associated with physical modifications to aquatic ecosystems, such as dredging and dam construction. These actions disrupt natural river flows, fragment habitats essential for sensitive species, and amplify pressures from predators while enabling invasive species to flourish (Moyle and Leidy 1992).

In the Ebro River, Spain, heavy metals and POP contamination have caused significant reductions in fish populations. Brominated flame retardants, for instance, pose a severe threat due to their persistence and bioaccumulation in aquatic environments (Eljarrat and Barceló 2008). Furthermore, nutrient runoff from agricultural activities has escalated eutrophication events, degrading habitat quality and reducing the survival rates of key aquatic species (Hald-Mortensen 2023).

The Reconquista River in Argentina exemplifies critical ecological degradation, with heavy metal and POP levels surpassing legal thresholds, drastically altering phytoplankton and zooplankton communities essential for ecosystem stability (Salibián 2006). Similarly, in Mexico, excessive use of the pesticide chlorpyrifos in agriculture has led to its accumulation in soils and water bodies, posing risks to aquatic biodiversity and the human populations depending on these resources (Ruiz-Arias et al. 2023).

Eutrophication and Dead Zones

Eutrophication is driven by the excessive enrichment of nutrients, primarily nitrogen and phosphorus, in water bodies. This process fosters unchecked growth of algae and aquatic plants, whose subsequent decomposition depletes oxygen levels, creating hypoxic or anoxic conditions known as “dead zones”, where aquatic life cannot survive (Jinglu et al. 2007).

For instance, the Baltic Sea is severely impacted by eutrophication, with expanding dead zones leading to significant reductions in fish populations and adversely affecting the regional fishing economy (Diaz and Rosenberg 2008). Similarly, Taihu Lake in China struggles with eutrophication due to excessive nutrient discharge, triggering toxic algal blooms and a marked decline in water quality (Jinglu et al. 2007). Additionally, the Changshou Reservoir in China has suffered from intensive aquaculture activities, which contribute to eutrophication and negatively affect local aquatic biodiversity (Zhang et al. 2006).

Eutrophication poses threats beyond biodiversity, with economic and public health implications. Algal blooms obstruct water intake systems, increase treatment costs, and produce toxins harmful to human health (Qin et al. 2013). Environmentally, algae reduce water transparency and limit light penetration, disrupting photosynthesis in submerged plants and degrading aquatic habitats (Sarkar 2018). These changes alter trophic chains, favour opportunistic species, and diminish phytoplankton diversity, affecting primary consumers and creating imbalances across the entire food web.

Ecologically, hypoxic conditions trigger the release of phosphorus from sediments, exacerbating eutrophication through a positive feedback loop. This process also fosters the accumulation of toxic compounds like hydrogen sulphide and ammonia, severely impacting sensitive organisms and further degrading water quality (Fei et al. 2024). These effects underscore the urgent need for integrated strategies to reduce nutrient loads and mitigate eutrophication impacts on aquatic ecosystems.

Impacts on Human Health and Terrestrial Ecosystems

Water pollution has severe repercussions for both human health and interconnected terrestrial ecosystems. Exposure to pollutants such as heavy metals and persistent organic compounds is closely linked to chronic diseases, including neurological disorders and cancer (Bellinger 2013). For example, lead is known for its neurotoxic effects, especially in children, while mercury is associated with neurological and renal damage (Patel et al. 2021). In India, wastewater discharges from hospitals, industries, and households have caused 21% of communicable diseases to be linked to the consumption of contaminated water (Meshram et al. 2021).

Pesticides and organochlorine compounds, such as lindane, affect both aquatic and terrestrial ecosystems. For instance, in certain regions, pesticide contamination has resulted in the destruction of up to 45% of crops and has also polluted nearby water bodies, threatening fish populations and the communities that rely on them as a food source, while also affecting adjacent ecosystems (Gupta and Sharma 2023). Furthermore, heavy metals from polluted water bodies can accumulate in soil, impacting plant health, reducing agricultural productivity, and degrading essential ecosystem services such as nutrient recycling and water filtration (Okereafor et al. 2020; Beduk et al. 2022).

In the Ganges River basin, water pollution has increased waterborne diseases among the local population and degraded soil quality, adversely affecting agriculture (Chakraborti et al. 2011). Arsenic accumulation in soils has significantly deteriorated the quality of rice produced in the region, exacerbating health and socioeconomic challenges for millions dependent on this staple food (Sonone et al. 2020).

A critical aspect of aquatic pollution is the presence of pharmaceutical residues, which have contributed to the development of antimicrobial resistance in aquatic microorganisms. For example, studies have detected antibiotic resistance genes in rivers near pharmaceutical facilities in China, illustrating how pharmaceutical waste exerts selective pressure on bacteria, promoting the proliferation of resistant strains (Berendonk et al. 2015). This phenomenon poses a serious public health threat as these resistant bacteria can transfer to humans and animals, complicating infection treatments (Kümmerer 2009).

Pollution and Ecosystem Services

Aquatic ecosystems provide essential ecosystem services, such as potable water supply, climate regulation, biodiversity support, and recreational opportunities. However, pollution degrades these services by impacting water quality, reducing biodiversity, and altering ecological functions, leading to significant economic and social implications for human communities (Reid et al. 2005). There is a direct relationship between the ecological status of water bodies and their ability to provide ecosystem services: ecosystems in better ecological condition have greater capacity for water purification, flood regulation, and sustainability for recreational activities. For instance, nitrogen retention efficiency, crucial for preventing eutrophication, is significantly higher in rivers and lakes with favourable ecological conditions, whereas this service is drastically reduced in degraded ecosystems, exacerbating water quality and biodiversity problems (Grizzetti et al. 2019).

For example, in Lake Erie, North America, eutrophication caused by nutrient pollution has resulted in toxic algal blooms that have affected the potable water supply for millions and reduced recreational and tourism activities in the region (Cooperrider et al. 2020). This pollution, worsened by intensive fertilizer use in surrounding watersheds, has also contributed to significant environmental and economic issues, underscoring the need to restore the lake's ecological state.

In Europe, it has been demonstrated that protecting and restoring riparian areas in watersheds improves key ecosystem services, such as sediment mitigation and nutrient retention. A notable example is the Danube River, where riparian vegetation plays a crucial role in preventing erosion and stabilizing adjacent ecosystems, enhancing the sustainability of the services offered by this fluvial system (Grizzetti et al. 2019).

Furthermore, improving ecological conditions not only benefits direct ecological services but also strengthens the attractiveness of water bodies for tourism and recreation. Policies that prioritize the conservation and restoration of aquatic ecosystems not only maximize these services but also ensure their long-term sustainability, fostering economic and social benefits (Grizzetti et al. 2019).

Continuous Monitoring as a Key Tool for Protecting Aquatic Ecosystems

Continuous water quality monitoring plays a crucial role in detecting and mitigating pollution impacts in real time, providing critical information for informed decision-making in the management and conservation of aquatic ecosystems. Advanced technologies such as online sensors, IoT, and GIS enable constant tracking of key parameters, facilitating early problem identification, ecological impact assessment, and timely implementation of corrective strategies (Thopanaiah et al. 2024).

For instance, in the Thames River, UK, the adoption of continuous monitoring systems has transformed the management of this iconic waterway. These systems have allowed for the early detection of pollution events, such as industrial spills or wastewater discharges, significantly improving authorities' response capabilities. As a result of this vigilance, the river has experienced remarkable recovery, with notable improvements in water quality and aquatic biodiversity, highlighting the positive impact of these technologies on ecological conservation (Sharma et al. 2021; Zhang et al. 2023).

Continuous monitoring not only aids in problem detection but also supports strategic decision-making based on data, enabling adaptation to changing conditions and promoting the long-term sustainability of water resources. These tools have become indispensable components of global aquatic conservation programs, helping to protect both biodiversity and the ecosystem services that rely on healthy water bodies.

Conclusions

This chapter has delved into the complexities of water body pollution, addressing its origins as well as its ecological, social, and economic consequences. The impacts of pollutants—including heavy metals, persistent organic pollutants, and emerging pollutants—extend beyond aquatic ecosystems, affecting essential ecosystem services such as the provision of drinking water, flood regulation, biodiversity support, and food supply. Eutrophication, bioaccumulation, and biomagnification stand out as key processes that amplify these effects, emphasizing the interconnectedness of aquatic and terrestrial ecosystems.

The chapter highlights the transformative role of continuous, precise, and intelligent monitoring in addressing these challenges. Tools such as real-time sensors, IoT, and GIS enable constant, real-time surveillance of water bodies, facilitating not only the early detection of pollutants but also the analysis of their behaviour. This approach allows for faster and more effective mitigation strategies, helping to prevent greater harm and restore the functionality of affected ecosystems.

The integration of emerging technologies such as AI and drones has revolutionized how aquatic ecosystems are monitored and managed. AI enables the analysis of large datasets, identification of pollution patterns, and predictive modelling, enhancing response and planning capabilities. Drones, in turn, expand monitoring reach, providing access to remote areas and collecting critical data quickly and accurately. These technologies not only optimize water resource management but also bolster ecosystem resilience against growing pressures, such as climate change and intensified human activities.

Beyond technological advancements, the chapter underscores the social implications of water pollution. The degradation of aquatic ecosystems affects drinking water quality, food security, and agricultural productivity, disproportionately impacting the most vulnerable communities. These realities reinforce the need for an integrated approach that combines biodiversity conservation, ecosystem service sustainability, and human well-being.

In the future, it will be essential to prioritize public policies that promote continuous and intelligent monitoring as a fundamental tool for sustainable water body management. Further research and innovation in monitoring technologies such as drones and AI will also be critical, improving their accessibility and applicability across diverse contexts. This, combined with interdisciplinary approaches and international cooperation, will enable the mitigation of pollution impacts, ensure ecosystem sustainability, and safeguard essential water resources for present and future generations.

References

Adimalla N, Li P, Qian H (2019) Evaluation of groundwater contamination for fluoride and nitrate in semi-arid region of Nirmal Province, South India: a special emphasis on human health risk assessment (HHRA). Hum Ecol Risk Assess Int J 25(5):1107–1124. https://doi.org/10.1080/10807039.2018.1460579

Akhtar N, Syakir Ishak MI, Bhawani SA, Umar K (2021) Various natural and anthropogenic factors responsible for water quality degradation: a review. Water 13(19):2660. https://doi.org/10.3390/w13192660

Alexander DE (1999) Bioaccumulation, bioconcentration, biomagnification. In: Environmental geology. Springer, Dordrecht, pp 43–44

Ali H, Khan E (2019) Trophic transfer, bioaccumulation, and biomagnification of non-essential hazardous heavy metals and metalloids in food chains/webs—concepts and implications for wildlife and human health. Hum Ecol Risk Assess Int J 25(6):1353–1376. https://doi.org/10.1080/10807039.2018.1469398

Alshami A, Ali E, Elsayed M, Eltoukhy AEE, Zayed T (2024) IoT innovations in sustainable water and wastewater management and water quality monitoring: a comprehensive review of advancements, implications, and future directions. IEEE Access 12:58427–58453. https://doi.org/10.1109/ACCESS.2024.3392573

Altenburger R, Brack W, Burgess RM, Busch W, Escher BI, Focks A, Mark Hewitt L, Jacobsen BN, de Alda ML, Ait-Aissa S, Backhaus T, Ginebreda A, Hilscherová K, Hollender J, Hollert H, Neale PA, Schulze T, Schymanski EL, Teodorovic I, Tindall AJ, de Aragão UG, Vrana B, Zonja B, Krauss M (2019) Future water quality monitoring: improving the balance between exposure and toxicity assessments of real-world pollutant mixtures. Environ Sci Eur 31(1):12. https://doi.org/10.1186/s12302-019-0193-1

Amatucci A, Ventura V, Simonetto A, Gilioli G (2024) The economic value of ecosystem services: meta-analysis and potential application of value transfer for freshwater ecosystems. Environ Resour Econ 87(11):3041–3061. https://doi.org/10.1007/s10640-024-00930-6

An X, Wang Y, Adnan M, Li W, Zhang Y (2024) Natural factors of microplastics distribution and migration in water: a review. Water 16(11):1595. https://doi.org/10.3390/w16111595

Arnold KE, Boxall ABA, Brown AR, Cuthbert RJ, Gaw S, Hutchinson TH, Jobling S, Madden JC, Metcalfe CD, Naidoo V, Shore RF, Smits JE, Taggart MA, Thompson HM (2013) Assessing the exposure risk and impacts of pharmaceuticals in the environment on individuals and ecosystems. Biol Lett 9(4):20130492. https://doi.org/10.1098/rsbl.2013.0492

Ashbolt NJ (2004) Microbial contamination of drinking water and disease outcomes in developing regions. Toxicology 198(1):229–238. https://doi.org/10.1016/j.tox.2004.01.030

Beduk F, Aydin S, Aydin ME, Bahadir M (2022) Consequences of heavy metals in water and wastewater for the environment and human health. In: Bahadir M, Haarstrick A (eds) Water and wastewater management: global problems and measures. Springer, Cham, pp 221–228

Bellinger DC (2013) Prenatal exposures to environmental chemicals and children's neurodevelopment: an update. Saf Health Work 4(1):1–11. https://doi.org/10.5491/SHAW.2013.4.1.1

Berendonk TU, Manaia CM, Merlin C, Fatta-Kassinos D, Cytryn E, Walsh F, Bürgmann H, Sørum H, Norström M, Pons M-N, Kreuzinger N, Huovinen P, Stefani S, Schwartz T, Kisand V, Baquero F, Martinez JL (2015) Tackling antibiotic resistance: the environmental framework. Nat Rev Microbiol 13(5):310–317. https://doi.org/10.1038/nrmicro3439

Berkowitz B, Dror I, Yaron B (2014) Transport of passive contaminants. In: Berkowitz B, Dror I, Yaron B (eds) Contaminant geochemistry: interactions and transport in the subsurface environment. Springer, Berlin, pp 255–266

Bertrand G, Hirata R, Pauwels H, Cary L, Petelet-Giraud E, Chatton E, Aquilina L, Labasque T, Martins V, Montenegro S, Batista J, Aurouet A, Santos J, Bertolo R, Picot G, Franzen M, Hochreutener R, Braibant G (2016) Groundwater contamination in coastal urban areas: anthropogenic pressure and natural attenuation processes. J Contam Hydrol 192:165–180. https://doi.org/10.1016/j.jconhyd.2016.07.008

Bhardwaj A, Dagar V, Khan MO, Aggarwal A, Alvarado R, Kumar M, Irfan M, Proshad R (2022) Smart IoT and machine learning-based framework for water quality assessment and device component monitoring. Environ Sci Pollut Res 29(30):46018–46036. https://doi.org/10.1007/s11356-022-19014-3

Carvalho L, Mackay EB, Cardoso AC, Baattrup-Pedersen A, Birk S, Blackstock KL, Borics G, Borja A, Feld CK, Ferreira MT, Globevnik L, Grizzetti B, Hendry S, Hering D, Kelly M, Langaas S, Meissner K, Panagopoulos Y, Penning E, Rouillard J, Sabater S, Schmedtje U, Spears BM, Venohr M, van de Bund W, Solheim AL (2019) Protecting and restoring Europe's waters: an analysis of the future development needs of the water framework directive. Sci Total Environ 658:1228–1238. https://doi.org/10.1016/j.scitotenv.2018.12.255

Ccanccapa A, Masiá A, Navarro-Ortega A, Picó Y, Barceló D (2016) Pesticides in the Ebro River basin: occurrence and risk assessment. Environ Pollut 211:414–424. https://doi.org/10.1016/j.envpol.2015.12.059

Chakraborti D, Das B, Murrill MT (2011) Examining India's groundwater quality management. Environ Sci Technol 45(1):27–33. https://doi.org/10.1021/es101695d

Chen L, Chen Z, Zhang Y, Liu Y, Osman AI, Farghali M, Hua J, Al-Fatesh A, Ihara I, Rooney DW, Yap P-S (2023) Artificial intelligence-based solutions for climate change: a review. Environ Chem Lett 21(5):2525–2557. https://doi.org/10.1007/s10311-023-01617-y

Cho KH, Pachepsky Y, Ligaray M, Kwon Y, Kim KH (2020) Data assimilation in surface water quality modeling: a review. Water Res 186:116307. https://doi.org/10.1016/j.watres.2020.116307

Chowdhary P, Bharagava RN, Mishra S, Khan N (2020) Role of industries in water scarcity and its adverse effects on environment and human health. In: Shukla V, Kumar N (eds) Environmental concerns and sustainable development: volume 1: Air, water and energy resources. Springer, Singapore, pp 235–256

Ciesielski TM, Pastukhov MV, Leeves SA, Farkas J, Lierhagen S, Poletaeva VI, Jenssen BM (2016) Differential bioaccumulation of potentially toxic elements in benthic and pelagic food chains in Lake Baikal. Environ Sci Pollut Res 23(15):15593–15604. https://doi.org/10.1007/s11356-016-6634-0

Cooperrider MC, Davenport L, Goodwin S, Ryden L, Way N, Korstad J (2020) Case studies on cultural eutrophication on watersheds around lakes that contribute to toxic blue-green algal blooms. In: Bauddh K, Kumar S, Singh RP, Korstad J (eds) Ecological and practical applications for sustainable agriculture. Springer, Singapore, pp 357–372

de Almeida Ribeiro Carvalho M, Botero WG, de Oliveira LC (2022) Natural and anthropogenic sources of potentially toxic elements to aquatic environment: a systematic literature review. Environ Sci Pollut Res 29(34):51318–51338. https://doi.org/10.1007/s11356-022-20980-x

Diaz RJ, Rosenberg R (2008) Spreading dead zones and consequences for marine ecosystems. Science 321(5891):926–929. https://doi.org/10.1126/science.1156401

Eljarrat E, Barceló D (2008) Occurrence, fate, and behavior of brominated flame retardants in Ebro River basin (Spain). In: Mehmetli E, Koumanova B (eds) The fate of persistent organic pollutants in the environment. Springer, Dordrecht, pp 283–299

Fei Q, Li W, Wang S, Zhou Z, Wang W, Li M (2024) Risk and mechanisms of phosphorus release at the sediment–water interface of lakes in cold and arid regions during non-frozen seasons. Environ Sci Pollut Res 31(16):23579–23590. https://doi.org/10.1007/s11356-024-32704-4

Fitriana F, Yudianto D, Kent S, Polisar A, Sanjaya S (2024) QUAL2Kw simulation for pollution control strategies: a case study of the Cikakembang River, West Java, Indonesia. Water Air Soil Pollut 235(11):722. https://doi.org/10.1007/s11270-024-07526-2

Gagneten AM (2011) Effects of contamination by heavy metals and eutrophication on zooplankton, and their possible effects on the trophic webs of freshwater aquatic ecosystems. In: Ansari AA, Singh Gill S, Lanza GR, Rast W (eds) Eutrophication: causes, consequences and control. Springer, Dordrecht, pp 211–223

Giani M, Djakovac T, Degobbis D, Cozzi S, Solidoro C, Umani SF (2012) Recent changes in the marine ecosystems of the northern Adriatic Sea. Estuar Coast Shelf Sci 115:1–13. https://doi.org/10.1016/j.ecss.2012.08.023

Giri S, Singh AK (2015) Human health risk assessment via drinking water pathway due to metal contamination in the groundwater of Subarnarekha River basin, India. Environ Monit Assess 187(3):63. https://doi.org/10.1007/s10661-015-4265-4

Giroux MS, Siddiqui S, Brander SM (2024) Classic contaminants in aquatic ecosystems: POPs, PFAS, heavy metals, and microplastics. In: Siddiqui S, Brander SM (eds) Aquatic ecotoxicology: understanding pollutants, aquatic organisms, and their environments. Springer, Cham, pp 43–58

Gogoi A, Mazumder P, Tyagi VK, Tushara Chaminda GG, An AK, Kumar M (2018) Occurrence and fate of emerging contaminants in water environment: a review. Groundw Sustain Dev 6:169–180. https://doi.org/10.1016/j.gsd.2017.12.009

Goodarzi S, Torabideh M, Parsaseresht G, Abdipour H, Kamani H, Zomorrodi Jangaee T (2024) Penicillin removal from the aqueous environment based on AOPs/challenges and outlook. A review. Appl Water Sci 14(7):164. https://doi.org/10.1007/s13201-024-02223-1

Grizzetti B, Liquete C, Pistocchi A, Vigiak O, Zulian G, Bouraoui F, De Roo A, Cardoso AC (2019) Relationship between ecological condition and ecosystem services in European rivers, lakes and coastal waters. Sci Total Environ 671:452–465. https://doi.org/10.1016/j.scitotenv.2019.03.155

Grizzetti B, Vigiak O, Udias A, Aloe A, Zanni M, Bouraoui F, Pistocchi A, Dorati C, Friedland R, De Roo A, Benitez Sanz C, Leip A, Bielza M (2021) How EU policies could reduce nutrient pollution in European inland and coastal waters. Glob Environ Chang 69:102281. https://doi.org/10.1016/j.gloenvcha.2021.102281

Gupta A, Sharma B (2023) A review on water pollution by γHCH (lindane) and its removal using nanomaterials. J Toxicol Stud 1(1):195–195. https://doi.org/10.59400/jts.v1i1.195

Hald-Mortensen C (2023) The main drivers of biodiversity loss: a brief overview. J Ecol Nat Resour 7:346

Hemdan EE-D, Essa YM, Shouman M, El-Sayed A, Moustafa AN (2023) An efficient IoT based smart water quality monitoring system. Multimed Tools Appl 82(19):28827–28851. https://doi.org/10.1007/s11042-023-14504-z

Horton AA, Svendsen C, Williams RJ, Spurgeon DJ, Lahive E (2017) Large microplastic particles in sediments of tributaries of the river Thames, UK—abundance, sources and methods for effective quantification. Mar Pollut Bull 114(1):218–226. https://doi.org/10.1016/j.marpolbul.2016.09.004

Huang F, Wang X, Lou L, Zhou Z, Wu J (2010) Spatial variation and source apportionment of water pollution in Qiantang River (China) using statistical techniques. Water Res 44(5):1562–1572. https://doi.org/10.1016/j.watres.2009.11.003

Jinglu W, Chengmin H, Haiao Z, Schleser GH, Battarbee R (2007) Sedimentary evidence for recent eutrophication in the northern basin of Lake Taihu, China: human impacts on a large shallow lake. J Paleolimnol 38(1):13–23. https://doi.org/10.1007/s10933-006-9058-x

John PM, Gopinath A (2024) An overview of heavy metal pollution in aquatic sediments around the world. In: Alshemmari H, Hashmi MZ, Kavil YN, Shu-hong W (eds) Contaminated land and water: remediation and management. Springer, Cham, pp 119–129

Kay P, Hiscoe R, Moberley I, Bajic L, McKenna N (2018) Wastewater treatment plants as a source of microplastics in river catchments. Environ Sci Pollut Res 25(20):20264–20267. https://doi.org/10.1007/s11356-018-2070-7

Khan MN, Mohammad F (2014) Eutrophication: challenges and solutions. In: Ansari AA, Gill SS (eds) Eutrophication: causes, consequences and control: volume 2. Springer, Dordrecht, pp 1–15

Khanam N, Singh AA, Singh AK, Hamidi MK (2022) Water quality characterization of industrial and municipal wastewater, issues, challenges, health effects, and control techniques. In: Madhav S, Singh P, Mishra V, Ahmed S, Mishra PK (eds) Recent trends in wastewater treatment. Springer, Cham, pp 1–30

Koelmans AA, Kooi M, Law KL, Sebille E (2017) All is not lost: deriving a top-down mass budget of plastic at sea. Environ Res Lett 12(11):114028. https://doi.org/10.1088/1748-9326/aa9500

Kristanti RA, Hadibarata T, Syafrudin M, Yılmaz M, Abdullah S (2022) Microbiological contaminants in drinking water: current status and challenges. Water Air Soil Pollut 233(8):299. https://doi.org/10.1007/s11270-022-05698-3

Kümmerer K (2009) Antibiotics in the aquatic environment—a review—part I. Chemosphere 75(4):417–434. https://doi.org/10.1016/j.chemosphere.2008.11.086

Kundzewicz ZW, Mata LJ, Arnell NW, Doll P, Kabat P, Jimenez B, Miller K, Oki T, Zekai S, Shiklomanov I (2007) Freshwater resources and their management. Cambridge University Press, Cambridge, pp 173–210

Lambert S, Sinclair CJ, Bradley EL, Boxall ABA (2013) Effects of environmental conditions on latex degradation in aquatic systems. Sci Total Environ 447:225–234. https://doi.org/10.1016/j.scitotenv.2012.12.067

Li J, See KF, Chi J (2019) Water resources and water pollution emissions in China's industrial sector: a green-biased technological progress analysis. J Clean Prod 229:1412–1426. https://doi.org/10.1016/j.jclepro.2019.03.216

Li J, Huang L, Zhu K (2023) Ecological health assessment of an Urban River: the case study of Zhengzhou City, China. Sustainability 15(10):8288. https://doi.org/10.3390/su15108288

López ADF, Fabiani M, Lassalle VL, Spetter CV, Severini MDF (2022) Critical review of the characteristics, interactions, and toxicity of micro/nanomaterials pollutants in aquatic environments. Mar Pollut Bull 174:113276. https://doi.org/10.1016/j.marpolbul.2021.113276

Ma J, Hung H, Macdonald RW (2016) The influence of global climate change on the environmental fate of persistent organic pollutants: a review with emphasis on the northern hemisphere and the Arctic as a receptor. Glob Planet Chang 146:89–108. https://doi.org/10.1016/j.gloplacha.2016.09.011

McGrane SJ (2016) Impacts of urbanisation on hydrological and water quality dynamics, and urban water management: a review. Hydrol Sci J 61(13):2295–2311. https://doi.org/10.1080/02626667.2015.1128084

Meshram VA, Singh AK, Verma SR (2021) Water's vital role: challenges and consequences of pollution: a review. Int J Health Sci Res 14:347

Mohanty S, Benya A, Hota S, Das AP, Singh S (2024) Impact of COVID-19 generated emerging pollutants on aquatic environment and its effect on fish population and its edibility: a review. In: Das AP, Mishra S (eds) Impact of COVID-19 waste on environmental pollution and its sustainable management: COVID-19 waste and its management. Springer, Cham, pp 17–33

Morin-Crini N, Lichtfouse E, Liu G, Balaram V, Ribeiro ARL, Lu Z, Stock F, Carmona E, Teixeira MR, Picos-Corrales LA, Moreno-Piraján JC, Giraldo L, Li C, Pandey A, Hocquet D, Torri G, Crini G (2022) Worldwide cases of water pollution by emerging contaminants: a review. Environ Chem Lett 20(4):2311–2338. https://doi.org/10.1007/s10311-022-01447-4

Moyle PB, Leidy RA (1992) Loss of biodiversity in aquatic ecosystems: evidence from fish faunas. In: Fiedler PL, Jain SK (eds) Conservation biology: the theory and practice of nature conservation preservation and management. Springer, Boston, pp 127–169

Najam L, Alam T (2023) Occurrence, distribution, and fate of emerging persistent organic pollutants (POPs) in the environment. In: Aftab T (ed) Emerging contaminants and plants: interactions, adaptations and remediation technologies. Springer, Cham, pp 135–161

Navarro-Ortega A, Barceló D (2011) Persistent organic pollutants in water, sediments, and biota in the Ebro river basin. In: Barceló D, Petrovic M (eds) The Ebro river basin. Springer, Berlin, pp 139–166

Nishan RK, Akter S, Sony RI, Hoque MM, Anee MJ, Hossain A (2024) Development of an IoT-based multi-level system for real-time water quality monitoring in industrial wastewater. Discov Water 4(1):43. https://doi.org/10.1007/s43832-024-00092-y

Novotny V (2002) Water quality: diffuse pollution and watershed management. Wiley, Hoboken

Noyes PD, McElwee MK, Miller HD, Clark BW, Van Tiem LA, Walcott KC, Erwin KN, Levin ED (2009) The toxicology of climate change: environmental contaminants in a warming world. Environ Int 35(6):971–986. https://doi.org/10.1016/j.envint.2009.02.006

Okereafor U, Makhatha M, Mekuto L, Uche-Okereafor N, Sebola T, Mavumengwana V (2020) Toxic metal implications on agricultural soils, plants, animals, aquatic life and human health. Int J Environ Res Public Health 17(7):2204. https://doi.org/10.3390/ijerph17072204

Ongley ED, Xiaolan Z, Tao Y (2010) Current status of agricultural and rural non-point source pollution assessment in China. Environ Pollut 158(5):1159–1168. https://doi.org/10.1016/j.envpol.2009.10.047

Patel N, Chauhan D, Shahane S, Rai D, Ali Khan MZ, Mishra U, Chaudhary VK (2021) Contamination and health impact of heavy metals. In: Inamuddin, Ahamed MI, Lichtfouse E (eds) Water pollution and remediation: heavy metals. Springer, Cham, pp 259–280

Porter EM, Bowman WD, Clark CM, Compton JE, Pardo LH, Soong JL (2013) Interactive effects of anthropogenic nitrogen enrichment and climate change on terrestrial and aquatic biodiversity. Biogeochemistry 114(1):93–120. https://doi.org/10.1007/s10533-012-9803-3

Potapowicz J, Szumińska D, Szopińska M, Polkowska Ż (2019) The influence of global climate change on the environmental fate of anthropogenic pollution released from the permafrost: part I. Case study of Antarctica. Sci Total Environ 651:1534–1548. https://doi.org/10.1016/j.scitotenv.2018.09.168

Qin B, Gao G, Zhu G, Zhang Y, Song Y, Tang X, Xu H, Deng J (2013) Lake eutrophication and its ecosystem response. Chin Sci Bull 58(9):961–970. https://doi.org/10.1007/s11434-012-5560-x
Qiu M, Liu L, Ling Q, Cai Y, Yu S, Wang S, Fu D, Hu B, Wang X (2022) Biochar for the removal of contaminants from soil and water: a review. Biochar 4(1):19. https://doi.org/10.1007/s42773-022-00146-1
Reid WV, Mooney HA, Cropper A, Capistrano D, Carpenter SR, Chopra K, Dasgupta P, Dietz T, Duraiappah AK, Hassan R, Kasperson R, Leemans R, May RM, McMichael AJ, Pingali P, Samper C, Scholes R, Watson RT, Zakri AH, Shidong Z, Ash NJ, Bennett E, Kumar P, Lee MJ, Raudsepp-Hearne C, Simons H, Thonell J, Zurek MB (2005) Ecosystems and human well-being - synthesis: a report of the millennium ecosystem assessment. Island Press, Washington
Ruhí A, Acuña V, Barceló D, Huerta B, Mor J-R, Rodríguez-Mozaz S, Sabater S (2016) Bioaccumulation and trophic magnification of pharmaceuticals and endocrine disruptors in a Mediterranean river food web. Sci Total Environ 540:250–259. https://doi.org/10.1016/j.scitotenv.2015.06.009
Ruiz-Arias MA, Medina-Díaz IM, Bernal-Hernández YY, Barrón-Vivanco BS, González-Arias CA, Romero-Bañuelos CA, Verdín-Betancourt FA, Herrera-Moreno JF, Ponce-Vélez G, Gaspar-Ramírez O, Bastidas-Bastidas PJ, González FB, Rojas-García AE (2023) The situation of chlorpyrifos in Mexico: a case study in environmental samples and aquatic organisms. Environ Geochem Health 45(8):6323–6351. https://doi.org/10.1007/s10653-023-01618-4
Salibián A (2006) Ecotoxicological assessment of the highly polluted Reconquista river of Argentina. In: Ware GW, Nigg HN, Doerge DR (eds) Reviews of environmental contamination and toxicology. Springer, New York, pp 35–65
Sanchez-Vila X, Vàzquez-Suñé E, Rodríguez-Escales P, Jurado A, Folch A, Carles-Brangarí A, Carrera J, Fernàndez-Garcia D (2016) Emerging organic contaminants in aquifers: sources, transport, fate, and attenuation. In: Petrovic M, Sabater S, Elosegi A, Barceló D (eds) Emerging contaminants in river ecosystems: occurrence and effects under multiple stress conditions. Springer, Cham, pp 47–75
Sarkar SK (2018) Algal blooms: basic concepts. In: Sarkar SK (ed) Marine algal bloom: characteristics, causes and climate change impacts. Springer, Singapore, pp 1–52
Schaffner M, Bader H-P, Scheidegger R (2009) Modeling the contribution of point sources and non-point sources to Thachin River water pollution. Sci Total Environ 407(17):4902–4915. https://doi.org/10.1016/j.scitotenv.2009.05.007
Schwabe KA (2001) Nonpoint source pollution, uniform control strategies, and the Neuse river basin. Rev Agric Econ 23(2):352–369
Schwarzenbach RP, Egli T, Hofstetter TB, Gunten U, Wehrli B (2010) Global water pollution and human health. Annu Rev Environ Resour 35:109–136. https://doi.org/10.1146/annurev-environ-100809-125342
Sharma V, Joshi H, Bowes MJ (2021) A tale of two rivers: can the restoration lessons of river Thames (southern UK) be transferred to river Hindon (northern India)? Water Air Soil Pollut 232(5):212. https://doi.org/10.1007/s11270-021-05152-w
Shukla N, Gupta S, Rai S (2023) Potential impacts of climatic changes and human activity on water quality. In: Shukla P, Singh P, Singh RM (eds) Environmental processes and management: tools and practices for groundwater. Springer, Cham, pp 103–111
Singh G, Sharma S (2024) Heavy metal contamination in fish: sources, mechanisms and consequences. Aquat Sci 86(4):107. https://doi.org/10.1007/s00027-024-01121-7
Singh A, Pratap SG, Raj A (2024a) Occurrence and dissemination of antibiotics and antibiotic resistance in aquatic environment and its ecological implications: a review. Environ Sci Pollut Res 31(35):47505–47529. https://doi.org/10.1007/s11356-024-34355-x
Singh PK, Kumar U, Kumar I, Dwivedi A, Singh P, Mishra S, Seth CS, Sharma RK (2024b) Critical review on toxic contaminants in surface water ecosystem: sources, monitoring, and its impact on human health. Environ Sci Pollut Res 31(45):56428–56462. https://doi.org/10.1007/s11356-024-34932-0

Skarphedinsdottir H, Gunnarsson K, Gudmundsson GA, Nfon E (2010) Bioaccumulation and biomagnification of organochlorines in a marine food web at a pristine site in Iceland. Arch Environ Contam Toxicol 58(3):800–809. https://doi.org/10.1007/s00244-009-9376-x

Sonone SS, Jadhav S, Sankhla MS, Kumar R (2020) Water contamination by heavy metals and their toxic effect on aquaculture and human health through food chain. Lett Appl NanoBioSci 10(2):2148–2166

Störmer O (2011) Climate change impacts on coastal waters of the Baltic Sea. In: Schernewski G, Hofstede J, Neumann T (eds) Global change and Baltic coastal zones. Springer, Dordrecht, pp 51–69

Talukdar P, Kumar B, Kulkarni VV (2023) A review of water quality models and monitoring methods for capabilities of pollutant source identification, classification, and transport simulation. Rev Environ Sci Biotechnol 22(3):653–677. https://doi.org/10.1007/s11157-023-09658-z

Tang KHD (2024) Microplastics and antibiotics in aquatic environments: a review of their interactions and ecotoxicological implications. Trop Aqua Soil Pollut 4(1):60–78. https://doi.org/10.53623/tasp.v4i1.446

Tarzia M, De Vivo B, Somma R, Ayuso RA, McGill RAR, Parrish RR (2002) Anthropogenic vs. natural pollution: an environmental study of an industrial site under remediation (Naples, Italy). Geochemistry 2(1):45–56. https://doi.org/10.1144/1467-787302-006

Thopanaiah CK, Gireesh Babu CN, Gurani V, Rajani T, Pavani A, Muthukumaran D, Deivasigamani S (2024) Advanced remote sensing and generative models for comprehensive water quality management in a changing climate. Remote Sens Earth Syst Sci. https://doi.org/10.1007/s41976-024-00149-5

Tiyasha, Tung TM, Yaseen ZM (2020) A survey on river water quality modelling using artificial intelligence models: 2000–2020. J Hydrol 585:124670. https://doi.org/10.1016/j.jhydrol.2020.124670

Tran NH, Reinhard M, Gin KY-H (2018) Occurrence and fate of emerging contaminants in municipal wastewater treatment plants from different geographical regions-a review. Water Res 133:182–207. https://doi.org/10.1016/j.watres.2017.12.029

Vasilachi IC, Asiminicesei DM, Fertu DI, Gavrilescu M (2021) Occurrence and fate of emerging pollutants in water environment and options for their removal. Water 13(2):181. https://doi.org/10.3390/w13020181

Verlicchi P, Al Aukidy M, Zambello E (2012) Occurrence of pharmaceutical compounds in urban wastewater: removal, mass load and environmental risk after a secondary treatment—a review. Sci Total Environ 429:123–155. https://doi.org/10.1016/j.scitotenv.2012.04.028

Vogelmann ES, Prevedello J, da Veiga KR, Awe GO (2022) Challenges and opportunities of water security in Latin America. In: Dubey SK, Jha PK, Gupta PK, Nanda A, Gupta V (eds) Soil-water, agriculture, and climate change: exploring linkages. Springer, Cham, pp 291–308

Walker CH, Sibly RM, Sibly RM, Peakall DB (2005) Principles of ecotoxicology, 3rd edn. CRC Press, Boca Raton

Weerasooriya RR, Liyanage LPK, Rathnappriya RHK, Bandara WBMAC, Perera TANT, Gunarathna MHJP, Jayasinghe GY (2021) Industrial water conservation by water footprint and sustainable development goals: a review. Environ Dev Sustain 23(9):12661–12709. https://doi.org/10.1007/s10668-020-01184-0

Weng H-X, Qin Y-C, Chen X-H (2007) Elevated iron and manganese concentrations in groundwater derived from the Holocene transgression in the hang-Jia-Hu plain, China. Hydrogeol J 15(4):715–726. https://doi.org/10.1007/s10040-006-0119-z

Witte W (1998) Medical consequences of antibiotic use in agriculture. Science 279(5353):996–997. https://doi.org/10.1126/science.279.5353.996

Xia Y, Zhang M, Tsang DCW, Geng N, Lu D, Zhu L, Igalavithana AD, Dissanayake PD, Rinklebe J, Yang X, Ok YS (2020) Recent advances in control technologies for non-point source pollution with nitrogen and phosphorous from agricultural runoff: current practices and future prospects. Appl Biol Chem 63(1):8. https://doi.org/10.1186/s13765-020-0493-6

Xu P, Peng G, Su L, Gao Y, Gao L, Li D (2018) Microplastic risk assessment in surface waters: a case study in the Changjiang estuary, China. Mar Pollut Bull 133:647–654. https://doi.org/10.1016/j.marpolbul.2018.06.020

Yaroshenko I, Kirsanov D, Marjanovic M, Lieberzeit PA, Korostynska O, Mason A, Frau I, Legin A (2020) Real-time water quality monitoring with chemical sensors. Sensors 20(12):3432. https://doi.org/10.3390/s20123432

Yaseen ZM, Sulaiman SO, Deo RC, Chau K-W (2019) An enhanced extreme learning machine model for river flow forecasting: state-of-the-art, practical applications in water resource engineering area and future research direction. J Hydrol 569:387–408. https://doi.org/10.1016/j.jhydrol.2018.11.069

Zhang S, Liu J, Wei S, Gao J, Wang D, Zhang K (2006) Impact of aquaculture on eutrophication in Changshou reservoir. Chin J Geochem 25(1):90–96. https://doi.org/10.1007/BF02894800

Zhang H, Shen N, Li Y, Hu C, Yuan P (2023) Source, transport, and toxicity of emerging contaminants in aquatic environments: a review on recent studies. Environ Sci Pollut Res 30(58):121420–121437. https://doi.org/10.1007/s11356-023-30869-y

Zhao S, Zhao Y, Cui Z, Zhang H, Zhang J (2024) Effect of pH, temperature, and salinity levels on heavy metal fraction in lake sediments. Toxics 12(7):494. https://doi.org/10.3390/toxics12070494

Zheng S, Chen B, Qiu X, Chen M, Ma Z, Yu X (2016) Distribution and risk assessment of 82 pesticides in Jiulong River and estuary in South China. Chemosphere 144:1177–1192. https://doi.org/10.1016/j.chemosphere.2015.09.050

Use of Omics Techniques for Assessing Water Quality: Omics to Evaluate Water Quality

José Alhama, Marina Barbudo-Lunar, and Carmen Michán

Contents

Abstract

Assessing chemical pollution in water environments is crucial to prevent or even reverse its negative effects on ecosystems, including the health of humans and other living beings. Chemical analysis of pollutants needs previous knowledge of the contaminating menaces and does not provide information on their biological effects. Thus, they are not useful for the detection of emerging pollutants or for evaluating the effects of compounds that do not present a monotonic dose-effect relationship. As an alternative, the presence, absence, or changes in the abundance of sentinel organisms (bioindicators) can be used to detect water

J. Alhama · M. Barbudo-Lunar · C. Michán (✉)
Departamento de Bioquímica y Biología Molecular, Campus de Excelencia Internacional Agroalimentario CeiA3, Universidad de Córdoba, Campus Universitario de Rabanales, Edificio Severo Ochoa, Córdoba, Spain
e-mail: bb2midoc@uco.es

D. Gutiérrez et al. (eds.), *Smart Water Quality Monitoring*,
https://doi.org/10.1007/978-3-032-03049-8_3

pollution. Furthermore, biological parameters (biomarkers) evaluated in these organisms at their biochemical, histological, immunological, physiological, or organismic levels are routinely used to detect and assess the negative effect of pollutants. Many of these parameters are related to oxidative stress, as pollutants commonly exert their toxicity and produce pathogenic damages by increasing the levels of reactive oxidative species. However, the number of biomarkers is limited, and their use requires prior knowledge of their mechanism of action, making them inefficient for the evaluation of new or unknown chemicals. Omics allow us to analyse all the biomolecules from a certain sample, what makes them ideal approaches for assessing pollution in an unbiased manner. They can be used to analyse DNA (genomics), mRNA (transcriptomics), proteins (proteomics), or metabolites (metabolomics). Moreover, they can also be utilised to study biomolecules with a spatial perspective (mass spectrometry imaging, MSI) to relate changes with morphological or physiological alterations in multicellular organisms. At the micro-scale, microbes are pivotal for the health of all different types of aquatic ecosystems (i.e. marine, lakes, rivers, wetlands, estuaries, and ponds) and are present even in the most extreme conditions. Analysing microbial alterations using metaomic techniques, mainly based on the taxonomic resolution by rDNA amplicon sequencing, which informs about the composition, structure, and diversity of the microbiome, is to date the most promising method to assess the health status of aquatic environments. Nevertheless, methodological improvements are needed to assure efficient and specific extraction/isolation protocols, and complete reference databases are also imperative for precise bacterial identification. Alterations in the functional capacity of the microbiome in response to pollution can easily be inferred using bioinformatic tools as FAPROTAX or Tax4Fun. This approach has proven to be particularly useful for wastewater-based epidemiology to identify health menaces in aquatic environments. Bacterial community biotic indexes (e.g. microgAMBI) can also be calculated to assess the ecological status of water bodies. In summary, metaomics are proving to be powerful tools for assessing the health of aquatic environments, also providing key information for the effective implementation of corrective bioremediation strategies. Moreover, the increasing development of artificial intelligence tools promises to generate ever more complete databases, faster and more precise identifications, and more accurate biological indexes to evaluate environmental alterations.

Keywords

Aquatic pollution · Bioindicators · Biomarkers · Metaomics · Microbiome functionality · Omics · Wastewater-based epidemiology

Since the Industrial Revolution, natural aquatic ecosystems are being affected by the release of thousands of chemicals and other stressors, including an increase in water temperature or the lowering of its pH in the present global climate change scenario (Wang et al. 2024). These rapid alterations are impacting the health and functioning of water ecosystems, as well as prompting the extinction of thousands of species. Common chemical pollutants that reach aquatic environments come from agriculture runoff, livestock farming, and untreated sewage and include metals, pesticides, herbicides, fertilisers, detergents, or oils. Coastal areas are particularly sensitive to pollution as around 30% of the world population lives within 50 km of coastlines and this proportion is increasing along the years (Cosby et al. 2024). The evaluation of the health status of aquatic ecosystems can help identify hotspots of weakness and assist in the management of interventions that may mitigate, or even reverse, the damage caused by stressors. Implementation of actions is not only important from an environmental point of view but also directly for human beings in the context of the "One Health" concept, which highlights the fact that environmental, animal, and human health are closely linked and interdependent (Wang et al. 2024).

Public awareness about pollution has led to actions being taken nationally and internationally and, as a consequence, the concentrations of many "legacy chemicals", those that have a lasting impact in communities, are decreasing (Hites 2019). However, on the other hand, the presence of "emerging pollutants" in aquatic ecosystems has become a growing and massive concern due to their serious risks to the environment and human health. Emerging pollutants refer to chemicals that were previously unknown, undetected, or unrecognised as harmful. Thus, their presence in the environment is not necessarily new, but, until now, there has been no concern about the possible consequences of their use, there is little information about their impact, and their regulation is also scarce. Moreover, they do not need to be persistent to cause negative effects, as they are currently being continuously introduced into the environment. Examples are medicines, recreational drugs, personal-care products, antibiotics, and hormones used in cosmetics and health (Geissen et al. 2015; Petrie et al. 2015; Escher et al. 2020; Mishra et al. 2023).

Traditionally, pollution has been assessed by chemical analysis, but it requires highly sophisticated and sensitive equipment due to the growing number of new molecules released into the environment and to the diversity of subsequent derivatives generated by physical, chemical, or biological transformations (Alhama Carmona et al. 2017). Although high-resolution mass spectrometry devices have been developed, prior information about the general chemical characteristics of the compounds to be analysed is also required, which is not always possible in the case of emerging pollutants. Furthermore, chemical analysis does not provide information about the biological effects of pollution, with some pollutants sharing very similar physical and chemical properties but having a different spatial configuration (enantiomers), what results in very diverse biological responses, e.g. many pharmaceuticals (Ribeiro et al. 2012). Also, chemical pollutants are not alone in the environment but in complex mixtures, where they interact with each other, having additive, synergistic, or antagonistic effects. Moreover, some compounds at very low concentrations, even below their analytical detection limit, can contribute to toxicity (Petrie et al. 2015; Escher et al. 2020). On top of that, although the effect of pollutants is always dose-dependent, they can follow a non-monotonic trend where dose and effect do not have a linear relationship (e.g. endocrine disruptors). Altogether, we can assume that the damaging effects to living beings cannot be inferred only by determining the structure and the concentration of the pollutants in the environment by chemical analyses. To overcome all these constrains, new approaches have emerged based on the evaluation of the biological effects of pollutants. Living organisms display several responses at different biological levels when exposed to chemicals, and these responses can be used as markers of pollution. Markers include antioxidative and biotransformation enzymes, or by-products resulting from oxidative damage to biomolecules. However, as with chemical analyses, their application requires a profound previous knowledge of the pollutant/s involved, plus information about their molecular toxicity mechanisms (Alhama Carmona et al. 2017).

To overcome these limitations, novel approaches are required to quickly detect new episodes of pollution, particularly those due to the release of emerging pollutants. Furthermore, the understanding of the underlying functional mechanisms altered by stressors in living organisms can help assess their health status and, ultimately, that of the whole aquatic ecosystems, and in the implementation of containment measures to stop or even reverse their negative effects. Both objectives could be fulfilled using modern molecular technologies that allow us to identify changes in all the biomolecules of a living organism or population in a global and unbiased manner.

Bioindicators and Biomarkers

Global environmental status can be assessed using "bioindicators", sentinel living organisms that can reflect the health status of ecosystems as they respond to environmental stresses through their occurrence, disappearance, or changes in their abundance. Limnologists have used the presence or absence of selected indicator species for assessing pollution since the second half of the twentieth century. At this time, different biotic indexes were developed to quantify the damage in aquatic environments based on the presence of certain taxonomic groups, e.g. the Trent Biotic Index or the Hilsenhoff Biotic Index, which consider the presence of selected macroinvertebrates or arthropods, respectively (Sumudumali and Jayawardana 2021). Common aquatic bioindicators were macroinvertebrates, bivalves, gastropods, fishes, zooplankton, phytoplankton, and macrophytes (Sumudumali and Jayawardana 2021). Those indexes are still applied today, and recent examples include the use of several species of plankton or invertebrates (Chandel et al. 2024), odonates (dragonflies and damselflies) (Datto-Liberato et al. 2024), or microalgae as bioindicators of water quality (Martínez-Burgos et al. 2024).

Living organisms, prior to sickness or death, respond to stressors by changing different parameters (the so-called "biomarkers") at the biochemical, histological, immunological, physiological, or organismic levels. These changes not only allow us to detect environmental aquatic pollution but also provide information on its impacts on the health of the exposed organisms (Lionetto et al. 2019). A biomarker could be defined as: "The measurement of body fluids, cells, tissues or whole animals that indicate in biochemical, cellular, physiological, behavioural or energetic terms the presence of contaminants or the extent of the host response" (López-Barea 1995). Biomarkers are based on changes in aquatic sentinel organisms at different levels of organisation. However, those that analyse lower levels of organisation (i.e. molecular, cellular) can reflect pollution events earlier and more significantly, allowing to identify alterations in ecosystems before they become irreversible (Alhama Carmona et al. 2017).

Oxidative stress is probably the most common mechanism by which pollutants exert their toxicity, leading to pathogenic damage (Alhama Carmona et al. 2017; Anetor et al. 2022; Zhou et al. 2023). Under environmental physiological conditions, organisms maintain a dynamic balance between prooxidants and antioxidants through three different levels of defence mechanisms: (1) avoiding the production of reactive oxygen species (ROS), (2) eliminating the produced ROS to avoid the damage of biomolecules, and (3) repairing damages once they occur. Many pollutants generate harmful free radicals or even impair the function of endogenous antioxidants, thereby inducing oxidative stress (Anetor et al. 2022). Common aquatic pollutants that generate oxidative stress include transition metals, polycyclic aromatic hydrocarbons, organochlorine and organophosphate pesticides, polychlorinated biphenyls and dioxins (Valavanidis et al. 2006). When oxidative stress occurs, cells attempt to counteract the oxidant effects by activating antioxidative defence mechanisms. For that reason, many primary antioxidative enzymes (e.g. superoxide dismutase, catalase, glutathione peroxidase, etc.) have been extensively used as biomarkers of response to pollution (Valavanidis et al. 2006; Alhama Carmona et al. 2017). Several studies have used these biomarkers for the assessment of natural environments, such as the Tunisian coastal lagoons (Bejaoui et al. 2020; Mansour et al. 2020), the Portuguese estuaries (Duarte et al. 2017), or the Southwest Spanish coast (Solé et al. 2009) or in controlled experiments after exposures to pharmaceuticals (Trombini et al. 2021), metal ions (Geret et al. 2002; Giarratano et al. 2023), or metal nanoparticles (Buffet et al. 2013, 2015; Gomes et al. 2014; Aouini et al. 2019).

Other common molecular biomarkers include the determination of proteins that also participate in the response to pollutants, including biotransformation enzymes (e.g. cytochrome P450, glutathione *S*-transferases, etc.), or those that capture or prevent their toxicity, as metallothioneins or heat shock proteins (Bebianno and Langston 1992; Cajaraville et al. 2000; Géret et al. 2002; Alhama et al. 2006; Choi et al. 2008; Pellerin and Amiard 2009; Solé et al. 2009; Fang et al. 2010; Rocha et al. 2016). Finally, low-molecular-weight molecules can also alter their levels after exposure to pollutants, such as nonenzymatic antioxidants (GSH, vitamin C, etc.) or cell damage by-products (malondialdehyde, 8-oxoguanine, etc.) (Viarengo et al. 1990; Géret et al. 2002; Alhama Carmona et al. 2017; Yuan et al. 2018; Aouini et al. 2019; Mansour et al. 2020; Santovito et al. 2021). These parameters are regarded as "conventional biomarkers" of pollution; however, their application is not without limitations, as it requires extensive prior molecular knowledge of toxicity mechanisms, which is not usually available, particularly for the so-called emerging pollutants. Thus, the detection of pollution events depends on the prior knowledge of the contaminants present, which may bias the results. To overcome these limitations, a new generation of stress/pollution biomarkers that involve the use of Omic unbiased approaches has emerged.

What Is Omics?

Omic techniques aim at the global characterisation and quantification of the different biological molecules of a/several given organism/s. Depending on the biomolecule analysed, we can distinguish: genomics (DNA), transcriptomics (mRNA), proteomics (proteins), or even metabolomics (metabolites). These approaches provide different and complementary information about the organism/s analysed. Genomics identifies the DNA sequences of the organism/s present in a sample and provides insights into their functional potential. Transcriptomics recognises the genes that are actively transcribed by sequencing and quantifying all mRNAs, thereby enabling the identification of those that are differentially expressed in response to different environmental conditions/stressors. Proteomics provides information on the expressed proteins and, thus, on the metabolic pathways functioning, typically using mass spectrometry (MS) techniques for peptide analysis. Finally, metabolomics identifies metabolites, usually by MS or nuclear magnetic resonance (NMR) spectroscopy, picturing the chemical processes that are happening in the organism/s. All these techniques reveal different aspects of the molecular responses in the organism/s exposed to pollution, providing a global, unbiased view without requiring prior knowledge of the nature of the alterations. Examples of their use include (1) the assessment of pollution by fish proteomics in French (Laurent et al. 2024) and Portuguese estuaries (Costa et al. 2012); (2) the responses to chronic exposure to wastewater treatment plant and stormwater effluents through the use of transcriptomics in the clam *Corbicula flumine* (Bertucci et al. 2018); (3) the evaluation of the effect of environmental pollution in natural ecosystems (e.g. Doñana National Park, the Tunisian coast), using proteomics in the clam *Scrobicularia plana* and the crabs *Carcinus maenas* and *Procambarus clarkii* (Romero-Ruiz et al. 2006; Fernández-Cisnal et al. 2014; Ghedira et al. 2016); (4) the effect of pollutants in exposure experiments, e.g. metals cadmium or arsenic, in the clams *Ruditapes philippinarum* or *Ruditapes decussates* (Chora et al. 2009; Wu et al. 2013; Zhang et al. 2023), crude oils in *Corbicula fluminea* clams (Miserazzi et al. 2020), or nanoplastics in the sea bass fish (Cuesta et al. 2023); (5) the evaluation of climate warming, e.g. in the fish *Paralichthys olivaceus* (Han et al. 2025) or the lobster *Homarus americanus* (Harrington et al. 2020); and (vi) to improve aquaculture systems, e.g. the resistance to infection in clams (Smits et al. 2020). We should highlight the fact that many of the biomonitoring studies have been performed in bivalve organisms because they are sessile, filter feeders, easily sampled, and capable of bioaccumulating high levels of pollutants.

Nevertheless, although databases for the identification of genes, transcripts, and proteins are quickly improving due to the increasing number of genome-sequenced organisms, they remain scarce and largely incomplete for many non-model aquatic organisms, which significantly limits the application of Omic approaches in environmental water assessment (Alhama Carmona et al. 2017). To address this issue, one of the first global advancements was the use of "protein expression signatures", sets of proteins' spots detected by 2D electrophoresis, without their associated identifications, that exhibited differences in intensity upon exposure to pollutants in sentinel organisms such as mussels or fish (Shepard et al. 2000; Apraiz et al. 2006; Jonsson et al. 2006; Roland et al. 2014). Some of those proteins were later identified by mass spectrometry protein sequencing, and, thus, functional alterations in the organisms could be associated to sources of pollution stress (Romero-Ruiz et al. 2006; Costa et al. 2012; Fernández-Cisnal et al. 2014; Jebali et al. 2014; Ghedira et al. 2016; Alhama Carmona et al. 2017). Recent advances in gel-free high-throughput (shotgun) proteomics have highly expanded the identification of signatures related to cellular perturbations due to environmental menaces (Laurent et al. 2024).

The scarcity of databases also hinders the potential of Omic approaches for the understanding of the molecular mechanisms underlying the response of organisms to xenobiotics. But, constructing DNA sequences databases from non-model sentinel organisms can be expensive and time-consuming; plus they usually have a high taxonomic variability, and so genomic sequences are highly variable. To overcome this limitation, one of the most effective approaches is sequencing the transcriptome rather than the genome, which reduces costs and considerably less time-consuming. By combining the obtained data with a shotgun mass spectrometry analysis of proteins, proteogenomics, it is possible to create transcriptome-informed protein databases at a significantly lower financial and temporal cost (Amil-Ruiz et al. 2021; Leprêtre et al. 2023). Furthermore, in addition to classic algorithmic approaches to construct genomic/transcriptomic databases, the use of artificial intelligence (AI), machine learning (ML), and deep learning (DL) tools is facilitating the development of automated pipelines that integrate data, dramatically improving both speed and accuracy. These tools can enhance the construction of databases for non-model organisms, reducing time and costs (Schmidt and Hildebrandt 2021; Ouanes 2024).

Additionally, another of the main problems of the aforementioned Omic techniques is that they do not provide spatially resolved information, and, thus, it is almost impossible to relate biomolecular alterations with morphological or physiological changes in multicellular organisms. To solve this problem, matrix-assisted laser desorption/ionisation (MALDI) mass spectrometry imaging (MALDI-MSI) generates MALDI data along the two dimensions of a thin tissue section. In this way, data for the identification of biomolecules as peptides or lipids can be obtained with a spatial perspective, when combined with optical microscopy techniques. From the combination of MSI and optical microscopy techniques, biomolecular changes can be assigned to specific morphological alterations (Fig. 1). MSI is currently being used mainly in cancer studies for the identification of peptides, lipids, or even metabolites associated with different tumour cells, or in the distribution of therapeutic drugs and their metabolites in the different organs or tissues. Just very

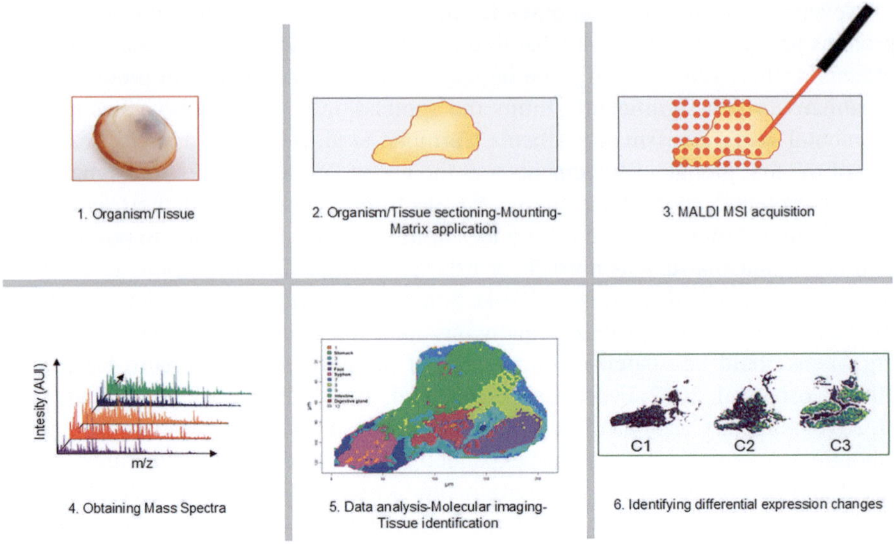

Fig. 1 General schematic workflow for a mass spectrometry imaging (MSI) analysis of biomolecules (e.g. proteins, peptides, lipids, etc.), from sample preparation to acquisition and subsequent analysis for evaluating molecular spatial changes in response to environmental stress

recently, this approach is emerging as a powerful technique for environmental analyses (Lagarrigue et al. 2016; Herruzo-Ruiz et al. 2024a). Several studies have used MSI to investigate the distribution and metabolism of pesticides such as organophosphates or pyrethroids in agricultural environments; however, to date, only a few have focused on assessing pollution in aquatic ecosystems (Liu et al. 2020; Shi et al. 2022a).

What Is the Microbiome? How Can the Microbiome of Aquatic Ecosystems Help to Assess their Quality?

In 1988, the term microbiome was used to describe "the collection of microbes and their activities within a given environment". In the last years, the interest in characterising the microbiomes of different environments has rapidly increased, mainly focusing on human-associated microbiomes, as they have unlocked new clinical treatments (Berg et al. 2020).

The evaluation of aquatic ecosystems regarding the control of drinking-water supplies and the cleaning of wastewaters to avoid public health scares has been a common practise since the dawn of human history. Among the parameters analysed to assess the quality of drinking water, microbial determinations were established in the 1980s to monitor faecal pollution, as most polluting events related to the presence of undesirable living organisms were due to unprocessed faecal discharges (McLellan and Eren 2014; Michán et al. 2021). Thus, specific enteric pathogens are currently used as indicators of faecal pollution, since they are very abundant in human and animal faeces. Standardised tests include the determination of heterotrophic plate counts (HPC) and of faecal coliforms that are used to estimate the global number of pathogens in the waters (https://iris.who.int/). To do this, slow and laborious routines are commonly used even nowadays, where standardised volumes of water must be centrifuged/filtered and the concentrated cells cultivated in selective agar media (Farhat et al. 2020). Therefore, the determination of microbial pathogens as markers of pollution in waters is neither optimal nor fast. Among the main limitations we should highlight that (1) not all pathogens can be easily detected by conventional microbiological techniques, (2) some pathogens are not even microorganisms, (3) microbes can be difficult-to-culture or can be concealed by other fast-growing organisms, and (4) the time between sampling and obtaining results can be significantly long (Chan et al. 2019; Farhat et al. 2020; Garner et al. 2021).

Water is the habitat of many different organisms, both at the macro- and micro-scales. Living organisms, particularly microorganisms, directly participate in many processes that contribute to biogeochemical cycles. These cycles refer to the natural pathways (transformations, transfers) of chemical elements and compounds between biotic (living organisms) and abiotic matrixes (water, sediments, air, etc.). They are fundamental to the health of environments, including waters, as they determine how chemical elements are stored and recycled (Graham et al. 2016). Hence, the activity of carbon, nitrogen, phosphorus, and sulphur cycles is closely related to the abundance and distribution of the different types of microorganisms within the environment. Microbes are also deeply involved in the formation of organic matter from inorganic compounds, probably constituting the most important group of organisms in the primary energy intake for the maintenance of life (Sehnal et al. 2021). When aquatic ecosystems are exposed to pollution (e.g. microplastics, nanoparticles, pharmaceuticals, etc.) or any other type of environmental stress (e.g. heat, salt, pH, etc.), the microbiome present in both the biotic and abiotic matrixes is affected (Seeley et al. 2020; Li et al. 2021; Lin et al. 2021; Shang et al. 2022; Duhamel 2024). Therefore, studying the microbiome to assess environmental aquatic changes is expected to have enormous potential applications.

It has been estimated that there are 10^{29} bacterial and archaeal cells in the oceans and that fresh waters contains 10^5 to 10^7 bacteria per millilitre. These prokaryotes are pivotal components of aquatic ecosystems as they play a key role in regulating most of their biogeochemical fluxes and processes (Graham et al. 2016; Overmann and Lepleux 2016; Grossart et al. 2020; Behera et al. 2021). Indigenous bacteria in natural ecosystems are capable of metabolising various classes of organic pollutants as polycyclic aromatic hydrocarbons or polychlorinated biphenyls. Thus, they directly participate in bioremediation processes, including the biodegradation of toxic pollutants related to oil and other chemical spills (Mapelli et al. 2017; Behera et al. 2021). Altogether, prokaryotic dynamics shape aquatic ecosystem processes, although, to date, the exact mechanisms remain unclear.

Microbes form complex communities in the environments (Angulo et al. 2019). The diversity and abundance of taxa and/or the global structure of the microbiome (e.g. organisms can compete or cooperate) in a certain ecosystem provide pivotal information on the potential functions and process rates within the community. Microorganisms are also particularly sensitive to environmental changes due to their high surface-to-volume ratio, which brings them into close, direct and immediate contact with the stressors (e.g. pollutants). Consequently, the abundance and distribution of the different species of microbes reflects the health status of the waters (Graham et al. 2016; Grossart et al. 2020; Michán et al. 2021; Fontaine et al. 2023). On top of that, contrary to macro-organisms, microbes are ubiquitous and inhabit all ecosystems, even those with the most extreme conditions, such as very low or high temperatures, high levels of radiation, high atmospheric pressures, extreme acidity or alkalinity conditions, absence of water, presence of high levels of salts, or exposure to high concentrations of toxic chemicals. The analysis of the water microbiome is considered one of the most promising, powerful, and reliable parameters for the environmental diagnosis of aquatic ecosystems since microbes are ubiquitous and respond to environmental changes in early and sensitive ways (Graham et al. 2016; Grossart et al. 2020; Michán et al. 2021).

Multicellular organisms also have host-associated microbiomes. Gut microbiota is heavily influenced by the diet and plays a crucial role in the health of marine organisms. As in mammals, the gut microorganisms from aquatic animals form a complex community that influences nutrition, immunity development, and metabolism. Several studies have linked changes in the microbial communities of the digestive tract with environmental stress, including aquatic pollution (Adamovsky et al. 2018; Fackelmann and Sommer 2019; Behera et al. 2021; Sehnal et al. 2021; Masanja et al. 2023). Also, the gills and skin of fish are coated with a mucus layer, which is also the habitat of numerous microbes. Furthermore, the microbiota from the respiratory system, the mouth, or the skin of marine mammals can be analysed using noninvasive techniques to provide information on their health status with the purpose of conserving these critically endangered animals. The microorganisms present in these tissues can not only influence and reflect their health but also quickly respond to environmental changes (Sehnal et al. 2021).

Alterations in the microbiome are due not only to the presence of unwanted/foreign chemicals or biological elements, e.g. pollutants, pathogens, etc., but also to changes in physical parameters, such as water temperature, salinity, or pH. All these factors may contribute to the ecological health status of the microbial community. Environmental changes shape the structure, composition, and functional activity of microbial communities and, therefore, can also be used for the early sensing of chemical and physical environmental stressors in a variety of aquatic environments, such as coastal waters, rivers, and marshes (e.g. Mustafa et al. 2016; Li et al. 2020, 2023; Behera et al. 2021; Michán et al. 2021; Wang et al. 2021).

How Can We Study the Microbiome?

One of the main hindrances for the study of the microbiome is that most environmental microorganisms are not detectable using cultivation-based methods, which have traditionally been the standard biological tools used in microbiological studies (Lloyd et al. 2018). Thus, to discover the relationships between environmental changes and alterations in the microbial composition, structure, and activity, cultivation-independent methodologies are required. Several technologies have been used to reveal the taxonomic composition of the microbiome in aquatic environments, including antibody-based methods that can be coupled with flow cytometry (e.g. Priyadarsini et al. 2023), quantitative polymerase chain reaction (qPCR) (e.g. Botes et al. 2013), or droplet digital PCR (ddPCR) (e.g. Picard et al. 2022). All of them are highly sensitive methodologies for the determination of specific species, although not optimal for the simultaneous analysis of numerous taxa. Furthermore, reverse qPCR could also be used for the identification of actively expressed genes in a certain condition (Botes et al. 2013; Tiwari et al. 2022). However, these techniques can analyse a limited number of microbial species. For that reason, as it happens with conventional biomarkers, they require prior knowledge of the microbes to be studied and their potential relationships with the stressors to be screened.

When all biomolecules from a complex environmental sample are simultaneously analysed without the previous segregation of the microbial living organisms, the Omic techniques are referred to as "Meta-Omics". Thus, we have metagenomics, metatranscriptomics, metaproteomics, or metabolomics, depending on if the global content of genes, transcripts, proteins, or metabolites of the microbiome, respectively, is analysed (Fig. 2). Generally, these techniques are versatile and very informative at a low cost. Their development has recently skyrocketed due to their intensive application in the clinical field with the discovery that dysbiosis in microbial communities of the human microbiome could be related to a broad spectrum of diseases. Accordingly, linking distinct bacterial taxa to specific diseases is opening new treatment targets and therapeutic approaches in many different clinical settings (Wang et al. 2022; Michán-Doña et al. 2024).

However, despite the importance of microorganisms in different types of environments, including aquatic ecosystems, the development of specific biomonitoring tools for assessing the status of water quality is still very scarce. The main challenges are related to technical limitations, such as (1) the high number of geographical genomic variations, (2) the difficulty in extracting biomolecules (e.g. nucleic acids, proteins) from complex abiotic and biotic matrixes, (3) the instability of biomolecules in complex environments, (4) the difficulty of deep nucleic acid/protein sequencing, or (5) the lacking of good databases for identifying the isolated biomolecules (Behera et al. 2021; Sehnal et al. 2021; Fontaine et al. 2023).

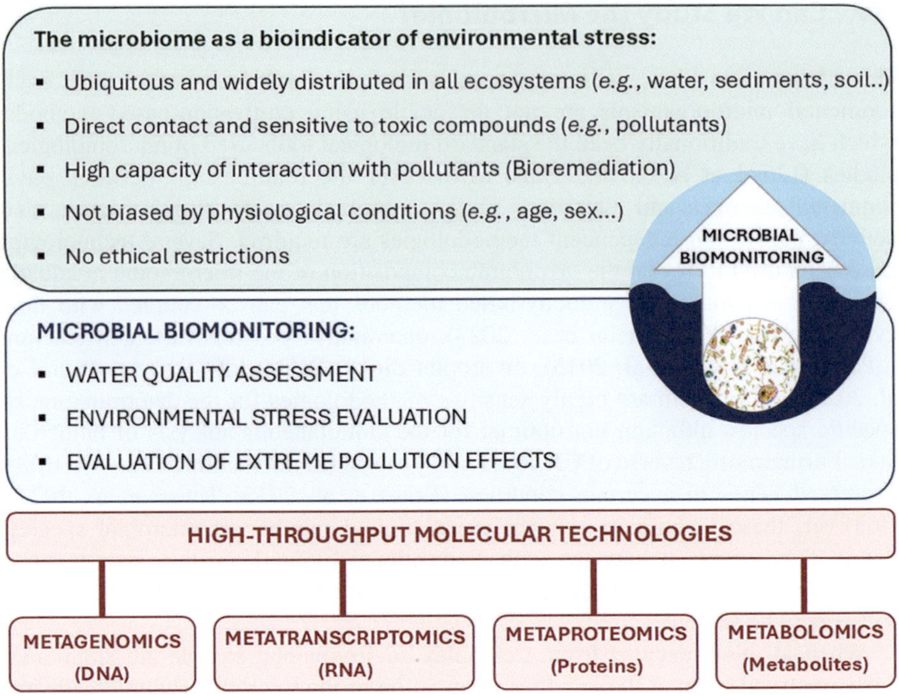

Fig. 2 Application of high-throughput technologies for the microbial biomonitoring of aquatic environments

Composition, Biodiversity, and Functionality of the Microbiome

The term environmental DNA (eDNA) refers to the genetic material isolated from environmental samples, including aquatic ones, that can be used to identify the different taxa within the ecosystems. This approach has been applied to evaluate water quality in riverine ecosystems as the Yangtze in China (Zhang et al. 2024), estuaries as the Gold Coast Broadwater in Australia (Downie et al. 2024) or to identify faecal pollution in coasts and beaches in Canada (Saleem et al. 2025).

Within eDNA, the bacteriome DNA is usually majoritarian and of great importance. The analysis of the 16S rDNA sequence is routinely used to infer the composition of the bacteriome (Fig. 3). This approximation only sequences segments of the hypervariable regions (V1–V9) of *rrs* genes and searches them against DNA databases (e.g. SILVA, Greengenes) (DeSantis et al. 2006; Quast et al. 2013). The strengths of this analysis lie in its simplicity, speed, and cost-effectiveness. Most importantly, it provides us pivotal information on bacterial composition and its evolution in response to environmental stress. However, among its weaknesses, it is worth highlighting that (1) only identifies bacteria and archaea, (2) family is the lowest taxonomic level most commonly identified, and (3) biased results can be obtained depending on the hypervariable region/s analysed (Johnson et al.

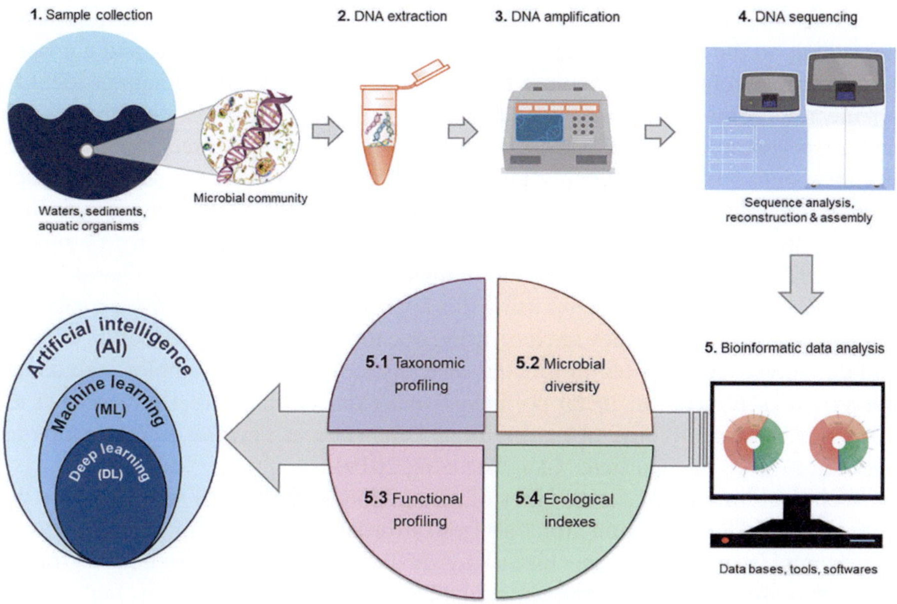

Fig. 3 General schematic workflow of a metagenomic analysis starting with the extraction and amplification of eDNA from the abiotic/biotic samples, followed by the high-throughput sequencing of DNA and ending with the bioinformatic analysis of the data. Relevant information about the composition, structure, and functional capacity of the environmental microbiome is obtained, allowing the ecological status of the aquatic ecosystem to be evaluated. The implementation of new powerful data integration technologies—AI, ML and DL—will provide a dramatic improvement in speed and accuracy

2019; Michán et al. 2021). Nevertheless, many studies have successfully used this approach to assess pollution effects in aquatic abiotic matrixes, e.g. metals and metallic nanoparticles (Tao et al. 2016; Guo et al. 2019; Herruzo-Ruiz et al. 2024b), drugs (Borsetto et al. 2021), mixtures of pollutants (Wang et al. 2021), and heavily contaminated sites (Liu et al. 2015; Mustafa et al. 2016; Chen et al. 2018; Jacquiod et al. 2018; Yin et al. 2022), and for the study of the bacteria present in marine organisms a.k.a. biotic matrixes, e.g. *Mytilus galloprovincialis* exposed to metal nanoparticles (Auguste et al. 2019) and *R. philippinarum* from polluted sites (Milan et al. 2018), etc.

To overcome the limitation of 16S rDNA sequencing, which only identifies bacteria and archaea, other DNA regions might be sequenced. Among them, it is worth noting the potential of the internal transcribed spacer (ITS) region to analyse the presence of fungi, the 18S rDNA sequence to characterise the presence of micro-eukaryotes, and the 23S rDNA region to report the occurrence of microalgae. Different databases are available to study these regions, such as UNITE, SILVA, or µgreen-db (Quast et al. 2013; Djemiel et al. 2020; Pérez-Cobas et al. 2020; Abarenkov et al. 2024). However, as it happens with 16S rDNA analysis, family is usually the lowest taxonomic level achieved, and the results could be biased depending on the sequenced regions. On the other hand, the complete and unbiased shotgun sequencing of all genomes from a certain environment can help us identify all the microorganisms present and their functionality (viruses, bacteria, fungi, and micro-eukaryotes) using next-generation sequencing (NGS) techniques. On the negative side, costs considerably increase, and the analysis of the results is much more challenging, as the genome fragments are difficult to assemble due to repetitive regions. In addition, environmental metagenomic databases are far for complete, particularly those from extreme environments (Behera et al. 2021; Garner et al. 2021). For these reasons, studies using this approach are fewer than those analysing the 16S rDNA, although they are increasing due to the progressive reduction of costs and the improvement of bioinformatics analysis tools. Recent examples include the analysis of sediments in metal polluted rivers (Yan et al. 2020; Dell'Anno et al. 2023), or the microbiome of marine organisms such as *Mytilus galloprovincialis* exposed to TiO_2 nanoparticles (Auguste et al. 2019).

Technical limitations for all types of metagenomic analysis of water include the lack of robust DNA isolation protocols from affordable volumes of water, as many of these samples contain very low biomass. Aquatic sediments can also be used for the analysis of the microbiome, but, although their biomass concentration is higher, they are very complex matrices containing elevated levels of mineral and organic matter. Minerals can absorb nucleic acids and make their extraction difficult, and organic phenolic compounds, as humic or fulvic acids, derived from degraded matter, can interfere with the extraction and inhibit enzymes required for the subsequent analysis, e.g. polymerases (Purdy 2005). All these facts can bias the analysis results as, ideally, for an accurate comparison among the different samples, DNA extraction methodologies should be 100% effective and consistent. Luckily, several DNA extraction kits with high yields are now available and can also be supplemented with good isolating reagents that avoid degradation, stabilising the DNA. Nevertheless, samples from highly polluted environments can be very challenging due to the presence of many and mostly unidentified, interfering compounds (authors' personal experience).

Taxon heterogenicity can also be used for the global assessment of water quality, yet this parameter is complex to determine. Several diversity/uniformity indexes based on the counts of individuals or their relative proportion among different taxa in a community are frequently used in a complementary manner (Fig. 3). Usually, these indexes are strongly correlated, though not interchangeable. Diversity or richness, e.g. indexes of Shannon-Wiener and Gini-Simpson, respectively, informs on the variety of species in a community. They are quantitative measures that reflect the number of different microbial species and how evenly the individuals are distributed among the community. The uniformity (evenness) of the microbial population can be estimated applying the Pielou's index. All of them can be intuitive mathematical tools to quantify and assess patterns and changes in biodiversity, which could reflect the health of the aquatic ecosystem (Chao et al. 2016; Thukral 2017). The interpretation of these indexes can also be complex, but a higher diversity is usually associated to a healthier environment, as stress tends to reduce the heterogenicity of the affected communities by selecting only the most stress-tolerant organisms (Francé et al. 2021). On the other hand, their main limitation is related not to the index itself but to the taxonomic resolution of the data available. As stated above, identification at low taxonomic levels (i.e. genus, species) is very difficult to achieve, what can constrain the ecological relevance of the calculated indexes.

Moreover, the functionality of the microbiome is intimately related to its taxonomy (Fig. 3), and who, how, and what is in the microbiome is connected to what it can do (Grossart et al. 2020). Actually, different taxonomic compositions may have similar functions as microorganisms may be functionally redundant (Louca et al. 2018). Thus, several algorithms have been developed to assess the potential functionality of the microbiome. These algorithms are based on previously available databases linking taxa to specific pathways/metabolic functions. Most of the currently available tools are based on 16S rDNA data and, thus, dedicated to bacteria/archaea, e.g. FAPROTAX, Tax4Fun, and PICRUSt (Aßhauer et al. 2015; Wemheuer et al. 2020; Djemiel et al. 2022). FAPROTAX matches prokaryotic taxa to ecological functions, using verified literature on previously cultured strains, covering over 7600 potential functions and 4600 taxa (Louca et al. 2016). Tax4Fun algorithm matches taxa with functions based on an association matrix built from a BLASTN analysis between the 16S rDNA gene sequences of the KEGG database prokaryotic organisms and the SILVA database, which allows to link KEGG pathways to prokaryotic taxa (Aßhauer et al. 2015). Other popular tool is PICRUSt2, which follows a more complex workflow, as it infers the genomic content of sampled 16S rDNA gene sequences based on 41,926 bacterial and archaeal genomes from the IMG database (version from 8 November 2017) (Douglas et al. 2020). Globally, the FAPROTAX database is focused on ecologically relevant potential pathways and functions, whereas Tax4Fun and PICRUSt2 support broader applications, including human microbiomes. These databases are a rapid and cheaper alternative to shotgun sequencing. However, they do not provide information about the real genomic content, as they only inform about potential functions, and, as with the heterogenicity indexes, taxonomic resolution is vital to obtain accurate functional predictions.

Based on these predictive approaches, the biological consequences of the stressors on ecosystems can be assessed. Recent examples of the use of microbiome diversity/uniformity indexes and potential functionalities for the assessment of their ecological status include the assessment of treatment innovations in drinking water quality (Kodera et al. 2023), or in the evaluation of the health of ponds (Stocker et al. 2024), corals (Zhu et al. 2023), or coastal areas (Galand et al. 2018).

On the other hand, using the relative abundance of putative indicator bacterial taxa, a biotic index (microgAMBI) was developed to assess the ecological status of aquatic environments (Fig. 3). This index is based on the assignation of each taxon to an ecological group (sensitive or not to disturbance) from a specific database that has kept growing (Aylagas et al. 2017). The microgAMBI index revealed a significant correlation with the traditional classification method based on macroinvertebrates, showing even better performance in some cases (Aylagas et al. 2017). The index has been validated by determining its response to different anthropogenic pressures (Aylagas et al. 2017; Borja 2018).

Recently, environmental genomics has been combined with machine learning tools to develop more reliable ecological quality status assessment routines (Cordier et al. 2018). AI technologies can improve taxonomic identifications by using metabarcoding sequencing in combination with morphological analyses. DNA barcoding is a method for species identifications that use short variable gene region flanked by highly conserved gene regions that can be amplified using shared primers. Metabarcoding refers to when those regions are used to specifically identify the different microbial organisms in eDNA. Genomic databases are usually incomplete and morphological taxa identification can be inaccurate. Nevertheless, the combination of both approaches, together with the use of AI, ML and DL tools to increase the speed and accuracy (Fig. 3), is improving the precision of the identifications (i.e. detecting potential bacterial bioindicators) and, consequently, the accuracy of the indexes for estimating the ecological status and water quality of water bodies (Fontaine et al. 2023). This approach is not only applied to bacteria but also to other taxonomic groups as invertebrates (e.g. (Kuntke et al. 2020; Macher et al. 2025)), algae (e.g. (Miyata et al. 2022)), or even fishes (e.g. (Corral-Lou and Doadrio 2023)).

Evaluating the Active Metabolism of the Microbial Communities: Metatranscriptomics, Metaproteomics, and Metabolomics

Although metagenomic studies can provide essential information about the composition, structure, biodiversity, and potential functionality of the microbiome, this technique also has important limitations since the data obtained cannot help us distinguish between microorganisms that are senescent, or even dead, from those fully alive. To accurately determine which functional pathways of the microbial communities are really operating in a certain environment and how they change upon exposure to stressful conditions, we need to know not only the microorganisms involved and their biodiversity but also the real active metabolic routes within the community. Thus, for that purpose, Omics that analyse transcripts, proteins, or metabolites should be implemented. Nevertheless, although the analysis of proteins or metabolites by metaproteomic or metabolomic techniques can provide information about the functions that are active, they are less effective in determining the precise taxonomic classification of the microorganisms present in the community, as they usually do not provide accurate information about the composition of the microbiome. Consequently, these studies should be complemented with Metagenomic studies (Shakya et al. 2019; Behera et al. 2021).

As previously mentioned, preceding transcriptomic analyses using reverse qPCR would identify and quantify active genes. New metatranscriptomic approaches use RNA sequencing (RNASeq) to determine all expressed transcripts within a microbiome at a given environmental condition. One of the strengths of this methodology lies in its ability to detect rare RNAs, including noncoding elements with regulatory functions, which may have a huge impact on the metabolic activity within the ecosystems even at very low concentrations. However, as the previously described techniques, metatranscriptomics also has several limitations that can affect the reproducibility and the efficiency of the analysis, such as the following: (1) transcripts can be very unstable, particularly those from prokaryote organisms; (2) transcript expression has a very large dynamic range, which can be affected by the quantification efficiency of the technique; and (3) transcript databases are scarce and incomplete, particularly for microorganisms from extreme environments (Shakya et al. 2019).

Mass spectrometry (MS) is used to analyse changes at the protein level in metaproteomic workflows. The non-amplifying nature of this technique allows accurate quantitative comparisons of the different protein levels among different ecosystems or environmental conditions. On the other hand, as already mentioned, its taxonomic resolution is low, as many homologous proteins from closely related microorganisms have near identical amino acid sequences. That is why metaproteomics can be used to identify global activity pathways, but it will not provide information on the specific microorganisms within the environments. Nevertheless, the main limitation of the metaproteomic approach is the lack of robust protocols for protein extraction from complex matrices (e.g. soil, water, sediments) (Nebauer et al. 2024). In addition, the scarcity of annotated genomes, particularly from non-model organisms, has hindered protein identification. Altogether, to date, only a few studies have used this approach to understand the molecular mechanisms underlying the response to pollutants (Herruzo-Ruiz et al. 2021).

Metabolomic methods, also based on MS techniques, are gradually being used to assess the health status of aquatic ecosystems. Metabolites in the ecosystems can not only provide insights into the active metabolic pathways, but they can also help to understand how microorganisms affect and "communicate" with each other (Thukral et al. 2023). As with the identification of other biomolecules, targeted metabolic analyses to identify specific chemicals are methodologically simpler and have been used in several environmental studies, e.g. to analyse aquatic ecological problems such as "cyanobacterial blooms" in many freshwater resources (Picard et al. 2022). However, non-targeted metabolomics is technically more challenging. As in the case of metatranscriptomic and metaproteomic analyses, the main obstacles are related to the lack of appropriate reference databases and to the low stability of the extracted biomolecules. One of the specific problems for the analysis of metabolites from marine samples is the interference of salt with the procedure, which needs to be removed prior MS analyses; hampering the results since several other metabolites may also be removed during this step (Sogin et al. 2019; Thukral et al. 2023).

Wastewater-Based Epidemiology (WBE): Identifying Health Menaces in Aquatic Environments

As mentioned earlier, a One Health approach, integrating human, animal, and environmental health, is vital for the effective control of pathogens. Furthermore, the COVID-19 pandemic and other viral outbreaks have highlighted the importance of assessing the surveillance of pathogens in waters to prevent epidemics in human and animal populations (Grassly et al. 2024). Wastewater surveillance has proven to be an ideal method for tracking a wide range of microbial pathogens, including viruses, bacteria, and microeukaryotes, due to its broad-spectrum detection and its cost-effectivity. Additionally, sampling is not invasive and, as community data are obtained, there are few ethical restrictions, in contrast to the analysis of human samples (Tiwari et al. 2024). Furthermore, considering that many pathogens are difficult to grow in laboratory conditions, Omic approaches can provide qualitative and quantitative insights into the composition of complex microbial and viral communities. In addition, these techniques identify many very different taxa, even those of low abundance, as they have a high dynamic range.

Recent studies have characterised the virome of different types of water samples, especially in untreated wastewaters, using Omic techniques. For examples of reviews, see McCall et al. (2020) and Schaeffer et al. (2023). Among the drawbacks of this approach, we should again mention those related to the scarcity of good databases, which can hinder the sequence analysis. In the case of viruses, and particularly for RNA viruses, this problem is exacerbated by the fact that their sequences usually evolve very quickly, and databases rapidly become outdated, leading to many unidentified sequences that need to be carefully analysed to prevent false negatives (McCall and Xagoraraki 2019; McCall et al. 2020). Also, wastewater samples can be very complex due to the presence of multiple chemical compounds, e.g. metals, pharmaceuticals, self-care products, etc. Thus, analytical protocols can be challenging, particularly for RNA samples due to their instability.

Pathogenic cellular organisms can also be found in wastewaters, although their levels can be very variable due to their high growth and decay rates in nutrient-rich waters (Tiwari et al. 2024). Nevertheless, the occurrence of bacteria harbouring antibiotic-resistant genes (ARGs) in wastewaters is dangerously increasing and represents a pivotal factor in the spread of antibiotic resistance (Behera et al. 2021). Furthermore, wastewaters have been pointed out as hotspots for antibiotic resistance transmission to new microorganisms by horizontal gene transfer events, which highlights the need to remove all ARGs in wastewater treatment plants (WWTP). The monitoring of ARGs-harbouring bacteria usually combines the quantification of the genes by qRT-PCR with metagenomic techniques to further associate the ARGs to specific taxa (Che et al. 2019; Nguyen et al. 2021). Dangerous bacteria that harbour resistance genes to several antibiotic classes as tetracyclines, beta-lactams, or macrolides, have been identified in contaminated and natural environments all over the world (Behera et al. 2021). The presence of pollutants (e.g. microplastics, metals) combined with other anthropogenic physicochemical (e.g. salinity or temperature) alterations of waters can contribute to increased rates of horizontal gene transfer, and thus to the dissemination of ARGs (Shi et al. 2022b; Tang et al. 2023; Yu et al. 2023).

Final Remarks and Future Directions

Unbiased biomarkers and biological indexes are urgently needed to efficiently evaluate the presence of pollutants and their biological effects in aquatic ecosystems. Microbes play pivotal functions in maintaining environmental health and, due to their ubiquity, are ideal candidates for environmental health evaluation. Omic techniques provide useful tools for this assessment, although there are some methodological problems that need to be solved. (1) The isolation of biomolecules as genes, transcripts, proteins, or metabolites from abiotic complex matrices, such as water or sediments, is difficult as they are scarce in nutrient-poor environments, (2) highly unstable, (3) may be bound to other compounds in complex matrices, and (4) the extraction efficiency of biomolecules from the microbial cells can be very low and variable, as some microbes have proven to be very reluctant to breakage, which can bias the results of the analyses. Therefore, new protocols need to be developed to overcome these limitations. On top of that, the absence of complete reference databases of aquatic organisms hinders the identification of their biomolecules and the understanding of metabolic processes altered by pollutants. This bottleneck can be overcome using improved and better trained AI, ML, and DL tools that, with greater speed and accuracy, allow a significant increase in the quantity and quality of databases, especially those of non-model organisms. The integrative capacity of these powerful tools is expected to generate new, more precise and reliable biological indexes for estimating the ecological quality status of water bodies in the short term.

Omic and particularly metaomic techniques can provide key information for the identification of impacted aquatic ecosystems, guiding the development of effective strategies to counteract the deleterious consequences of pollution.

An integrated Omic approach will facilitate the deep understanding of the role of microbial communities in aquatic ecosystems. This knowledge will certainly help in the discovery of microorganisms/genes/pathways that could be used as biomarkers of stress in aquatic ecosystems. A recent example is the use of certain pathogenic viruses as indicators of faecal pollution (Bibby et al. 2019). The final goal is the identification of low-level organisational molecular biomarkers that respond sensitively, specifically, and early to environmental perturbations. Their implementation in biosensors coupled to automatic autonomous vehicles may allow the continuous monitoring of water courses to assess environmental threats.

Acknowledgments The authors environmental research related to this chapter has been funded by projects CTM2016-75908-R, PID2019-110049RB, and PID2022-139807OB-I00 by "Ministerio de Ciencia e Innovación" (MCIN/AEI) and "European Regional Development Fund" (ERDF) "A way of making Europe", by projects PCM_00118 and 1262384-R "Junta de Andalucía", and by funds to group BIO187 "Junta de Andalucía" and the University of Cordoba. Project PCM_00118 is financed with resources from the "Next Generation EU" Recovery Fund through the Recovery, Transformation and Resilience Plan and co-financed by the Ministry of University, Research and Innovation of the Administration of the Junta de Andalucía. The views and opinions expressed are solely those of the authors and do not necessarily reflect those of the European Union or the European Commission. Neither the European Union nor the European Commission is responsible for them. M. Barbudo-Lunar received a predoctoral contract from the University of Cordoba ("Plan Propio") and the "Ministerio de Ciencia e Innovación" (FPU).

Competing Interests The authors declare that they have no known competing financial interests or personal relationships that could have appeared to influence the work reported in this paper.

References

Abarenkov K, Nilsson RH, Larsson K-H, Taylor AFS, May TW, Frøslev TG et al (2024) The UNITE database for molecular identification and taxonomic communication of fungi and other eukaryotes: sequences, taxa and classifications reconsidered. Nucleic Acids Res 52:D791–D797

Adamovsky O, Buerger AN, Wormington AM, Ector N, Griffitt RJ, Bisesi JH, Martyniuk CJ (2018) The gut microbiome and aquatic toxicology: an emerging concept for environmental health. Environ Toxicol Chem 37:2758–2775

Alhama Carmona J, Michán Doña C, López-Barea J (2017) New trends in aquatic pollution monitoring: from conventional biomarkers to environmental proteomics. In: Gomez-Ariza JL, García-Barrena T (eds) Environmental problems in marine biology: methodological aspects and applications. CRC Press, Boca Raton

Alhama J, Romero-Ruiz A, López-Barea J (2006) Metallothionein quantification in clams by reversed-phase high-performance liquid chromatography coupled to fluorescence detection after monobromobimane derivatization. J Chromatogr A 1107:52–58

Amil-Ruiz F, Herruzo-Ruiz AM, Fuentes-Almagro C, Baena-Angulo C, Jiménez-Pastor JM, Blasco J et al (2021) Constructing a de novo transcriptome and a reference proteome for the bivalve *Scrobicularia plana:* comparative analysis of different assembly strategies and proteomic analysis. Genomics 113:1543–1553

Anetor GO, Nwobi NL, Igharo GO, Sonuga OO, Anetor JI (2022) Environmental pollutants and oxidative stress in terrestrial and aquatic organisms: examination of the total picture and implications for human health. Front Physiol 13:1–13

Angulo MT, Moog CH, Liu YY (2019) A theoretical framework for controlling complex microbial communities. Nat Commun 10:1–12

Aouini F, Trombini C, Sendra M, Blasco J (2019) Biochemical response of the clam *Ruditapes philippinarum* to silver (AgD and AgNPs) exposure and application of an integrated biomarker response approach. Mar Environ Res 152:104783

Apraiz I, Mi J, Cristobal S (2006) Identification of proteomic signatures of exposure to marine pollutants in mussels (*Mytilus edulis*). Mol Cell Proteomics 5:1274–1285

Aßhauer KP, Wemheuer B, Daniel R, Meinicke P (2015) Tax4Fun: predicting functional profiles from metagenomic 16S rRNA data. Bioinformatics 31:2882–2884

Auguste M, Lasa A, Pallavicini A, Gualdi S, Vezzulli L, Canesi L (2019) Exposure to TiO_2 nanoparticles induces shifts in the microbiota composition of *Mytilus galloprovincialis* hemolymph. Sci Total Environ 670:129–137

Aylagas E, Borja Á, Tangherlini M, Dell'Anno A, Corinaldesi C, Michell CT et al (2017) A bacterial community-based index to assess the ecological status of estuarine and coastal environments. Mar Pollut Bull 114:679–688

Bebianno MJ, Langston WJ (1992) Cadmium induction of metallothionein synthesis in *Mytilus galloprovincialis*. Compar Biochem Physiol Part C Compar Pharmacol 103:79–85

Behera BK, Dehury B, Rout AK, Patra B, Mantri N, Chakraborty HJ et al (2021) Metagenomics study in aquatic resource management: recent trends, applied methodologies and future needs. Gene Rep 25:101372

Bejaoui S, Michán C, Telahigue K, Nechi S, Cafsi M, Soudani N et al (2020) Metal body burden and tissue oxidative status in the bivalve *Venerupis decussata* from Tunisian coastal lagoons. Mar Environ Res 159:105000

Berg G, Rybakova D, Fischer D, Cernava T, Vergès M-CC, Charles T et al (2020) Microbiome definition re-visited: old concepts and new challenges. Microbiome 8:103

Bertucci A, Pierron F, Gourves PY, Klopp C, Lagarde G, Pereto C et al (2018) Whole-transcriptome response to wastewater treatment plant and stormwater effluents in the Asian clam, *Corbicula fluminea*. Ecotoxicol Environ Saf 165:96–106

Bibby K, Crank K, Greaves J, Li X, Wu Z, Hamza IA, Stachler E (2019) Metagenomics and the development of viral water quality tools. NPJ Clean Water 2:9

Borja A (2018) Testing the efficiency of a bacterial community-based index (microgAMBI) to assess distinct impact sources in six locations around the world. Ecol Indic 85:594–602

Borsetto C, Raguideau S, Travis E, Kim DW, Lee DH, Bottrill A et al (2021) Impact of sulfamethoxazole on a riverine microbiome. Water Res 201:117382

Botes M, de Kwaadsteniet M, Cloete TE (2013) Application of quantitative PCR for the detection of microorganisms in water. Anal Bioanal Chem 405:91–108

Buffet P-E, Pan J-F, Poirier L, Amiard-Triquet C, Amiard J-C, Gaudin P et al (2013) Biochemical and behavioural responses of the endobenthic bivalve *Scrobicularia plana* to silver nanoparticles in seawater and microalgal food. Ecotoxicol Environ Saf 89:117–124

Buffet PE, Zalouk-Vergnoux A, Poirier L, Lopes C, Risso-de-Faverney C, Guibbolini M et al (2015) Cadmium sulfide quantum dots induce oxidative stress and behavioral impairments in the marine clam *Scrobicularia plana*. Environ Toxicol Chem 34:1659–1664

Cajaraville MP, Bebianno MJ, Blasco J, Porte C, Sarasquete C, Viarengo A (2000) The use of biomarkers to assess the impact of pollution in coastal environments of the Iberian Peninsula: a practical approach. Sci Total Environ 247:295–311

Chan AW, Naphtali J, Schellhorn HE (2019) High-throughput DNA sequencing technologies for water and wastewater analysis. Sci Prog 102:351–376

Chandel P, Mahajan D, Thakur K, Kumar R, Kumar S, Brar B et al (2024) A review on plankton as a bioindicator: a promising tool for monitoring water quality. World Water Policy 10:213–232

Chao A, Chiu C-H, Jost L (2016) Phylogenetic diversity measures and their decomposition: a framework based on hill numbers. In: Pellens R, Grandcolas P (eds) Biodiversity conservation and phylogenetic systematics: preserving our evolutionary heritage in an extinction crisis. Springer, Cham, pp 141–172

Che Y, Xia Y, Liu L, Li AD, Yang Y, Zhang T (2019) Mobile antibiotic resistome in wastewater treatment plants revealed by nanopore metagenomic sequencing. Microbiome 7:1–13

Chen Y, Jiang Y, Huang H, Mou L, Ru J, Zhao J, Xiao S (2018) Long-term and high-concentration heavy-metal contamination strongly influences the microbiome and functional genes in Yellow River sediments. Sci Total Environ 637:1400–1412

Choi YK, Jo PG, Choi CY (2008) Cadmium affects the expression of heat shock protein 90 and metallothionein mRNA in the Pacific oyster, *Crassostrea gigas*. Compar Biochem Physiol Part C Toxicol Pharmacol 147:286–292

Chora S, Starita-Geribaldi M, Guigonis J-M, Samson M, Roméo M, Bebianno MJ (2009) Effect of cadmium in the clam *Ruditapes decussatus* assessed by proteomic analysis. Aquat Toxicol 94:300–308

Cordier T, Forster D, Dufresne Y, Martins CIM, Stoeck T, Pawlowski J (2018) Supervised machine learning outperforms taxonomy-based environmental DNA metabarcoding applied to biomonitoring. Mol Ecol Resour 18:1381–1391

Corral-Lou A, Doadrio I (2023) Environmental DNA metabarcoding of water samples as a tool for monitoring Iberian freshwater fish composition. PLoS One 18:e0283088

Cosby AG, Lebakula V, Smith CN, Wanik DW, Bergene K, Rose AN et al (2024) Accelerating growth of human coastal populations at the global and continent levels: 2000–2018. Sci Rep 14:1–10

Costa PM, Chicano-Gálvez E, Caeiro S, Lobo J, Martins M, Ferreira AM et al (2012) Hepatic proteome changes in *Solea senegalensis* exposed to contaminated estuarine sediments: a laboratory and in situ survey. Ecotoxicology 21:1194–1207

Cuesta A, Espinosa C, Esteban MA, González-Fernández C (2023) Application of transcriptomic profiling to investigate the toxicity mechanisms caused by dietary exposure of nanoplastics in fish. Aquat Toxicol 264:106712

Datto-Liberato FH, Lopez VM, Quinaia T, Valle RF, Samways MJ, Juen L et al (2024) Total environment sentinels: dragonflies as ambivalent/amphibiotic bioindicators of damage to soil and freshwater. Sci Total Environ 934:173110

Dell'Anno F, Joaquim van Zyl L, Trindade M, Buschi E, Cannavacciuolo A, Pepi M et al (2023) Microbiome enrichment from contaminated marine sediments unveils novel bacterial strains for petroleum hydrocarbon and heavy metal bioremediation. Environ Pollut 317:120772

DeSantis TZ, Hugenholtz P, Larsen N, Rojas M, Brodie EL, Keller K et al (2006) Greengenes, a chimera-checked 16S rRNA gene database and workbench compatible with ARB. Appl Environ Microbiol 72:5069–5072

Djemiel C, Plassard D, Terrat S, Crouzet O, Sauze J, Mondy S et al (2020) Mgreen-db: a reference database for the 23S rRNA gene of eukaryotic plastids and cyanobacteria. Sci Rep 10:5915

Djemiel C, Maron PA, Terrat S, Dequiedt S, Cottin A, Ranjard L (2022) Inferring microbiota functions from taxonomic genes: a review. Gigascience 11:1–30

Douglas GM, Maffei VJ, Zaneveld JR, Yurgel SN, Brown JR, Taylor CM et al (2020) PICRUSt2 for prediction of metagenome functions. Nat Biotechnol 38:685–688

Downie AT, Bennett WW, Wilkinson S, de Bruyn M, DiBattista JD (2024) From land to sea: environmental DNA is correlated with long-term water quality indicators in an urbanized estuary. Mar Pollut Bull 207:116887

Duarte IA, Reis-Santos P, França S, Cabral H, Fonseca VF (2017) Biomarker responses to environmental contamination in estuaries: a comparative multi-taxa approach. Aquat Toxicol 189:31–41

Duhamel S (2024) The microbial phosphorus cycle in aquatic ecosystems. Nat Rev Microbiol 23:239–255

Escher BI, Stapleton HM, Schymanski EL (2020) Tracking complex mixtures of chemicals in our changing environment. Science 392:388–392

Fackelmann G, Sommer S (2019) Microplastics and the gut microbiome: how chronically exposed species may suffer from gut dysbiosis. Mar Pollut Bull 143:193–203

Fang Y, Yang H, Wang T, Liu B, Zhao H, Chen M (2010) Metallothionein and superoxide dismutase responses to sublethal cadmium exposure in the clam *Mactra veneriformis*. Compar Biochem Physiol Part C Toxicol Pharmacol 151:325–333

Farhat N, Kim LH, Vrouwenvelder JS (2020) Online characterization of bacterial processes in drinking water systems. NPJ Clean Water 3:16

Fernández-Cisnal R, Alhama J, Abril N, Pueyo C, López-Barea J (2014) Redox proteomics as biomarker for assessing the biological effects of contaminants in crayfish from Doñana National Park. Sci Total Environ 490:121–133

Fontaine L, Pin L, Savio D, Friberg N, Kirschner AKT, Farnleitner AH, Eiler A (2023) Bacterial bioindicators enable biological status classification along the continental Danube river. Commun Biol 6:1–11

Francé J, Varkitzi I, Stanca E, Cozzoli F, Skejić S, Ungaro N et al (2021) Large-scale testing of phytoplankton diversity indices for environmental assessment in Mediterranean sub-regions (Adriatic, Ionian and Aegean seas). Ecol Indic 126:107630

Galand PE, Pereira O, Hochart C, Auguet JC, Debroas D (2018) A strong link between marine microbial community composition and function challenges the idea of functional redundancy. ISME J 12:2470–2478

Garner E, Davis BC, Milligan E, Blair MF, Keenum I, Maile-Moskowitz A et al (2021) Next generation sequencing approaches to evaluate water and wastewater quality. Water Res 194:116907

Geissen V, Mol H, Klumpp E, Umlauf G, Nadal M, van der Ploeg M et al (2015) Emerging pollutants in the environment: a challenge for water resource management. Int Soil Water Conserv Res 3:57–65

Géret F, Jouan A, Turpin V, Bebianno MJ, Cosson RP (2002) Influence of metal exposure on metallothionein synthesis and lipid peroxidation in two bivalve mollusks: the oyster (*Crassostrea gigas*) and the mussel (*Mytilus edulis*). Aquat Living Resour 15:61–66

Geret F, Serafim A, Barreira L, Bebianno MJ (2002) Effect of cadmium on antioxidant enzyme activities and lipid peroxidation in the gills of the clam *Ruditapes decussatus*. Biomarkers 7:242–256

Ghedira J, Chicano-Gálvez E, Fernández-Cisnal R, Jebali J, Banni M, Chouba L et al (2016) Using environmental proteomics to assess pollutant response of *Carcinus maenas* along the Tunisian coast. Sci Total Environ 541:109–118

Giarratano E, Sturla Lompré J, Malanga G (2023) Evidences of metabolic alterations and cellular damage in different tissues of scallops *Aequipecten tehuelchus* exposed to cadmium. Mar Environ Res 188:106011

Gomes T, Pereira CG, Cardoso C, Sousa VS, Teixeira MR, Pinheiro JP, Bebianno MJ (2014) Effects of silver nanoparticles exposure in the mussel *Mytilus galloprovincialis*. Mar Environ Res 101:208–214

Graham EB, Knelman JE, Schindlbacher A, Siciliano S, Breulmann M, Yannarell A et al (2016) Microbes as engines of ecosystem function: when does community structure enhance predictions of ecosystem processes? Front Microbiol 7:214

Grassly NC, Shaw AG, Owusu M (2024) Global wastewater surveillance for pathogens with pandemic potential: opportunities and challenges. Lancet Microbe 6:100939

Grossart H-P, Massana R, McMahon KD, Walsh DA (2020) Linking metagenomics to aquatic microbial ecology and biogeochemical cycles. Limnol Oceanogr 65:S2–S20

Guo Y, Cichocki N, Schattenberg F, Geffers R, Harms H, Müller S (2019) AgNPs change microbial community structures of wastewater. Front Microbiol 10:1–12

Han P, Yuan M, Sun Z, Xue Y, Liu X, Chen J et al (2025) Effect of heat exposure on histology, transcriptomics and co-expression network: a synthetic study in Japanese flounder (*Paralichthys olivaceus*). Aquaculture 595:741490

Harrington AM, Clark KF, Hamlin HJ (2020) Expected Ocean warming conditions significantly alter the transcriptome of developing postlarval American lobsters (*Homarus americanus*): implications for energetic trade-offs. Comp Biochem Physiol Part D Genomics Proteomics 36:100716

Herruzo-Ruiz AM, Fuentes-Almagro CA, Jiménez-Pastor JM, Pérez-Rosa VM, Blasco J, Michán C, Alhama J (2021) Meta-omic evaluation of bacterial microbial community structure and activity for the environmental assessment of soils: overcoming protein extraction pitfalls. Environ Microbiol 23:4706–4725

Herruzo-Ruiz AM, Peralbo-Molina Á, López CM, Michán C, Alhama J, Chicano-Gálvez E (2024a) Mass spectrometry imaging in environmental monitoring: from a scarce existing past to a promising future. Trends Environ Anal Chem 42:e00228

Herruzo-Ruiz AM, Trombini C, Moreno-Garrido I, Blasco J, Alhama J, Michán C (2024b) Ions and nanoparticles of Ag and/or Cd metals in a model aquatic microcosm: effects on the abundance, diversity and functionality of the sediment bacteriome. Mar Pollut Bull 204:116525

Hites RA (2019) Statistical approach for assessing the Stockholm convention's effectiveness: great lakes atmospheric data. Environ Sci Technol 53:8585–8590

Jacquiod S, Cyriaque V, Riber L, Al-soud WA, Gillan DC, Wattiez R, Sørensen SJ (2018) Long-term industrial metal contamination unexpectedly shaped diversity and activity response of sediment microbiome. J Hazard Mater 344:299–307

Jebali J, Chicano-Gálvez E, Fernández-Cisnal R, Banni M, Chouba L, Boussetta H et al (2014) Proteomic analysis in caged Mediterranean crab (*Carcinus maenas*) and chemical contaminant exposure in Téboulba harbour, Tunisia. Ecotoxicol Environ Saf 100:15–26

Johnson JS, Spakowicz DJ, Hong B-Y, Petersen LM, Demkowicz P, Chen L et al (2019) Evaluation of 16S rRNA gene sequencing for species and strain-level microbiome analysis. Nat Commun 10:5029

Jonsson H, Schiedek D, Grøsvik BE, Goksøyr A (2006) Protein responses in blue mussels (*Mytilus edulis*) exposed to organic pollutants: a combined CYP-antibody/proteomic approach. Aquat Toxicol 78:49–56

Kodera SM, Sharma A, Martino C, Dsouza M, Grippo M, Lutz HL et al (2023) Microbiome response in an urban river system is dominated by seasonality over wastewater treatment upgrades. Environ Microbiome 18:10

Kuntke F, de Jonge N, Hesselsøe M, Lund Nielsen J (2020) Stream water quality assessment by metabarcoding of invertebrates. Ecol Indic 111:105982

Lagarrigue M, Caprioli RM, Pineau C (2016) Potential of MALDI imaging for the toxicological evaluation of environmental pollutants. J Proteome 144:133–139

Laurent J, Le Berre I, Armengaud J, Waeles M, Sturbois A, Durand G et al (2024) Environmental signatures and fish proteomics: a multidisciplinary study to identify the major stressors in estuaries located in French agricultural watersheds. Environ Pollut 363:124876

Leprêtre M, Degli Esposti D, Sugier K, Espeyte A, Gaillard J-C, Delorme N et al (2023) Organ-oriented proteogenomics functional atlas of three aquatic invertebrate sentinel species. Sci Data 10:643

Li C, Quan Q, Gan Y, Dong J, Fang J, Wang L, Liu J (2020) Effects of heavy metals on microbial communities in sediments and establishment of bioindicators based on microbial taxa and function for environmental monitoring and management. Sci Total Environ 749:141555

Li C, Wang L, Ji S, Chang M, Wang L, Gan Y, Liu J (2021) The ecology of the plastisphere: microbial composition, function, assembly, and network in the freshwater and seawater ecosystems. Water Res 202:117428

Li W, Luo D, Adyel TM, Wu J, Miao L, Kong M, Hou J (2023) Dynamic responses of carbon metabolism of sediment microbial communities to Ag nanoparticles: effects of the single and repeated exposure scenarios. Sci Total Environ 870:161891

Lin W, Lu J, Yao H, Lu Z, He Y, Mu C et al (2021) Elevated pCO_2 alters the interaction patterns and functional potentials of rearing seawater microbiota. Environ Pollut 287:117615

Lionetto MG, Caricato R, Giordano ME (2019) Pollution biomarkers in environmental and human biomonitoring. Open Biomark J 9:1–9

Liu J, He XX, Lin XR, Chen WC, Zhou QX, Shu WS, Huang LN (2015) Ecological effects of combined pollution associated with e-waste recycling on the composition and diversity of soil microbial communities. Environ Sci Technol 49:6438–6447

Liu W, Nie H, Liang D, Bai Y, Liu H (2020) Phospholipid imaging of zebrafish exposed to fipronil using atmospheric pressure matrix-assisted laser desorption ionization mass spectrometry. Talanta 209:120357

Lloyd KG, Steen AD, Ladau J, Yin J, Crosby L (2018) Phylogenetically novel uncultured microbial cells dominate earth microbiomes. mSystems 3:10.1128/msystems.00055-18

López-Barea J (1995) Biomarkers in ecotoxicology: an overview. Arch Toxicol Suppl 17:57–79

Louca S, Parfrey LW, Doebeli M (2016) decoupling function and taxonomy in the global ocean microbiome. Science 353:1272–1277

Louca S, Polz MF, Mazel F, Albright MBN, Huber JA, O'Connor MI et al (2018) Function and functional redundancy in microbial systems. Nat Ecol Evol 2:936–943

Macher T-H, Beermann AJ, Arle J, Foerster J, Greyer M, Mora D et al (2025) Fit for purpose? Evaluating benthic invertebrate DNA metabarcoding for ecological status class assessment in streams under the water framework directive. Water Res 272:122987

Mansour C, Guibbolini M, Rouane Hacene O, Saidane Mosbahi D, Risso-de Faverney C (2020) Oxidative stress and damage biomarkers in clam *Ruditapes decussatus* exposed to a polluted site: the reliable biomonitoring tools in hot and cold seasons. Arch Environ Contam Toxicol 78:478–494

Mapelli F, Scoma A, Michoud G, Aulenta F, Boon N, Borin S et al (2017) Biotechnologies for marine oil spill cleanup: indissoluble ties with microorganisms. Trends Biotechnol 35:860–870

Martínez-Burgos WJ, Pozzan R, de Carvalho JC, Cavali M, Mariano AB, Vargas JVC et al (2024) The role of microalgae as bioindicators of aquatic contamination. In: Kurniawan TA, Anouzla A (eds) Algae as a natural solution for challenges in water-food-energy nexus: toward carbon neutrality. Springer, Singapore, pp 323–347

Masanja F, Yang K, Xu Y, He G, Liu X, Xu X et al (2023) Bivalves and microbes: a mini-review of their relationship and potential implications for human health in a rapidly warming ocean. Front Mar Sci 10:1–9

McCall C, Xagoraraki I (2019) Metagenomic approaches for detecting viral diversity in water environments. J Environ Eng 145:4019039

McCall C, Wu H, Miyani B, Xagoraraki I (2020) Identification of multiple potential viral diseases in a large urban center using wastewater surveillance. Water Res 184:116160

McLellan SL, Eren AM (2014) Discovering new indicators of fecal pollution. Trends Microbiol 22:697–706

Michán C, Blasco J, Alhama J (2021) High-throughput molecular analyses of microbiomes as a tool to monitor the wellbeing of aquatic environments. Microb Biotechnol 14:870–885

Michán-Doña A, Vázquez-Borrego MC, Michán C (2024) Are there any completely sterile organs or tissues in the human body? Is there any sacred place? Microb Biotechnol 17:2–5

Milan M, Carraro L, Fariselli P, Martino ME, Cavalieri D, Vitali F et al (2018) Microbiota and environmental stress: how pollution affects microbial communities in Manila clams. Aquat Toxicol 194:195–207

Miserazzi A, Perrigault M, Sow M, Gelber C, Ciret P, Lomenech AM et al (2020) Proteome changes in muscles, ganglia, and gills in *Corbicula fluminea* clams exposed to crude oil: relationship with behavioural disturbances. Aquat Toxicol 223:105482

Mishra RK, Mentha SS, Misra Y, Dwivedi N (2023) Emerging pollutants of severe environmental concern in water and wastewater: a comprehensive review on current developments and future research. Water-Energy Nexus 6:74–95

Miyata K, Inoue Y, Amano Y, Nishioka T, Nagaike T, Kawaguchi T et al (2022) Comparative environmental RNA and DNA metabarcoding analysis of river algae and arthropods for ecological surveys and water quality assessment. Sci Rep 12:19828

Mustafa GA, Abd-Elgawad A, Ouf A, Siam R (2016) The Egyptian Red Sea coastal microbiome: a study revealing differential microbial responses to diverse anthropogenic pollutants. Environ Pollut 214:892–902

Nebauer DJ, Pearson LA, Neilan BA (2024) Critical steps in an environmental metaproteomics workflow. Environ Microbiol 26:e16637

Nguyen AQ, Vu HP, Nguyen LN, Wang Q, Djordjevic SP, Donner E et al (2021) Monitoring antibiotic resistance genes in wastewater treatment: current strategies and future challenges. Sci Total Environ 783:146964

Ouanes K (2024) The role of artificial intelligence and machine learning in NGS. In: Dash HR, Elkins KM, Al Snan A (eds) Next generation sequencing (NGS) technology in DNA analysis. Academic Press, New York, pp 531–538

Overmann J, Lepleux C (2016) Marine bacteria and archaea: diversity, adaptations, and culturability. In: Stal LJ, Cretoiu MS (eds) The marine microbiome: an untapped source of biodiversity and biotechnological potential. Springer, Cham, pp 21–55

Pellerin J, Amiard J-C (2009) Comparison of bioaccumulation of metals and induction of metallothioneins in two marine bivalves (*Mytilus edulis* and *Mya arenaria*). Comp Biochem Physiol C Toxicol Pharmacol 150:186–195

Pérez-Cobas AE, Gomez-Valero L, Buchrieser C (2020) Metagenomic approaches in microbial ecology: an update on whole-genome and marker gene sequencing analyses. Microb Genom 6:1–22

Petrie B, Barden R, Kasprzyk-Hordern B (2015) A review on emerging contaminants in wastewaters and the environment: current knowledge, understudied areas and recommendations for future monitoring. Water Res 72:3–27

Picard M, Pochon X, Atalah J, Pearman JK, Rees A, Howarth JD et al (2022) Using metabarcoding and droplet digital PCR to investigate drivers of historical shifts in cyanobacteria from six contrasting lakes. Sci Rep 12:1–17

Priyadarsini M, Kushwaha J, Pandey KP, Rani J, Dhoble AS (2023) Application of flow cytometry for rapid, high-throughput, multiparametric analysis of environmental microbiomes. J Microbiol Methods 214:106841

Purdy KJBT-M (2005) Nucleic acid recovery from complex environmental samples. In: Environmental microbiology. Academic Press, New York, pp 271–292

Quast C, Pruesse E, Yilmaz P, Gerken J, Schweer T, Yarza P et al (2013) The SILVA ribosomal RNA gene database project: improved data processing and web-based tools. Nucleic Acids Res 41:D590–D596

Ribeiro AR, Castro PML, Tiritan ME (2012) Chiral pharmaceuticals in the environment. Environ Chem Lett 10:239–253

Rocha TL, Gomes T, Durigon EG, Bebianno MJ (2016) Subcellular partitioning kinetics, metallothionein response and oxidative damage in the marine mussel *Mytilus galloprovincialis* exposed to cadmium-based quantum dots. Sci Total Environ 554:130–141

Roland K, Kestemont P, Loos R, Tavazzi S, Paracchini B, Belpaire C et al (2014) Looking for protein expression signatures in European eel peripheral blood mononuclear cells after in vivo exposure to perfluorooctane sulfonate and a real world field study. Sci Total Environ 468:958–967

Romero-Ruiz A, Carrascal M, Alhama J, Gómez-Ariza JL, Abian J, López-Barea J (2006) Utility of proteomics to assess pollutant response of clams from the Doñana bank of Guadalquivir estuary (SW Spain). Proteomics 6(Suppl 1):245–255

Saleem F, Jiang JL, Li E, Tran KL, Schellhorn HE, Edge TA (2025) eDNA metabarcoding-based source attribution of fecal indicator bacteria exceedances in urban freshwater beaches, sand and rivers. Environ Sci Adv 4:456–468

Santovito G, Trentin E, Gobbi I, Bisaccia P, Tallandini L, Irato P (2021) Non-enzymatic antioxidant responses of *Mytilus galloprovincialis*: insights into the physiological role against metal-induced oxidative stress. Comp Biochem Physiol C Toxicol Pharmacol 240:108909

Schaeffer J, Desdouits M, Besnard A, Le Guyader FS (2023) Looking into sewage: how far can metagenomics help to detect human enteric viruses? Front Microbiol 14:1161674

Schmidt B, Hildebrandt A (2021) Deep learning in next-generation sequencing. Drug Discov Today 26:173–180

Seeley ME, Song B, Passie R, Hale RC (2020) Microplastics affect sedimentary microbial communities and nitrogen cycling. Nat Commun 11:1–10

Sehnal L, Brammer-Robbins E, Wormington AM, Blaha L, Bisesi J, Larkin I et al (2021) Microbiome composition and function in aquatic vertebrates: small organisms making big impacts on aquatic animal health. Front Microbiol 12:567408

Shakya M, Lo C-C, Chain PSG (2019) Advances and challenges in metatranscriptomic analysis. Front Genet 10:904

Shang Y, Wu X, Wang X, Wei Q, Ma S, Sun G et al (2022) Factors affecting seasonal variation of microbial community structure in Hulun Lake, China. Sci Total Environ 805:150294

Shepard JL, Olsson B, Tedengren M, Bradley BP (2000) Protein expression signatures identified in *Mytilus edulis* exposed to PCBs, copper and salinity stress. Mar Environ Res 50:337–340

Shi Q, Fang C, Yan C, Chang XL, Zhang X, Zhang H (2022a) Visualization of fullerenol nanoparticles distribution in *Daphnia magna* using laser ablation-isotope ratio mass (LA-IRMS) and matrix-assisted laser desorption/ionization imaging mass spectrometry (MALDI-IMS). Ecotoxicol Environ Saf 232:113226

Shi X, Xia Y, Wei W, Ni BJ (2022b) Accelerated spread of antibiotic resistance genes (ARGs) induced by non-antibiotic conditions: roles and mechanisms. Water Res 224:119060
Smits M, Artigaud S, Bernay B, Pichereau V, Bargelloni L, Paillard C (2020) A proteomic study of resistance to Brown ring disease in the Manila clam, *Ruditapes philippinarum*. Fish Shellfish Immunol 99:641–653
Sogin EM, Puskás E, Dubilier N, Liebeke M (2019) Marine metabolomics: a method for non-targeted measurement of metabolites in seawater by gas chromatography-mass spectrometry. mSystems 4(6):e00638–e00619
Solé M, Kopecka-Pilarczyk J, Blasco J (2009) Pollution biomarkers in two estuarine invertebrates, *Nereis diversicolor* and *Scrobicularia plana*, from a marsh ecosystem in SW Spain. Environ Int 35:523–531
Stocker MD, Smith JE, Pachepsky YA, Blaustein RA (2024) Fine-scale spatiotemporal variations in bacterial community diversity in agricultural pond water. Sci Total Environ 915:170143
Sumudumali RGI, Jayawardana JMCK (2021) A review of biological monitoring of aquatic ecosystems approaches: with special reference to macroinvertebrates and pesticide pollution. Environ Manag 67:263–276
Tang M, Chen Q, Zhong H, Ju F, Wu Y, Hu J et al (2023) Exploring diversity patterns and driving mechanisms of the antibiotic resistome and microbiome in saline groundwater. J Hazard Mater 446:130734
Tao W, Chen G, Zeng G, Yan M, Chen A, Guo Z et al (2016) Influence of silver nanoparticles on heavy metals of pore water in contaminated river sediments. Chemosphere 162:117–124
Thukral AK (2017) A review on measurement of alpha diversity in biology. Agric Res J 54:1
Thukral M, Allen AE, Petras D (2023) Progress and challenges in exploring aquatic microbial communities using non-targeted metabolomics. ISME J 17:2147–2159
Tiwari A, Ahmed W, Oikarinen S, Sherchan SP, Heikinheimo A, Jiang G et al (2022) Application of digital PCR for public health-related water quality monitoring. Sci Total Environ 837:155663
Tiwari A, Radu E, Kreuzinger N, Ahmed W, Pitkänen T (2024) Key considerations for pathogen surveillance in wastewater. Sci Total Environ 945:173862
Trombini C, Kazakova J, Montilla-López A, Fernández-Cisnal R, Hampel M, Fernández-Torres R et al (2021) Assessment of pharmaceutical mixture (ibuprofen, ciprofloxacin and flumequine) effects to the crayfish *Procambarus clarkii*: a multilevel analysis (biochemical, transcriptional and proteomic approaches). Environ Res 200:111396
Valavanidis A, Vlahogianni T, Dassenakis M, Scoullos M (2006) Molecular biomarkers of oxidative stress in aquatic organisms in relation to toxic environmental pollutants. Ecotoxicol Environ Saf 64:178–189
Viarengo A, Canesi L, Pertica M, Poli G, Moore MN, Orunesu M (1990) Heavy metal effects on lipid peroxidation in the tissues of *Mytilus galloprovincialis* clam. Compar Biochem Physiol Part C Compar Pharmacol 97:37–42
Wang L, Li Y, Zhao Z, Cordier T, Worms IA, Niu L et al (2021) Microbial community diversity and composition in river sediments contaminated with tetrabromobisphenol A and copper. Chemosphere 272:129855
Wang L, Li F, Gu B, Qu P, Liu Q, Wang J et al (2022) Metaomics in clinical laboratory: potential driving force for innovative disease diagnosis. Front Microbiol 13:883734
Wang F, Xiang L, Sze-Yin Leung K, Elsner M, Zhang Y, Guo Y et al (2024) Emerging contaminants: a one health perspective. Innovation 5:100612
Wemheuer F, Taylor JA, Daniel R, Johnston E, Meinicke P, Thomas T, Wemheuer B (2020) Tax4Fun2: prediction of habitat-specific functional profiles and functional redundancy based on 16S rRNA gene sequences. Environ Microbiome 15:11
Wu H, Liu X, Zhang X, Ji C, Zhao J, Yu J (2013) Proteomic and metabolomic responses of clam Ruditapes philippinarum to arsenic exposure under different salinities. Aquat Toxicol 136:91–100
Yan C, Wang F, Geng H, Liu H, Pu S, Tian Z et al (2020) Integrating high-throughput sequencing and metagenome analysis to reveal the characteristic and resistance mechanism of microbial community in metal contaminated sediments. Sci Total Environ 707:136116

Yin X, Wang W, Wang A, He M, Lin C, Ouyang W, Liu X (2022) Microbial community structure and metabolic potential in the coastal sediments around the Yellow River estuary. Sci Total Environ 816:151582

Yu Q, Han Q, Shi S, Sun X, Wang X, Wang S et al (2023) Metagenomics reveals the response of antibiotic resistance genes to elevated temperature in the Yellow River. Sci Total Environ 859:160324

Yuan L, Richardson CJ, Ho M, Willis CW, Colman BP, Wiesner MR (2018) Stress responses of aquatic plants to silver nanoparticles. Environ Sci Technol 52:2558–2565

Zhang X, Li F, Ji C, Wu H (2023) Toxicological mechanism of cadmium in the clam *Ruditapes philippinarum* using combined ionomic, metabolomic and transcriptomic analyses. Environ Pollut 323:121286

Zhang Y, Qiu Y, Liu K, Zhong W, Yang J, Altermatt F, Zhang X (2024) Evaluating eDNA and eRNA metabarcoding for aquatic biodiversity assessment: from bacteria to vertebrates. Environ Sci Ecotechnol 21:100441

Zhou X, Gao S, Yue M, Zhu S, Liu Q, Zhao XE (2023) Recent advances in analytical methods of oxidative stress biomarkers induced by environmental pollutant exposure. TrAC Trends Anal Chem 160:116978

Zhu W, Liu X, Zhang J, Zhao H, Li Z, Wang H et al (2023) Response of coral bacterial composition and function to water quality variations under anthropogenic influence. Sci Total Environ 884:163837

Hyperspectral Technology to Monitor Marine Pollution

Ámbar Pérez-García, Adrián Rodríguez-Molina, Emma Hernández-Suárez, Alba Martín-Lorenzo, and José F. López

Contents

Á. Pérez-García (✉) · A. Rodríguez-Molina · E. Hernández-Suárez · A. Martín-Lorenzo · J. F. López
Institute for Applied Microelectronics, University of Las Palmas de Gran Canaria, Las Palmas, Spain
e-mail: ambar.perez@ulpgc.es; armolina@iuma.ulpgc.es; ehernandez@iuma.ulpgc.es; alba.martin@ulpgc.es; lopez@iuma.ulpgc.es

D. Gutiérrez et al. (eds.), *Smart Water Quality Monitoring*,
https://doi.org/10.1007/978-3-032-03049-8_4

Abstract

Marine pollution is a pressing environmental issue with profound implications for ecosystems, biodiversity, and human livelihood. This chapter delves into the transformative role of hyperspectral imaging in monitoring and mitigating pollution in marine environments. Pollutants such as oil spills and plastics are particularly impactful, causing severe damage in the short term or remaining persistently in the marine environment. Hyperspectral images provide unparalleled precision in detecting these contaminants by capturing detailed spectral information across hundreds of narrow bands. The chapter introduces key advancements, including specialised spectral indices that allow real-time detection. Band selection methodologies are highlighted as tools to identify the most relevant spectral features, improve pollutant classification, and facilitate machine learning (ML) model transfer across diverse environments. Emerging technologies, such as specialised multispectral sensors integrated into cost-effective platforms like CubeSats, demonstrate the potential for scalable environmental monitoring. Additionally, cloud computing platforms enable efficient processing of vast datasets, enhancing the global accessibility of hyperspectral imaging (HSI) solutions. While these innovations address critical challenges in data complexity and model generalisation, the chapter underscores the need for interdisciplinary collaboration to refine these methods and integrate them into policy frameworks. These tools offer a path towards more effective marine pollution management and ocean stewardship.

Marine Pollution and the Need for Advanced Monitoring

Marine pollution is one of the twentieth century's most significant environmental challenges, with devastating implications for ecosystems, biodiversity, and the human communities dependent on healthy oceans (Häder et al. 2020). Over the past few decades, intensifying human activities have turned oceans into repositories for pollutants such as oil spills, plastic debris, and industrial chemicals. These contaminants can persist for extended periods, travel vast distances via ocean currents, and inflict widespread ecological harm (Kershaw and Protection 2015).

This section examines the key aspects of marine pollution, including its impact on coastal ecosystems, human health, and socio-economic systems. Additionally, it underscores the urgency of deploying advanced monitoring technologies to mitigate pollution's detrimental effects and safeguard marine ecosystems for future generations (van Emmerik et al. 2023).

Environmental and Social Impact

Coastal areas are particularly vulnerable due to their proximity to human activities such as urbanisation and industry (Burak et al. 2004). When pollutants such as oil and plastics enter coastal waters, they disrupt critical habitats, including coral reefs, mangroves, and estuaries, which are critical for biodiversity and ecosystem services (Paches et al. 2019). Coral reefs, for example, support a wide variety of marine life but are particularly sensitive to oil contamination, which can smother corals and inhibit their ability to grow and reproduce (Negri et al. 2016). Additionally, plastics can physically damage reefs, which further contributes to their decline (Burke et al. 2011).

Mangroves and estuaries serve as nurseries for many marine species and act as natural barriers against storm surges and coastal erosion (Asari et al. 2021). However, when these ecosystems are polluted, they can become degraded, leading to a reduction in biodiversity and the loss of critical ecosystem services (Botterell et al. 2019). The loss of biodiversity weakens the resilience of marine ecosystems, making them less able to withstand environmental stressors such as climate change (Bernhardt and Leslie 2013). Furthermore, pollution disrupts the food chain, from plankton to large predators, ultimately affecting fish populations that many coastal communities rely on for food and economic stability (Koelmans et al. 2015).

The impacts of pollution are not limited to wildlife, as shown in Fig. 1. Coastal pollution also presents severe threats to human health. Pollutants can accumulate in marine organisms, increasing toxin levels as they ascend the food chain (Avio

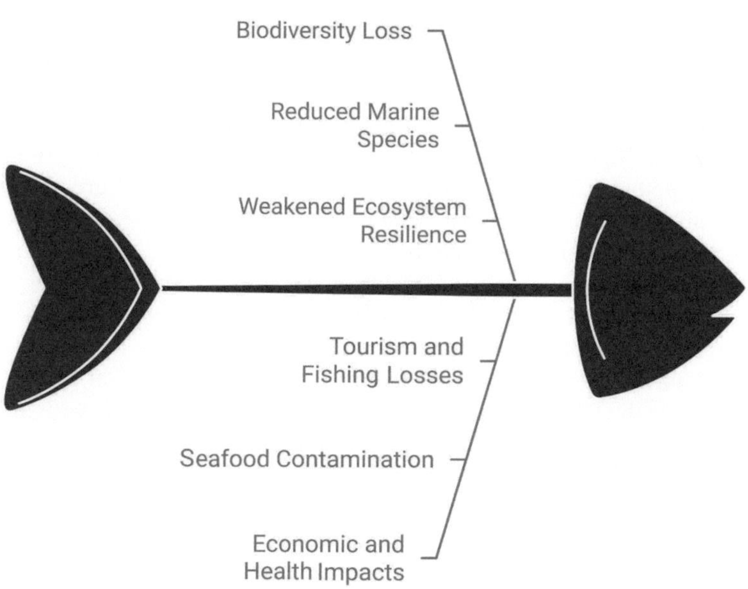

Fig. 1 Coastal pollution impact

et al. 2017). This creates a direct risk for humans who consume contaminated seafood. In addition, pollution degrades coastal water quality, affecting tourism and fishing (Spalding et al. 2017).

Pollutants

This study mainly focuses on two pollutants: oil spills and plastics. Both contaminants can be widespread easily and are persistent in marine environments. Plastic pollution presents a widespread challenge that impacts ecosystems on land and in water worldwide (Morales-Caselles et al. 2021). Over time, plastic breaks down into tiny fragments that persist for hundreds of years, creating enduring ecological risks (Meijer et al. 2021). Though less frequent, oil spills cause immediate and severe damage to marine ecosystems, coating wildlife, disrupting food chains, and polluting coastlines (Häder et al. 2020). These pollutants are particularly harmful due to their capacity to spread across vast areas via ocean currents, making them critical targets for monitoring efforts.

Oil Spills

Oil spills, often caused by accidents on tankers, refineries, and offshore platforms during drilling, transportation, transfer, or oil extraction, are among the most damaging forms of marine pollution (Angelova et al. 2011). The physical properties of oil, which makes it float on the water's surface, allow it to spread quickly, covering large areas and forming slicks that block sunlight from penetrating the water (Mishra and Kumar 2015). Marine invertebrates and aquatic vegetation exposed to oil spill pollution can experience varying levels of initial mortality and prolonged sublethal effects, depending on the persistence of exposure, the type and amount of oil, and environmental conditions such as habitat type, temperature, and depth (Keesing et al. 2018). Gradually, oil can sink to the seafloor, contaminating and harming deep-sea ecosystems (Fisher et al. 2016). The Deepwater Horizon (DWH) disaster in 2010 is one of the most infamous oil spills, where nearly 800,000 cubic meters of oil was released into the Gulf of Mexico over 3 months, causing catastrophic environmental damage (Liu et al. 2011; Beyer et al. 2016). Some consequences of the DWH disaster included severe harm to wildlife, with thousands of birds, mammals, and sea turtles coated in oil (Beyer et al. 2016). It also led to significant economic losses, including tens of thousands of jobs and $3.5 billion in industry output, gross regional product, and tax revenues (Court et al. 2020).

Plastic waste

Plastic waste is another critical form of marine pollution. Tons of plastic enter the oceans annually, breaking down into smaller particles known as microplastics, which can persist for centuries (Hermsen et al. 2018). These plastics are non-biodegradable, accumulating in marine ecosystems and harming wildlife (Green 2020). Animals frequently confuse plastic debris with food, resulting in ingestion and entanglement that can lead to injuries, malnutrition, and mortality (Markic et al. 2018). Birds, fish, and even large marine mammals have been found with plastic residues in their stomachs (Markic et al. 2018). Due to their small size, microplastics are ingested by plankton and small fish, entering the food chain and ultimately reaching humans through seafood consumption (Abbasi et al. 2018). Plastics also carry chemical pollutants that attach to their surfaces and are subsequently absorbed by organisms, causing prolonged toxic effects (Gallo et al. 2020). The highest concentrations of plastic pollution are seen along shorelines, water, and seafloors, particularly near densely populated regions and industrialised areas (Jambeck et al. 2015).

! Attention

Oil spills and plastic pollution pose significant threats to marine ecosystems, making early detection essential to mitigate their impacts (Fig. 2). Due to the severe short-term consequences of certain pollutants, such as oil spills, early identification is crucial. Exposure to oil can quickly cause widespread mortality or lasting harm to marine flora and fauna. In contrast, plastic pollution primarily exerts long-term effects, persisting in the environment, dispersing across the globe, and accumulating in wildlife over time.

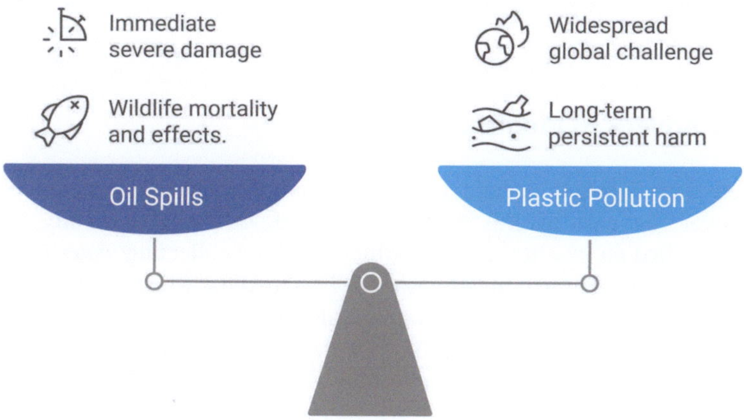

Fig. 2 Impact comparison between pollutants

The Role of Hydrodynamics

Hydrodynamics, the study of liquids in motion, is fundamental to understanding how water movement influences the distribution and transport of water pollutants (Martin et al. 2018). Ocean currents, influenced by factors such as wind, tides, temperature variations, and the Earth's rotation, carry pollutants over extensive distances (Okubo 1971). These currents frequently cause pollutants to settle in regions far from their original source (Legleiter et al. 2022). For example, the dispersion of oil spills is influenced by factors such as the physical properties of the water and the viscosity of the oil (Kistovich and Chaplina 2021). Another well-known case is the Great Pacific Garbage Patch, a massive concentration of plastic debris in the North Pacific Ocean. The hydrodynamic forces in oceanic gyres trap plastic waste in a continuous vortex that pulls in and contains millions of tons of floating plastic and other waste (Van Sebille et al. 2020). Similarly, physical processes such as wave action and friction with other materials affect the degradation of plastics, which causes plastics to break down into smaller pieces (Hermsen et al. 2018). Therefore, understanding how contaminants move within an aquatic environment is essential for developing effective monitoring and mitigation strategies (Ji et al. 2020).

Scientists often use dyes such as rhodamine to gain deeper insights into local water movement and better predict how pollutants will spread (Clark et al. 2014). Rhodamine, a fluorescent dye, is commonly released into water systems to trace the flow and velocity of currents in a given area (Smart and Laidlaw 1977). By studying the dispersal patterns of these dyes, researchers can create accurate models of water movement, helping to determine how contaminants might be transported (Köppl et al. 2023). This local understanding is crucial for effective monitoring, particularly in coastal zones where pollution accumulates.

Studying how water movement influences the transport of contaminants in specific regions improves pollution pattern predictions, helps identify vulnerable marine ecosystems, and designs more targeted mitigation efforts (Van Sebille et al. 2020). This information is critical to developing efficient cleanup strategies and preventing pollutants from causing further harm to marine life and coastal environments.

Need for Advanced Monitoring

The need for advanced aquatic pollution monitoring has become increasingly evident over the past several decades (Ji et al. 2020). Historically, pollution monitoring relied on in situ sampling, which involved collecting water or sediment samples and analysing them in laboratories. This method, while accurate, is labour-intensive, time-consuming, and limited in scale (Armitage et al. 2022). As a result, early efforts to monitor marine pollution were often reactive instead of proactive, responding to visible pollution events rather than preventing them (van Emmerik et al. 2023). Effectively monitoring hazardous agents in aquatic environments is crucial for addressing geographical and resource constraints. It helps establish baseline conditions, enables rapid response to emerging threats, evaluates the success of mitigation efforts with reduced operational costs, tracks environmental changes, and pinpoints pollution hotspots (Tasseron et al. 2024). In this sense, remote sensing analysis is a powerful tool to increase the level of confidence when dealing with monitoring pollutants in aquatic environments.

Overview
Marine pollution is a multifaceted problem with both visible and invisible consequences. Two of the most pressing pollutants are oil spills and plastic waste. Each has distinct characteristics, but both can cause significant disruptions to marine ecosystems. Therefore, efficient monitoring of marine pollution is crucial for understanding its extent, mitigating its effects, and protecting marine biodiversity.

Fundaments of Remote Sensing

As environmental challenges intensify globally, advanced monitoring technologies have become indispensable for managing pollution (Lebreton et al. 2018). Among these, remote sensing (RS) and hyperspectral imaging (HSI) have proven to be transformative in monitoring aquatic ecosystems and detecting marine pollution (Fingas and Brown 2014). By providing data over vast and often inaccessible areas, these technologies offer unparalleled insights into the state of our environment (Andriolo et al. 2023).

Remote sensing refers to acquiring information about the Earth's surface without direct physical contact (Elachi and Van Zyl 2021). Typically conducted using satellites, drones, or aircraft, this method detects and analyses reflected or emitted energy from various surfaces (Emery and Camps 2017). Observing large-scale phenomena or areas complex to reach by traditional methods is precious for Earth observation (Fingas and Brown 2014). Remote sensing's versatility has led to its adoption across diverse fields, including agriculture, forestry, geology, and disaster management (Horstrand et al. 2019; Jordan 2019; Haut et al. 2024). Its ability to track natural and anthropogenic changes over time enables researchers and policymakers to make informed decisions about sustainable development and resource management (Kramer et al. 2002).

Remote sensing technologies can be broadly categorised into active and passive systems, each with distinct mechanisms and applications. Active remote sensing generates energy through lasers or radio waves and measures the return signal after interacting with the surface (Garcia-Pineda et al. 2013a). Examples include light detection and ranging (LiDAR), which uses laser pulses to map terrain, and

(a) (b) (c)

Fig. 3 Examples of remote sensing systems. (**a**) LiDAR YellowScan Vx20-300. (**b**) Multispectral Silios CMS-V. (**c**) Hyperspectral Specim FX10

radar systems, which can penetrate clouds and operate effectively in low-light or nighttime conditions (Garcia-Pineda et al. 2017). In contrast, passive remote sensing relies on natural energy sources, primarily sunlight, to detect and measure the reflected radiation from the Earth's surface (Camps-Valls et al. 2011). Common examples include optical sensing systems such as multispectral and hyperspectral imaging (Gonçalves and Andriolo 2022). Multispectral systems capture data in a limited number of discrete spectral bands, offering generalised information suitable for large-scale assessments, while hyperspectral systems measure hundreds of continuous spectral bands, enabling highly detailed material identification and discrimination (Emery and Camps 2017; Zhang et al. 2017) (Fig. 3).

In this section, we will explore the foundational concepts of remote sensing, focusing on the principles and applications of hyperspectral imaging in marine pollution monitoring. The following subsections will delve into optical sensing technologies, examining the differences between multispectral and hyperspectral approaches and the platforms used for data collection. Subsequently, we will provide an in-depth discussion of hyperspectral imaging, outlining the scientific principles that underpin this cutting-edge technology.

Optical Sensing Technologies

Optical sensors are essential to RS, enabling the conversion of light into measurable signals for analysing diverse biophysical parameters (Satriano et al. 2019). These technologies operate within specific ranges of the electromagnetic spectrum, primarily the visible (VIS, 400–700 nm), near-infrared (NIR, 700–1400 nm), visible and near-infrared (VNIR, 400–1000 nm), and short-wave-infrared (SWIR, 1000–2500 nm) regions (Camps-Valls et al. 2011). Their non-destructive nature allows researchers to study ecosystems without causing any disruption, making them indispensable for environmental monitoring.

These optical sensors can be used to measure the solar radiation reflected from the Earth's surface. The reflected radiation is quantified using radiant flux intensity, expressed in units of watts per square meter steradian $[W/(sr \cdot \mathrm{m}^2)]$. This metric considers the energy flux at a specific wavelength over a surface area and within

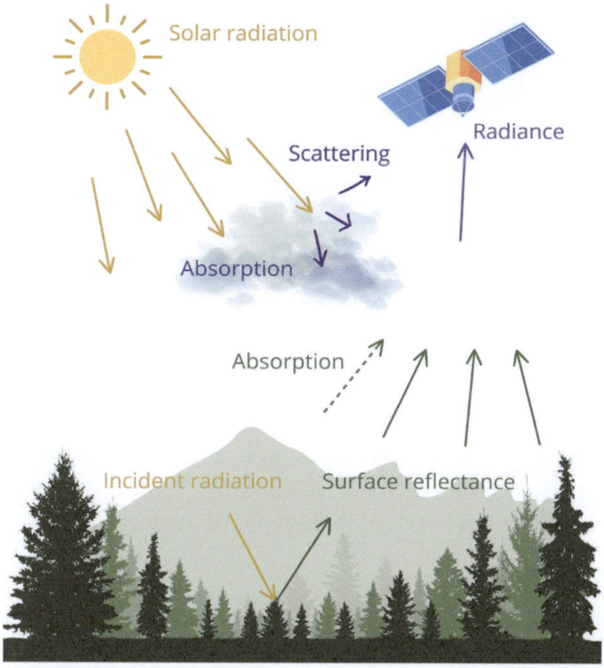

Fig. 4 Interaction of solar radiation with Earth's atmosphere and surface. The sun emits solar radiation, which encounters clouds, causing scattering and absorption. Some radiation reaches the Earth's surface as incident radiation, where it is either absorbed or reflected. Reflected radiation, known as surface reflectance, travels back through the atmosphere, suffering absorption and scattering again. A satellite detects the radiance on top of the atmosphere (Pérez-García 2024)

a defined angular field of view (Camps-Valls et al. 2011). However, the raw measurements collected by optical sensors are significantly affected by atmospheric conditions, influenced by interactions with gases, aerosols, and water vapour. These effects can distort the spectral signals, as illustrated in Fig. 4, leading to inaccuracies in interpreting surface properties. Reflectance is used as a corrected measure to account for these distortions.

Reflectance is defined as the ratio of the radiant flux emitted from a surface to the incident solar flux. This parameter provides a normalised, dimensionless value that better compares the spectral characteristics of a surface, independently of the atmospheric conditions (Tempfli et al. 2009). Achieving accurate reflectance values requires applying atmospheric correction algorithms, which remove the influence of atmospheric components from the raw data. This correction ensures the reflectance values more accurately represent surface properties.

Based on their spectral capabilities, optical sensors can be categorised into panchromatic, multispectral, and hyperspectral systems, each suited for specific applications (Camps-Valls et al. 2011). These systems differ in spectral resolution

Table 1 Comparison of optical sensors based on their characteristics

Characteristic	Panchromatic	Multispectral	Hyperspectral
Bands	Single broad band	3–10 discrete bands	100–300 contiguous narrow bands
Spatial Resolution	High (grayscale images)	Moderate to high	Lower than multispectral
Spectral resolution	Low	Moderate	High
Uses	Mapping, surveillance	Vegetation, land cover	Material ID, pollution
Examples sensors	WorldView-2 Panchromatic	Sentinel-2, Micasense	Specim FX10, AVIRIS
Advantages	High detail, low cost	Good balance, efficient	Precise analysis
Limitations	No spectral info	Limited bands	Expensive, complex

and data collection capabilities, offering distinct advantages and limitations that are critical to consider in marine pollution monitoring. Table 1 presents a summary of the general characteristics of each of these types of sensors, which will be discussed in more detail below.

Panchromatic Sensors

Panchromatic sensors capture data in a single broad spectral band, encompassing the entire visible spectrum and sometimes extending into the NIR. Panchromatic sensors' simpler manufacturing process makes it accessible to acquire devices with a higher spatial resolution, making them ideal for detailed imaging and feature extraction (Camps-Valls et al. 2011). They are commonly used in satellite-based systems, such as the WorldView-2 Panchromatic Imager (Corporation 2024) or Lansat 8 Panchromatic band (Morfitt et al. 2015), which provides detailed imagery for urban planning, infrastructure monitoring, and land cover mapping.

Panchromatic sensors are also widely employed in security and surveillance due to their ability to produce sharp, high-resolution grayscale images, even in low-light conditions (Corbane et al. 2008). Despite their advantages in spatial resolution, panchromatic sensors lack spectral differentiation, making them less effective for identifying specific materials or pollutants. However, their data is often combined with multispectral or hyperspectral imagery through a process known as pansharpening. This process merges the high spatial resolution of panchromatic data with the rich spectral detail of multispectral or hyperspectral imagery (Ehlers et al. 2010). This fusion enhances the interpretability and utility of the combined dataset, making panchromatic sensors an essential component of many remote sensing applications.

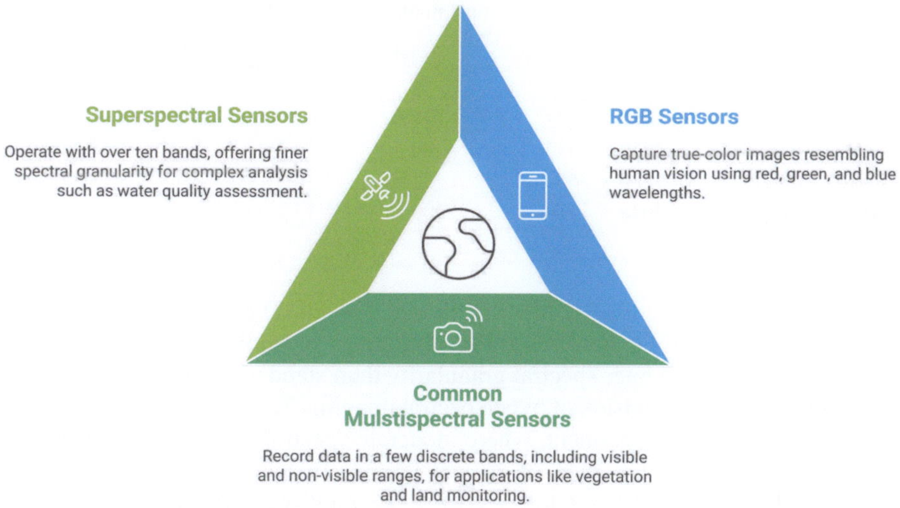

Fig. 5 Multispectral sensor comparison

Multispectral Sensors

Multispectral sensors capture data in a limited number of broad spectral bands, typically ranging from three to around ten bands, with each band covering a wide range of wavelengths (Camps-Valls et al. 2011). Depending on the characteristics of the sensor, they can be subclassified into RGB, common multispectral sensors, and superspectral sensors, as shown in Fig. 5.

The most common example is RGB sensors, which record data in the red (R), green (G), and blue (B) ranges of the spectrum to produce true-colour images that resemble natural human vision. RGB is widely used in consumer devices like smartphones and security cameras and is essential in applications such as facial recognition and traffic monitoring (Smith et al. 2015).

In environmental monitoring, multispectral sensors can include additional bands beyond the visible spectrum, such as NIR, to provide vegetation maps and land cover classification and monitor general ecosystem health. For instance, cameras with a NIR band, such as the Micasense RedEdge, are widely used in agriculture to assess vegetation health (Micasense 2019). Parrot (2023) and SILIOS Technologies (2023) include red-edge and NIR bands, enabling advanced plant growth and stress analysis. Table 2 presents a comparison of different characteristics such as number of bands or pixels captured by different commercial multispectral sensors.

Other multispectral sensors operate with more than ten spectral bands but lack the narrow, contiguous wavelength intervals that define hyperspectral sensors. These

Table 2 Comparison of characteristics of some multispectral sensors

Sensor name	Resolution (pixels)	Weight (g)	Number of bands	Price (€)
Silios CMS-V	426 × 339	75	8	7000
Mini-MCA6	1280 × 1024	700	6	5200
Sentera Double 4K	4056 × 3040	80	4	3000
Parrot Sequoia	1280 × 960	72	4	3000
ADC Micro	2048 × 1536	200	3	3000

systems, such as the Sentinel-2 MSI (Pahlevan et al. 2017) and the Envisat MERIS sensor (Signature 2021), are often called "bridging" technologies or superspectral sensors. They provide finer spectral granularity than standard multispectral sensors but do not match the precision of hyperspectral imaging. These intermediate systems benefit water quality assessment, where moderate spectral detail is needed (Pérez-García et al. 2024d).

While their simplicity and lower cost make multispectral sensors accessible, their relatively low spectral resolution can limit their ability to differentiate between materials with similar reflectance characteristics (Tempfli et al. 2009). For example, detecting specific pollutants or subtle changes in water quality may exceed the capabilities of multispectral imaging (Köppl et al. 2023). Despite these limitations, their efficiency in processing smaller data volumes and their cost-effectiveness make multispectral sensors an attractive option for many routine environmental assessments (Pérez-García et al. 2023b).

Hyperspectral Sensors

HSI is a specialised form of optical sensing that builds on remote sensing principles by capturing detailed spectral information. Hyperspectral sensors record data across hundreds of narrow, contiguous spectral bands, typically ranging from 100 to 300 (Zhang et al. 2017). The high spectral resolution can detect subtle differences in similar surfaces, enabling material identification and discrimination and making HSI particularly advantageous for complex analyses such as marine pollution monitoring (Hafeez et al. 2019).

HSI generates data organised into three-dimensional data cubes. In the hyperspectral cubes, each pixel contains an N-dimensional vector corresponding to the number of spectral bands and representing its spectral signature (García et al. 2022). The foundation of HSI lies in the interaction of solar radiation with materials. Solar radiation is absorbed or reflected at specific wavelengths depending on a surface's atomic structure and composition, resulting in a unique spectral response for each material (Jackson and Alpers 2010). This response, or spectral signature, is influenced by the material's physical and chemical properties and surface

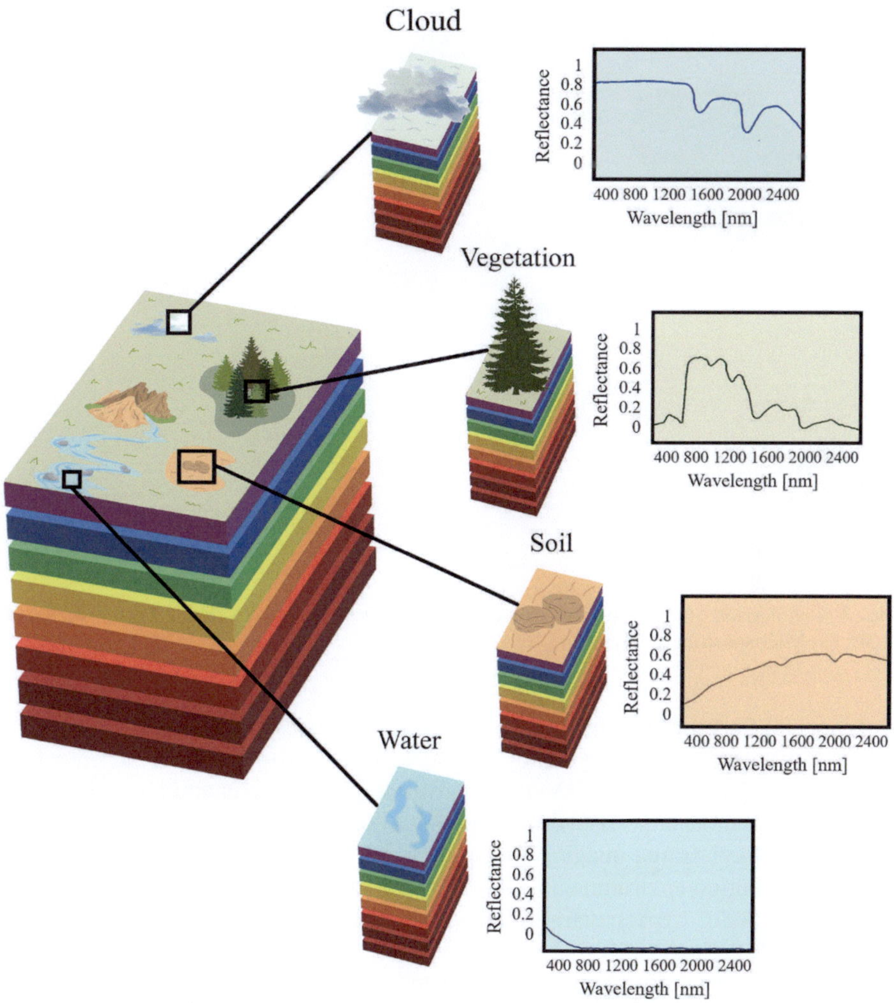

Fig. 6 Hyperspectral image cube with cloud, vegetation, soil, and water sections. Each section is connected to a graph showing its specific spectral signature across wavelengths from 400 to 2400 nm. The cloud graph shows low reflectance with minor fluctuations. Vegetation has higher reflectance with distinct peaks. Soil shows moderate, steady reflectance. Water has low reflectance, decreasing sharply. The layers are colour-coded, representing different spectral bands (Pérez-García 2024)

characteristics like roughness (Fig. 6) (Pérez-García et al. 2023a). By processing these spectral signatures, HSI can differentiate and identify various substances, including pollutants, with high accuracy (Brekke and Solberg 2005).

Several hyperspectral sensors are renowned for their advanced capabilities. Specim offers high-performance hyperspectral sensors such as the Specim FX10 and FX17, widely used for environmental monitoring and industrial quality con-

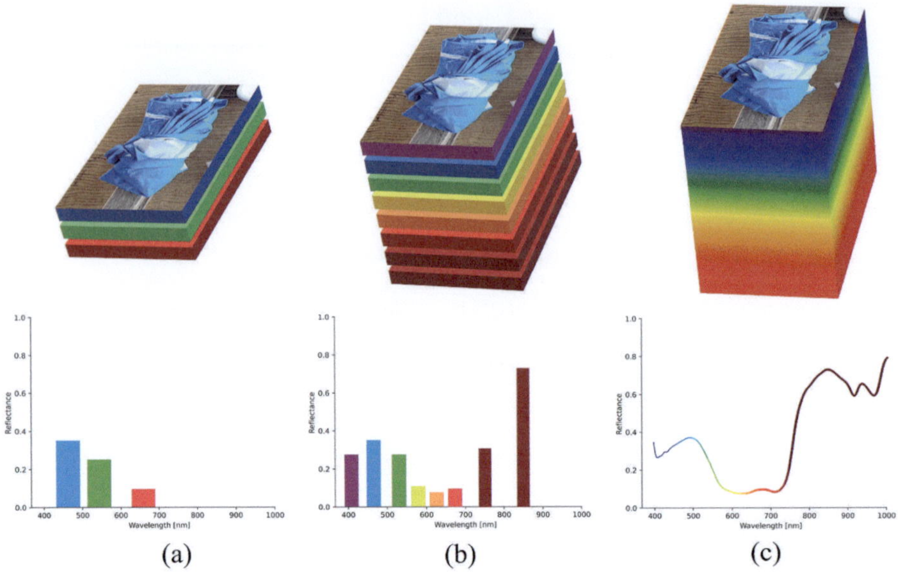

Fig. 7 Comparison of the volume of data captured for each sensor type (Pérez-García 2024). (**a**) RGB. (**b**) Multispectral. (**c**) Hyperspectral

trol (Specim Spectral Imaging 2022). Headwall Photonics produces versatile hyperspectral systems, such as the Nano-Hyperspec and Micro-Hyperspec, designed for uncrewed aerial vehicles (UAVs) and airborne platforms, enabling flexible data collection in remote areas (Headwall Photonics 2020). Another notable system is the airborne visible/infrared imaging spectrometer (AVIRIS), a NASA sensor designed for high-resolution environmental assessments (Vane et al. 1993).

The benefits of hyperspectral imaging come with significant challenges. These systems are more expensive—three to five times the cost of multispectral sensors—and generate large data volumes (Fig. 7), often 10–20 times greater than their multispectral counterparts (Pérez-García et al. 2023b). This data complexity necessitates advanced processing techniques and significant computational resources (Zhang et al. 2017). Despite these challenges, hyperspectral sensors' ability to deliver detailed and accurate insights makes them invaluable for monitoring and managing marine pollution.

! Attention

Optical sensors are pivotal in remote sensing, offering diverse capabilities tailored to specific applications. HSI is a transformative tool, offering unparalleled spectral precision for material identification and environmental analysis. Despite its potential, the large data volumes it generates present significant challenges in storage and processing, necessitating advanced solutions to fully harness its capabilities—a topic explored in a later section.

Platforms

Optical sensors deployed across various platforms, including satellites, airborne systems, and uncrewed aerial vehicles, provide a versatile framework for environmental monitoring. Each platform offers unique capabilities and limitations, making it essential to match its strengths to specific monitoring needs.

Satellite-Based Systems

Satellites are a cornerstone of remote sensing, providing extensive coverage and continuous data over large areas (Fingas and Brown 2014). These platforms are particularly well-suited for long-term environmental monitoring, such as tracking oceanic currents, mapping large-scale oil spills, or observing plastic patches (Garcia-Pineda et al. 2017; Papageorgiou and Topouzelis 2024). For instance, the Sentinel-2 series of satellites, equipped with multispectral sensors, offers a revisit time of 5 days and spatial resolutions ranging from 10 to 60 meters, making it ideal for broad-scale environmental applications (Pahlevan et al. 2017).

Satellites can host a variety of sensors, including multispectral systems such as the MERIS sensor onboard the Envisat mission (Signature 2021) and hyperspectral instruments like Hyperion on board the EO-1 spacecraft (Survey and NASA 2021), allowing them to capture detailed spectral information across diverse wavelengths. However, spatial resolution may be insufficient for detecting localised pollution events, such as small oil slicks or isolated debris clusters, especially in marine environments (Andriolo et al. 2023). Additionally, satellite data acquisition is hindered by cloud cover, which obstructs optical imaging in certain weather conditions (Gómez-Chova et al. 2007).

Despite these limitations, satellite-based systems remain indispensable due to their global reach and ability to provide consistent, repeatable observations over time (Rußwurm et al. 2023). This makes them a critical tool for monitoring large-scale trends in marine and terrestrial ecosystems.

Airborne Systems

Airborne platforms offer higher spatial resolution than satellites, making them ideal for localised and detailed assessments of pollution events (Vane et al. 1993). For example, systems like AVIRIS can capture hyperspectral data at resolutions as precise as a few meters, enabling accurate mapping of oil spills, sediment plumes, or industrial discharges (Jet Propulsion Laboratory, NASA 2021).

Airborne systems provide greater data collection flexibility, as they can be deployed on demand and adjusted to focus on areas of interest (Green et al. 1998). They are commonly used in post-disaster scenarios, such as oil spill assessments after maritime accidents (Liu et al. 2011). However, these systems have notable limitations, including higher operational costs, limited flight duration, and constraints on coverage area due to the physical range of the aircraft.

Airborne platforms are versatile in terms of the sensors they can carry, accommodating hyperspectral, multispectral, thermal, and LiDAR sensors (Leifer et al. 2012). This adaptability allows them to address diverse monitoring needs, from detecting marine pollutants to mapping coastal erosion.

Uncrewed Aerial Vehicle-Based Systems

Uncrewed aerial vehicles represent a rapidly growing technology in remote sensing. Their high spatial resolution and flexibility in deployment make them ideal for localised and short-term monitoring (Odonkor et al. 2019). UAVs can carry advanced sensors, including compact hyperspectral imagers such as the Specim AFX10 (Specim Spectral Imaging 2022) or multispectral cameras like the SILIOS Technologies (2023), enabling detailed pollution assessments at a fraction of the cost of traditional airborne systems (Horstrand et al. 2019).

The key advantages of UAVs include their quick deployment and ability to operate in areas inaccessible to satellites or aircraft, such as nearshore environments, rivers, or urban waterways (Baek et al. 2019; Cortesi et al. 2022). UAVs are particularly effective for mapping small oil spills, monitoring illegal waste dumping, identifying plastic items, and tracking water quality in restricted zones (Gonçalves and Andriolo 2022). Moreover, UAVs are far more affordable than satellite platforms. Figure 8 presents two UAV models commonly used in remote

(a) (b)

Fig. 8 Examples of UAVs used in remote sensing applications. (**a**) DJI Matrice 600. (**b**) ETRAIR Akila eOcto

sensing applications. Commercial models offer ready-to-use solutions for various remote sensing applications. For instance, the DJI Matrice 600 (Fig. 8a) is widely used for environmental monitoring due to its stability and high payload capacity. Custom-built UAVs, such as those provided by professional manufacturers like ETRAIR (Etrair 2024), can be commissioned for specialised requirements, ensuring the platform is customised to meet specific operational needs. The ETRAIR Akila eOcto (Fig. 8b) is an example of a highly adaptable UAV designed for scientific and industrial applications requiring advanced payload integration.

Despite their benefits, UAVs face operational limitations, including short flight times, limited range, and the need for skilled operators to manage both flight and data processing (Qiao et al. 2022). Moreover, regulatory restrictions in many regions can pose challenges for UAV deployment, particularly in sensitive or crowded areas (Köppl et al. 2023).

> **Important**

Each platform—satellites, airborne systems, and UAVs—plays a critical role in optical remote sensing, offering distinct advantages and trade-offs. Satellites provide broad-scale, repeatable coverage for long-term monitoring; airborne systems deliver high-resolution data for medium-scale applications; and UAVs excel in localised, high-precision assessments. By leveraging the strengths of these platforms, RS facilitates informed decision-making across a range of spatial and temporal scales.

Marine Pollution Monitoring Techniques

HSI provides detailed spectral information, but its high dimensionality often poses challenges in data analysis due to noise and redundancy produced by the high correlation between adjacent bands (Emery and Camps 2017). Various techniques have been developed to address these complexities, including spectral indices and artificial intelligence (AI) methods. These approaches enhance the interpretability and utility of HSI data across different applications.

Spectral Indices

Spectral indices simplify the analysis of complex HSI datasets by combining multiple spectral bands through arithmetic operations (Pérez-García et al. 2024d). These indices highlight specific spectral behaviours while suppressing noise or irrelevant information (Bijeesh and Narasimhamurthy 2019). The effectiveness of spectral indices depends on selecting suitable bands and operations based on the phenomena's spectral characteristics (Kolokoussis and Karathanassi 2018). For instance, pollutants in aquatic environments often exhibit distinct spectral responses at specific wavelengths, enabling the creation of indices tailored to emphasise these features (Hafeez et al. 2019).

A widely used example is the normalised difference vegetation index (NDVI) (Carlson 1997) in Eq. 1, which assesses plant health by comparing near-infrared (NIR) and red bands, and the chlorophyll content index (CHL) (Hu et al. 2012) in Eq. 2, which measures chlorophyll-a concentration with bands corresponding to different wavelengths ($R_{wavelength}$):

$$NDVI = \frac{NIR - Red}{NIR + Red} \tag{1}$$

$$CHL = log(max(R_{433,490,510})/R_{555}) \tag{2}$$

Spectral indices for water characteristics are the water absorption feature (WAF) (Lu et al. 2007) in Eq. 3 that assesses water content, the coloured dissolved organic matter (CDOM) (Kutser et al. 2005) in Eq. 4 that identifies concentrations of organic matter in water, and the normalised difference water index (NDWI) (cai Gao 1996) in Eq. 5 that highlights water features:

$$WAF = \frac{R_{1343} + R_{1563}}{2} - R_{1453} \tag{3}$$

$$CDOM = R_{565}/R_{660} \tag{4}$$

$$NDWI = \frac{R_{860} - R_{1240}}{R_{860} + R_{1240}} \tag{5}$$

Some indices identify specific pollutants. Some standard indices for crude oil are the fluorescence index (FI) (Loos et al. 2012) in Eq. 6 that identifies the source of dissolved organic matter; the rotation-absorption index (RAI) (b_i is the pixel radiance for the image band) (Loos et al. 2012) in Eq. 7 that detects shifts in light absorption; the hydrogen index (HI) (Kühn et al. 2004) in Eq. 8 that assesses hydrocarbon content, where $\lambda_{wavelength}$ refers to a specific wavelength; and the oil slope index (OSI) (Li et al. 2012) in Eq. 9 that differentiates oil pollution from natural substances:

$$FI = \frac{Blue - Red}{Blue + Red} \tag{6}$$

$$RAI = \frac{Blue - IR}{Blue + IR} \cdot \sqrt{\sum_{i=1}^{N} {b_i}^2} \tag{7}$$

$$HI = (\lambda_{1729} - \lambda_{1705})\frac{R_{1741} - R_{1705}}{\lambda_{1741} - \lambda_{1705}} + (R_{1705} - R_{1729}) \tag{8}$$

$$OSI = \frac{R_{\lambda_{Red}} - R_{\lambda_{Yellow}}}{\lambda_{Red} - \lambda_{Yellow}} \tag{9}$$

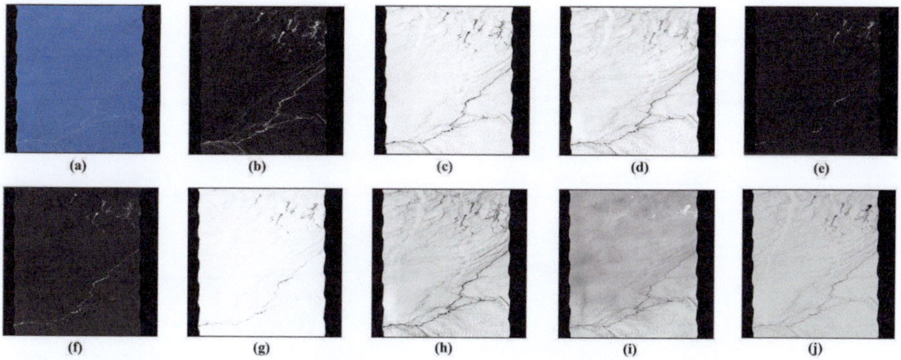

Fig. 9 Index performance in Deepwater Horizon oil spill (Pérez-García et al. 2024d). (**a**) RGB. (**b**) NDVI. (**c**) RAI. (**d**) FI. (**e**) HI. (**f**) OSI. (**g**) WAF. (**h**) CDOM. (**i**) CHL. (**j**) NDWI

Both the specific pollutant detection indices and the water characteristics and vegetation indices can be used to identify oil spills. Figure 9 shows the performance of these indices in part of the Deepwater Horizon oil spill in Mexico in 2010. RAI, FI, and CDOM are the best at identifying oil.

For plastic items, the floating debris index (FDI) (Biermann et al. 2020) presented in Eq. 10 and the plastic index (PI) (Themistocleous et al. 2020) in Eq. 11 are proposed to identify debris and plastic patches in the ocean:

$$FDI = R_{800} - R_{700} + (R_{1600} - R_{730}) \cdot \frac{\lambda_{800} - \lambda_{665}}{\lambda_{1600} - \lambda_{665}} \cdot 10 \tag{10}$$

$$PI = R_{860}/(R_{650} + R_{860}) \tag{11}$$

Artificial Intelligence

AI techniques have become fundamental for addressing the challenges of HSI, offering solutions for efficient surface classification, material detection, and data interpretation (Lary et al. 2016; Mou et al. 2018; Zhang and Zhang 2022). These techniques mimic human cognitive functions, such as learning and problem-solving, to enhance hyperspectral data analysis (Campesato 2020). Machine learning (ML) enables data-driven predictions, deep learning extracts high-level features through neural networks, dimensionality reduction optimises data processing by eliminating redundancies, and transfer learning allows models to adapt knowledge from related domains. Figure 10 provides an overview of these techniques. AI's ability to process and analyse massive HSI datasets efficiently makes it a cornerstone of modern marine environment monitoring. By leveraging these techniques, researchers can

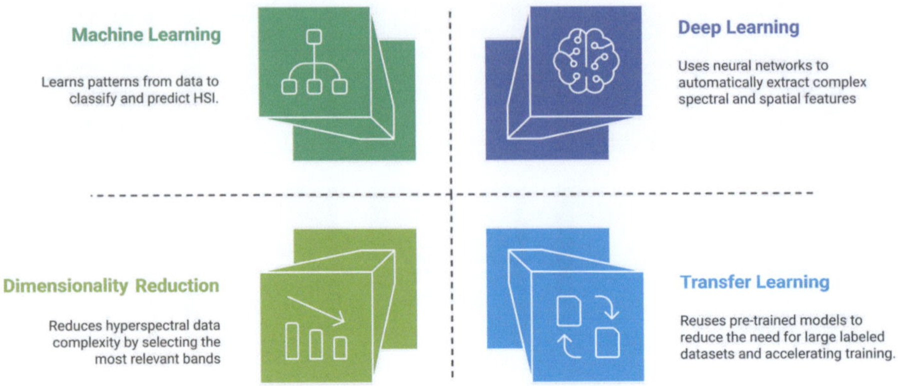

Fig. 10 Overview of AI techniques used in hyperspectral image analysis

achieve higher accuracy in identifying pollutants, tracking their spread, and informing strategies for mitigating environmental impacts (Pérez-García 2024). In this section, we focus on the aspects of AI most relevant to HSI and its applications in marine pollution monitoring.

Machine Learning

Machine learning (ML) is a branch of AI that enables computers to learn patterns from data and make predictions or decisions without explicit programming (Maxwell et al. 2018). It has become a vital tool in analysing HSI data for environmental applications, including the detection of marine pollutants such as oil spills, plastics, and harmful algal blooms (Pérez-García 2024). In marine pollution detection, ML algorithms classify HSI data into meaningful categories based on spectral signatures. ML techniques can be subdivided into supervised learning and unsupervised learning.

Supervised Learning

Supervised learning, the most commonly used approach, requires labelled datasets to train models that accurately predict outcomes for new data (Campesato 2020). Supervised learning methods include classification algorithms, with support vector machines (SVM) (Hearst et al. 1998) and random forests (RF) (Ho 1995) standing out for their effectiveness in marine applications. SVM separates data into categories by finding the best boundary (called the hyperplane) that divides them, aiming to maximise the distance between the groups for better classification (Hearst et al. 1998). RF combines the results of several decision trees (flowchart with yes-or-no questions based on the wavelength values) to make more accurate and reliable predictions, reducing the chance of errors or overfitting. These techniques have been applied to identify oil spill extents by distinguishing spectral patterns of hydrocarbons from surrounding water (Hua et al. 2021).

Unsupervised Learning

Unsupervised learning techniques identify patterns and groupings in unlabelled datasets, making it useful when labelled data is scarce (Campesato 2020). K-means clustering groups data points into clusters based on their similarity, making it easier to find patterns (Hartigan and Wong 1979). Methods like clustering are effective for segmenting hyperspectral images into regions of similar spectral characteristics, facilitating tasks such as detecting plastic debris or mapping algae concentrations in water bodies (Barbato et al. 2022).

Deep Learning

Deep learning (DL) is a subset of ML that uses multilayered neural networks to analyse complex patterns in large datasets (Fitton et al. 2022). A neural network is a computer system inspired by the human brain, designed to process information by learning patterns and relationships in data (Garcia-Pineda et al. 2013b). This approach excels in image recognition tasks, making it particularly suitable for interpreting hyperspectral imagery (Yuan et al. 2020). However, these models are also prone to overfitting, especially when dealing with limited training data or highly complex architectures, which can reduce their generalisation capability. DL methods have been widely adopted in marine pollution detection due to their ability to extract features automatically from raw data, reducing reliance on manual preprocessing.

For instance, DL models have been utilised to identify macroplastics in aquatic environments by detecting subtle spectral differences that traditional methods might miss (Armitage et al. 2022). These models also play a crucial role in recognising oil spills, leveraging spectral and spatial features to improve accuracy in challenging conditions, such as turbid waters or areas with mixed pollutants (Yuan et al. 2020).

DL model's performance is heavily influenced by loss functions, which measure the discrepancy between the predicted and the actual outcomes during training (Pérez-García et al. 2023a). Optimising the loss function ensures that models achieve higher accuracy, even in complex datasets like hyperspectral cubes (Bottou et al. 2018).

Dimensionality Reduction

HSI datasets are inherently high dimensional, with hundreds of spectral bands per pixel (Emery and Camps 2017). While the rich spectral detail provides valuable information for training models, it also poses challenges, such as noise and computational inefficiency arising from the high correlation between adjacent bands (Zhang et al. 2017). These limitations make real-time analysis and large-scale applications more demanding, requiring advanced optimisation techniques. Dimensionality reduction techniques simplify datasets by retaining only the most relevant information, improving model performance and reducing processing time (Bajcsy and Groves 2004).

In marine pollution studies, dimensionality reduction is crucial for identifying key spectral bands associated with pollutants (Sun and Du 2019). Techniques like sequential feature selection (SFS) iteratively identify bands that contribute most to predictive accuracy (Kramer and Kramer 2016). Other techniques, such as the principal component analysis (PCA), reduce dimensions by transforming variables to capture the greatest variance (Wold et al. 1987). By focusing on critical wavelengths, researchers can optimise models for detecting specific pollutants, such as oil or plastics, while minimising redundancy and computational costs (Pérez-García et al. 2024a).

Transfer Learning

Transfer learning is a technique or strategy that can be applied across various ML paradigms (Manas et al. 2021). Transfer learning leverages pre-trained models developed for one task and adapts them to related tasks, enabling faster training and improved accuracy (Hosna et al. 2022). This strategy is precious in marine pollution detection, where datasets may be limited or expensive to obtain.

Transfer learning enables the use of pre-trained models, allowing them to be adapted to new tasks with minimal modification (Hosna et al. 2022). For instance, a DL model trained on hyperspectral data to detect oil spills in one region can be adapted to another geographic area by fine-tuning only the final layers of the neural network, which define the specific classes (Manas et al. 2021). By adapting only the later layers to refine the model for the specific task, transfer learning enhances performance across diverse conditions, such as variations in water salinity, temperature, or pollutant types.

Overview

Integrating HSI with spectral indices and AI techniques offers robust solutions for monitoring marine pollution. Spectral indices help highlight specific features in the data, making it easier to detect pollutants like oil spills and plastics. Meanwhile, ML and DL techniques provide robust tools for classifying, segmenting, and analysing complex datasets. Dimensionality reduction optimises the efficiency of these methods by focusing on the most relevant data, while transfer learning enables models to be adapted across different regions and reduces the computational cost.

Current Challenges

Despite its advantages, HSI faces several challenges limiting its large-scale marine pollution monitoring deployment. These challenges include issues related to data processing, the availability of labelled datasets, and the integration with policy, as shown in Fig. 11. Addressing these obstacles is critical to realising HSI's full potential in marine pollution applications. Section "Cutting-Edge Technology" presents some innovative solutions to overcome these obstacles.

Data Volume and Complexity

HSI detailed spectral information provides unique insights into material properties, including the identification of marine pollutants, but also presents significant challenges in terms of data management and analysis (Sun and Du 2019). The sheer size of HSI datasets demands substantial storage capacities, advanced computational resources, and efficient algorithms to extract meaningful information (Donoho

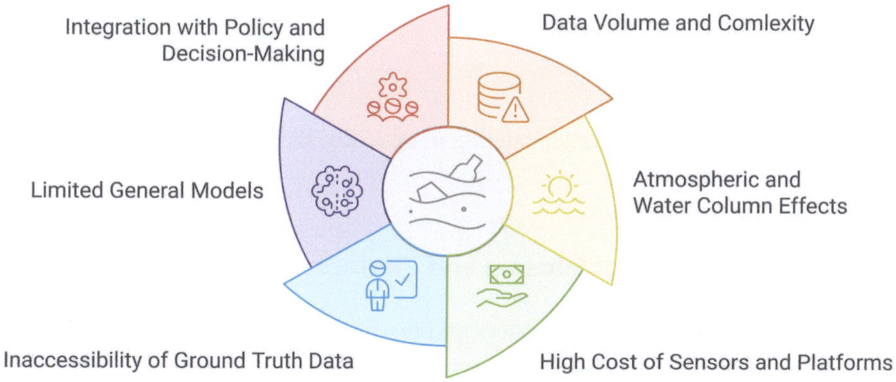

Fig. 11 Main HSI challenges for marine pollution monitoring

et al. 2000). Traditional data processing techniques often falter under the "curse of dimensionality", where the increasing number of features complicates the detection of relevant patterns and increases the risk of overfitting (Bajcsy and Groves 2004). Additionally, noise and redundancy in hyperspectral data can obscure critical signals, complicating pollutant detection and classification tasks.

While spectral indices and dimensionality reduction techniques, such as principal component analysis (PCA) or band selection, have been developed to address these issues, they often involve trade-offs between complexity and interpretability (Qiao et al. 2022). For instance, spectral indices may simplify analysis by focusing on specific wavelengths but struggle differentiating between multiple pollutants with overlapping spectral features (Karnieli et al. 2010). Dimensionality reduction methods risk discarding essential information if not carefully applied.

? Question

What methods can simplify the analysis of high-dimensional HSI data while maintaining its accuracy and utility?

Atmospheric and Water Column Effects

Accurately interpreting HSI data from marine environments is challenging due to atmospheric interference and water column variability (Xiong et al. 2020). Factors such as light absorption by gases and aerosols, scattering by air molecules, and turbidity significantly impact data quality (Liang 2005). Sunglint, the intense reflection of sunlight off the water's surface, poses additional difficulties, particularly for high-resolution imaging (Harmel et al. 2018).

Current atmospheric correction techniques, such as those assuming zero water-leaving radiance in specific bands, are not always effective in turbid or optically shallow waters, where additional noise can distort results (Li et al. 2017). Addressing these challenges requires developing more robust and adaptable correction methods tailored to marine environments.

? Question

How can we improve HSI understanding in optically shallow and turbid waters?

Cost and Limitations of Sensors and Platforms

The high cost of hyperspectral sensors and the platforms that support them—such as satellites, drones, and aircraft—presents a significant barrier to the widespread adoption of HSI in marine pollution monitoring (Pérez-García et al. 2023b).

Hyperspectral sensors require precise manufacturing, driving up costs compared to traditional multispectral sensors. Compared to their multispectral counterparts, hyperspectral systems' larger size and weight further restrict their deployment on lightweight drones and nanosatellites (Andriolo et al. 2023). Satellites require significant investment in launch and maintenance, while small platforms face challenges related to payload limitations and operational costs (Grøtte et al. 2022).

The expense of these systems limits their accessibility, particularly for low-resource settings or developing countries (Emery and Camps 2017). Reducing the cost of hyperspectral platforms requires advancements in sensor miniaturisation, efficient manufacturing processes, and the integration of lower-cost platforms like nanosatellites and consumer-grade drones. Collaborative funding models and public-private partnerships could also help make HSI technology more affordable and accessible for marine pollution monitoring.

? Question

How can technological advances reduce the costs of deploying optical sensors and platforms for marine applications?

Inaccessibility of Ground Truth Data

The scarcity of labelled data, or "ground truth" data, is a critical challenge in applying HSI to marine pollution monitoring (Maxwell et al. 2018). Ground truth data is the foundation for training supervised ML models, which rely on extensive, accurately labelled datasets to classify and detect pollutants effectively. However, acquiring such data for marine environments is time-consuming and resource-intensive, often requiring expert interpretation and field validation under challenging conditions (Hua et al. 2021).

Remote or hazardous locations, such as open oceans or oil spill sites, compound the difficulty of obtaining ground truth data due to logistical constraints, high costs, and safety concerns. For example, measuring spectral signatures of pollutants in situ involves deploying specialised equipment and personnel to areas that may be difficult to access or monitor consistently (Pérez-García et al. 2024c). This reliance on manual data collection limits the scalability and speed at which new labelled datasets can be generated, hindering the development of more accurate ML models (Xu et al. 2022).

To address this challenge, researchers are increasingly exploring semi-supervised and unsupervised learning approaches that can identify patterns and groupings in HSI data without requiring extensive labelled datasets. Synthetic data generation using simulation models or data augmentation techniques also shows promise in filling the gap by creating realistic training datasets that mimic the spectral properties of pollutants (Michau and Fink 2021). Simple classification tasks may require hundreds to a few thousand labelled images, while more complex tasks, such as detailed pollutant mapping across diverse conditions, often demand tens of thousands of high-quality, labelled samples (Liu et al. 2020). Collaborations between research institutions, governments, and private entities are vital to establishing shared repositories of labelled datasets, ensuring broader access to critical training resources and enabling the development of more generalisable models.

? Question

What strategies can reduce reliance on extensive HSI field validation datasets?

Limited General Models

General models can also reduce the need for extensive labelled data and expert supervision in new environments, making them ideal for broad-scale implementation. Most models are designed and trained for specific scenarios, relying heavily on localised data that captures the spectral characteristics of pollutants under specific environmental settings (Legleiter et al. 2019). However, variations in water salinity, temperature, turbidity, and other factors can significantly alter spectral signatures, making applying models trained in one region to another difficult without retraining (Ma et al. 2024). For instance, detecting pollutants like rhodamine or oil spills often involves site-specific calibration, limiting the generalisation of existing techniques (Legleiter et al. 2022).

Transfer learning has emerged as a promising strategy to enhance model generalisation by adapting pre-trained models to new environments with minimal additional training (Pan and Yang 2009). This approach leverages shared features across datasets, reducing the need for extensive labelled data and enabling models to perform reliably in different contexts. Additionally, band selection methods that identify the most influential spectral ranges for detecting specific pollutants can help improve model adaptability by focusing on the most relevant features (Sun and Du 2019). Developing standardised workflows and open-access repositories for model training and validation can further support general models, fostering collaboration among researchers and practitioners (Legleiter et al. 2024).

? Question

What techniques can improve the generalisation of pollutant detection methods across different environments?

Integration with Policy and Decision-Making

The effective use of HSI in marine pollution monitoring requires its integration into policy frameworks and decision-making processes. There is a gap between the generation of HSI data and its utilisation by policymakers and environmental managers (Martínez-Vicente et al. 2019). Complex HSI datasets often need to be translated into actionable insights, such as maps of pollution hotspots or risk assessments, which requires collaboration between scientists, data analysts, and decision-makers (Tasseron et al. 2024).

To bridge this gap, standardised data interpretation and reporting protocols must be developed to ensure that HSI outputs are precise, reliable, and aligned with regulatory frameworks (Cózar et al. 2024). Additionally, there is a need for capacity-building initiatives to train policymakers and stakeholders in interpreting and applying HSI-derived information. Incorporating HSI into environmental monitoring policies can also benefit from real-time analytics and user-friendly tools, which enable timely responses to pollution events and long-term management strategies (Mahrad et al. 2020).

? Question

What frameworks are needed to translate hyperspectral data into actionable insights for marine pollution management?

Cutting-Edge Technology

Emerging technologies are shaping the future of marine pollution monitoring by partially addressing the current significant HSI challenges. This chapter highlights recent advancements, including novel spectral indices, generalisable ML methods, next-generation sensors, and cloud computing platforms.

Specialised Spectral Indices for Real-Time Monitoring

Spectral indices simplify the analysis of high-dimensional HSI data while maintaining its utility by extracting specific information from a subset of spectral bands (Hernández-Suárez et al. 2025). They are usually calculated using simple mathematical operations (Kolokoussis and Karathanassi 2018). By design, spectral indices emphasise the differences in reflectance properties of specific targets, enabling precise detection of pollutants such as oil spills, algae, and plastics. Spectral indices offer rapid, computationally efficient solutions that are ideal for real-time applications where decisions must be made quickly, and resources are limited. Their low computational time makes them ideal for integration into UAV onboard systems, facilitating real-time monitoring (Odonkor et al. 2019).

Case Study: Normalised Difference Oil Index

A notable example of a specialised spectral index is the normalised difference oil index (NDOI) (Pérez-García et al. 2024d), proposed for coastal oil spill detection. NDOI, in Eq. 12, is a novel index that addresses the unique challenges of sand and suspended particles, which often confound standard spectral indices (Pérez-García et al. 2022):

$$NDOI = \frac{R_{599} - R_{870}}{R_{599} + R_{870}} \tag{12}$$

By leveraging two widely available spectral bands (599 nm and 870 nm), NDOI provides a computationally efficient tool for distinguishing oil spills from surrounding features. As illustrated in Fig. 12, features that appear as reddish filaments to the human eye (RGB) are distinctly identified by the NDOI as an oil spill. In addition, NDOI reveals other contaminated areas that are not discernible to the naked eye. Furthermore, because the index is designed to work with commonly used bands, such as those of Landsat 8 (Morfitt et al. 2015), its implementation is broadly accessible and cost-effective. For instance, in emergency oil spill responses, using NDOI could expedite the deployment of containment measures with real-time data. However, like most spectral indices, NDOI is sensitive to environmental variability and relies on manual thresholding, underscoring the need for complementary ML techniques in complex scenarios (Karnieli et al. 2010).

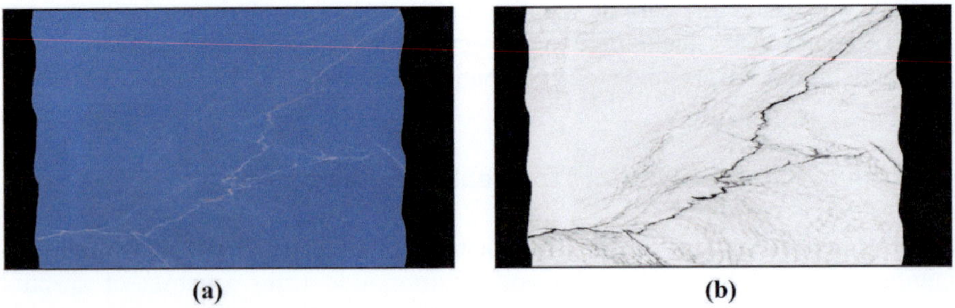

Fig. 12 Visualisation of an oil spill in RGB and applying the NDOI. (**a**) RGB. (**b**) NDOI

Generalisable Methods

Generalisable methods focus on extending the utility of detection models to diverse environments and datasets (Ma et al. 2024). Such methods are essential for overcoming challenges associated with limited labelled data, variable environmental conditions, and the need for transferable solutions across environments. There are several techniques to improve model generalisation in different environments, including unsupervised algorithms, highly transferable models, and data augmentation techniques.

Unsupervised methods have gained attention for their ability to extract patterns from unlabelled data, reducing reliance on costly fieldwork and manual annotations (Barbato et al. 2022). These techniques are particularly beneficial for monitoring hazardous or remote areas, such as the open oceans, due to logistical challenges (Xu et al. 2022). Unsupervised methods still trail behind supervised models in precision and need more refinement to handle environmental variability.

Transferable models represent another breakthrough, enabling pre-trained algorithms to generalise across diverse scenarios (Ma et al. 2024). Band selection algorithms play a pivotal role in identifying the most influential spectral bands, allowing the development of refined spectral indices and ML classifiers with improved generalisation across diverse environmental contexts. By focusing on critical spectral features, these algorithms hold significant promise for reducing dependence on extensive labelled datasets (Pan and Yang 2009). The most promising band selection algorithm that optimises ML classifiers has been presented by (Pérez-García et al. 2024b). It has been tested with standard HSI datasets to discriminate between different types of plastics and rhodamine concentrations in optically shallow waters (Pérez-García et al. 2024a).

Another effective strategy for enhancing models' generalisation involves expanding the training datasets by collecting more real-world data or generating synthetic datasets. HSI data augmentation techniques generate synthetic data that enrich the training datasets (Li et al. 2018). These methods improve the ML model's robustness by simulating realistic variations, thereby reducing overfitting and enhancing their performance across diverse settings (Bartolini et al. 2018).

Next-Generation Optical Sensors

Advancements in optical sensors are critical for enabling high-precision monitoring while addressing constraints such as cost, weight, and spectral resolution (Grøtte et al. 2022). Multispectral sensors, such as MAIA (2024) and SILIOS Technologies (2023), offer a promising alternative to hyperspectral systems. While their broader spectral bands lack the finer resolution of hyperspectral sensors, they are often sufficient for detecting pollutants with gradual spectral variations (Legleiter et al. 2022). This advancement paves the way for a new generation of specialised sensors tailored to detect marine pollutants (Pérez-García 2024). These sensors, built with cost-effective materials, can be seamlessly integrated into low-cost platforms like UAVs and CubeSats, enabling widespread and accessible deployment for environmental monitoring.

Case Study: Bentayga-I and PANDORA

Bentayga-I (Rodríguez-Molina et al. 2024), a CubeSat integrated with the PANDORA multispectral camera, represents a milestone in remote sensing technology. PANDORA is a modular multispectral sensor with open-source architecture. Its interchangeable filters and narrow 10 nm bandwidths enable it to capture spectral signatures with hyperspectral-like precision while maintaining the compact form factor of a multispectral sensor (Pérez-García et al. 2023b). PANDORA successfully differentiated coastal features during its first field validation, although further refinement is needed to reduce false positives.

The ongoing development of Bentayga-I underscores the potential of specialised sensors to expand the accessibility and affordability of RS technologies. The CubeSat was validated from 20 km of altitude using a helium balloon of the company B2Space (B2Space 2024), in which the subsystems were tested on near-space conditions. Figure 13 illustrates two perspectives of the first validation mission. Figure 13a shows the CubeSat mounted on the validation structure before the launch, while Fig. 13b captures the helium balloon as it ascends, providing a view of the near-space testing environment.

Cloud Computing

Cloud computing is revolutionising the processing and analysis of RS data by providing scalable infrastructure and computational power (Haut et al. 2019). Cloud-based workflows optimise computational efficiency, supporting large-scale environmental monitoring initiatives (Gorelick et al. 2017). In this context, imple-

(a) (b)

Fig. 13 Bentayga first validation mission. (**a**) View from the validation structure. (**b**) Helium balloon carrying the CubeSat during its ascent

menting the NDOI on cloud platforms has significantly reduced processing time, making it feasible to analyse massive HSI datasets almost instantly (Haut et al. 2024). Platforms such as Google Earth Engine (GEE) already have direct access to satellite imagery, including Copernicus data, as soon as it becomes available, eliminating users' need to upload large datasets manually, therefore allowing researchers to analyse vast datasets without extensive local resources (Wilder 2012). Furthermore, integrating cloud computing with Python workflows has facilitated the global identification of macroplastic hotspots, showcasing the potential for scalable environmental solutions (Pérez-García 2024). However, transmitting large datasets from remote sensing platforms to cloud servers can be challenging due to the data volume and the high latency, particularly for UAV or airborne systems. Therefore, cloud-based processing is typically performed after data acquisition rather than in real time.

Cloud platforms reduce computational barriers, enabling researchers to focus on developing advanced detection methods and improving model transferability. As satellite missions continue to generate increasing volumes of data, cloud computing will remain a cornerstone of RS innovation, enhancing the scalability and impact of environmental monitoring systems (Gorelick et al. 2017).

Conclusion

This chapter explores the transformative potential of HSI and related remote sensing technologies for environmental monitoring, focusing on detecting and addressing marine pollution. From introducing foundational concepts to presenting cutting-edge innovations, it highlights how advancements in technology and methodology are unlocking new possibilities for understanding and managing complex environmental challenges.

Key contributions include the development of specialised spectral indices and more robust, transferable, and general AI models for pollutant identification and classification. These advancements underscore the importance of tailoring solutions to specific environmental contexts, bridging the gap between the complex handling of high-dimensional data and actionable insights. Emerging tools, such as specialised multispectral sensors and scalable cloud computing platforms, further extend the reach and impact of these methods. Cost-effective technologies can deliver high-performance data collection while decreasing the cost. The efficient analysis of large datasets reduces the barriers to global-scale implementation.

Despite these advancements, challenges remain. Data complexity, the limited availability of ground truth data, and generalisable models highlight the ongoing need for research and innovation. Addressing these gaps will require interdisciplinary collaboration, leveraging the combined expertise of environmental scientists, remote sensing specialists, and policymakers. These tools provide the means to monitor and respond to pollution and offer insights to drive preventive measures, ensuring a healthier, more sustainable future for our planet.

Competing Interests This work was completed while Ámbar Pérez-García was beneficiary of a predoctoral grant (ID: TESIS2021010059) given by the "Agencia Canaria de Investigacion, Innovacion y Sociedad de la Informacion (ACIISI)" of the "Consejería de Universidades, Ciencia e Innovación y Cultura", part-financed by the European Social Fund Plus (FSE+) "Programa Operativo Integrado de Canarias 2021–2027, Eje 3 Tema Prioritario 74 (85%)". Adrián Rodríguez-Molina and Emma Hernández-Suárez have benefited from a predoctoral grant given by the "Vicerrectorado de Investigación y Transferencia de la Universidad de las Palmas de Gran Canaria" during the research. We also would like to thank the OASIS-HARMONIE project, under contract PID2023-148285OB-C43 of "Proyectos de Generación de Conocimiento" 2023, and the PERSEO project, under contract CPP2021-008527, of "Programa de Colaboración Pblico-Privada" of "Gobierno de Espaa", 2023–2025. The authors have no conflict of interest to declare relevant to this chapter's content.

Ethics Approval No ethics approval was required.

References

Abbasi S, Soltani N, Keshavarzi B, Moore F, Turner A, Hassanaghaei M (2018) Microplastics in different tissues of fish and prawn from the musa estuary, persian gulf. Chemosphere 205:80–87

Andriolo U, Topouzelis K, van Emmerik TH, Papakonstantinou A, Monteiro JG, Isobe A, Hidaka M, Kako S, Kataoka T, Gonçalves G (2023) Drones for litter monitoring on coasts and rivers: suitable flight altitude and image resolution. Marine Pollut Bullet 195:115521

Angelova D, Uzunov I, Uzunova S, Gigova A, Minchev L (2011) Kinetics of oil and oil products adsorption by carbonized rice husks. Chem Eng J 172(1):306–311

Armitage S, Awty-Carroll K, Clewley D, Martinez-Vicente V (2022) Detection and classification of floating plastic litter using a vessel-mounted video camera and deep learning. Remote Sensing 14(14), 3425

Asari N, Suratman MN, Mohd Ayob NA, Abdul Hamid NH (2021) Mangrove as a natural barrier to environmental risks and coastal protection. In: Mangroves: ecology, biodiversity and management, pp 305–322

Avio CG, Gorbi S, Regoli F (2017) Plastics and microplastics in the oceans: from emerging pollutants to emerged threat. Marine Environ Res 128:2–11

Baek D, Seo IW, Kim JS, Nelson JM (2019) UAV-based measurements of spatio-temporal concentration distributions of fluorescent tracers in open channel flows. Adv Water Resour 127:76–88

Bajcsy P, Groves P (2004) Methodology for hyperspectral band selection. Photogram Eng Remote Sensing 70(7):793–802

Barbato MP, Napoletano P, Piccoli F, Schettini R (2022) Unsupervised segmentation of hyperspectral remote sensing images with superpixels. Remote Sensing Appl Soc Environ 28:100823

Bartolini S, Mecocci A, Pozzebon A, Zoppetti C, Bertoni D, Sarti G, Caiti A, Costanzi R, Catani F, Ciampalini A et al (2018) Augmented virtuality for coastal management: a holistic use of in situ and remote sensing for large scale definition of coastal dynamics. ISPRS Int J Geo-Inform 7(3):92

Bernhardt JR, Leslie HM (2013) Resilience to climate change in coastal marine ecosystems. Annu Rev Marine Sci 5(1):371–392

Beyer J, Trannum HC, Bakke T, Hodson PV, Collier TK (2016) Environmental effects of the deepwater horizon oil spill: a review. Marine Pollut Bull 110(1):28–51

Biermann L, Clewley D, Martinez-Vicente V, Topouzelis K (2020) Finding plastic patches in coastal waters using optical satellite data. Sci Rep 10(1):5364

Bijeesh T, Narasimhamurthy K (2019) A comparative study of spectral indices for surface water delineation using landsat 8 images. In: 2019 international conference on data science and communication (IconDSC). IEEE, Piscataway, pp 1–5

Botterell ZL, Beaumont N, Dorrington T, Steinke M, Thompson RC, Lindeque PK (2019) Bioavailability and effects of microplastics on marine zooplankton: a review. Environ Pollut 245:98–110

Bottou L, Curtis FE, Nocedal J (2018) Optimization methods for large-scale machine learning. SIAM Rev 60(2):223–311

Brekke C, Solberg AH (2005) Oil spill detection by satellite remote sensing. Remote Sensing Environ 95(1):1–13

Burak S, Dogan E, Gazioglu C (2004) Impact of urbanization and tourism on coastal environment. Ocean Coastal Manag 47(9-10):515–527

Burke L, Reytar K, Spalding M, Perry A (2011) Reefs at risk revisited. World Resources Institute (WRI), Washington

cai Gao B (1996) Ndwi—a normalized difference water index for remote sensing of vegetation liquid water from space. Remote Sensing Environ 58(3):257–266

Campesato O (2020) Artificial intelligence, machine learning, and deep learning. Mercury Learning and Information

Camps-Valls G, Tuia D, Gómez-Chova L, Jiménez S, Malo J (2011) Remote sensing image processing. Synthesis Lectures Image Video Multimedia Process 5(1):1–192

Carlson TN, Ripley DA (1997) On the relation between NDVI, fractional vegetation cover, and leaf area index. Remote Sensing of Environment 62(3):241–252

Clark DB, Lenain L, Feddersen F, Boss E, Guza R (2014) Aerial imaging of fluorescent dye in the near shore. J Atmos Ocean Technol 31(6):1410–1421

Corbane C, Marre F, Petit M (2008) Using spot–5 hrg data in panchromatic mode for operational detection of small ships in tropical area. Sensors 8(5):2959–2973

Corporation SI (2024) Worldview-2 satellite sensor. https://www.satimagingcorp.com/satellite-sensors/worldview-2/

Cortesi I, Masiero A, Tucci G, Topouzelis K (2022) UAV-based river plastic detection with a multispectral camera. Int Arch Photogram Remote Sensing Spatial Inform Sci 43:855–861

Court C, Hodges AW, Coffey K, Ainsworth CH, Yoskowitz D (2020) Effects of the deepwater horizon oil spill on human communities: catch and economic impacts. In: Deep oil spills: facts, fate, and effects, pp 569–580

Cózar A, Arias M, Suaria G, Viejo J, Aliani S, Koutroulis A, Delaney J, Bonnery G, Macías D, de Vries R et al (2024) Proof of concept for a new sensor to monitor marine litter from space. Nat Commun 15(1):4637

Donoho DL et al (2000) High-dimensional data analysis: the curses and blessings of dimensionality. AMS Math Challenges Lecture 1(2000):32

Ehlers M, Klonus S, Johan Åstrand P, Rosso P (2010) Multi-sensor image fusion for pansharpening in remote sensing. Int J Image Data Fusion 1(1):25–45

Elachi C, Van Zyl JJ (2021) Introduction to the physics and techniques of remote sensing. John Wiley & Sons, London

Emery W, Camps A (2017) Introduction to satellite remote sensing: atmosphere, ocean, land and cryosphere applications. Elsevier, Amsterdam

Fingas M, Brown C (2014) Review of oil spill remote sensing. Marine Pollut Bull 83(1):9–23

Fisher CR, Montagna PA, Sutton TT (2016) How did the deepwater horizon oil spill impact deep-sea ecosystems? Oceanography 29(3):182–195

Fitton D, Laurens E, Hongkarnjanakul N, Schwob C, Mezeix L (2022) Land cover classification through convolutional neural network model assembly: a case study of a local rural area in thailand. Remote Sensing Appl Soc Environ 26:100740

Gallo F, Fossi C, Weber R, Santillo D, Sousa J, Ingram I, Nadal A, Romano D (2020) Marine litter plastics and microplastics and their toxic chemicals components: the need for urgent preventive measures. In: Analysis of nanoplastics and microplastics in food, CRC Press, Boca Raton, pp 159–179

García JL, Paoletti ME, Jiménez LI, Haut JM, Plaza A (2022) Efficient semantic segmentation of hyperspectral images using adaptable rectangular convolution. IEEE Geosci Remote Sensing Lett 19:1–5

Garcia-Pineda O, Holmes J, Rissing M, Jones R, Wobus C, Svejkovsky J, Hess M (2017) Detection of oil near shorelines during the deepwater horizon oil spill using synthetic aperture radar (SAR). Remote Sensing 9(6):567

Garcia-Pineda O, MacDonald I, Hu C, Svejkovsky J, Hess M, Dukhovskoy D, Morey SL (2013a) Detection of floating oil anomalies from the deepwater horizon oil spill with synthetic aperture radar. Oceanography 26(2):124–137

Garcia-Pineda O, MacDonald IR, Li X, Jackson CR, Pichel WG (2013b) Oil spill mapping and measurement in the gulf of Mexico with textural classifier neural network algorithm (TCNNA). IEEE J Selected Topics Appl Earth Observ Remote Sensing 6(6):2517–2525

Gonçalves G, Andriolo U (2022) Operational use of multispectral images for macro-litter mapping and categorization by unmanned aerial vehicle. Marine Pollut Bull 176:113431

Gorelick N, Hancher M, Dixon M, Ilyushchenko S, Thau D, Moore R (2017) Google earth engine: planetary-scale geospatial analysis for everyone. Remote Sensing Environ 202:18–27

Grøtte ME, Birkeland R, Honoré-Livermore E, Bakken S, Garrett JL, Prentice EF, Sigernes F, Orlandić M, Gravdahl JT, Johansen TA (2022) Ocean color hyperspectral remote sensing with high resolution and low latency—the hypso-1 cubesat mission. IEEE Trans Geosci Remote Sensing 60(1000619):1–19

Green DS (2020) Biological and ecological impacts of plastic debris in aquatic ecosystems. In: Plastics in the aquatic environment—Part I: current status and challenges. Springer, Berlin, pp 111–133

Green RO, Eastwood ML, Sarture CM, Chrien TG, Aronsson M, Chippendale BJ, Faust JA, Pavri BE, Chovit CJ, Solis M et al (1998) Imaging spectroscopy and the airborne visible/infrared imaging spectrometer (AVIRIS). Remote Sensing Environ 65(3):227–248

Gómez-Chova L, Camps-Valls G, Calpe-Maravilla J, Guanter L, Moreno J (2007) Cloud-screening algorithm for envisat/meris multispectral images. IEEE Trans Geosci Remote Sensing 45(12):4105–4118

Häder DP, Banaszak AT, Villafañe VE, Narvarte MA, González RA, Helbling EW (2020) Anthropogenic pollution of aquatic ecosystems: emerging problems with global implications. Sci Total Environ 713:136586

Hafeez S, Wong MS, Abbas S, Kwok CY, Nichol J, Lee KH, Tang D, Pun L (2019) Detection and monitoring of marine pollution using remote sensing technologies. In: Fouzia HB (ed) Monitoring of marine pollution, Chapters, IntechOpen

Harmel T, Chami M, Tormos T, Reynaud N, Danis PA (2018) Sunglint correction of the multi-spectral instrument (MSI)-SENTINEL-2 imagery over inland and sea waters from SWIR bands. Remote Sensing of Environment 204:308–321

Hartigan JA, Wong MA (1979) Algorithm as 136: a k-means clustering algorithm. J Roy Stat Soc Series C (Applied Statistics) 28(1):100–108

Haut JM, Gallardo JA, Paoletti ME, Cavallaro G, Plaza J, Plaza A, Riedel M (2019) Cloud deep networks for hyperspectral image analysis. IEEE Trans Geosci Remote sensing 57(12):9832–9848

Haut JM, Moreno-Alvarez S, Pastor-Vargas R, Pérez-García Á, Paoletti ME (2024) Cloud-based analysis of large-scale hyperspectral imagery for oil spill detection. IEEE J Selected Topics Appl Earth Observ Remote Sensing 17:2461–2474

Hearst M, Dumais S, Osuna E, Platt J, Scholkopf B (1998) Support vector machines. IEEE Intell Syst Appl 13(4):18–28

Hermsen E, Mintenig SM, Besseling E, Koelmans AA (2018) Quality criteria for the analysis of microplastic in biota samples: a critical review. Environ Sci Technol 52(18):10230–10240

Hernández-Suárez E, Rodríguez-Molina A, Pérez-García Á, Mirza-Rosca J, López JF (2025) Corrosion severity index (CSI) for spectral characterization of corroded steel and iron samples. IEEE Trans Instrum Measur 74(5006113)

Ho TK (1995) Random decision forests. In: Proceedings of 3rd international conference on document analysis and recognition, vol 1, pp 278–282

Horstrand P, Guerra R, DÍaz M, Morales A, Jiménez A, López S, López JF (2019) A spectral imaging system for precision agriculture: from its inception till a pre-commercial prototype. In: 2019 XXXIV conference on design of circuits and integrated systems (DCIS). IEEE, Piscataway, pp 1–6

Hosna A, Merry E, Gyalmo J, Alom Z, Aung Z, Azim MA (2022) Transfer learning: a friendly introduction. J. Big Data 9(1):102

Hu C, Lee Z, Franz B (2012) Chlorophyll algorithms for oligotrophic oceans: a novel approach based on three-band reflectance difference. J Geophys Res Oceans 117(C1)

Hua Y, Marcos D, Mou L, Zhu XX, Tuia D (2021) Semantic segmentation of remote sensing images with sparse annotations. IEEE Geosci Remote Sensing Lett 19:1–5

Jackson CR, Alpers W (2010) The role of the critical angle in brightness reversals on sunglint images of the sea surface. J Geophys Res Oceans 115(C9)

Jambeck JR, Geyer R, Wilcox C, Siegler TR, Perryman M, Andrady A, Narayan R, Law KL (2015) Plastic waste inputs from land into the ocean. Science 347(6223):768–771

Jet Propulsion Laboratory, NASA (2021) Airbone visible infrared imaging spectrometer website. In: AVIRIS. https://aviris.jpl.nasa.gov/

Ji C, Beegle-Krause CJ, Englehardt JD (2020) Formation, detection, and modeling of submerged oil: a review. J Marine Sci Eng 8(9):642

Jordan BR (2019) Collecting field data in volcanic landscapes using small UAS (SUAS)/drones. J Volcanol Geothermal Res 385:231–241. Geological data in volcanology: collection, organisation and applications

Kühn F, Oppermann K, Hörig B (2004) Hydrocarbon index—an algorithm for hyperspectral detection of hydrocarbons. Int J Remote Sensing 25(12):2467–2473

Karnieli A, Agam N, Pinker RT, Anderson M, Imhoff ML, Gutman GG, Panov N, Goldberg A (2010) Use of NDVI and land surface temperature for drought assessment: merits and limitations. J Climate 23(3):618–633

Keesing JK, Gartner A, Westera M, Edgar GJ, Myers J, Hardman-Mountford NJ, Bailey M (2018) Impacts and environmental risks of oil spills on marine invertebrates, algae and seagrass: a global review from an Australian perspective. Oceanogr Marine Biol 56:311–370

Kershaw P, Protection J (2015) Sources, fate and effects of microplastics in the marine environment: a global assessment. UNESCO: United Nations Educational, Scientific and Cultural Organisation

Kistovich A, Chaplina T (2021) Analytical and experimental modeling of the hydrocarbon slick form and its spreading on the water surface. Phys Fluids 33(7):076605

Koelmans AA, Besseling E, Shim WJ (2015) Nanoplastics in the aquatic environment. critical review. In: Marine anthropogenic litter, pp 325–340

Kolokoussis P, Karathanassi V (2018) Oil spill detection and mapping using sentinel 2 imagery. J Marine Sci Eng 6(4):1–12

Kramer O, Kramer O (2016) Scikit-learn. In: Machine learning for evolution strategies, pp 45–53

Kramer HJ et al (2002) Observation of the earth and its environment: survey of missions and sensors, vol 1982. Springer, Berlin

Kutser T, Pierson DC, Kallio KY, Reinart A, Sobek S (2005) Mapping lake CDOM by satellite remote sensing. Remote Sensing Environ 94(4):535–540

Köppl CJ, McKnight US, Lemaire GG, Nørregaard AM, Thiim TC, Bjerg PL, Bauer-Gottwein P, García M (2023) Tracer concentration mapping in a stream with hyperspectral images from unoccupied aerial systems. Adv Water Resour 182:104567

Lary DJ, Alavi AH, Gandomi AH, Walker AL (2016) Machine learning in geosciences and remote sensing. Geosci Front 7(1):3–10

Lebreton L, Slat B, Ferrari F, Sainte-Rose B, Aitken J, Marthouse R, Hajbane S, Cunsolo S, Schwarz A, Levivier A et al (2018) Evidence that the great pacific garbage patch is rapidly accumulating plastic. Sci Rep 8(1):1–15

Legleiter CJ, McDonald RR, Nelson JM, Kinzel PJ, Perroy RL, Baek D, Seo IW (2019) Remote sensing of tracer dye concentrations to support dispersion studies in river channels. J Ecohydraulics 4(2):131–146

Legleiter CJ, Sansom BJ, Jacobson RB (2022) Remote sensing of visible dye concentrations during a tracer experiment on a large, turbid river. Water Resour Res 58(4):e2021WR031396

Legleiter CJ, Scholl VM, Sansom BJ, Burgess MA (2024) Hyperspectral image transects during transient events in rivers (hitter): framework development and application to a tracer experiment on the Missouri river, USA. Remote Sensing 16(3743):1–33

Leifer I, Lehr WJ, Simecek-Beatty D, Bradley E, Clark R, Dennison P, Hu Y, Matheson S, Jones CE, Holt B, Reif M, Roberts DA, Svejkovsky J, Swayze G, Wozencraft J (2012) State of the art satellite and airborne marine oil spill remote sensing: application to the BP deepwater horizon oil spill. Remote Sensing Environ 124:185–209

Li Q, Lu L, Zhang B, Tong Q (2012) Oil slope index: An algorithm for crude oil spill detection with imaging spectroscopy. In: 2012 second international workshop on earth observation and remote sensing applications, pp 46–49

Li J, Hu C, Shen Q, Barnes BB, Murch B, Feng L, Zhang M, Zhang B (2017) Recovering low quality modis-terra data over highly turbid waters through noise reduction and regional vicarious calibration adjustment: a case study in Taihu Lake. Remote Sensing Environ 197:72–84

Li W, Chen C, Zhang M, Li H, Du Q (2018) Data augmentation for hyperspectral image classification with deep CNN. IEEE Geosci Remote Sensing Lett 16(4):593–597

Liang S (2005) Quantitative remote sensing of land surfaces. John Wiley & Sons, London

Liu Y, MacFadyen A, Ji ZG, Weisberg RH (2011) Introduction to monitoring and modeling the deepwater horizon oil spill. Geophys Monogr Series 195:1–7

Liu S, Shi Q, Zhang L (2020) Few-shot hyperspectral image classification with unknown classes using multitask deep learning. IEEE Trans Geosci Remote Sensing 59(6):5085–5102

Loos E, Brown L, Borstad G, Mudge T, Álvarez M (2012) Characterization of oil slicks at sea using remote sensing techniques. In: 2012 oceans. IEEE, Piscataway, pp 1–4

Lu WZ, Yuan H, Xu G, et al. (2007) Modern near-infrared spectroscopy analytical technology, 2nd edn. China Petrochemical Press, Beijing

Ma Y, Chen S, Ermon S, Lobell DB (2024) Transfer learning in environmental remote sensing. Remote Sensing Environ 301:113924

Mahrad BE, Newton A, Icely JD, Kacimi I, Abalansa S, Snoussi M (2020) Contribution of remote sensing technologies to a holistic coastal and marine environmental management framework: a review. Remote Sensing 12(14):2313

Manas O, Lacoste A, Giró-i Nieto X, Vazquez D, Rodriguez P (2021) Seasonal contrast: unsupervised pre-training from uncurated remote sensing data. In: Proceedings of the IEEE/CVF international conference on computer vision, pp 9414–9423

Markic A, Niemand C, Bridson JH, Mazouni-Gaertner N, Gaertner JC, Eriksen M, Bowen M (2018) Double trouble in the south pacific subtropical gyre: increased plastic ingestion by fish in the oceanic accumulation zone. Marine Pollut Bull 136:547–564

Martínez-Vicente V, Clark JR, Corradi P, Aliani S, Arias M, Bochow M, Bonnery G, Cole M, Cózar A, Donnelly R, Echevarría F, Galgani F, Garaba SP, Goddijn-Murphy L, Lebreton L, Leslie HA, Lindeque PK, Maximenko N, Martin-Lauzer FR, Moller D, Murphy P, Palombi L, Raimondi V, Reisser J, Romero L, Simis SG, Sterckx S, Thompson RC, Topouzelis KN, van Sebille E, Veiga JM, Vethaak AD (2019) Measuring marine plastic debris from space: initial assessment of observation requirements. Remote Sensing 11(20):2443

Martin JL, Schottman RW, McCutcheon SC (2018) Hydrodynamics and transport for water quality modeling. CRC Press, Boca Raton

Maxwell AE, Warner TA, Fang F (2018) Implementation of machine-learning classification in remote sensing: an applied review. Int J Remote Sensing 39(9):2784–2817

Meijer LJ, Van Emmerik T, Van Der Ent R, Schmidt C, Lebreton L (2021) More than 1000 rivers account for 80% of global riverine plastic emissions into the ocean. Sci Adv 7(18):eaaz5803

Micasense (2019) Rededge-MX camera. https://www.micasense.com/rededge-mx. Accessed 06 Oct 2019

Michau G, Fink O (2021) Unsupervised transfer learning for anomaly detection: application to complementary operating condition transfer. Knowl-Based Syst 216:106816

Mishra AK, Kumar GS (2015) Weathering of oil spill: modeling and analysis. Aquatic Procedia 4:435–442. International Conference on water resources, coastal and ocean engineering (ICWRCOE'15)

Morales-Caselles C, Viejo J, Martí E, González-Fernández D, Pragnell-Raasch H, González-Gordillo JI, Montero E, Arroyo GM, Hanke G, Salvo VS et al (2021) An inshore–offshore sorting system revealed from global classification of ocean litter. Nat Sustain 4(6):484–493

Morfitt R, Barsi J, Levy R, Markham B, Micijevic E, Ong L, Scaramuzza P, Vanderwerff K (2015) Landsat-8 operational land imager (OLI) radiometric performance on-orbit. Remote Sensing 7(2):2208–2237

Mou L, Ghamisi P, Zhu XX (2018) Unsupervised spectral–spatial feature learning via deep residual conv–deconv network for hyperspectral image classification. IEEE Trans Geosci Remote Sensing 56(1):391–406

Negri AP, Brinkman DL, Flores F, Botté ES, Jones RJ, Webster NS (2016) Acute ecotoxicology of natural oil and gas condensate to coral reef larvae. Sci Rep 6(1):21153

Odonkor P, Ball Z, Chowdhury S (2019) Distributed operation of collaborating unmanned aerial vehicles for time-sensitive oil spill mapping. Swarm Evol Comput 46:52–68

Okubo A (1971) Oceanic diffusion diagrams, vol 18. Elsevier, Amsterdam

Paches M, Aguado D, Martínez-Guijarro R, Romero I (2019) Long-term study of seasonal changes in phytoplankton community structure in the western mediterranean (Valencian community). Environ Sci Pollut Res 26:14266–14276

Pahlevan N, Sarkar S, Franz B, Balasubramanian S, He J (2017) Sentinel-2 multispectral instrument (MSI) data processing for aquatic science applications: demonstrations and validations. Remote Sensing Environ 201:47–56

Pan SJ, Yang Q (2009) A survey on transfer learning. IEEE Trans Knowl Data Eng 22(10):1345–1359

Papageorgiou D, Topouzelis K (2024) Experimental observations of marginally detectable floating plastic targets in sentinel-2 and planet super dove imagery. Int J Appl Earth Observ Geoinform 134:104245

Pérez-García Á (2024) Towards transferable pollution detection methods in aquatic environments using hyperspectral technology. Ph.D. thesis, Universidad de Las Palmas de Gran Canaria

Pérez-García Á, Horstrand P, López JF (2022) A novel spectral index for ocean oil spill detection. In: ESA living planet symposium

Pérez-García Á, Paoletti ME, Haut JM, López JF (2023a) Novel spectral loss function for unsupervised hyperspectral image segmentation. IEEE Geosci Remote Sensing Lett 20:1–5

Pérez-García Á, Rodríguez-Molina A, Hernández E, Vera L, López JF (2023b) Development of low-cost multi-spectral cameras for precision agriculture. In: IGARSS 2023—2023 IEEE international geoscience and remote sensing symposium, pp 3466–3469

Pérez-García Á, van Emmerik TH, Mata A, Tasseron PF, López JF (2024a) Efficient plastic detection in coastal areas with selected spectral bands. Marine Pollut Bull 207:116914

Pérez-García Á, Lorenzo AM, López J (2024b) Spectral band selection methodology for future sensor development. In: 2024 47th MIPRO ICT and electronics convention (MIPRO), pp 152–156

Pérez-García Á, Martín Lorenzo A, Hernández E, Rodríguez-Molina A, van Emmerik THM, López JF (2024c) Developing a generalizable spectral classifier for rhodamine detection in aquatic environments. Remote Sensing 16(3090):1–18

Pérez-García Á, Rodríguez-Molina A, Hernández E, López Feliciano JF (2024d) Spectral indices survey for oil spill detection in coastal areas. IEEE J Selected Topics Appl Earth Observ Remote Sensing 17:15359–15372

Qiao L, Tang W, Gao D, Zhao R, An L, Li M, Sun H, Song D (2022) UAV-based chlorophyll content estimation by evaluating vegetation index responses under different crop coverages. Comput Electron Agric 196:106775

Rodríguez-Molina A, Santana A, Machado F, Barrios Y, Hernández-Suárez E, Pérez-García Á, Díaz M, Santana R, Sánchez AJ, López JF (2024) Bentayga-I: development of a low-cost and open-source multispectral cubesat for marine environment monitoring and prevention. Sensors 24(23):7648

Rußwurm M, Venkatesa SJ, Tuia D (2023) Large-scale detection of marine debris in coastal areas with sentinel-2. iScience

Satriano V, Ciancia E, Lacava T, Pergola N, Tramutoli V (2019) Improving the RST-oil algorithm for oil spill detection under severe sun glint conditions. Remote Sensing 11(23):2762

Signature ESA (2021) Meris. In: Earth online, Envisat. https://earth.esa.int/eogateway/instruments/meris

Smart PL, Laidlaw I (1977) An evaluation of some fluorescent dyes for water tracing. Water Resour Res 13(1):15–33

Smith DF, Wiliem A, Lovell BC (2015) Face recognition on consumer devices: reflections on replay attacks. IEEE Trans Inform Forensics Secur 10(4):736–745

Spalding M, Burke L, Wood SA, Ashpole J, Hutchison J, Zu Ermgassen P (2017) Mapping the global value and distribution of coral reef tourism. Marine Policy 82:104–113

Sun W, Du Q (2019) Hyperspectral band selection: a review. IEEE Geosci Remote Sensing Mag 7(2):118–139

Survey UG, NASA (2021) Hyperion website. In: Earth Observing 1. https://eo1.usgs.gov/sensors/hyperion

Tasseron PF, van Emmerik TH, Vriend P, Hauk R, Alberti F, Mellink Y, van der Ploeg M (2024) Defining plastic pollution hotspots. Sci Total Environ 934:173294

Tempfli K, Huurneman G, Bakker W, Janssen LL, Feringa W, Gieske A, Grabmaier K, Hecker C, Horn J, Kerle N, et al. (2009) Principles of remote sensing: an introductory textbook. International Institute for Geo-Information Science and Earth Observation

Themistocleous K, Papoutsa C, Michaelides S, Hadjimitsis D (2020) Investigating detection of floating plastic litter from space using sentinel-2 imagery. Remote Sensing 12(16):2648

van Emmerik TH, Kirschke S, Schreyers LJ, Nath S, Schmidt C, Wendt-Potthoff K (2023) Estimating plastic pollution in rivers through harmonized monitoring strategies. Marine Pollut Bull 196:115503

Vane G, Green RO, Chrien TG, Enmark HT, Hansen EG, Porter WM (1993) The airborne visible/infrared imaging spectrometer (AVIRIS). Remote Sensing Environ 44(2):127–143

Van Sebille E, Aliani S, Law KL, Maximenko N, Alsina JM, Bagaev A, Bergmann M, Chapron B, Chubarenko I, Cózar A et al (2020) The physical oceanography of the transport of floating marine debris. Environ Res Lett 15(2):023003

Wilder B (2012) Cloud architecture patterns: using microsoft azure. O'Reilly Media

Wold S, Esbensen K, Geladi P (1987) Principal component analysis. Chemometrics Intell Lab Syst 2(1-3):37–52

Xiong Y, Ran Y, Zhao S, Zhao H, Tian Q (2020) Remotely assessing and monitoring coastal and inland water quality in China: progress, challenges and outlook. Critical Rev Environ Sci Technol 50(12):1266–1302

Xu M, Wu M, Chen K, Zhang C, Guo J (2022) The eyes of the gods: a survey of unsupervised domain adaptation methods based on remote sensing data. Remote Sensing 14(17):4380

Yuan Q, Shen H, Li T, Li Z, Li S, Jiang Y, Xu H, Tan W, Yang Q, Wang J et al (2020) Deep learning in environmental remote sensing: achievements and challenges. Remote Sensing Environ 241:111716

Zhang L, Zhang L (2022) Artificial intelligence for remote sensing data analysis: a review of challenges and opportunities. IEEE Geosci Remote Sensing Mag 10(2):270–294

Zhang S, Li S, Fu W, Fang L (2017) Multiscale superpixel-based sparse representation for hyperspectral image classification. Remote Sensing 9(2):139

B2Space (2024) Taking your Business to Space. https://b2-space.com/ Accessed 12 Dec 2024

Etrair (2024) Sistemas Aéreos no Tripulados (UAS). https://etraair.com/. Accessed 12 Dec 2024

Headwall Photonics (2020) Spectral imaging instruments. https://www.headwallphotonics.com. Accessed 22 Jun 2020

Horstrand P, Guerra R, Rodríguez A, Díaz M, López S, López JF (2019) A UAV platform based on a hyperspectral sensor for image capturing and on-board processing. IEEE Access 7:66919–66938

MAIA (2024) MAIA WV model. https://www.spectralcam.com/maia-tech/. Accessed 06 Oct 2019

Parrot (2023) Sequoia+ camera. https://www.parrot.com/soluciones-business/profesional/parrot-sequoia. Accessed 6 Oct 2019

SILIOS Technologies (2023) Multispectral cameras. CMS series. https://www.silios.com/cms-series. Accessed 3 Jan 2023

Specim Spectral Imaging (2022) FX series hyperspectral cameras. http://www.specim.fi/fx/. Accessed 6 Oct 2019

Machine and Deep Learning Approaches for Water Pollution Detection Using Hyperspectral Imaging

María Gema Carrasco-García,
María Inmaculada Rodríguez-García, Javier González-Enrique,
Juan Jesús Ruiz-Aguilar, and Ignacio José Turias Domínguez

Contents

Abstract

Hydrocarbon contamination in aquatic environments poses a major environmental challenge, which requires the development of advanced tools for its early detection. In this chapter, hyperspectral images captured with high spatial resolution local cameras are presented to identify the presence of hydrocarbons

M. G. Carrasco-García (✉) · J. J. Ruiz-Aguilar
Department of Industrial and Civil Engineering, Algeciras School of Engineering and Technology (ASET), University of Cádiz, Algeciras, Spain
e-mail: maria.carrasco@uca.es; juanjesus.ruiz@uca.es

M. I. Rodríguez-García · J. González-Enrique · I. J. Turias Domínguez
Department of Computer Science Engineering, Algeciras School of Engineering and Technology (ASET), University of Cádiz, Algeciras, Spain
e-mail: inma.rodriguezgarcia@uca.es; javier.gonzalezenrique@uca.es; ignacio.turias@uca.es

D. Gutiérrez et al. (eds.), *Smart Water Quality Monitoring*,
https://doi.org/10.1007/978-3-032-03049-8_5

in water. Two main approaches are compared: classifiers trained with hyperspectral signatures using machine learning techniques and classifiers trained with full hyperspectral images (HSI) using deep learning methods, especially convolutional neural networks (CNNs). The comparative analysis provides critical insights into the performance of both methodologies, outlining their strengths and limitations in terms of accuracy, robustness, and generalization. The findings underscore the potential of deep learning approaches for improving the automated detection of hydrocarbon contamination, while also recognising the reliability of classical machine learning techniques in specific scenarios. This study advances the development of more efficient monitoring strategies for aquatic ecosystems, reinforcing the role of artificial intelligence in hyperspectral data analysis for environmental protection.

Keywords

HSI · Hyperspectral · Deep learning · Machine learning · ANNs · CNN · AE · PCA

Introduction

The concern over oil spills in the past few decades has driven the development of international plans to address them, ranging from spill prevention to response capabilities once an incident has occurred. The adverse effects of oil spills are well documented: their toxicity, high dispersion capacity, cleaning challenges, and persistence in the environment make oil spills a major environmental hazard, endangering both aquatic and coastal ecosystems (Yildiz et al. 2021; González et al. 2022; Kababu et al. 2022; French-McCay et al. 2023) and posing risks to human health (Stroope et al. 2023; Hwang et al. 2024). Recognizing their devastating effects on ecosystems, human health, and the blue economy, oil spills remain one of the 32 key scientific challenges to be addressed by 2030, as identified by the American Academy of Sciences (Yang et al. 2022a).

To mitigate their impact, international agreements such as the International Convention for the Prevention of Pollution from Ships (MARPOL), developed by the International Maritime Organization (IMO), and regional regulatory measures such as Directive 2005/35/EC, which governs pollution caused by ships and imposes penalties for illegal discharges, and the Marine Strategy Framework Directive (2008/56/EC) have been established. However, once a spill occurs, none of these regulatory frameworks are effective unless supported by an efficient monitoring system that ensures a timely and adequate response.

Satellite imagery based on spectral technology, such as HSI, has been emerging as a cost-effective method for monitoring oceans and coastlines, facilitating decision-making, enhancing mitigation efforts, and aiding in identifying responsible parties. These systems, managed by space agencies, operate continuously, mapping the entire planet. Additionally, the hyperspectral data is often available for free, making satellite-borne HSI one of the most widely used techniques for planetary-scale environmental monitoring, particularly in the event of major oil spills (Leifer et al. 2012; Deepthi 2020). Despite these advantages, satellite-based monitoring has inherent limitations. The frequency of data acquisition for a specific location is constrained by orbital periodicity. For instance, the Sentinel-1 mission of the European Space Agency (ESA) consists of two satellites with a 6-day repeat cycle when both are operational (Sentiel 2024). This means that a given location will not be captured from the same angle until the cycle repeats. Second, spaceborne HSI systems are affected by weather conditions, as cloud cover can alter or degrade the quality of acquired data, reducing detection capabilities (Yang et al. 2022b; Haut et al. 2024).

To address these challenges, airborne hyperspectral and multispectral imaging missions on manned aircraft, such as AVIRIS, provide a viable alternative. Extensive research has demonstrated the effectiveness of airborne HSI in oil spill detection (Leifer et al. 2012; Yang et al. 2022b; Menezes and Poojary 2020; Pelta et al. 2019). However, the logistical complexity of flight planning and the high operational costs of manned aircraft make this solution impractical for continuous monitoring.

The limitations of both satellite-based and manned airborne monitoring are particularly critical in areas with high maritime traffic, such as port waters, where the detection of small-scale oil spills (<7 tons) poses a significant environmental challenge. Along European coasts, these small spills constitute a substantial portion of total hydrocarbon pollution and are among the most frequent occurrences, as reported by the EMSA and EEA (2021). Early detection of these spills is crucial to mitigating environmental consequences and assigning liability.

HSI systems on unmanned aerial vehicles (UAVs) or drones are emerging as a leading solution (Jiang et al. 2022; Asadzadeh et al. 2022; Kieu and Law 2021), offering triple benefits: (1) they allow continuous and real-time monitoring of local areas such as harbour zones with a high spatial-spectral resolution, (2) they retain complete spectral information of the target by eliminating the effect of clouds cover, and (3) they allow the detection of not only macro oil spills but also small ones. The superior spatial-spectral resolution of HSI of small dimensions is particularly beneficial in applications related to the early detection of small spills, where accurate and timely detection is crucial for effective response and for minimizing environmental damage (Mohammadiun et al. 2021).

Hyperspectral technology such as any optical sensor is based on capturing the unique response with which bodies reflect, transmit, emit, or scatter light, known as the spectral signature. The differentiating aspect of this technology is its spectral resolution of less than 20 nm, which allows quasi-continuous spectral signatures (hyperspectral signatures) (Fig. 1) to be obtained and also to identify small differences between spectral responses and discriminate between different materials (Menezes and Poojary 2020; Dalponte et al. 2013; Shan et al. 2018). Therefore, hyperspectral images are three-dimensional data cubes captured instantaneously composed of spatial (2D: x-y) and spectral (1D: wavelength, λ) information. In other words, each pixel (x-y) of the image contains reflectance values across the visible to infrared range, unlike conventional Red-Green-Blue (RGB) cameras, which record only three broad spectral bands. Each pixel (*x*-*y*) in a hyperspectral image is associated with a unique hyperspectral profile, providing a rich dataset for early oil spill detection and enhanced material discrimination.

HSI has proven to be highly effective in detecting water pollutants by analysing how specific spectral bands correlate with the presence of various chemical substances, enabling early detection and continuous monitoring of contaminants in aquatic ecosystems (Table 1).

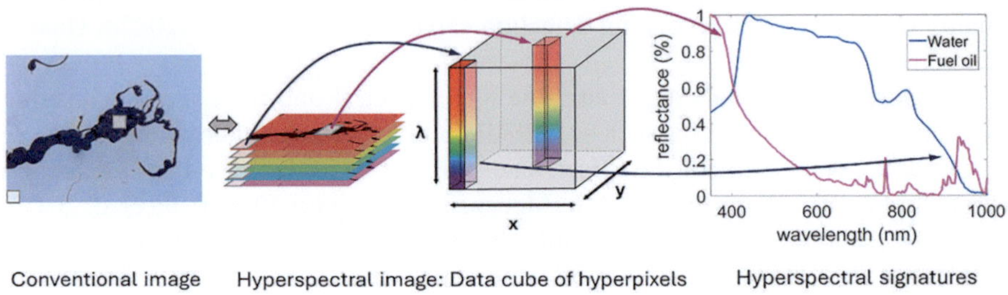

Fig. 1 Representation of a hyperspectral image. The sensor acquires the spectral response of each pixel across hundreds of wavelengths spanning from the visible to the infrared range, creating a data cube of spatial-spectral information. Each pixel in the image is a hyperpixel with its unique hyperspectral signature

Table 1 Application areas of HSI for water pollution detection, highlighting the machine learning techniques employed and the corresponding references

Application areas	Machine learning techniques	References
Oil spill monitoring	ANNs, DT, AE, PCA, CNN	Salam and Saxena (2024), Carrasco-García et al. (2023), Carrasco-García et al. (2024), Zhou (2020), Zhang et al. (2016), Zhang et al. (2021)
Oil emulsion detection in waters		Sicot et al. (2014), Viallefont-Robinet et al. (2018)
Water quality indexes: suspended particulate matter, chlorophyll-a, turbidity, etc.	MLR, ANNs, DFM	Goyens et al. (2022), Shengfang et al. (2019), Torres-Vera (2023), Krizhevsky et al. (2017)
Detection of chemical pollution of water		Aira et al. (2024), Chevalier et al. (2018)
Detection of heavy metals in water	CNN	Sun et al. (2024)
Detection of biological contaminants in water		Zhao et al. (2024), Fingas and Brown (2018)
Soil and water conservation	RM, ANNs	Kim et al. (2024), Cruz-Retana et al. (2023), Lu et al. (2021)

The spectral differences between hydrocarbon and water, particularly in the near-infrared region, facilitate their detection in aquatic environments (Salam and Saxena 2024). This capability extends beyond large-scale spills to more complex scenarios, such as emulsions on water surfaces. The role of HSI in identifying oil emulsions in water is crucial for environmental monitoring and pollutant detection. By capturing the detailed spectral signatures of oil emulsions, including the impact of bubble size distribution on the spectral profile, HSI contributes to a more comprehensive approach to water quality assessment and pollution control (Sicot et al. 2014; Viallefont-Robinet et al. 2018). Furthermore, parameters like chlorophyll-a concentration and suspended particulate matter (Goyens et al. 2022) are key water quality indicators that can be effectively assessed using HSI. Similarly, in (Shengfang et al. 2019; Torres-Vera 2023), the authors investigated the use of remote sensing to monitor water turbidity and the presence of suspended particles. The detection of chemical contaminants in water bodies is considered another field of application for remote sensing and spectral analysis. In (Aira et al. 2024), they present a spectrometer using IoT technology to detect glyphosate contamination, offering fast, field-based results without lab processing and providing a promising solution for timely monitoring and prevention of health risks. In (Chevalier et al. 2018), a quantum cascade laser-based sensor is demonstrated for detecting chemicals in water, with its sensitivity analysed and real-time spectroscopic detection validated through ethanol injections. Additionally, monitoring the impact of industrial activities on nearby water bodies is another field of HSI application. The authors in (Sun et al. 2024) focus on identifying pollutants such as heavy metals and chemical waste. The study emphasizes the importance of timely and accurate water quality assessment to mitigate environmental risks in industrial regions, underscoring the need for effective monitoring systems to protect aquatic ecosystems from industrial contamination.

Further expanding the use of remote sensing for water quality monitoring, the research in (Zhao et al. 2024; Fingas and Brown 2018) explores methods for detecting biological contaminants, including microorganisms, in water. The rapid growth of phytoplankton and microalgae poses environmental challenges, and (Zhao et al. 2024; Fingas and Brown 2018) proposes using hyperspectral data to calculate water chromatic indices (WCIs) for algal bloom detection, overcoming satellite limitations. A global dataset confirmed significant chromatic differences, and successful applications in the Bohai Sea and Taihu Lake highlight its potential for satellite-based monitoring. This capability of HSI to acquire the subtle spectral difference responses to distinguish between polluted and non-polluted water is significantly enhanced through the application of two powerful methodologies: machine learning (ML) and deep learning (DL). These approaches enable the efficient processing and analysis of the vast amounts of data generated by HSI, maximizing the use of its rich spectral and spatial information for accurate water quality assessment.

The combination of spectral analysis with ML methods has proven to be a powerful tool approach for soil and water conservation, enabling advanced analyses and more efficient resource management, as evidenced in (Kim et al.

2024; Cruz-Retana et al. 2023). Furthermore, studies comparing ML methods have demonstrated their high accuracy in detecting and classifying water quality indicators derived from spectral information (Lu et al. 2021). A novel approach has been developed by Carrasco-García et al. (Carrasco-García et al. 2023; Carrasco-García et al. 2024), who propose classical ML models for oil spill detection based on hyperspectral signatures as inputs for the classifier training. Hyperspectral signatures offer the advantage of being point data, which significantly reduces the data management requirements and eliminates the costly task of image labelling required by hyperspectral imagery. To address the curse of dimensionality, feature selection techniques such as principal component analysis (PCA) (Hotelling 1933; Jolliffe 1986) and autoencoders (AEs) (Kramer 1991) were successfully employed in combination with artificial neural networks (ANNs) and decision tree (DTs).

On the other hand, advanced DL techniques have been increasingly employed to process and extract relevant information from HSI data, achieving success in classification tasks (Paoletti et al. 2019; Kumar et al. 2020). CNNs stand out as the most powerful deep learning techniques for developing predictive classification models (Zhou 2020; Krizhevsky et al. 2017). The strength of CNNs lies in their exceptional ability to extract relevant spatial-spectral information and local contextual details from HSI, facilitating more accurate image classification. While traditional 2D-CNNs excel in spatial processing, they fall short when it comes to handling spectral data. This limitation drives the adoption of 3D-CNNs, which effectively process both the spectral and spatial dimensions, offering a more comprehensive approach to HSI classification (Zhang et al. 2016). The use of HSI and DL is explored in Zhang et al. (2021) for quantifying water quality parameters, including phosphorus, nitrogen, COD, BOD, and chlorophyll-A. Through a deep learning-based model (HF-DFM) that integrates UAV remote sensing, it outperforms other models in accuracy, offering a reliable tool for real-time biochemical monitoring of urban water resources. CNNs are also outperforming the classical models for the detection of oil spills. Hu et al. (2022) conducted a study where they integrated one-dimensional and two-dimensional CNN (1D-CNN and 2D-CNN) to extract spectral and spatial features, respectively. In a similar approach, Yang et al. (2019) applied CNN to detect oil spills on the sea surface using airborne HSI, achieving an accuracy exceeding 85%. A deep encoder-decoder model for detecting oil spills in SAR images is developed by Satyanarayana and Dhali (2023). The authors develop a deep learning model using CNNs to detect oil spills from remote sensing satellite imagery. The model is trained to distinguish between oil spills and other environmental features, offering an automated and more accurate approach compared to traditional methods. The results show that the deep learning model performs better in detecting oil spills under various conditions, making it a promising tool for real-time monitoring of oil spills using remote sensing data.

According to the state of the art regarding HSI classification, this research developed a CNN model. The innovation of this research lies in the integration of small hyperspectral technology that allows it to be installed in UAVs for the continuous acquisition of local hyperspectral images. Using a fixed CNN architecture, the impact of sample size on classification performance was evaluated.

However, considering that CNNs, along with all other DL methods, face the drawback of needing a favourable ratio of available training samples to the number of features to achieve optimal classifier performance, a classical approach was also developed. Shallow ANNs, trained on the hyperspectral signatures of targets, were designed to reduce the complexity of the problem. Various classifiers were assessed based on the dimensionality reduction technique employed-either PCA or AE-as well as the extent of dimensionality reduction applied. While hyperspectral data provides valuable information, a significant portion of it is redundant. To enhance classification efficiency, it is essential to extract the most relevant features from the hyperspectral signatures. The optimal dimensionality reduction was identified by training classifiers with input datasets of different dimensions and subsequently assessing their classification performance.

The aim of this research was to demonstrate the suitability of CNNs and Shallow ANNs in hyperimage classification. A prior spectral characterization of hyperspectral signatures was performed using PCA and AEs, specifically for classical Shallow ANNs. Section "Materials: Data Acquisition and Preprocessing" describes the materials, detailing the data acquisition process and preprocessing steps. Section "Methods" explains the methodologies applied and the workflow followed throughout the research. Section "Results and Discussion" presents the results along with their analysis and discussion. Finally, section "Conclusion" summarizes the conclusions drawn based on the findings.

Materials: Data Acquisition and Preprocessing

The initial phase of the research consisted of sample preparation, as well as the acquisition of the hyperspectral signatures and the hyperspectral images. This process was conducted on a laboratory scale to establish an initial understanding of the issue under controlled environmental conditions.

Sample Preparation

Samples of water and water-hydrocarbon mixtures were prepared for the subsequent acquisition of the hyperspectral signatures and hyperspectral images that constituted the dataset.

Vessels and bunkering are potential sources of significant oil inputs into the marine and river environment. So, real hydrocarbons involved in these activities were selected to assess the effectiveness of the research: (a) diesel, (b) high-sulphur fuel oil, and (c) C10.

The samples were prepared in plastic trays as a mixture of hydrocarbons and fresh water, ensuring the entire surface was covered with hydrocarbons at varying concentrations (Fig. 2). This approach serves a dual purpose. Firstly, it guarantees that all pixels in the acquired hyperspectral images are properly labelled, facilitating accurate model training and testing. Secondly, the presence of different

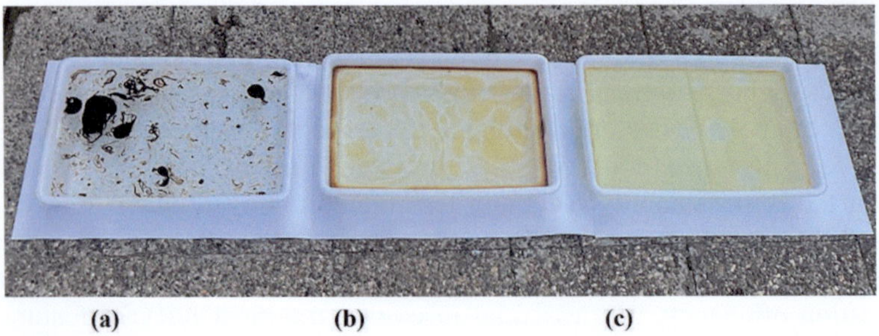

Fig. 2 Sample preparation: (a) fuel oil, (b) C10, and (c) diesel

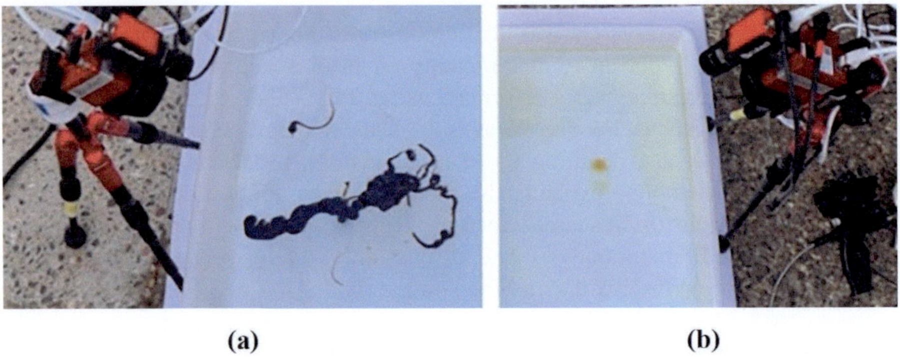

Fig. 3 Acquisition process of hyperspectral images of (a) water-fuel oil and (b) water-C10

concentration levels enables the development of models capable of detecting oil pollution even when it is not visually apparent. This highlights the added value of hyperspectral technology, which can identify subtle spectral variations undetectable to the naked eye.

Hyperimage Acquisition

The hyperimages of the different samples: water-diesel, water-fuel oil, and water-C10, were taken with the Cubert Ultris X20 Plus hyperspectral snapshot camera (Fig. 3). This camera provides the spectral information in the VNIR (350–1000 nm) of a complete scene in a single capture, with a bandwidth of 4 nm and a spatial resolution of 1880 × 1880 px. The high spatial and spectral resolution becomes the scene in a very detailed information cube. The small dimensions of the hyperspectral camera, 86 × 121 × 105 mm and 630 g, are also another remarkable characteristic that allows it to be mounted on a UAV.

Reflectance, representing the percentage of light reflected by the target, is the optical property measured in this study. To obtain the reflectance of each hyperpixel in the images, a precise protocol needs to be followed to set the references needed. First of all, it is necessary to adjust the integration time of the camera to the light of the scene so as not to over-illuminate any wavelength. Then, white and black references are taken to set the maximum and minimum light exposure that can be reflected. The white reference was taken using a certified 100% reflectance panel provided by the camera manufacturer. The black reference requires the lens caps to be attached to the lenses.

As the hyperspectral camera, like a spectroradiometer, is a passive device, the sun is used as a light source. Thus, the integration time and the white reference must be repeated according to the illumination changes.

A collection of 80 hyperimages was acquired consisting of 20 images of water-diesel, 20 images of water-fuel oil, 20 images of water-C10, and 20 images of water. Before training the CNNs, the hyperspectral images underwent a preprocessing stage. DL methods require large amounts of training examples to achieve optimal performance. Additionally, smaller image sizes, in this study, are advantageous for CNN training, as they help to preserve spatial information across the different convolutional layers and improve the computational speed. To address these needs, the original hyperspectral images were divided into 81 smaller patches of 25 $\times$ 25 pixels each (Fig. 4). This approach resulted in a final dataset consisting of 6480 hyperspectral images, providing a sufficient number of samples while maintaining critical spatial details.

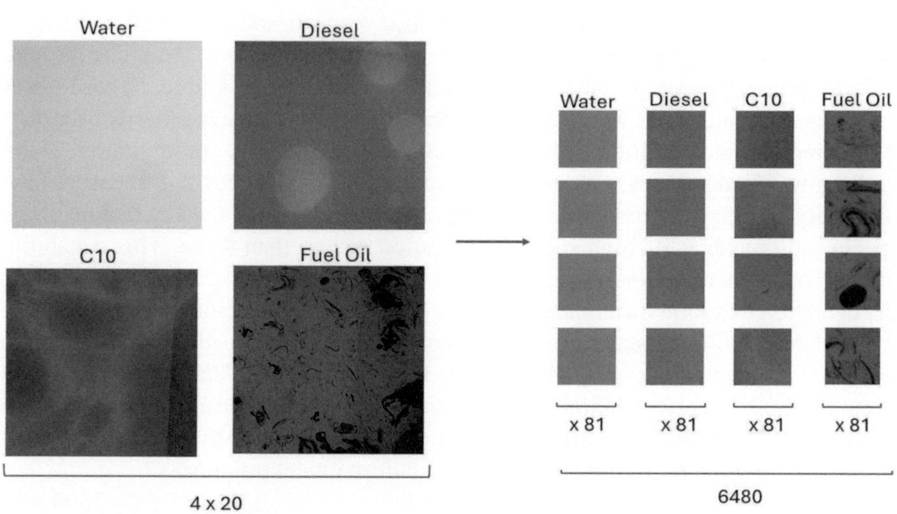

Fig. 4 Representation of the preprocessing done. Each hyperimage was divided into 80 new hyperimage of 25 $\times$ 25, resulting in a total of 6480 patches

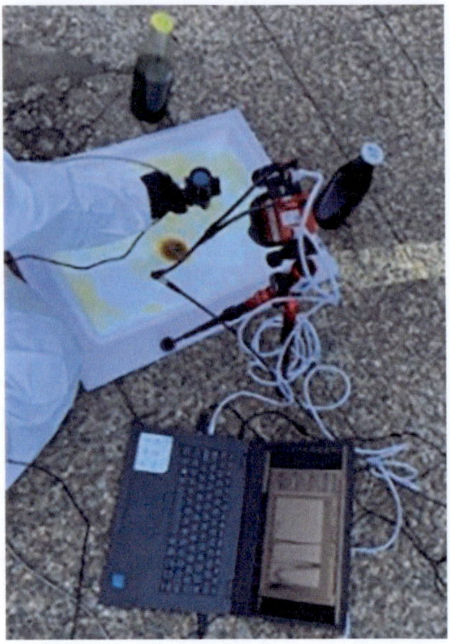

Fig. 5 Hyperspectral signatures acquisition. The fibre optic cable of the spectroradiometer is handled in the gun shown in the picture and pointed at the target

Hyperspectral Signature Acquisition

The acquisition of the hyperspectral signatures of water, diesel, fuel oil, and C10 on water was performed using the ASD FieldSpec 4 spectroradiometer (Fig. 5). This device captures the electromagnetic response of specific points in the scene through a fibre optic cable composed of 57 randomly distributed glass fibres. These fibres are organized into three groups: one for the VNIR spectrum, another for the first SWIR region, and a third for the second SWIR region. This configuration, along with built-in redundancy, ensures maximum signal throughput and measurement reliability. Beyond its broad spectral range and high reliability, the key feature of the spectroradiometer is its high spectral resolution of less than 4 nm. This capability allows for the acquisition of near-continuous hyperspectral signatures, facilitating the detection of subtle differences between highly similar spectral profiles.

Similar to the hyperspectral camera, the experimental procedure requires white and black reference measurements prior to data acquisition. These references were taken multiple times throughout the experiment to minimize the effects of variations in solar tilt. Additionally, any residual illumination variations were further mitigated by normalizing the hyperspectral signatures before model training.

Furthermore, since the spectral range of the spectroradiometer (350–2500 nm) exceeds that of the hyperspectral camera (350–1000 nm) and has a lower spectral resolution, prior alignment of the hyperspectral signatures is required. This was achieved by downsampling the reflectance data of the spectroradiometer to match the specific wavelengths captured by the hyperspectral camera.

A set of 80 samples was acquired and distributed as follows: 20 hyperspectral signatures of water, 20 hyperspectral signatures of diesel, 20 hyperspectral signa-

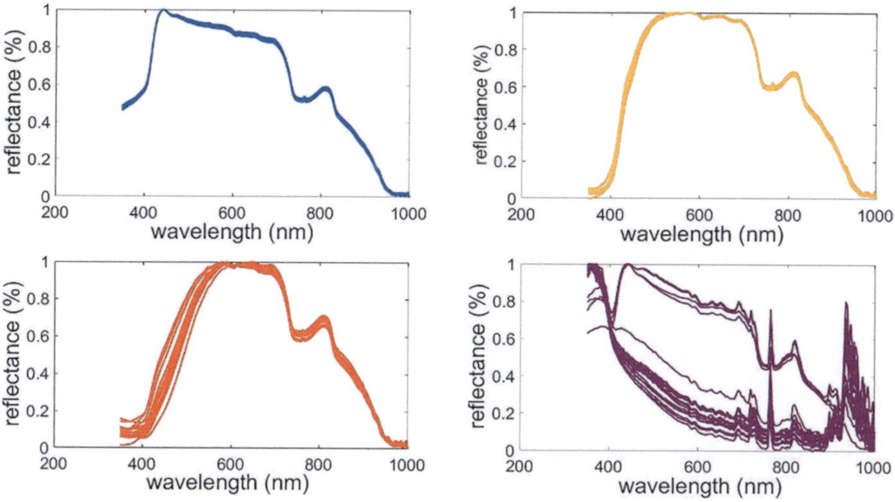

Fig. 6 Representation of the hyperspectral signatures of water (blue), diesel (yellow), C10 (orange), and fuel oil (purple) after reducing the spectral range and resolution and the normalization. Note that the signatures of water and hydrocarbons in their low concentrations were influenced by the background where the samples were placed

tures of fuel oil, and 20 hyperspectral signatures of C10 (Fig. 6). All the curves presented the shape according to their nature. In the case of fuel oil, four signatures were identified whose first maximum is shifted towards the water zone. However, as the remaining maxima match in wavelength, it was decided to keep them to provide information on the cases of less concentrated fuel oil.

Methods

The development of predictive classification models can be approached through two distinct methodological frameworks. The first follows a classical scheme, which relies on the hyperspectral signatures, the use of dimensionality reduction techniques, and the application of traditional classifiers which classify the hyperimages according to hypersignatures. In contrast, the second framework adopts a state-of-the-art scheme, leveraging hyperimages as complete data structures for model development using CNNs. This advanced methodology exploits both the spectral and spatial information embedded in hyperspectral data, allowing for automatic feature extraction and enhanced classification performance through deep learning techniques.

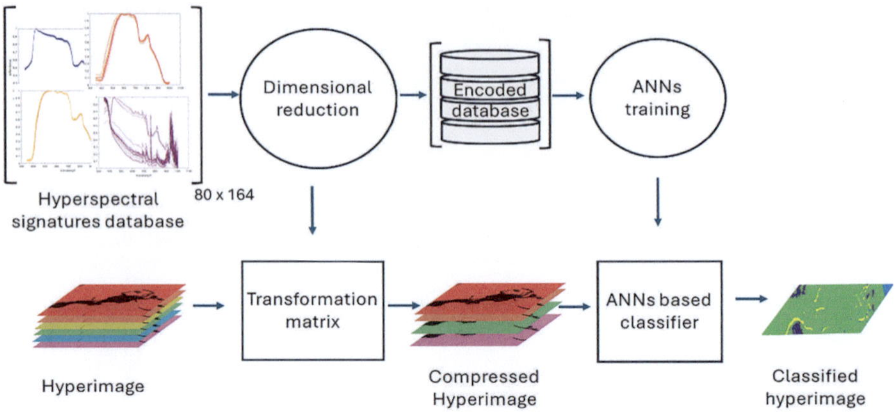

Fig. 7 Workflow of the classical machine learning approach for hyperimage classification. The process includes dimensionality reduction of hyperspectral signatures, followed by the training of an ANN classifier. The hyperspectral image is then transformed into the reduced-dimensional space using the transformation matrix and classified at the hyperpixel level

Classical Machine Learning Approach

The classical machine learning approach for detecting oil spills (or any other pollutants) in water relies on hyperspectral signatures. A visual representation of the classification workflow is shown in Fig. 7. Given the high dimensionality inherent to hyperspectral data, a preliminary dimensionality reduction is required to optimize computational efficiency and model performance. The resulting encoded dataset serves as the input for an ANN classifier. Once trained, the model is used to classify the hyperpixels of the hyperspectral image, which are first transformed into the reduced-dimensional space using the corresponding transformation matrix derived during the dimensionality reduction process of the signatures. Thus, to accomplish the compression of each hyperimage, the data cube was arranged into a two-dimensional matrix, where each row represented each hyperpixel (Fig. 8). Using this arrangement, the spatio-spectral information of the pixels was not lost.

Dimensional Reduction: PCA and AE

Hyperspectral technology provides detailed and valuable information about materials but at the same time a lot of redundant information. A feature selection is convenient for discerning the most relevant variables from the unnecessary ones. This is a key step to reduce the dimensionality of the data to the most informative characteristics and success in the development of efficient classifier models. For this purpose, principal component analysis (PCA) and autoencoders (AEs) were employed independently.

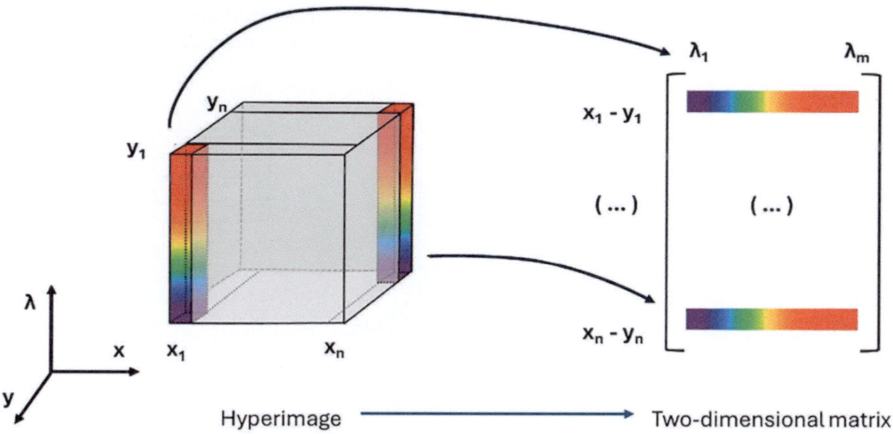

Fig. 8 Data cube rearrange for hyperpixel compression. Each row of the two-dimensional matrix represents a hyperpixel of the hyperimage. For instance, the first row of the matrix corresponds to the reflectance values of the first hiperpixel (x_1-y_1), the second row of the matrix to the reflectance values of the second pixel (x_1-y_2), and so on, until the last row of the matrix represents the reflectance values of last pixel of the hyperimage (x_n-y_n)

PCA

PCA is a statistical technique developed by Hotelling (1933) and Jolliffe (1986) in which the original data, x, of a P-dimensional space, are projected in a new Q-dimensional space of orthonormal principal components (PCs). The transformed data, y, are determined as a linear combination of the original data, being the coefficients, a_{mn}, the weights in each PC.

$$y_n = \sum_{m=1}^{n} x_m \cdot a_{mn} \tag{1}$$

These PCs are the axes of the new space and are defined according to the directions of maximum variance of the original data. The first PC is established in the direction in which the data show the most variation. The same criterion is followed for the calculation of the rest of PCs but with the extra condition of the orthogonality between axes. Thus, the first PC explains the most information of the original data while the last one the least. The ranking of the PCs allows us to select only the PCs that accumulate a significant percentage of explained variance, thus reducing the dimensionality of the data ($P > Q$).

PCA was applied to the dataset of hyperspectral signatures of water and different oil pollutants. The objective is to create new datasets of different dimensions in order to evaluate their performance in the development of classifiers. For this purpose, 14 datasets were organized as follows: 1 dataset containing the values only from the first PC, another dataset with the values from the first and second PCs, a dataset comprising values from the first to the third PCs, and so on, until datasets with

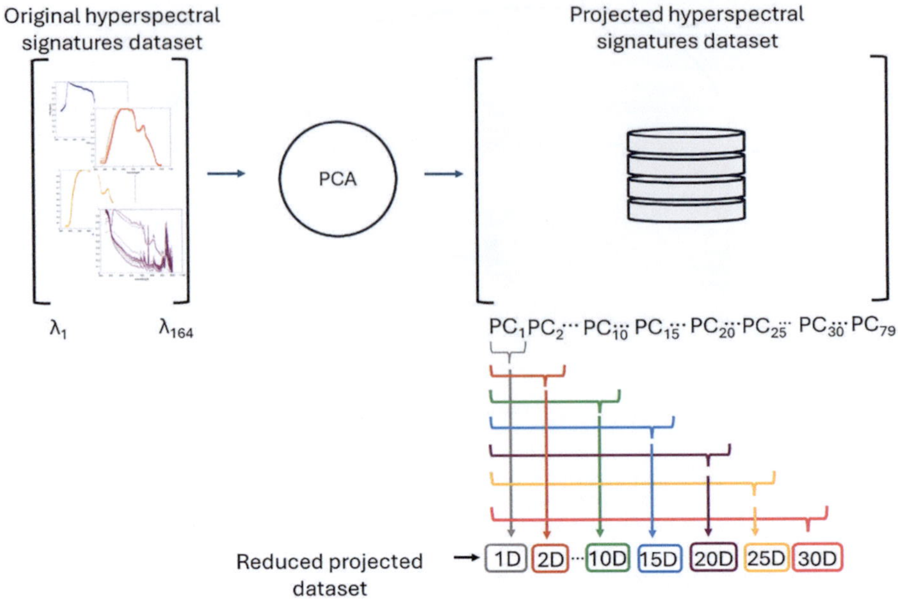

Fig. 9 Dimensionality reduction flow with PCA of the hyperspectral signature dataset. The resulting new dataset is projected into a space in which the PCs are ordered by decreasing the explained original variance. Subsequently, 14 reduced datasets are organized as follows: 1 dataset containing the values only from the first PC, 1 dataset with the values from the first and second PCs, 1 dataset comprising values from the first to the third PCs, and so on, until datasets with values spanning from the first to the tenth PCs, 15th PCs, 20th PCs, 25th PCs, and 30th PCs were reached

values spanning from the first to the tenth PCs, 15th PCs, 20th PCs, 25th PCs, and 30th PCs were reached (Fig. 9). The selection of these specific dimensions was made based on the cumulative variance explained. We aimed to assess datasets with cumulative explained variance values below 99% and others with values above 99%. Considering that the variance explained by PC beyond 30th was negligible, we do not to evaluate datasets with higher dimensions.

The transformation matrix obtained during the dimensionality reduction process is used to project the hyperpixels of hyperspectral images into the new reduced-dimensional space. Prior to this projection, the 3D data cube is rearranged into a 2D matrix, where each row represents a hyperpixel. The resulting projected matrix maintains the same structure as the encoded hyperspectral signatures, with the most relevant information concentrated in the first PCs and the least relevant information distributed across the later PCs.

AE

An autoencoder (AE) (Kramer 1991) is a type of neural network designed for analysing variable relevance and learning complex non-linear relationships. This method effectively captures the intrinsic non-linear structure of spectral data

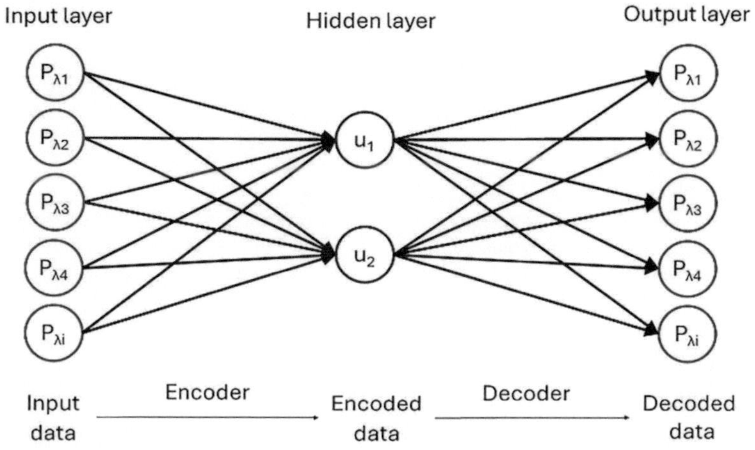

Fig. 10 Representation of an AE with a hidden layer of two neurons. The data are compressed in two dimensions, corresponding to the outputs of these neurons

(Abiodun et al. 2018), making it a valuable tool for interpreting and processing hyperspectral information. AEs help identify the most significant spectral features, enhancing the efficiency of subsequent modelling tasks.

What sets autoencoders apart is that their input and output layers are identical (Fig. 10). The AE compresses the data through hidden layers, encoding the original information into a more compact form. The encoding process is mathematically represented in Eq. (2), where *P* denotes the original variables, *IW* represents the weights between the input layer and hidden neurons, *b* refers to the biases, and *f* is the activation function applied during encoding. After encoding, the AE decodes the compressed data to reconstruct the original inputs, as shown in Eq. (3). Here, the decoder applies its transfer function to the sum of the decoding bias (*b*) and the product of the encoded data with the output layer weights (*OW*). This encoding-decoding process enables the AE to filter out redundant information, effectively reducing the dimensionality of the dataset.

$$f\ (b + P \cdot IW) \tag{2}$$

$$f\,(b + f\ (b + P \cdot IW) \cdot OW\ ) \tag{3}$$

Multiple autoencoder (AE) configurations were evaluated to compress the hyperspectral water-oil dataset, varying the number of neurons in the hidden layer from 1 to 15. This approach aimed at creating 15 different projections of the data, each corresponding to a specific dimensionality, in order to identify the optimal level of dimensional reduction for subsequent classification. To ensure maximum generalization and robustness, each AE configuration was trained 20 times (Fig. 11). Before each training session, the dataset of hyperspectral signatures was randomized

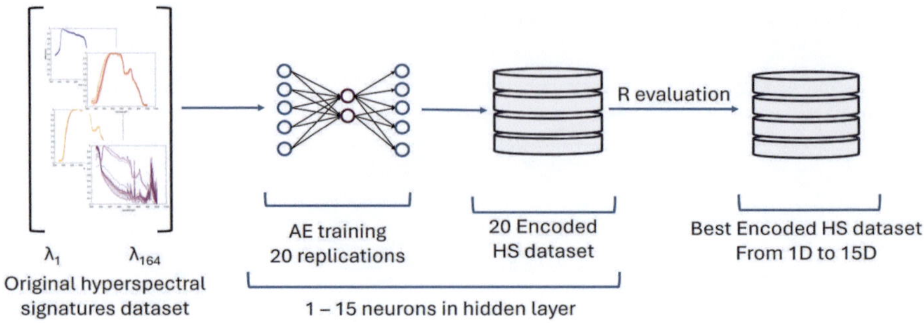

Fig. 11 Representation of the AE encoder scheme

to prevent overfitting and improve model reliability. The selection of the most suitable AE for each dimensionality was based on the correlation coefficient and the mean squared error (MSE) between the original input data and the reconstructed output, allowing for an objective assessment of the performance.

The AE is applied to hypercube that has been reshaped from 3D into 2D matrix as was previously described in Fig. 8. The encoding function (Eq. (1)) is then applied to these matrices, achieving significant data compression.

Classification Methods: ANNs

Classification aims to solve the problem of assigning a predefined class label to a given input. In this research, the objective is to develop a classifier to predict whether a spectral response (input) corresponds to water, diesel, fuel oil, or C10 (labels or classes).

ANNs are highly effective machine learning techniques for classification tasks due to their ability to learn and model complex patterns (LeCun et al. 2015). Inspired by the neural architecture of the human brain, Shallow ANNs consist of a fully connected structure that consists of an input layer for receiving data, an output layer for generating predictions or labels, and one or more hidden layers made up of neurons or nodes. Information flows through the network from the input to the output layer via weighted connections between neurons. These weights are iteratively adjusted using the backpropagation algorithm (Rumelhart et al. 1986), a simple yet powerful method that enables the architecture of the backpropagation neural network (BPNN) to function as universal approximators (Hornik et al. 1980).

Multiple classifiers for the determination of water and oils were trained using datasets of varying dimensions as inputs. The transformed dataset of hyperspectral signatures of water and the hydrocarbons was taken in different ways, keeping different levels of the original information in each new reduced dataset. The idea behind the research was to evaluate different ANN-based classifiers in order to determine the optimal dimension of the inputs.

For the classifiers trained with datasets reduced using PCA, a total of 14 models were developed. These classifiers were built using inputs ranging from 1 to 30

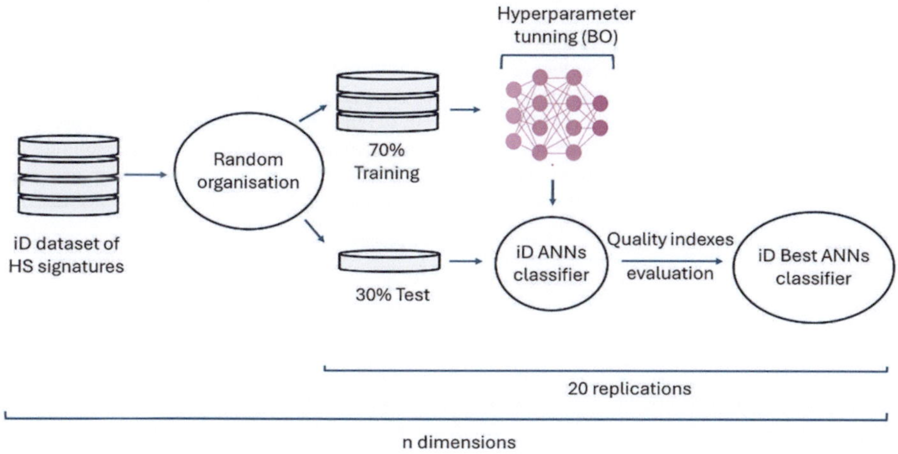

Fig. 12 Computational flow of ANN-based classifiers. For each i-dimensional projected dataset of hyperspectral signatures of water and oils, 20 networks with different inputs and different hyperparameters were trained. Bayesian optimization was used for the selection of the hyperparameters, conducted up to 200 times as the maximum limit. The resulting classifier was tested each time with different test sets, and the accuracy, recall, and F-score were determined as parameters against which to evaluate the performance of the classifier

dimensions, specifically considering the first 1 to 10 principal component PCs, along with additional models using 15, 20, 25, and 30 PCs as inputs. This approach aimed to assess how the retention of different amounts of spectral variance impacts classification performance.

In parallel, classifiers were also trained with datasets compressed using AEs. Various AE architectures were tested by adjusting the number of neurons in the hidden layer, ranging from 1 to 15. Each configuration produced a new dataset projected into a corresponding number of dimensions, enabling the evaluation of 15 distinct classifiers. This process was designed to determine the optimal level of dimensionality reduction achieved through AEs for subsequent classification tasks.

The sequence shown in Fig. 12 was followed for the obtention of the ANN-based classifiers. Each 1 of the classifiers was trained 20 times using different hyperparameters in each iteration, aiming to obtain the architecture that demonstrates the best generalization capability and robustness. To ensure this, in addition, the inputs were also different each time by randomly arranging the dataset before starting each calculation. Thus, for each dimension, each of the 20 networks was trained using different hyperparameters and training data, comprising 70% of the inputs, and then tested using different test sets, representing 30% of the inputs. The optimal architecture for each network was determined using BO (Mockus 1975), conducted up to 200 times as the maximum limit. Table 2 presents the hyperparameters that were tuned to optimize the architecture of the classification model.

The optimal classifier for each input dimension was selected based on the evaluation of performance metrics, such as accuracy (Eq. (4)), recall (Eq. (5)),

Table 2 Hyperparameters tuned using Bayesian optimization to determine the most effective architecture for the classification model. The optimized hyperparameters include (1) the number of neurons per hidden layer, capped at ten; (2) the number of hidden layers, with a maximum of two; (3) the weight initialization method for fully connected layers, using either He initializer (He et al. 2015) or Glorot initializer (Glorot and Bengio 2010); (4) the choice of activation functions, such as tanh, ReLU, identity, or sigmoid (with softmax applied in the final layer for output generation); (5) the application of standardization to numeric predictors or not; (6) the adjustable regularization strength; and (7) the initialization of biases in fully connected layers, set as either zeros or ones

Hyperparameters to optimize	Configuration options
N° of neurons per hidden layer	Up to ten neurons
Maximum hidden layers	Up to two hidden layers
Weight initialization method	He et al. (2015) or Glorot and Bengio (2010) initialization
Activation functions	Tanh, ReLU, identity, or sigmoid functions
Standardization of numeric inputs	Standardized or non-standardized
Regularization strength	Adjustable parameter
Initial bias values for dense layers	Initialized with zeros or ones

and F-score (Eq. (6)). These indicators offer a thorough evaluation of the model's predictive performance, highlighting both its strengths and potential weaknesses. They play a crucial role in assessing the model's ability to accurately classify previously unseen data. The formulas used are presented in Eqs. (3)–(6), where true positives (TP) and true negatives (TN) denote correctly identified instances, while false negatives (FN) and false positives (FP) indicate misclassification errors:

$$\text{Accuracy} = \frac{TP + TN}{TP + TN + FP + FN} \tag{4}$$

$$\text{Recall} = \frac{TP}{TP + FN} \tag{5}$$

$$F - \text{score} = \frac{2 \times \text{Precision} \times \text{Sensitivity}}{\text{Precision} + \text{Sensitivity}} \tag{6}$$

Once the optimal classifiers for each input dimension were identified, they were applied to the corresponding compressed hyperimages (Fig. 6). Each classifier was responsible for labelling the compressed hyperpixels within these images as water, C10, fuel oil, or diesel. Specifically, the classifier trained with the one-dimensional hyperspectral signature dataset was used to classify the one-dimensional compressed hyperimage, with the same approach applied to all other dimensions. The quantitative evaluation of the classifiers was conducted by analysing the confusion matrix and key classification metrics. This evaluation was streamlined by the experimental design, in which each hyperspectral image contained hyperpixels belonging exclusively to a single, known class.

Convolutional Neural Networks Approach

Convolution is a mathematical operation that merges two functions to represent the degree of overlap between them. It involves "sliding" one function over the other, multiplying their values at each point of intersection, and summing the resulting products to generate a new function. This process produces a function that reflects how the original functions interact with each other.

Convolution plays a crucial role in CNNs. CNNs (LeCun et al. 1990) are a type of deep neural network commonly used for image analysis. Unlike traditional methods that require manual feature extraction, CNNs automatically learn relevant features, making them highly effective for complex tasks such as image recognition and medical image analysis. In hyperspectral data processing, the ability of CNNs to operate on full images provides a significant advantage. By analysing both spectral information and the spatial arrangement of pixels simultaneously, CNNs can better preserve contextual relationships within the data. This approach enhances their robustness to noise and local variability, allowing them to extract more meaningful patterns.

CNNs are inspired by the biological structure of the visual cortex, which consists of arrangements of simple and complex cells (Hubel and Wiesel 1959). These cells are known to activate in response to specific subregions within a visual field, referred to as receptive fields. Based on these findings, neurons in a convolutional layer are connected only to localized subregions of the preceding layer. As a result, neurons are responsive only to the information within their designated receptive fields and do not react to areas outside these regions. Therefore, the number of parameter are minimized through a reduced number of connections, shared weights, and subsampling techniques.

Convolutional layers act as feature extractors, where each layer applies a specific filter (or kernel) to identify distinct patterns within the input data. Mathematically, the convolution between a filter W and an input image X is defined in Eq. (7), where $X(i,j)$ represents the pixel value at position (i,j) in the input image, $W(m,n)$ denotes the filter with dimensions $M \times N$, and $Y(i,j)$ is the resulting feature map. This operation is performed across all image channels, with the outputs summed to produce the final feature map. During training, the network learns the optimal values for these filters. The initial convolutional layers tend to capture simple, low-level features such as edges or textures, while deeper layers progressively extract more complex, high-level representations, enabling the model to recognize abstract patterns:

$$Y(i, j) = \sum_{m=0}^{M-1} \sum_{n=0}^{N-1} X(i + m, j + n) W(m, n) \tag{7}$$

CNNs consist of several layers, including convolutional layers, ReLU layers, and fully connected layers. The neurons in the convolutional layers learn non-linear combinations of the original inputs, extracting key features from the data. These learned features, also referred to as activations, are passed on as inputs to the next layer. ReLU layers introduce non-linearity, allowing the network to learn more complex representations. In fully connected layers, each neuron performs matrix multiplication, connecting all neurons from the previous layer to the next. Finally, the learned features are used as inputs for the classifier or regression function at the end of the network. The architecture varies based on the complexity of the problem. In this research, the hyperparameters were fixed before a grid search in the space of the parameters.

Here, the CNN consists of four 3D-CNN layers, each followed by a ReLU activation layer. After these convolutional layers, there are two fully connected layers, each followed by a dropout layer, and, finally, a fully connected layer and a softmax layer for classification. The first 3D-CNN layer uses 8 kernels of size $3 \times 3 \times 130$, the second has 16 kernels of size $3 \times 3 \times 20$, the third has 32 kernels of size $3 \times 3 \times 15$, and the fourth has 8 kernels of size $3 \times 3 \times 1$. This last convolutional layer filters only the spatial dimensions without affecting the spectral dimension, reducing the amount of information. The fully connected layers have 256, 128, and 4 units, respectively. The number of neurons of the last one corresponds to the number of classes. The dropout layers are set to 40%, meaning 40% of the connections are randomly eliminated during each iteration to prevent overfitting. The softmax layer provides the classification probabilities. Figure 13 presents a visual representation of architecture.

This research aims to evaluate how sample size impacts the performance of the CNN model. To achieve this, three different training set sizes were defined as percentages of the original dataset: 8%, 15%, and 30%. The same percentages were applied to the internal validation sets, while the remaining patches were used as test data. Before dividing the dataset into these groups, the data was randomly shuffled to prevent any dependency on the order in which the network saw the patterns. To further ensure robustness, each experiment was repeated ten times. The performance of the CNN was evaluated based on classification quality metrics for the test set, including accuracy, recall, and F-score.

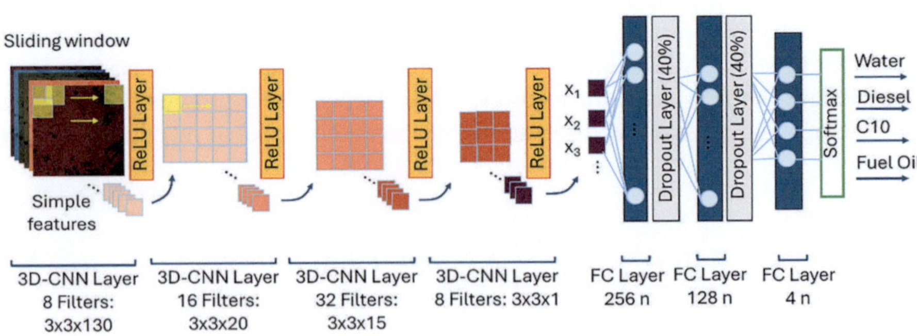

Fig. 13 Visual representation of CNN architecture used in the research. Four 3D-CNN, followed by a RELU layer each one, two fully connected layers followed by a dropout layer each one, and a final fully connected layer and a softmax layer

Results and Discussion

This section presents the obtained results based on the experimental procedure outlined below, where both the classical and CNN-based approaches were developed to perform the classification of hyperimages, specifically to determine the presence or absence of oils in water. The results and discussion follow the structure outlined in the methodology. First, the results obtained from the different stages within the classical approach are presented. This includes the outcomes from PCA, AE, and their respective classifiers. On the other hand, the results for the CNN-based approach are also presented separately, highlighting the performance of the convolutional neural networks in the classification task.

Classical Machine Learning Approach

After acquiring the experimental data, the preprocessed hyperspectral signatures of water and hydrocarbons—diesel, fuel oil, and C10— were projected into a new PCA space. This projection reduces the dimensionality of the hyperspectral signature dataset, with the information mapped onto the principal components (PCs) ordered by the amount of original variance they explain, from highest to lowest. Table 3 presents the variance explained by each PC (up to 10 PCs), along with the cumulative variance.

The analysis of the results from applying PCA demonstrates the effectiveness of dimensionality reduction. By considering only the first three principal components (PCs), over 99% of the original information is preserved. This is further highlighted by visualizing the hyperspectral signatures in these three PCs, as shown in Fig. 14, where the high degree of variance explained by the first three PCs is clearly evident. The different classes are easily distinguishable, and notably, the four fuel oil hyperspectral signatures that deviate from the rest of their class are also identified. Examining the variance explained by each PC allows for the selection of dimensions of interest for further study. Given that the first three PCs accumulate 99% of the explained variance, databases with one and two dimensions are constructed to assess the impact of retaining a reduced amount of original information. Similarly, databases with an increasing number of dimensions, incremented by one up to ten, where 99.95% of the explained variance is accumulated, are created. To evaluate the effects of higher levels of information retention, databases with 15, 20, 25, and

Table 3 Variance explained (EV) for each PC of the original dataset of hyperspectral signatures of water, diesel, fuel oil, and C10, and cumulative variance (CV). The first three PCs retained 99% of the original variance of the data

PCs	1	2	3	4	5	6	7	8	9	10
EV (%)	80.93	16.70	1.48	0.70	0.05	0.03	0.02	0.01	0.01	0.01
CV	80.93	97.63	99.11	99.81	99.87	99.90	99.92	99.94	99.96	99.97

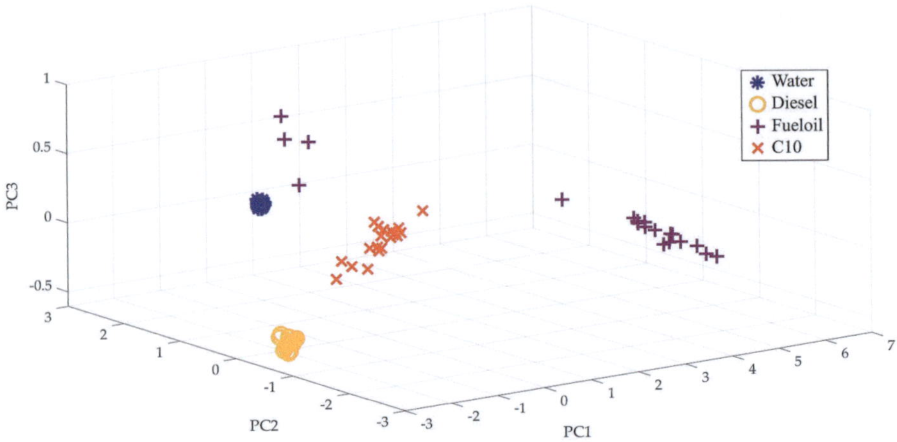

Fig. 14 Visual representation of the projected hyperspectral signatures into the PCA space. The first three PCs have been represented

30 dimensions are also generated, reaching an accumulated explained variance of 99.999%. In summary, the evaluation of the results allows for the critical creation of the projected databases of different dimensions. This is essential to reach accurate conclusions in the subsequent study of the optimal classifier.

The application of AEs with different numbers of neurons in the hidden layer for feature selection produced notable results (Table 4). The analysis focuses on the evaluation of the correlation coefficient and MSE across 20 replicates for each AE configuration. Both the mean values, their standard deviations, and the best-performing results are considered to assess performance. The worst results were observed when reducing the dataset to a single dimension, with a correlation coefficient of 0.956 and an MSE of 0.0102. This decline underscores the challenge of maintaining data integrity under extreme dimensionality reduction. However, performance improves as the number of neurons in the hidden layer increases. Overall, the correlation coefficients between the original and encoded data exceed 0.990, while the MSE remains below 0.003, indicating the effectiveness of AEs in preserving critical information during dimensionality reduction.

The most effective encoding was observed with a 15-dimensional reduction, achieving the highest correlation coefficient of 0.9984 and the minimum MSE of less than 5E-04. This indicates that 15 dimensions provide an optimal balance, retaining critical information while minimizing the risks of overfitting or data loss during the encoding process. However, this trend does not hold consistently, as seen in the encoding results for 11, 13, and 14 dimensions. Interestingly, increasing the number of dimensions does not always enhance encoding performance. Reducing the data to 11, 13, and 14 dimensions led to noticeably lower correlation coefficients and higher MSE compared to reductions to 10 and 12 dimensions. This suggests that higher dimensionality can introduce unnecessary complexity or

Table 4 AE results. The first column shows the dimensionality of the new dataset, corresponding with the number of neurons in the hidden layer of the AE. Columns 2, 3, and 4 display the results related to correlation coefficients, reporting the mean, the standard deviation, and the best correlation coefficient. Columns 5, 6, and 7 present the results relative to the MSE, being the mean, the standard deviation, and the minimum value

	r			MSE		
Dim	Mean	SD	Max	Mean	SD	Min
1	0.9560	3.74E-07	0.9560	0.0102	9.54E-08	0.0102
2	0.9887	4.75E-04	0.9901	0.0027	1.09E-04	0.0024
3	0.9940	3.14E-03	0.9965	0.0015	7.21E-04	9.13E-04
4	0.9912	2.68E-03	0.9975	0.0021	6.20E-04	6.54E-04
5	0.9920	3.15E-03	0.9978	0.0019	7.30E-04	5.73E-04
6	0.9935	3.44E-03	0.9979	0.0016	7.97E-04	5.59E-04
7	0.9932	3.39E-03	0.9979	0.0016	7.86E-04	5.59E-04
8	0.9948	2.91E-03	0.9979	0.0013	6.75E-04	5.61E-04
9	0.9961	1.65E-03	0.9979	9.77E-04	3.84E-04	5.68E-04
10	0.9967	2.91E-04	0.9979	8.52E-04	6.98E-05	5.63E-04
11	0.9966	4.32E-05	0.9967	8.67E-04	1.16E-05	8.46E-04
12	0.9967	4.05E-04	0.9979	8.34E-04	9.70E-05	5.54E-04
13	0.9966	3.02E-05	0.9967	8.61E-04	8.64E-06	8.41E-04
14	0.9966	3.98E-05	0.9967	8.60E-04	1.05E-05	8.41E-04
15	0.9968	5.58E-04	0.9984	8.08E-04	1.33E-04	4.37E-04

noise, negatively impacting the encoding quality. These findings underscore that dimensionality reduction does not inherently improve with more variables; instead, fewer dimensions can often achieve a better trade-off between compression and information retention.

Furthermore, the evaluation of standard deviations offers key insights into the stability and reliability of the AE models. The mean correlation and MSE values closely align with the maximum values obtained in the 20 replications, and the consistently low standard deviations indicate strong model stability. This suggests that the dimensionality reduction process is robust, showing minimal variability across different training iterations. The low standard deviations confirm the repeatability of the encoding process, demonstrating that results are not significantly affected by the stochastic nature of AE training.

Once the classifiers have been trained using the described experimental procedure, which involves conducting 20 repetitions with random inputs and different hyperparameters for training the 14 classifiers with different dimensions in the inputs, it becomes imperative to select the optimal one for each dimension. The criteria of selection attend to the best quality index classification for the test set and the simplest architecture of the neural network (Ockham's razor criteria). Table 5 presents the main findings in terms of accuracy.

Interestingly, a more complex architecture was not required to achieve perfect classifications. All optimal classifiers consisted of a single hidden layer. Starting

Table 5 Results from a 20-replication experiment when PCA/AE is used as the dimensional reduction technique for ANN input. Columns 2 and 6 show the mean accuracy, columns 3 and 7 display the standard deviation, columns 4 and 8 present the maximum accuracy, and columns 5 and 10 detail the optimal architecture for each ANN (number of hidden layers and number of neurons in each hidden layer). The accuracy refers to the overall performance of the model

ANNs—PCA inputs					ANNs—AE inputs				
	Accuracy					Accuracy			
Dim	Mean	SD	Max	Optimal ANNs	Dim	Mean	SD	Max	Optimal ANNs
1	0.974	0.0338	1	6	1	0.925	0.042	1	[9, 10]
2	0.998	0.0097	1	3	2	1	0	1	4
3	0.985	0.0255	1	3	3	0.977	0.035	1	3
4	0.976	0.0330	1	3	4	0.984	0.034	1	3
5	0.980	0.0222	1	3	5	0.961	0.045	1	2
6	0.987	0.0248	1	3	6	0.982	0.027	1	[2, 2]
7	0.996	0.0134	1	3	7	0.961	0.047	1	4
8	0.991	0.0227	1	3	8	0.980	0.040	1	3
9	0.980	0.0359	1	3	9	0.970	0.050	1	1
10	0.985	0.0324	1	3	10	0.998	0.010	1	4
15	0.994	0.0213	1	3	11	1	0	1	2
20	0.989	0.0193	1	3	12	0.980	0.038	1	4
25	0.985	0.0213	1	3	13	1	0	1	3
30	0.991	0.0178	1	3	14	1	0	1	3
					15	1	0	1	3

from two principal components (PCs) as inputs, the optimal hidden layer consistently contained just three neurons. Even though the maximum architecture parameters allowed up to two hidden layers and ten neurons per layer, the best-performing models in all cases relied on a simpler structure: one hidden layer with fewer than ten neurons. These results suggest that simple ANN architectures can achieve perfect classification of hyperspectral signatures, which holds promising potential for effectively identifying water and hydrocarbons among the hyperpixels. Concerning the use of AE to encode the ANN input data, once again, classifiers with perfect accuracy were obtained for all dimensions (Table 5, columns 6–10). There are several attempts where all the replications performed a perfect classification, since the mean accuracies of 1 and their standard deviations of 0.

ANNs trained with the 2D dataset as input demonstrated superior classification performance compared to models using datasets with a greater number of original features. This suggests that, in this case, the 2D dataset effectively retained the critical information necessary for optimal classification, resulting in better generalization than models trained on more complex data. A similar trend is observed in classifiers trained with datasets containing more than 11 dimensions (excluding the 12D dataset). The consistently high classification accuracy, even with just 2D inputs, highlights the ability of AEs to preserve key features for accurate classification. Additionally, the low standard deviation values across all dimensions underscore the stability and reliability of the classifiers. The minimal variability in performance across replications indicates consistent and robust behaviour. Notably, the mean accuracy exceeds 0.96 for all dimensions, except the 1D reduction, where accuracy drops to 0.925. This slight decline suggests that although dimensionality reduction can enhance performance, excessive reduction may lead to the loss of critical information, negatively impacting classification results.

In terms of network architecture, most classifiers exhibit simple structures, typically featuring a single hidden layer with fewer than five neurons. This simplicity reflects the efficiency of well-encoded datasets, minimizing the need for complex architectures to capture underlying data patterns. The only exception is the classifier trained on the 1D encoded hyperspectral signature dataset, which requires a more intricate architecture with two hidden layers containing nine and ten neurons, respectively. This added complexity likely compensates for the limited information available in the 1D dataset, enabling the model to better capture the inherent relationships within the data.

Convolutional Neural Networks Approach

The performance of the CNN models in distinguishing water and polluted water was evaluated for different training set sizes: 518 (8%), 972 (15%), and 1944 (30%) patches, based on accuracy, recall, and F-score for the test set. As shown in Table 6, an increase in the training set size resulted in a noticeable improvement in classification performance across all classes—water, diesel, C10, and fuel oil. For the smallest training set (8%), the water class achieved the highest F-score (0.917 ± 0.186), while diesel and fuel oil exhibited lower performance, with F-scores of 0.720 ± 0.322 and 0.732 ± 0.314, respectively. When the training set increased to 15%, improvements were observed, particularly for water (F-score $= 0.898 \pm 0.315$) and C10 (F-score $= 0.923 \pm 0.190$). The best results were obtained with the largest training set (30%), where all classes showed high accuracy and F-scores, with water and C10 reaching nearly perfect scores (F-score $= 0.998 \pm 0.003$ and 0.999 ± 0.001, respectively). These results highlight the positive impact of larger training sets on model performance, reducing variability and enhancing classification accuracy, especially for more challenging classes like diesel and fuel oil.

Conclusion

This study demonstrates the effectiveness of hyperspectral imaging (HSI) combined with machine learning (ML) and deep learning (DL) techniques for detecting hydrocarbons in water. By comparing classical ML approaches using hyperspectral signatures with deep learning models trained on full hyperspectral images, several key insights were obtained:

Table 6 CNN classification results: accuracy, recall, and F-score for the test sets across different training set sizes: 518, 972, and 1944 samples, representing 8%, 15%, and 30% of the total dataset (6480 samples), respectively. The table presents the mean values and standard deviations of these performance metrics. Recall and F-score values are reported to classify water, diesel, C10, and fuel oil, respectively. The accuracy refers to the overall performance of the model

		Accuracy		Recall		F-score	
Training set size	Class	Mean	SD	Mean	SD	Mean	SD
518 (8%)	Water	0.816	0.232	0.991	0.330	0.917	0.186
	Diesel			0.725	0.329	0.720	0.322
	C10			0.814	0.324	0.826	0.308
	Fuel oil			0.736	0.324	0.732	0.314
972 (15%)	Water	0.857	0.224	0.899	0.317	0.898	0.315
	Diesel			0.746	0.008	0.762	0.300
	C10			0.995	0.286	0.923	0.190
	Fuel oil			0.793	0.286	0.782	0.280
1944 (30%)	Water	0.978	0.052	0.998	0.022	0.998	0.003
	Diesel			0.984	0.002	0.964	0.083
	C10			0.999	0.184	0.999	0.001
	Fuel oil			0.930	0.184	0.945	0.137

1. Performance of Classical Machine Learning Methods.

   - PCA and AEs successfully reduced the dimensionality of hyperspectral data while preserving crucial spectral information.
   - ANNs trained on hyperspectral signatures achieved high classification accuracy, with some configurations reaching nearly perfect results.
   - The optimal dimensionality for classification was found to be around 2 for both PCA and AEs, highlighting that excessive dimensionality does not necessarily improve classification performance.

2. Effectiveness of Deep Learning Approaches.

   - CNNs outperformed classical ML models in classifying full hyperspectral images, benefiting from their ability to extract spatial-spectral features. By effectively capturing spatial and spectral correlations, CNNs enable more robust feature extraction and exhibit greater resilience to local noise and variability compared to models based solely on spectral signatures.
   - The performance of CNNs is highly dependent on the size of the training dataset, with larger datasets significantly improving classification accuracy. This aligns with existing literature on DL models, where larger training samples contribute to better generalization.
   - The CNN achieved near-perfect classification when trained with a sufficiently large dataset, confirming its robustness for automated oil spill detection.

3. Finally, it is worth mentioning that:

   - HSI combined with ML/DL techniques enables early detection of oil spills and water contamination, offering a valuable tool for environmental protection.
   - The findings emphasize the importance of selecting the appropriate classification method depending on the application, where classical ML models offer computational efficiency, while CNNs provide superior accuracy in image-based detection tasks.

4. Challenges and future directions
   Although the results are promising, several challenges remain:

   - Generalization to real-world conditions: The models need to be tested with real oil spills and varying environmental conditions (e.g. different water types, lighting conditions, and oil compositions).
   - Integration with operational frameworks: The deployment of UAV-mounted hyperspectral sensors for continuous monitoring requires optimization in terms of sensor calibration, data transmission, and real-time processing capabilities.
   - The results obtained in this study reinforce the potential of hyperspectral imaging as a powerful tool for oil spill detection in aquatic environments. The combination of HSI with ML and DL methods offers high accuracy, automation, and scalability, making it a promising approach for environmental protection and sustainable resource management.

Acknowledgements This work is part of the research projects 2023-057/PV/AYUDCCRRTT/AY–REINFORCED, "Research Network on Autonomous Vehicles and Artificial Intelligence for Water Resources Monitoring (REINFORCED): Technology in Action" and 2022-113/PV/FUND-CEIMAR-TRANSF-INNOV-2022/PT, "Hyperspectral monitoring of oil spills in marine and river waters with machine learning" supported by CEIMAR.

References

Abiodun OI, Jantan A, Omolara AE, Dada KV, Mohamed NA, Arshad H (2018) State-of-the-art in artificial neural network applications: a survey. Heliyon 4:938. https://doi.org/10.1016/j.heliyon.2018

Aira J, Olivares T, Delicado FM (2024) SpectroGLY: a low-cost IOT-based ecosystem for the detection of glyphosate residues in waters. arXiv:2401.16009. https://doi.org/10.48550/arXiv.2401.16009

Asadzadeh S, Oliveira WJ, Souza Filho CR (2022) UAV-based remote sensing for the petroleum industry and environmental monitoring: state-of-the-art and perspectives. J Pet Sci Eng 208:109633. https://doi.org/10.1016/j.petrol.2021.109633

Carrasco-García MG, Rodríguez-García MI, González-Enrique J, Cubillas-Fernández PR, Ruiz-Aguilar JJ, Turias-Domínguez IJ (2023) Hyperspectral technology for oil spills detection by

using artificial neural network classifier. In: 18th international conference on soft computing models in industrial and environmental applications (SOCO 2023). Springer, Cham, pp 82–91. https://doi.org/10.1007/978-3-031-42529-5_8
Carrasco-García MG, Rodríguez-García MI, Ruiz-Aguilar JJ, Deka L, Elizondo D, Turias-Domínguez IJ (2024) Oil spill classification using an autoencoder and hyperspectral technology. J Mar Sci Eng 12(3):495. https://doi.org/10.3390/jmse12030495
Chevalier P, Piccardo M, de Naurois G-M, Gabay I, Katzir A, Capasso F (2018) In-water chemical sensing by fiber-optic evanescent waves spectroscopy using mid-infrared quantum cascade lasers. Sensors Actuators B Chem 262:195–199. https://doi.org/10.1016/j.snb.2018.01.159
Cruz-Retana A, Becerril-Piña R, Fonseca CR, Gómez-Albores MA, Gaytán-Aguilar S, Hernández-Téllez M, Mastachi-Loza CA (2023) Assessment of regression models for surface water quality modeling via remote sensing of a water body in the Mexican highlands. Water 15(21):3828
Dalponte M, Ørka HO, Gobakken T, Gianelle D, Næsset E (2013) Tree species classification in boreal forests with hyperspectral data. IEEE Trans Geosci Remote Sens 51(5):2632–2645. https://doi.org/10.1109/TGRS.2012.2216272
Deepthi TT (2020) Spectral similarity algorithm-based image classification for oil spill mapping of hyperspectral datasets. J Spectr Imaging 9:1–17. https://doi.org/10.1255/jsi.2020.a14
EMSA & EEA (2021) European maritime transport environmental report 2021. Publications Office of the European Union, Maastricht
Fingas M, Brown CE (2018) Review of oil spill remote sensing. Sensors 18(1):91. https://doi.org/10.3390/s18010091
French-McCay DP, Parkerton TF, de Jourdan B (2023) Bridging the lab to field divide: advancing oil spill biological effects models requires revisiting aquatic toxicity testing. Aquat Toxicol 256:106389. https://doi.org/10.1016/j.aquatox.2022.106389
Glorot X, Bengio Y (2010) Understanding the difficulty of training deep feedforward neural networks. Artif Intell Stat 9:249–256
González J, Fernández E, Figueiras FG (2022) Assessing the effect of oil spills on the dynamics of the microbial plankton community using a NPZD model. Estuar Coast Shelf Sci 265:107734. https://doi.org/10.1016/j.ecss.2021.107734
Goyens C, Lavigne H, Dille A, Vervaeren H (2022) Using hyperspectral remote sensing to monitor water quality in drinking water reservoirs. Remote Sens 14(21):5607. https://doi.org/10.3390/rs14215607
Haut JM, Moreno-Alvarez S, Pastor-Vargas R, Perez-Garcia A, Paoletti ME (2024) Cloud-based analysis of large-scale hyperspectral imagery for oil spill detection. IEEE J Select Top Appl Earth Observ Remote Sens 17(62):2461–2474. https://doi.org/10.1109/JSTARS.2023.3344022
He K, Zhang X, Ren S, Sun J (2015) Delving deep into rectifiers: surpassing human-level performance on ImageNet classification. In: Proceedings of the IEEE international conference on computer vision. Piscataway, IEEE, pp 1026–1034. https://doi.org/10.1109/ICCV.2015.123
Hornik K, Stinchcombe M, White H (1980) Multilayer feedforward networks are universal approximators. Neural Netw 2(5):359–366. https://doi.org/10.1016/0893-6080(89)90020-8
Hotelling H (1933) Analysis of a complex statistical variables into principal components 8. Determination of principal components for individuals. J Educ Psychol 24:498–520. https://doi.org/10.1037/H0071325
Hu T, Yuan J, Wang X, Yan C, Ju X (2022) Spectral-spatial features extraction of hyperspectral remote sensing oil spill imagery based on convolutional neural networks. IEEE Access 10:127969–127983. https://doi.org/10.1109/ACCESS.2022.3194260
Hubel HD, Wiesel TN (1959) Receptive fields of single neurones in the cat's striate cortex. J Physiol 148:574–591
Hwang SH, Lee YJ, Choi YH, Huh DA, Kang MS, Moon KW (2024) Long-term effects of the Hebei Spirit oil spill on the prevalence and incidence of allergic disorders. Sci Total Environ 912:168801. https://doi.org/10.1016/j.scitotenv.2023.168801
Jiang Z, Zhang J, Ma Y, Mao X (2022) Hyperspectral remote sensing detection of marine oil spills using an adaptive long-term moment estimation optimizer. Remote Sens 14(1):157. https://doi.org/10.3390/rs14010157

Jolliffe IT (1986) Principal component analysis. Springer-Verlag, Cham

Kababu E, Angel DL, Sisma-Ventura G, Belkin N, Rubin-Blum M, Rahav E (2022) Effects of crude oil and gas condensate spill on coastal benthic microbial populations. Front Environ Sci 10:1051460. https://doi.org/10.3389/fenvs.2022.1051460

Kieu HT, Law AWK (2021) Remote sensing of coastal hydro-environment with portable unmanned aerial vehicles a state-of-the-art. J Hydro Environ Res 37:32–45. https://doi.org/10.1016/j.jher.2021.04.003

Kim YI, Park WH, Shin Y, Park J-W, Engel B, Yun Y-J, Jang WS (2024) Applications of machine learning and remote sensing in soil and water conservation. Hydrology 11(11):183. https://doi.org/10.3390/hydrology11110183

Kramer MA (1991) Nonlinear principal component analysis using autoassociative neural networks. AICHE J 32(2):233–243. https://doi.org/10.1002/AIC.690370209

Krizhevsky A, Sutskever I, Hinton GE (2017) ImageNet classification with deep convolutional neural networks. Commun ACM 60:84–60. http:/\penalty\z@/\penalty\z@code.google.com/\penalty\z@p/\penalty\z@cuda-\penalty\z@convnet/\penalty\z@

Kumar B, Dikshit O, Gupta A, Singh MK (2020) Feature extraction for hyperspectral image classification: a review. Int J Remote Sens 41(16):6248–6287. https://doi.org/10.1080/01431161.2020.1736732

LeCun Y et al (1990) Handwritten digit recognition with a back-propagation network. In: Proc. advances in neural information processing systems. IEEE, Piscataway, pp 396–404

LeCun Y, Bengio Y, Hinton G (2015) Deep learning. Nature 521:436–444. https://doi.org/10.1038/nature14539

Leifer I, Lehr WJ, Simecek-Beatty D, Bradley E, Clark R, Dennison P, Hu Y, Matheson S, Jones CE, Holt B, Reif M, Roberts DA, Svejkovsky J, Swayze G, Wozencraft J (2012) State of the art satellite and airborne marine oil spill remote sensing: application to the BP Deepwater horizon oil spill. Remote Sens Environ 124:185–209. https://doi.org/10.1016/j.rse.2012.03.024

Lu Q, Si W, Wei L, Li Z, Xia Z, Ye S, Xia Y (2021) Retrieval of water quality from UAV-borne hyperspectral imagery: a comparative study of machine learning algorithms. Remote Sens 13(19):3928. https://doi.org/10.3390/rs13193928

Menezes J, Poojary N (2020) A fusion approach to classify hyperspectral oil spill data. Multimed Tools Appl 79(7–8):5399–5418. https://doi.org/10.1007/s11042-018-6709-7

Mockus J (1975) On the Bayes methods for seeking the extremal point. IFAC Proc 8(1):428–431. https://doi.org/10.1016/S1474-6670(17)67769-3

Mohammadiun S, Hu G, Gharahbagh AA, Li J, Hewage K, Sadiq R (2021) Intelligent computational techniques in marine oil spill management: a critical review. J Hazard Mater 419:126425. https://doi.org/10.1016/j.jhazmat.2021.126425

Paoletti ME, Haut JM, Plaza J, Plaza A (2019) Deep learning classifiers for hyperspectral imaging: a review. ISPRS J Photogramm Remote Sens 158:279–317. https://doi.org/10.1016/j.isprsjprs.2019.09.006

Pelta R, Carmon N, Ben-Dor E (2019) A machine learning approach to detect crude oil contamination in a real scenario using hyperspectral remote sensing. Int J Appl Earth Obs Geoinf 82:101901. https://doi.org/10.1016/j.jag.2019.101901

Rumelhart DE, Hinton GE, Williams RJ (1986) Learning representations by back-propagating errors. Nature 323:533–536. https://doi.org/10.1038/323533a0

Salam ML, Saxena RS (2024) Comparative analysis of SVM and CNN for hyperspectral image classification. Non-Peer Rev. https://doi.org/10.31223/X5811M

Satyanarayana AR, Dhali MA (2023) Oil spill segmentation using deep encoder-decoder models. Comput Vis Pattern Recogn. arXiv:2305.01386. https://doi.org/10.48550/arXiv.2305.01386

Sentiel (2024) https:/\penalty\z@/\penalty\z@sentinels.copernicus.eu/\penalty\z@web/\penalty\z@sentinel/\penalty\z@copernicus/\penalty\z@sentinel-\penalty\z@1. Accessed 15 Dec 2024

Shan J, Zhao J, Liu L, Zhang Y, Wang X, Wu F (2018) A novel way to rapidly monitor microplastics in soil by hyperspectral imaging technology and chemometrics. Environ Pollut 238:121–129. https://doi.org/10.1016/j.envpol.2018.03.026

Shengfang M, Yuting Z, Prasanna HG, Jinwei D, Geli Z, Vijaya GK, Pradeep W, Liangfu C, Colton F, Weiguo J (2019) Application of the water-related spectral reflectance indices: a review. Ecol Indic 98:68–79. https://doi.org/10.1016/j.ecolind.2018.10.049

Sicot G, Lennon M, Miegebielle V, Dubucq D (2014) Analysis of the reflectance spectra of oil emulsion spilled on the sea surface. SPIE Proc 2014:924007. https://doi.org/10.1117/12.2067204

Stroope S, Slack T, Kroeger RA, Keating KS, Beedasy J, Sury JJ, Brooks J, Chandler T (2023) Deepwater horizon oil spill exposures and long-term self-rated health effects among parents in coastal Louisiana. Disaster Med Public Health Prep 17(3-4):223. https://doi.org/10.1017/dmp.2022.223

Sun Y, Wang D, Li L, Ning R, Yu S, Gao N (2024) Application of remote sensing technology in water quality monitoring: from traditional approaches to artificial intelligence. Water Res 267:122546. https://doi.org/10.1016/j.watres.2024.122546

Torres-Vera MA (2023) Mapping of total suspended solids using Landsat imagery and machine learning. Int J Environ Sci Technol 20:11877–11890. https://doi.org/10.1007/s13762-023-04787-y

Viallefont-Robinet F, Moussous A, Déliot P, Roupioz L, Miegebielle V (2018) Analysis of water-in-oil emulsion hyperspectral signature: contribution of pool experiment. Proc SPIE 107:107840. https://doi.org/10.1117/12.2324865

Yang JF, Wan JH, Ma Y, Zhang J, Hu YB, Jiang ZC (2019) Oil spill hyperspectral remote sensing detection based on DCNN with multi-scale features. J Coast Res 90:332–339. https://doi.org/10.2112/SI90-042.1

Yang J et al (2022a) Decision fusion of deep learning and shallow learning for marine oil spill detection. Remote Sens 14(3):666. https://doi.org/10.3390/rs14030666

Yang J, Hu Y, Ma Y, Li Z, Zhang J (2022b) Research on oil spill pollution type identification using RPNET deep learning model and airborne hyperspectral image. International Geoscience and Remote Sensing Symposium (IGARSS), pp 807–810. https://doi.org/10.1109/IGARSS46834.2022.9883530

Yildiz S, Sönmez VZ, Uğurlu Ö, Sivri N, Loughney S, Wang J (2021) Modelling of possible tanker accident oil spills in the Istanbul Strait in order to demonstrate the dispersion and toxic effects of oil pollution. Environ Monit Assess 193:538. https://doi.org/10.1007/s10661-021-09339-w

Zhang L, Zhang L, Du B (2016) Deep learning for remote sensing data: a technical tutorial on the state of the art. IEEE Geosci Remote Sens Magaz 4:22–40. https://doi.org/10.1109/MGRS.2016.2540798

Zhang Y, Wu L, Deng L, Ouyang B (2021) Retrieval of water quality parameters from hyperspectral images using a hybrid feedback deep factorization machine model. Water Res 204:117618. https://doi.org/10.1016/j.watres.2021.117618

Zhao D, Luo Q, Qiu Z (2024) Chromaticity-based discrimination of algal bloom from inland and coastal waters using in situ hyperspectral remote sensing reflectance. Water 16(16):2276. https://doi.org/10.3390/w16162276

Zhou DX (2020) Theory of deep convolutional neural networks: Downsampling. Neural Netw 124:319–327. https://doi.org/10.1016/j.neunet.2020.01.018

Intelligent Real-Time Anomaly Detection for Optimization of Water Monitoring Systems

Adrian Bazan-Muñoz, Guadalupe Ortiz, Alfonso Garcia-de-Prado, Manuel Cano-Crespo, Juan Boubeta-Puig, Enrique Daneri-Vias, and Juan-Manuel Mariscal-Chavinet

Contents

A. Bazan-Muñoz · G. Ortiz · A. Garcia-de-Prado · M. Cano-Crespo · J. Boubeta-Puig (✉)
UCASE Software Engineering Research Group, University of Cadiz, Cádiz, Spain
e-mail: juan.boubeta@uca.es

E. Daneri-Vias · J.-M. Mariscal-Chavinet
Development Area - Grupo Energético de Puerto Real S.A., Cádiz, Spain

D. Gutiérrez et al. (eds.), *Smart Water Quality Monitoring*,
https://doi.org/10.1007/978-3-032-03049-8_6

Abstract

The growing emergence of the Internet of Things and the proliferation of smart cities have created significant challenges for software engineering. One of the key challenges in smart sustainable cities is the efficient management of water resources, which has become a global priority. Smart water meters, equipped with sensors and communication capabilities, offer a promising solution for optimizing water management. However, the successful deployment of smart water metering systems requires a robust and scalable software architecture capable of efficiently processing and analyzing large volumes of data to improve decision making. Complex event processing (CEP) has emerged as a powerful technology to address such a challenge since CEP enables real-time definition and detection of complex patterns within data streams, facilitating the identification of anomalies, trends, and critical events. In this chapter we present an effective and efficient CEP-based software architecture for intelligent water management which facilitates an early detection of leaks and frauds, as well as further anomalies. In addition, we have configured and made use of an intuitive graphical user interface that allows domain experts to define custom patterns without the need for programming skills, as well as a RESTful API to manage the patterns deployed within the architecture at runtime.

Keywords

Complex event processing · Internet of things · Smart cities · Intelligent water-supply networks · Model-driven development · Real-time anomaly detection · Decision making · Performance evaluation

Introduction

The advent of the Internet of Things (IoT) and the rise of smart cities has presented a significant challenge for engineers, particularly software engineers, especially concerning the capacity to ingest, process, and correlate heterogeneous data in real time. This trend is driven by several factors: the ubiquitous nature of the internet, the decreasing cost of communication technologies, and the widespread adoption of powerful Internet-connected devices such as smartphones and tablets. Additionally, the proliferation of sensors and other data sources for IoT applications has further promoted this growth. The potential applications of these advancements are vast, enabling improved decision-making, enhanced urban sustainability, and elevated quality of life for citizens. Numerous publications have explored the challenges and opportunities in this field, as well as presented specific solutions for various application domains (Khan et al. 2020; Gavrilović and Mishra 2021).

On the other hand, environmental degradation and water scarcity pose significant challenges to sustainable development (UNEP 2025; UN 2025). The obvious link between clean air, water, and healthy ecosystems underscores the importance of environmental management in smart cities. Particularly, water is a finite and indispensable resource which is increasingly consumed by growing domestic and industrial demands. Groundwater depletion, a consequence of unsustainable water use, threatens the long-term viability of our societies. This situation is further exacerbated by undetected water leaks and frauds.

To address these challenges, strategic and efficient water resource management is imperative. Traditional water metering systems, often reliant on manual readings, hinder effective management and real-time monitoring. Smart water meters, equipped with advanced sensors and communication technologies, offer a promising solution. While the initial investment in smart meter infrastructure may be substantial, the long-term benefits, including reduced water loss, improved billing accuracy, and enhanced operational efficiency, outweigh the costs. However, the deployment of smart meters is only the first step toward intelligent water management. It is essential to complement the acquisition of these devices with the appropriate software architecture. This architecture must be designed to efficiently collect, process, and analyze in real time the large datasets generated by smart meters, and at the same time it is crucial to balance the frequency of data collection with battery life considerations. While frequent data updates provide valuable insights, excessive data transmission can drain battery power, requiring frequent replacements.

In our experience, the use of complex event processing (CEP) engines, integrated with the appropriate software architecture according to the particular needs of the specific domain, provides an effective and efficient solution to support real-time decision-making in the scope of smart cities (Ortiz et al. 2022a, b; Bazan-Muñoz et al. 2024). CEP enables the real-time definition and detection of situations of interest, referred to as patterns, based on the correlation of incoming data over time. This capability allows for the timely identification of anomalies, trends, and other significant occurrences within the system, facilitating proactive decision-making and enabling effective response to changing conditions (Luckham 2011).

In this sense, within the domain of intelligent water management, we have proposed, in collaboration with the GEN company (GEN 2025) in Puerto Real, Spain, which manages the water supply networks of the locality, a software architecture that enables real-time processing and correlation of data obtained from smart meters in the municipality's water supply networks. As we will explain throughout this chapter, this architecture comprises various software artifacts and components, with a CEP engine playing a key role. Our proposed architecture represents a significant advancement in the field of water management by providing an efficient and scalable solution for monitoring and optimizing water management systems. Complementing the core data processing architecture, we have incorporated a graphical tool that empowers domain experts to define custom patterns without the need for coding expertise. Moreover, a RESTful Application Programming Interface (API) facilitates dynamic management of deployed patterns at runtime, eliminating the need for system downtime. These features promote customization and adaptability, enabling the architecture to be tailored to specific use cases and evolving requirements.

The rest of the chapter is organized as follows. After the Introduction, background on CEP and message brokers is provided. Afterward, related work is examined and described. The following section provides the software architecture proposed followed by the section describing the implemented CEP patterns. Then, the modeling tool is explained together with the graphical definition of some of the previous implemented CEP patterns. Afterward, the evaluation performed to the proposed architecture and patterns is described, including the description of the used resources, the method followed, and the analysis of the performance test results. Finally, conclusions and future work are provided.

Background

This section provides background on CEP and message brokers.

Complex Event Processing

CEP (Luckham 2011) is a cutting-edge technology that allows large amounts of data to be processed, analyzed, and correlated in real time to detect situations of interest to the application domain.

In CEP, the data received is referred to as simple events. These simple events are processed by the CEP engine to detect situations of interest, known as complex events. For the CEP engine to be able to detect these complex events, it must first have defined both an event schema and a set of event patterns. While the event schema allows the type and structure of events to be received by the CEP engine to be defined, the event patterns are a set of rules that allow simple events, and even other already generated complex events, to be analyzed and correlated for real-time detection of relevant situations.

Today, there are a large number of CEP engines, each with its own event processing language (EPL) for defining event patterns. In this case, Esper CEP engine was used (Espertech 2024) due to its already proven high performance (Corral-Plaza et al. 2021; Ortiz et al. 2022c) and high scalability (Ortiz et al. 2023).

Message Brokers

A message broker is a middleware that allows communication between message producers and consumers to be managed in distributed systems. There are currently many models for messaging broker communication. However, in the IoT domain, the most common model is the publish/subscribe model. In this model, messages are published in topics to which it is possible to subscribe in order to receive these messages. In this model, the message remains in the topic until all active subscriptions have extracted the message for processing. Although there are many message brokers, this proposal uses RabbitMQ, one of the best rated for performance (Ionescu 2015).

RabbitMQ (2024) is an open-source message broker that acts as middleware, enabling communication between producers and consumers using Advanced Message Queuing Protocol (AMQP) and the publish/subscribe model. In our system, messages from the water supply network are stored in a RabbitMQ topic until they are immediately extracted by the CEP engine for processing. The generated complex events are then sent to another RabbitMQ topic to be consumed by the interested parties.

Related Work

Regarding the state of the art from the academia, existent proposals related to efficient management of water supply networks do not usually provide the capacity to make a real-time analysis of the data obtained from such a network but tend to work with stored data (Ma et al. 2022; Hu et al. 2021; Axiotidis et al. 2024). This prevents appropriate action from being taken at the time a leak is detected and not several hours or days later, when it is decided to process the stored data.

Other proposals focus on predicting events, especially in terms of water consumption, but do not detect deviations that would allow us to assume that there is a water leak or a fraud in real time (Preciado et al. 2019). Nevertheless, several innovative solutions have recently emerged to address this issue. For instance, (Farah and Shahrour 2024) proposes the comparison of flow pattern distribution (CFPD) method that analyses real-time water supply and consumption data to identify leak events and consumption patterns, contributing to water conservation and operational efficiency. By predicting future water usage and comparing it with real-time data, the methods proposed in (Serban et al. 2024) can identify deviations indicative of leaks or abnormal consumption.

On the other hand, companies either do not invest in meters that allow remote access, and this is an issue that will be solved little by little when it becomes necessary to replace meters; or, if they are innovative and decide to invest in commercial smart meters, they are tied to the proprietary software that comes with these meters. These systems usually lack flexibility and interoperability with devices from different manufacturers, limiting the utility's ability to integrate diverse technologies and adapt to changing needs.

The work presented in this chapter faces to solve a large part of such problems, by being able to make use of the proposed software architecture to process the measurements of smart meters through a series of patterns in real time for both industrial and residential subscribers and check whether consumption is in line with expected consumption or whether the flow of water entering the sector does not coincide with consumption, among other scenarios. This would allow us to detect leaks, fraud, and other anomalous consumption in real time and take immediate action to prevent further damage.

Software Architecture for Intelligent Water Management

The software architecture, shown in Fig. 1, consists of the following five components:

- Firstly, there is a series of water measurement sensors in different sectors of the supply network.
- Secondly, there is a message broker, RabbitMQ, with three message queues configured in it.
- Thirdly, there is a CEP engine, Esper CEP, in which an event schema and the different event patterns that will perform the analysis of the information have been deployed.
- Fourthly, the MEdit4CEP (Boubeta-Puig et al. 2015) event pattern modeling tool will permit to define the CEP schemas and patterns graphically and to automatically generate their code from the graphical design.

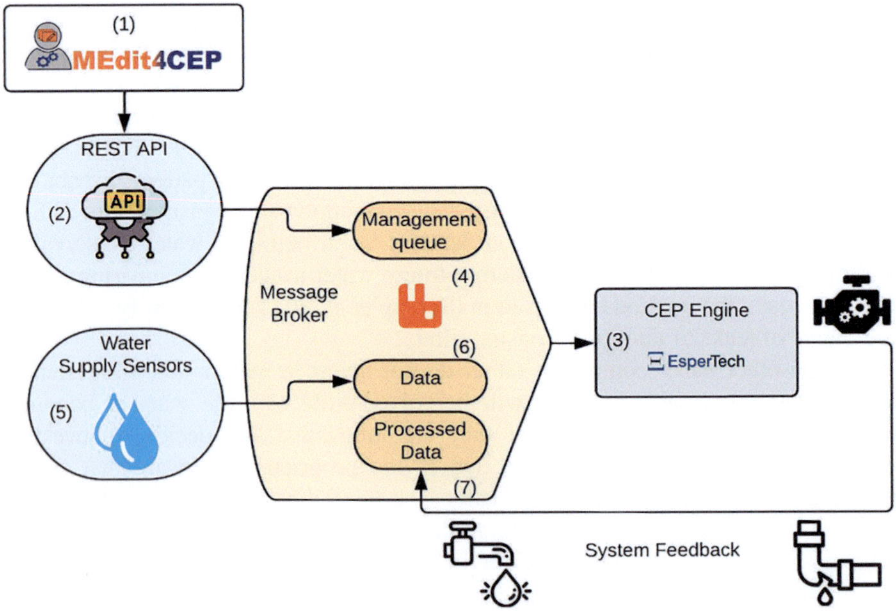

Fig. 1 Software architecture proposed

- Fifthly, API4CEP is a REST API that can be used to manage the CEP schemas and patterns at runtime in the CEP engine, without the need to stop the system.

The proposed software architecture works as follows:

- Firstly, the schema and event patterns are modeled using the MEdit4CEP tool (see (1) in Fig. 1). Then, at any time during execution, the event patterns can be redesigned or new event patterns can be created. The code for such schemas and patterns can be automatically generated from MEdit4CEP.
- Secondly, a REST API (see (2) in Fig. 1) is used to implement and/or modify the event schemas and patterns. When using MEdit4CEP, the REST API is invoked with the generated event pattern code; otherwise, the API can be invoked through any REST client. The REST API communicate with the CEP engine through a queue located in the RabbitMQ message broker dedicated to the management of the CEP engine.
- Thirdly, the CEP engine (see (3) in Fig. 1) receives requests to deploy schemas and event patterns through the subscription to the RabbitMQ management queue (see (4) in Fig. 1). The CEP engine thus deploys the event schemas and patterns defined to analyze and detect situations of interest. During execution, new event patterns can be added and existing ones modified or deleted. Dependencies between existing event patterns are managed internally.
- Fourthly, the data detected in the sensors (see (5) in Fig. 1) distributed along the water supply network are published in a queue (see (6) in Fig. 1) in the RabbitMQ message broker dedicated to the data to be processed by the CEP engine.
- Finally, the CEP engine receives the simple events with the sensor data from the supply network, subscribing to the RabbitMQ queue with the data. The CEP engine will process these simple events according to the deployed event patterns, detecting situations of interest for the application domain. The complex events generated will be sent to a new message queue (see (7) in Fig. 1) in the RabbitMQ message broker.

Smart Water Event Patterns

This section introduces the schema and event patterns used in the proposed architecture. All simple input events will arrive to the CEP engine in a common JSON format. They are always identified by the serial number of the device, *serialNumber*. This is followed by the date the measurement was taken, *measureDateTime*, and the date the measurement was sent, *processingDateTime*. This is followed by the sector of the supply network in which the device is located in the variable *sector*, the commercial classification of the meter *contractType*, and the classification of the meter according to its consumption mode *consumptionMode*. This is followed by the variables *q1* or minimum flow, *q2* or temporary flow, *q3* or permanent flow, and *q4* or maximum flow. Finally, there is the quantity to be measured *magnitude*, the type of measurement *measureType*, the numerical value of the measurement *measureValue*, and the unit in which the measurement is carried out *measureUnit*. In order to simplify the messages received, the variables *magnitude* and *contractType* use integers that refer to the different magnitudes measured and to the different types of contracts that these devices have respectively. The equivalences of the numerical values of the *magnitude* variable are as follows:

1. Daily consumption
2. Hourly consumption
3. Chlorine
4. Pressure
5. Start-up
6. Sleep time
7. Normal time
8. Leakage time
9. Maximum daily flow rate
10. Minimum daily flow rate
11. Inverse consumption
12. Battery level

The equivalences of the *contractType* variable are as follows:

1. Domestic
2. Commercial
3. Industrial
4. Municipal dependencies
5. Municipal irrigation
6. Works
7. Public agencies
8. Industrial fire-fighting
9. Industrial garage
10. Other uses
11. Fire-fighting
12. Domestic garage
13. Irrigation

The event scheme is as follows:

```
Measure (serialNumber String, measureDateTime long,
   processingDateTime long, sector int, deviceType
   String, contractType int, consumptionMode String, q1
   double, q2 double, q3 double, q4 double, magnitude
   int, type String, value double, unit String)
```

When a simple event arrives in this format, it is analyzed by the CEP engine according to the following event patterns, which are detailed below and the code for which is provided in the appendix:

Excessive night minimum flow rate: This pattern generates an alarm if, at any time during the night-time consumption period between 01:00 and 06:00, the maximum threshold of 3000 liters per hour (l/h) is exceeded. This situation can occur due to interruptions in the supply network or fraud in the network. To do this, it is necessary to check that the simple event occurs temporarily between 01:00 and 06:00, that the magnitude of the measurement corresponds to hourly consumption, that it is an input device in the network, and that the measured value is greater than 3000 l/h.

Excessive average daily flow: An alert is generated if the average daily consumption between 09:00 and 13:00 exceeds 80,000 l on weekdays, including Saturdays, or 55,000 l on Sundays. This would be a symptom of a break in the network or fraud. The first step is to calculate the cumulative total of the consumption of the same appliance between 09:00 and 13:00.

After each cumulative sum, it is checked whether the consumption value, depending on the day of the week, is higher than the maximum value set for that day.

High pressure in the supply network: An alarm is generated if a pressure measurement exceeds the maximum set value of 40 m of water column (mca), which would indicate a fault in the pressure regulator. This is done by checking that, in the simple event received, the type of device is the control one, the magnitude is pressure, and the reading is greater than 40.

Low pressure in the network: An alarm is generated if two consecutive pressure measurements are less than 35 mca, which could mean an increase in flow due to a break in the network or a fault in the pressure regulator. For this purpose, it is checked that a first simple event is received with a control type device, a pressure magnitude, and a value lower than 35. It is then checked that the next pressure reading from this device is not less than 35.

Excess of unregistered water (ANR): An alert is generated when the difference between the amount of water entering a sector and the amount of water leaving the sector, over the water injected in the same sector, is greater than 0.30. This daily indicator would indicate poor network performance. To do this, the first step is to calculate the cumulative sum of the values of inflow, ANRin, and outflow, ANRout, of the same sector.

After each cumulative sum of sector input and output measurements, the ANR during the day is calculated.

High chlorine levels: An alarm is generated if the chlorine measurements are higher than the set maximum. This is done by checking that the reading is from a control type device with a chlorine level magnitude and the reading is greater than 1.2 parts per million (ppm).

Low chlorine levels. An alarm is generated if chlorine measurements are below the set minimum. This is done by checking that the measurement is from a control type device with a chlorine magnitude and the reading is less than 0.2 ppm.

Excessive leak time: An alarm is generated when a possible internal leak is detected. This is done by checking that the device is a network output device, that the magnitude is of the leak time type, and that the value of the measurement in percent per one over the daily seconds duration is greater than 0.05.

No consumption (commercial or industrial): An alarm is generated if, for more than one consecutive day, the number of starts of a device is zero, its consumption is zero, and it is classified as a commercial-industrial type. To do this, it is checked that a first simple event is of type output, magnitude starts, reading zero, and contract type commercial or industrial. It is then checked that, for more than one consecutive day, another simple event is received from the same device, in which the consumption and the number of starts are also zero.

Low battery: An alarm is generated if the battery level of the device is low. To do this, it is checked that it is an output device, that the magnitude is the battery level, and that the value of the measurement is 1, which would indicate that the level is low.

Excessive flow rate: If the device has been incorrectly sized and the maximum daily flow rate exceeds 90% of the device's under-load flow rate q4, an alarm is generated. This is done by checking that the device is of the output type, that the value of the magnitude is of the maximum daily flow, and that the value of the measurement is greater than 90% of the maximum flow of this device.

Reverse consumption: If the customer's network is poorly designed, a reverse consumption alarm must be generated. In this case, it is checked that the device is an output device, that the magnitude is reverse consumption, and that the value of this measurement is greater than 0.

Internal leak: A possible internal leak alarm is generated when a device has a number of zero starts but a consumption greater than zero. This is done by checking that a first simple event is of type output, with a magnitude of type starts and zero value. And that during the same day, no second reading is received from the same device, indicating that despite the absence of starts, the daily consumption of this device is greater than zero.

Fire protection with continuous daily consumption: An alarm is generated when more than five consecutive hourly consumption readings are taken from a device that is commercially classified as fire protection. To do this, it first checks that the device is of the output type, that the magnitude is of hourly consumption type, that the contract type is fire or industrial fire, and that the consumption value is greater than zero. It then checks that, in the next hours, four more consecutive simple events are received, corresponding to the same device, which also have a consumption greater than zero.

Devices without communication: A communication failure alert is generated if there are no messages from the devices for 2 days. This is done by checking if a single event with the same serial number has not been received for two consecutive days.

Modeling Tool for Pattern Definition and Code Generation

A modeling tool should be provided to enable domain experts to adapt the system to their needs without the need of being Esper EPL experts. With this tool they should easily create and modify their own schemas and event patterns. For this reason, a modeling tool called MEdit4CEP (Boubeta-Puig et al. 2015) has been used to configure the particular domain of water supply network. MEdit4CEP is a model-based system that allows domain matter experts to define both event schemas, patterns, and alerts for real-time notification in a graphical mode, without having to be familiar with the Esper EPL language. Below we show how the scheme used in this proposal and some of the patterns described can be implemented with MEdit4CEP:

The event diagram is shown in Fig. 2, which illustrates how the graphical interface can be used to easily define all the variables that conform the schema. In order to do this, an object of the type *Schema(E)* has to be selected and as many objects of the type *Property(P)* have to be added to it as there are variables in the Schema. Once the schema is complete, MEdit4CEP allows the EPL code corresponding to the graphical representation to be generated. Thus, from the diagram in Fig. 2, the following schema code is generated:

```
@public @buseventtype create json schema Measure
   (serialNumber String, measureDateTime long,
   processingDateTime long,  sector int, deviceType
   String, contractType int, consumptionMode String, q1
   double,q2 double, q3 double, q4 double, magnitude int,
   type String, value double, unit String)
```

After defining the schema, it is also possible to define the event patterns using MEdit4CEP. Figure 3 shows the pattern for a high chlorine level. To apply this event pattern, a simple event corresponding to the defined schema must first be received.

This would be represented in MEdit4CEP by first placing a schema object like the one previously defined in Fig. 2. Then all the conditionals of this pattern must be applied. In MEdit4CEP, this is done by associating the variable to be analyzed with the corresponding condition or operator, such as the variable *deviceType*, which must be equal to the *CONTROL* type, or the variable *value*, which must be greater than 1.2. To generate the complex event, we must select an object of type *New Complex Event* and add as many properties as we wish and then associate these properties with the corresponding property of the input event.

Figure 4 shows a more complex pattern: fire protection pattern with continuous daily consumption. The procedure to create such an event pattern would be the same as for the previous pattern. Starting by including as many event schemas as simple events have to occur in the event, assigning to each simple event the relevant conditionals, and linking all events with the "followed by" clause. Once all event patterns have been created and their respective EPL codes generated, the REST API can be invoked to add them to the CEP engine or to modify already existing CEP patterns.

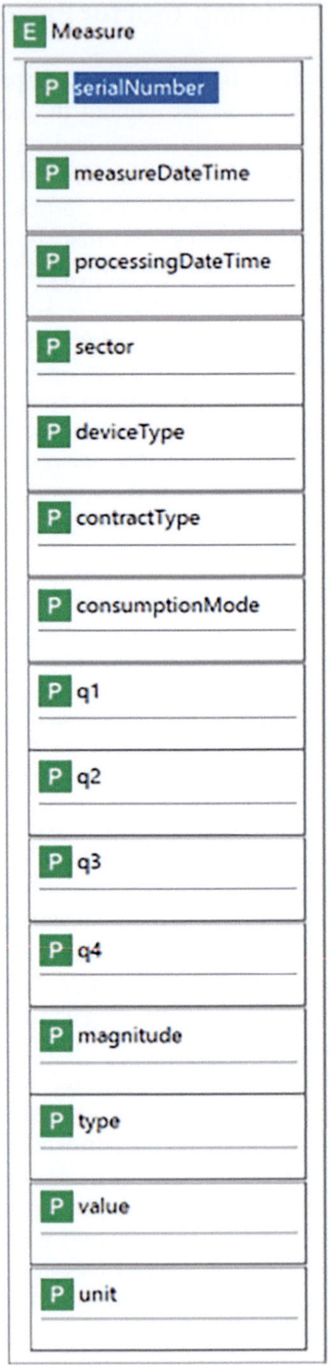

Fig. 2 MEdit4CEP event schema

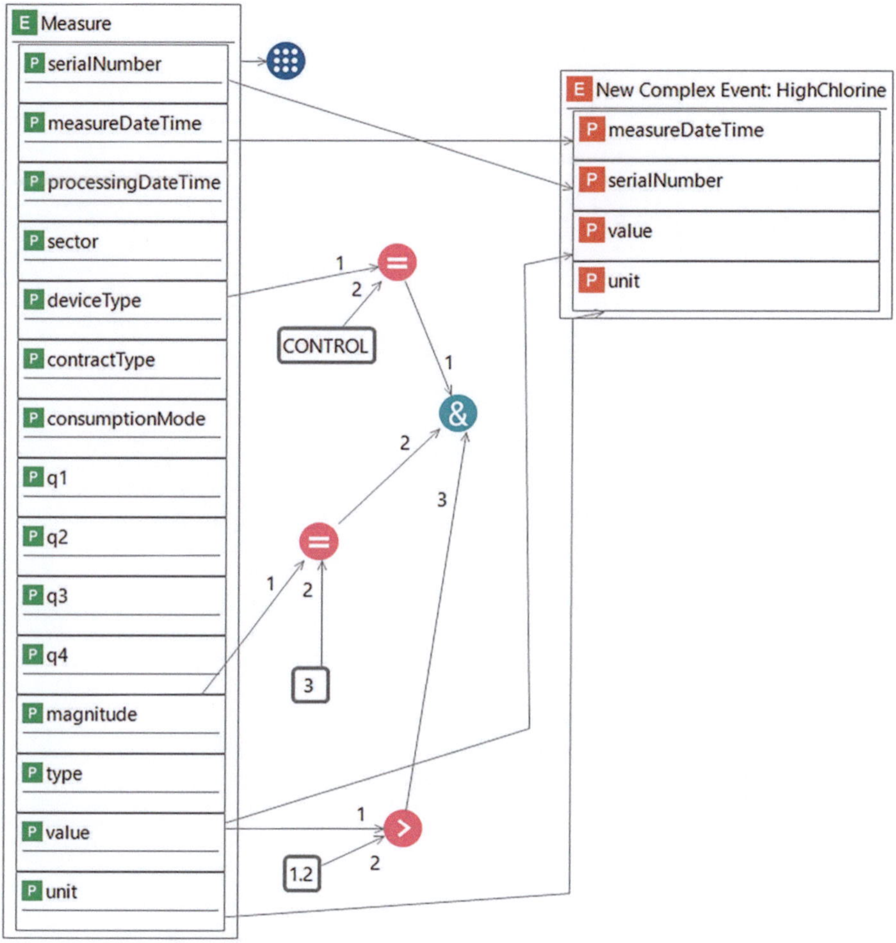

Fig. 3 MEdit4CEP HighChlorine pattern

Evaluation

Every abovementioned event pattern has been tested over an implementation of the proposed software architecture in order to evaluate the performance of the system.

The data flows within the implemented software architecture are as follows (see Fig. 5):

1. Event patterns and schemas are implemented through the REST API. Moreover, the data from the different sensors are generated by a Java-based simulator, which emulates the real periodicity in which the data is generated by the deployed smart

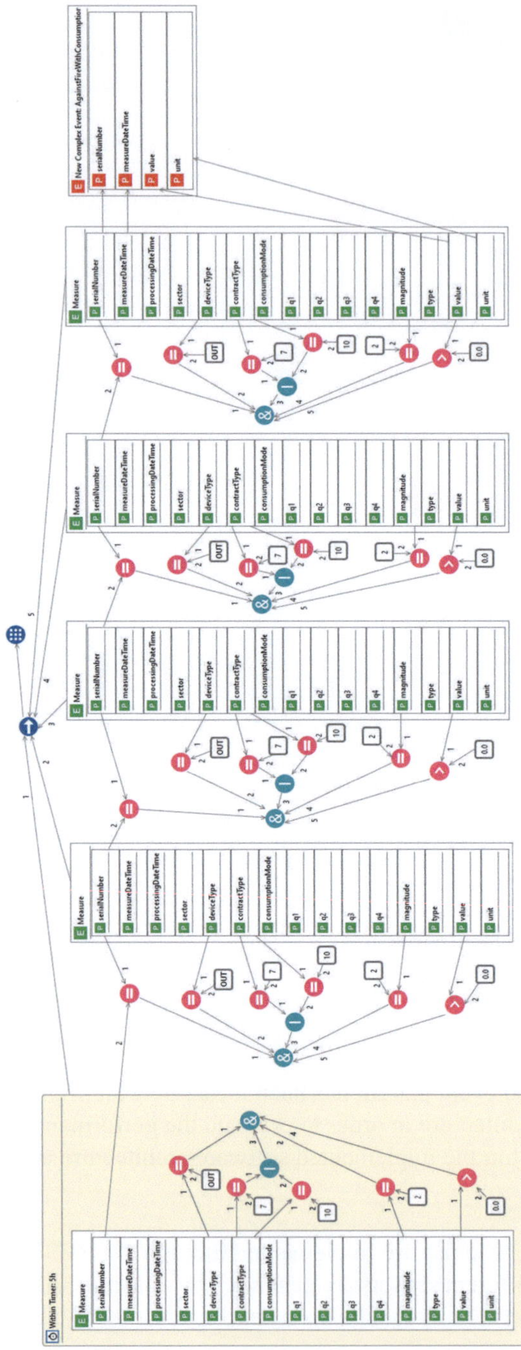

Fig. 4 MEdit4CEP AgainstFireWithConsumption pattern

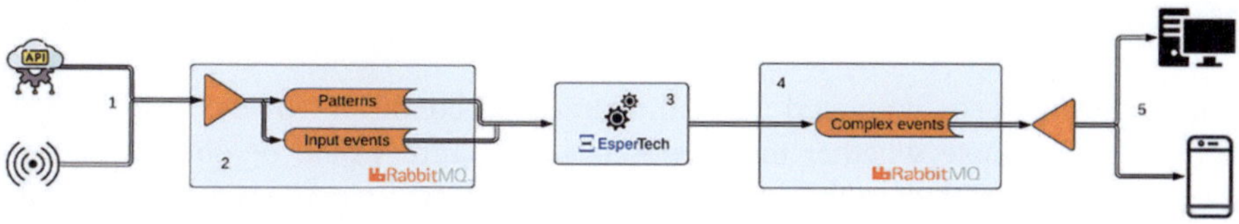

Fig. 5 Software architecture implemented for evaluation

meters. The data values have been emulated to ensure pattern compliance for the purpose of verifying their proper operation.
2. Both patterns and input data (simple events) are published on the corresponding queue of the RabbitMQ server.
3. An application executing the Esper CEP engine is responsible for (1) consuming the messages from these mentioned queues, (2) deploying, modifying or deleting the event patterns and schemas, and (3) processing the simple events received.
4. Complex events generated by the Esper engine are published on another RabbitMQ queue.
5. These complex events are received on the different devices subscribed to the output RabbitMQ queue.

Note that the mentioned software architecture was used to perform different tests, which will be explained below, to prove that our system could be implemented in a real-world water supply network.

Computer Resources

The evaluation tests were performed within the University of Cadiz network, using the following computer resources:

- A PC with an Intel i3-3220T 2.80GHz and 4GB of RAM. This machine was used as the external source, from which our system received the incoming simple events. All data has been emulated using an ad hoc Java-based simulator.
- Another PC identical to the previous one, with an Intel i3-3220T 2.80GHz and 4GB of RAM. This machine hosts a RabbitMQ instance which is used as a broker for the CEP engine input and output events.
- A PC with an Intel i7-3770 3.40GHz and 8GB of RAM. The CEP engine was deployed in this machine. All the performance measurements were taken in this PC.

Methods

The performed tests emulated the real operational conditions of the water supply network to evaluate whether the proposed architecture would be able to support the huge amount of data that it would receive in a real-world scenario. For that, every event pattern described in the Event Patterns section was deployed in the CEP engine.

The water supply network which we emulated is composed of 20,000 water meters. Each water meter sent daily data from nine different magnitudes: (1) daily consumption, (2) starts, (3) sleep time, (4) normal time, (5) leak time, (6) maximum daily flow, (7) minimum daily flow, (8) reverse consumption, and (9) battery level.

Hourly, each water meter sent the value of the hourly consumption, and in the case of the global or section meter (there is only one in the network), the value of the chlorine and the pressure were sent.

This leads to a total of 180,000 events sent every day once a day and 20,002 events sent every hour per day.

Three tests were conducted, with different durations, with the aim of comparing the results and proving the scalability and stability of the system:

1. Stress test: The maximum data load that our system would have to handle was sent only once.
2. Resistance test: The system was running for 3 to 4 hours, with the dispatch of the daily events being followed by several hourly dispatches.
3. Stability test: The system was running for a week to demonstrate that it can be maintained over time without suffering major alterations in resource consumption.

In all three tests, the following performance indicators were measured:

- CPU usage: Percentage of CPU usage by the CEP engine process during the test execution.
- RAM usage: Megabytes (MB) of memory consumed by the CEP engine process along the test execution.

In the last test, the stability one, additionally to the CPU and RAM usage, time performance was also measured, defined by two metrics:

- Processing time: Time taken by complex events to be processed by the CEP engine from the moment the data reaches the CEP engine, measured in ms. This variable is also calculated making use of the *DummyEvents, which are sent* periodically every second interlaced with the rest of the simple events and for which we have a CEP rule that processes them. In such a rule, we compare the timestamp of when the *DummyEvent* reached the CEP engine to the timestamp of when the *DummyEvent* CEP rule ended processing it.

Results

After executing the tests described in the previous subsections, we have analyzed the results as explained below. Please note that all the results are referred to the machine where the CEP engine is deployed, which is the one which might be overloaded by the processing task.

Regarding the stress tests, as illustrated in Figs. 6 and 7, the results obtained are satisfactory. In this particular instance, a single transmission containing all the data the system would typically receive was simulated. Figure 6 reveals a CPU consumption peak of approximately 15 s upon data reception; however, this peak remains below 25% of total CPU capacity. Subsequently, CPU usage drops to near-zero levels. Concerning memory usage, we observe a spike to around 750 Kb at the moment of data reception. This elevated memory consumption is sustained due

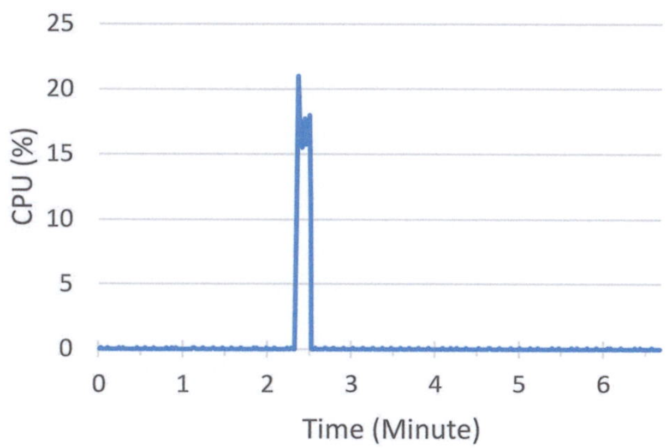

Fig. 6 CPU consumption per second for stress tests

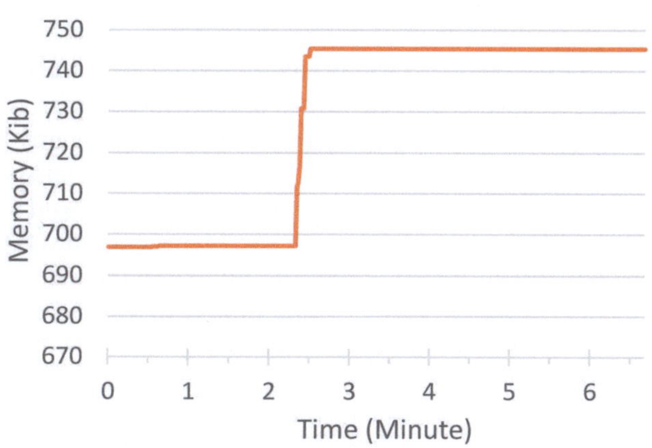

Fig. 7 Memory consumption per second for stress tests

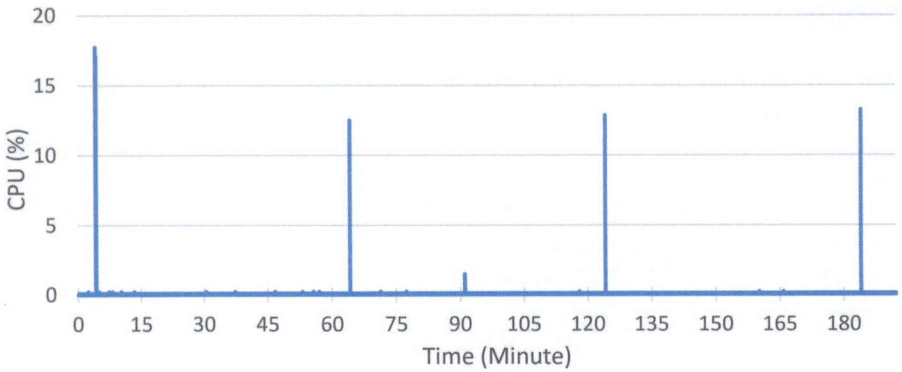

Fig. 8 CPU consumption per second for resistance tests

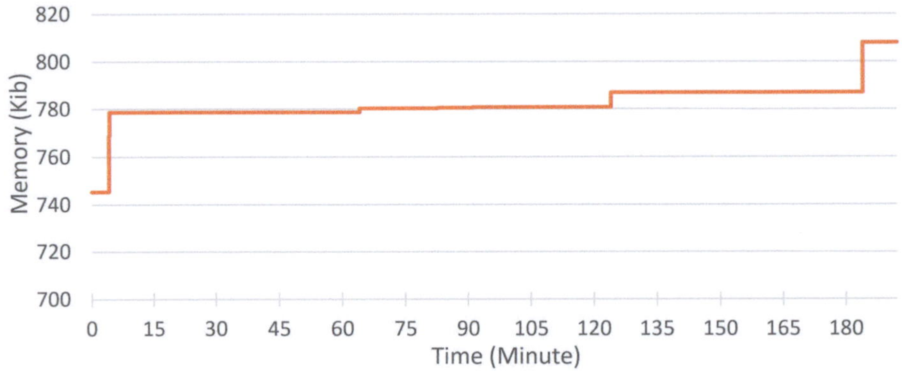

Fig. 9 Memory consumption per second for resistance tests

to the defined temporal windows within the patterns, necessitating the retention of multiple data points in memory for extended periods. Nevertheless, the trend remains stable, as further demonstrated in subsequent longer-duration tests.

Regarding the resistance endurance tests, where a larger data load is initially received (comprising the daily dataset) followed by smaller hourly loads, Figs. 8 and 9 demonstrate once again satisfactory CPU and memory consumption results. Figure 8 reveals several memory peaks primarily associated with the initial data transmission (18% CPU usage) and then with hourly transmissions (up to 13% CPU usage), all of which remain within acceptable limits. Moreover, Fig. 9 illustrates a gradual increase in memory consumption with each hourly interval, as more temporal windows are opened upon receiving hourly data, reaching a maximum of 810 Kb, still within reasonable bounds.

The results of the stability tests can be seen in Figs. 10, 11, and 12, which represents the average CPU consumption per minute, the average RAM consumption per minute, and the time taken to process each event received in the CEP engine, all of

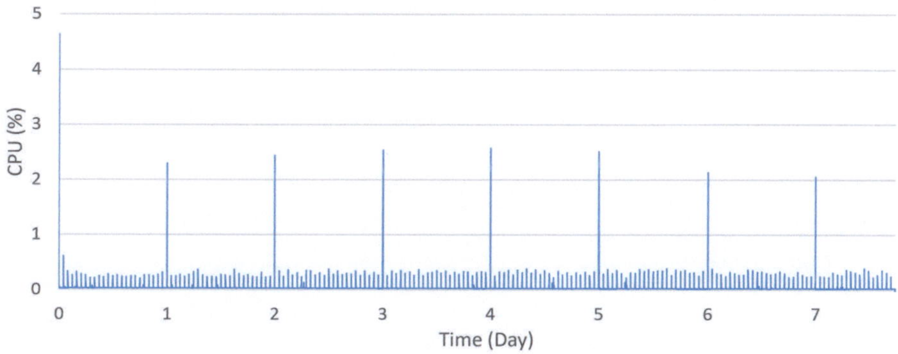

Fig. 10 Average CPU consumption per minute for stability tests

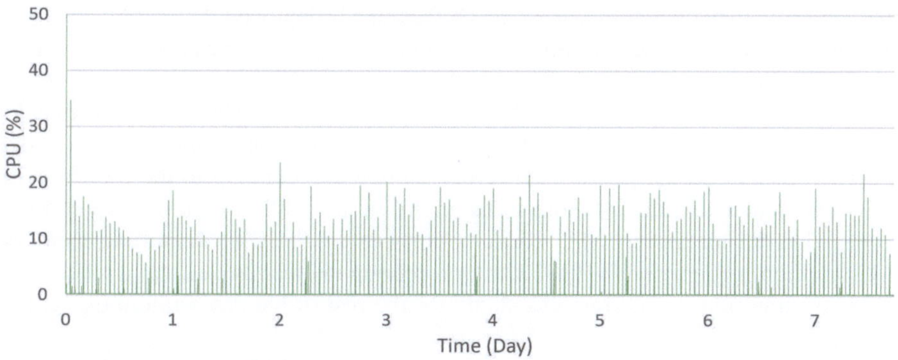

Fig. 11 Maximum CPU consumption per minute for stability tests

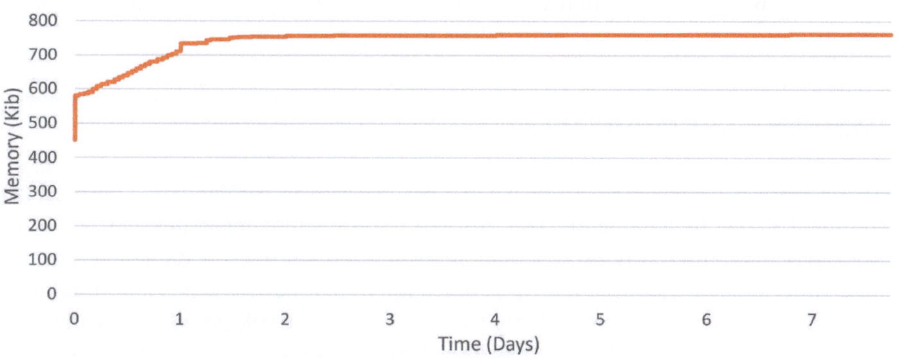

Fig. 12 Memory consumption per minute for stability tests

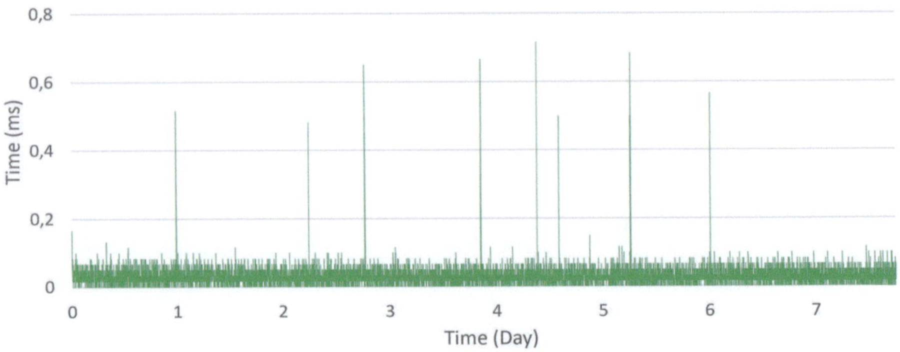

Fig. 13 Average processing time per second for stability tests

them referred to the machine where the CEP engine is deployed. In this occasion, we have grouped the average consumption per minute to provide a clearer visualization of a larger data range.

As shown in Fig. 10, average CPU consumption exhibits a series of smaller peaks corresponding to hourly data reception throughout the 24-h period, not exceeding 1% CPU usage. Larger peaks occur every 24 h, corresponding to the reception of daily accumulated data, reaching a maximum of 3% CPU usage. As we can see in Fig. 11, the maximum CPU consumption can reach 20%. Regarding memory consumption, as depicted in Fig. 12, after receiving the first day's data, it reaches its peak consumption, which remains stable for the rest of the day without exceeding 800 KiB.

On the other hand, with a processing time consistently below 1 ms, the system demonstrates its ability to handle events in real time. As Fig. 13 shows, the average processing time is well below this threshold, with peak values reaching a maximum of 0.7 ms. Considering the total values, rather than the mean, processing time rarely exhibits any significant peaks, with values that are negligible.

Conclusion and Future Work

This chapter has presented a novel CEP-based software architecture tailored for intelligent water management. The proposed architecture leverages the power of CEP engines to efficiently process and analyze large volumes of data generated by smart water meters, enabling real-time detection of anomalies, leaks, and fraudulent activities. A key contribution of this work lies in the development of an intuitive graphical user interface that empowers domain experts to define complex event patterns without requiring programming expertise. This user-friendly approach significantly reduces the time and effort needed to deploy and adapt the system to evolving requirements. Furthermore, the integration of a RESTful API provides a flexible and scalable mechanism for managing and controlling the CEP engine, enabling seamless integration with other systems and applications. The successful implementation and the evaluation of the proposed architecture emulating the amount and frequency of data for a city of 180,000 citizens demonstrate its efficiency and, therefore, guarantees an improvement in water management efficiency, decrease in water loss, and enhancement of overall sustainability.

Future research directions may include providing a tool to model water distribution in the city and exploring advanced machine learning techniques for more sophisticated anomaly detection, as well as investigating the integration of additional data sources, such as weather data and demographic information, to further refine the analysis and decision-making process.

Acknowledgment This work was partially supported by I+D+i grant PID2021-122215NB-C33 (AwESOMe) funded by MICIU/AEI/10.13039/501100011033 and by ERDF/EU, by grant PDC2022-133522-I00 (ASSENTER) funded by MCIN/AEI /10.13039/501100011033 and by the European Union Next GenerationEU/ PRTR, and by I+D+i grant REINFORCED funded by Research Plan of the University of Seville.

Appendix

Pattern ExcessiveNetworkMinimumNightTimeFlow

```
insert into ExcessiveNetworkMinimumNightTimeFlow
select measureDateTime as measureDateTime, serialNumber
   as serialNumber, value as value, unit as unit
from Measure(measureDateTime.between
   (measureDateTime.withTime(1, 0, 0, 0),
   measureDateTime.withTime(6, 0, 0, 0))
   and magnitude=2 and deviceType='IN' and value>3000);
```

Patterns for ExcessiveAverageDailyFlow

```
insert into Addition
select a1.measureDateTime as measureDateTime,
   a1.serialNumber as serialNumber, sum(value) as value,
   unit as unit
from Measure(measureDateTime.between
   (measureDateTime.withTime(9, 0, 0, 0),
   measureDateTime.withTime(13, 0, 0, 0))
   and magnitude=2 and deviceType= 'IN')
   .win:time(4 hour) a1
group by a1.serialNumber;

insert into ExcessiveAverageDailyFlow
select measureDateTime as measureDateTime, serialNumber
   as serialNumber, value as value, unit as unit
from Addition(value > Main.getBoundForWeekDay
   (measureDateTime.getDayOfWeek()));
```

```
public static Integer getBoundForWeekDay (Integer day) {
   if (day == 7)
   {
      return 55000;
   }
   return 80000;
}
```

Pattern HighPressure

```
insert into HighPressure
select measureDateTime as measureDateTime, serialNumber
   as serialNumber, value as value, unit as unit
from Measure(deviceType='CONTROL' and magnitude=4
   and value>40);
```

Pattern LowPressure

```
insert into LowPressure
select a2.measureDateTime as measureDateTime,
   a2.serialNumber as serialNumber, a2.value as value,
   a2.unit as unit
from pattern [every (a1 = Measure(
   a1.deviceType='CONTROL' and a1.magnitude=4
   and a1.value < 35) where timer:within(2 hours)
   -> a2 = Measure(a1.serialNumber = a2.serialNumber
   and a2.deviceType='CONTROL' and a2.magnitude=4
   and a2.value < 35))];
```

Patterns for ExcessiveANR

```
insert into ANRin
select measureDateTime as measureDateTime, serialNumber
   as serialNumber, sector as sector, sum(value) as
   value, unit as unit
from Measure(magnitude=2 and
   deviceType= 'IN').win:time(1 day)
group by sector;

insert into ANRout
select measureDateTime as measureDateTime, serialNumber
   as serialNumber, sector as sector, sum(value) as
   value, unit as unit
from Measure(magnitude=2
   and deviceType= 'OUT').win:time(1 day)
group by sector;

insert into ExcessiveANR
```

```
select a1.serialNumber as serialNumber, a1.sector as
   sector, a2.measureDateTime as measureDateTime,
   1-(a2.value/a1.value) as value, a2.unit as unit
from pattern [every a1 = ANRin()
   where timer:within(1 days)
   -> a2 = ANRout(a1.sector=a2.sector
   and (1-(a2.value/a1.value))>0.30)];
```

Pattern HighChlorine

```
insert into HighChlorine
select measureDateTime as measureDateTime, serialNumber
   as serialNumber, value as value, unit as unit
from Measure(deviceType='CONTROL' and magnitude=3
   and value>1.2);
```

Pattern LowChlorine

```
insert into LowChlorine
select measureDateTime as measureDateTime, serialNumber
   as serialNumber, value as value, unit as unit
from Measure(deviceType='CONTROL' and magnitude=3
   and value<0.2);
```

Pattern LeakTime

```
insert into LeakTime
select measureDateTime as measureDateTime, serialNumber
   as serialNumber, value as value, unit as unit
from Measure(deviceType='OUT' and magnitude=8
   and value/86400>0.05);
```

Pattern NoConsumption

```
insert into NoConsumption
select a2.measureDateTime as measureDateTime,
   a2.serialNumber as serialNumber, a1.value as value,
   a2.value as volume
from pattern [every a1 = Measure(a1.deviceType='OUT'
   and a1.magnitude= 5  and a1.value= 0
   and (a1.contractType=1 or a1.contractType=2))
   where timer:within(2 days)
   -> a2 = Measure (a1.serialNumber = a2.serialNumber
   and a2.deviceType='OUT' and a2.magnitude=1
   and a2.value=0 and (a2.contractType=1
   or a2.contractType=2))];
```

Pattern LowBattery

```
insert into LowBattery
select measureDateTime as measureDateTime, serialNumber
   as serialNumber, value as value, unit as unit
from Measure(deviceType='OUT' and magnitude=12
   and value = 1);
```

Pattern ExcessiveFlow

```
insert into ExcessiveFlow
select measureDateTime as measureDateTime, serialNumber
   as serialNumber, value as value, 0.9*q4 as maxAllowed,
   unit as unit
from Measure(deviceType='OUT' and magnitude=9
   and value>0.9*q4);
```

Pattern ReverseConsumption

```
insert into ReverseConsumption
select measureDateTime as measureDateTime, serialNumber
   as serialNumber, value as value, unit as unit
from Measure(deviceType='OUT' and magnitude=11
   and value > 0);
```

Pattern InternalLeak

```
insert into InternalLeak
select a2.measureDateTime as measureDateTime,
   a2.serialNumber as serialNumber, a1.value as value,
   a2.value as volume
from pattern [every a1 = Measure(a1.deviceType='OUT' and
   a1.magnitude=5  and a1.value = 0)
   where timer:within(1 day)
   -> a2 = Measure(a1.serialNumber = a2.serialNumber and
   a2.deviceType='OUT' and a2.magnitude= 1
   and a2.value > 0)];
```

Pattern AgainstFireWithConsumption

```
insert into AgainstFireWithConsumption
select a5.serialNumber as serialNumber,
   a5.measureDateTime as measureDateTime,
   a5.value as value, a5.unit as unit
from pattern [every (a1 = Measure(a1.deviceType='OUT'
   and a1.magnitude= 2 and (a1.contractType=7
```

```
or a1.contractType =10) and a1.value> 0 )
where timer:within(5 hours)
-> a2 = Measure(a1.serialNumber = a2.serialNumber and
a2.deviceType='OUT' and  a2.magnitude= 2 and
(a2.contractType=7 or a2.contractType =10)
and a2.value> 0)
-> a3 = Measure(a1.serialNumber = a3.serialNumber and
a3.deviceType='OUT' and  a3.magnitude= 2 and
(a3.contractType=7 or a3.contractType =10)
and a3.value> 0)
-> a4 = Measure(a1.serialNumber = a4.serialNumber and
a4.deviceType='OUT' and  a4.magnitude= 2 and
(a4.contractType=7 or a4.contractType =10)
and a4.value> 0)
-> a5 = Measure(a1.serialNumber = a5.serialNumber and
a5.deviceType='OUT' and  a5.magnitude= 2 and
(a5.contractType=7 or a5.contractType =10)
and a5.value> 0))];
```

Pattern CommunicationError

```
insert into CommunicationError
select a1.serialNumber as serialNumber
from pattern [every a1 = Measure()
   -> (timer:interval(2 day) and not
   a2 = Measure (a1.serialNumber = a2.serialNumber))];
```

References

Axiotidis C, Konstantopoulou E, Sklavos N (2024) A wireless sensor network IoT platform for consumption and quality monitoring of drinking water. Discov Appl Sci 7:15. https://doi.org/10.1007/s42452-024-06384-1

Bazan-Muñoz A, Ortiz G, Augusto JC, Garcia-de-Prado A (2024) Taxonomy and software architecture for real-time context-aware collaborative smart environments. Internet Things 26:101160. https://doi.org/10.1016/j.iot.2024.101160

Boubeta-Puig J, Ortiz G, Medina-Bulo I (2015) MEdit4CEP: a model-driven solution for real-time decision making in SOA 2.0. Knowl-Based Syst 89:97–112. https://doi.org/10.1016/j.knosys.2015.06.021

Corral-Plaza D, Ortiz G, Medina-Bulo I, Boubeta-Puig J (2021) MEdit4CEP-SP: a model-driven solution to improve decision-making through user-friendly management and real-time processing of heterogeneous data streams. Knowl-Based Syst 213:106682. https://doi.org/10.1016/j.knosys.2020.106682

Espertech (2024) Esper documentation. https://www.espertech.com/esper/

Farah E, Shahrour I (2024) Use of data-driven methods for water leak detection and consumption analysis at microscale and macroscale. Water 16:2530. https://doi.org/10.3390/w16172530

Gavrilović N, Mishra A (2021) Software architecture of the internet of things (IoT) for smart city, healthcare and agriculture: analysis and improvement directions. J Ambient Intell Humaniz Comput 12:1315–1336. https://doi.org/10.1007/s12652-020-02197-3

GEN (2025) Grupo Energético de Puerto Real. https://www.grupoenergetico.es/. Accessed 11 January 2025

Hu X, Han Y, Yu B, Geng Z, Fan J (2021) Novel leakage detection and water loss management of urban water supply network using multiscale neural networks. J Clean Prod 278:123611. https://doi.org/10.1016/j.jclepro.2020.123611

Ionescu VM (2015) The analysis of the performance of RabbitMQ and ActiveMQ. In: 2015 14th roedunet international conference - networking in education and research (RoEduNet NER), pp 132–137. https://doi.org/10.1109/RoEduNet.2015.7311982

Khan IH, Khan MI, Khan S (2020) Challenges of IoT implementation in smart city development. In: Ahmed S, Abbas SM, Zia H (eds) Smart cities—opportunities and challenges. Springer, Singapore, pp 475–486. https://doi.org/10.1007/978-981-15-2545-2_40

Luckham DC (2011) Event processing for business: organizing the real-time enterprise. Wiley, Hoboken

Ma Z, Liu S, Guo G, Yu X (2022) Hybrid attention networks for flow and pressure forecasting in water distribution systems. IEEE Geosci Remote Sens Lett 19:1–5. https://doi.org/10.1109/LGRS.2020.3030839

Ortiz G, Zouai M, Kazar O, Garcia-de-Prado A, Boubeta-Puig J (2022a) Atmosphere: context and situational-aware collaborative IoT architecture for edge-fog-cloud computing. Comput Standards Interf 79:103550. https://doi.org/10.1016/j.csi.2021.103550

Ortiz G, Boubeta-Puig J, Criado J, Corral-Plaza D, Garcia-de-Prado A, Medina-Bulo I, Iribarne L (2022b) A microservice architecture for real-time IoT data processing: a reusable Web of things approach for smart ports. Comput Standards Interf 81:103604. https://doi.org/10.1016/j.csi.2021.103604

Ortiz G, Castillo I, Garcia-de-Prado A, Boubeta-Puig J (2022c) Evaluating a flow-based programming approach as an alternative for developing CEP applications in IoT. IEEE Internet Things J 9:11489–11499. https://doi.org/10.1109/JIOT.2021.3130498

Ortiz G, Bazan-Muñoz A, Lamersdorf W, Garcia-de-Prado A (2023) Evaluating the integration of Esper complex event processing engine and message brokers. PeerJ Comput Sci 9:e1437. https://doi.org/10.7717/peerj-cs.1437

Preciado JC, Prieto ÁE, Benitez R, Rodríguez-Echeverría R, Conejero JM (2019) A high-frequency data-driven machine learning approach for demand forecasting in smart cities. Sci Program 2019:8319549. https://doi.org/10.1155/2019/8319549

RabbitMQ (2024) One broker to queue them all. RabbitMQ. https://www.rabbitmq.com/. Accessed 18 October 2024

Serban CM, Sebestyen G, Hangan A (2024) Anomaly detection in water consumption patterns using prediction and clustering approaches. In: 2024 IEEE international conference on automation, quality and testing, robotics (AQTR). IEEE, Piscataway, pp 1–6. https://doi.org/10.1109/AQTR61889.2024.10554265

UN (2025) Martin: water and Sanitation. https://www.un.org/sustainabledevelopment/water-and-sanitation/. Accessed 12 January 2025

UNEP (2025) Water action | Countries hunt for novel sources of water. https://www.unep.org/news-and-stories/story/shortages-mount-countries-hunt-novel-sources-water. Accessed 12 January 2025

Smart Sensors for Water Quality Monitoring in Aquaculture Systems

María del Valle Zurita-Lozano, Ángela Écija-Arenas, and María Paz Aguilar-Caballos

Contents

Abstract

The growing population has provoked the need for alternative food sources, increasing the demand for aquaculture products to supplement wild fishing. Traditional aquaculture has evolved to precision aquaculture, where monitoring water quality is essential to maintain adequate production by keeping toxic substances to a minimum. These substances come from animal or plant excretion or feed residues, which may result in high concentration levels in ponds or tanks, thus causing disease outbreaks and reducing farm productivity. This chapter aims to provide insight into the most recent contributions of artificial intelligence and the Internet of Things to sensing platforms for water quality monitoring in the aquaculture industry. The control of some parameters, such as

M. del Valle Zurita-Lozano · Á. Écija-Arenas · M. P. Aguilar-Caballos (✉)
Facultad de Ciencias, Departamento de Química Analítica, Instituto Químico Para la Energía y el Medio Ambiente (IQUEMA), Campus Universitario de Rabanales, Universidad de Córdoba, Córdoba, Spain
e-mail: qa1agcam@uco.es

D. Gutiérrez et al. (eds.), *Smart Water Quality Monitoring*,
https://doi.org/10.1007/978-3-032-03049-8_7

pH, temperature, electrical conductivity, salinity, and dissolved oxygen, seems adequate nowadays, but there are still some challenges regarding pollutant sensing. Only a few examples of smart sensors for hormones and antimicrobials have been applied to aquaculture. Furthermore, only a few examples of on-site sensors for inexperienced end-users have been developed within this field.

Keywords

Precision aquaculture · Water quality monitoring · Sensors · Machine learning · Internet of things

Introduction

United Nations World Population Prospects 2022 foresee a world global population to grow from 8.5 billion people in 2030 to reach 9.7 billion by 2050. These figures have led to looking at the aquaculture industry as an alternative sector to yield plant and animal food to increase production and meet the requirements of the increasing population. Aquaculture increased by 5.1% yearly in the 2000–2020 period, with 87.5 million tons in 2020 (Biazi and Marques 2023).

The strategic guidelines of the European Union (EU) to achieve a more sustainable and competitive aquaculture from 2021 to 2030 (Communication from the Commission to the European Parliament, the Council, the European Economic and Social Committee and the Committee of the Regions 2021) start from the premises of the European Green Deal and the Farm to Fork strategy. EU aquaculture products only represent 10% of the EU consumption, so there is huge potential for a sector that contributes to mitigating climate change, owing to the economy's decarbonization and reducing pollution. This EU strategy has four main objectives: (1) building resilience and competitiveness, (2) participating in the green transition, (3) ensuring social acceptance and consumer information, and (4) increasing knowledge and innovation. To achieve the objective (1), the development of new aquaculture areas is needed, with the subsequent impact assessment on ecosystems, in such a way that monitoring water quality should accompany these designations of new aquaculture areas. The Sustainable Development 2030 Agenda has also established objectives related to the contribution of fishing and aquaculture to empower food safety and nutrition, as well as with the use of natural resources by the main actors of the aquaculture sector so that sustainable development is guaranteed from the economic, social and environmental point of view (Consejería de Agricultura, Pesca, Agua y Desarrollo Rural 2022). Sustainable Development Goals (SDGs) are useful for defining policies, planning, and managing aquaculture. There is a clear relationship between the aquaculture field and SDGs, as established in the current Andalusian aquaculture policies.

Figure 1 shows SDG1 (no poverty), SDG2 (zero hunger), SDG3 (good health and well-being), SDG5 (gender equality), SD8 (decent work and economic growth), SDG12 (responsible production and production), SDG13 (climate action), SDG14

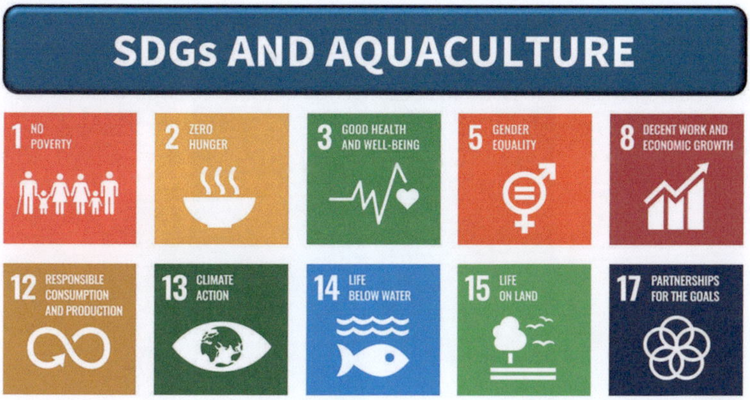

Fig. 1 Main SDGs used to develop Andalusian policies in the aquaculture sector (Consejería de Agricultura, Pesca, Agua y Desarrollo Rural 2022)

(life underwater), and SDG15 (life on land) and SDG17 (partnership for the goals). The Andalusian strategy promotes corporate social responsibility and gender equality, through the mentioned SDGs, specifically SDG1, SDG2, SDG3, and SDG5. Developing new technologies is essential to achieve a constantly increasing competitive and innovative sector that contributes to economic growth (SDG8). The commitment to achieving a sustainable sector result in taking actions to reduce environmental impacts and mitigate climate change, which is aligned with SDG2, SDG13, SDG14, and SDG15. Creating networks among producers, scientists, and consumers provides the framework to establish new alliances to comply with the 2030 Agenda (SDG17). The FAO's strategic framework for 2022–2031 seeks to support this Agenda with the *four betters*: *better* production, *better* nutrition, *better* environment, and a *better* life, leaving no one behind (FAO Strategic Framework 2021). Some attempts have been made to provide observing networks at the European level, such as the JERICO Research Infrastructure (Farcy et al. 2019). This network is composed of research infrastructures, national monitoring programs, and marine industries through their monitoring activities. This network acknowledges the relevance of the investment in monitoring physical, chemical, and biological parameters of complex processes and interactions in the coastal ocean. Recently, a scientific tool has been developed to identify and apply quality indicators to assess the impact on marine Natura 2000 areas (Birkeland et al. 2024). The applied methodology relies on measurements of dynamic physical, chemical, and biogeochemical processes and changes in marine ecosystems to simplify their management and quantify the environmental impact of the aquaculture processes.

Wild fish populations have plummeted due to the overexploitation of global marine reserves. Aquaculture provides a constant food source to millions of people worldwide, thus reducing the pressure on those overexploited areas where conventional fishing may be limited. Water scarcity has driven the switch from traditional low-input, land-based pond systems to open-sea aquaculture, where controlling

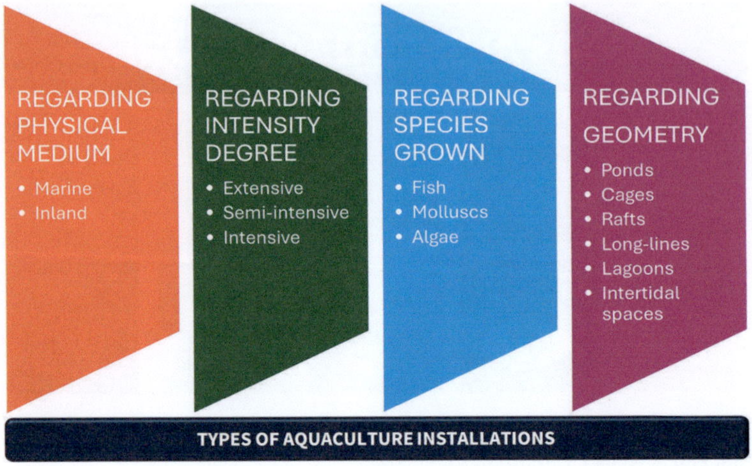

Fig. 2 Categorization criteria for current aquaculture installations

variables is crucial. Thus, the search for sustainability in intensive aquaculture production is leading to a shift from the traditional management of aquaculture systems to intelligent or smart aquaculture farms. Currently, aquaculture farms can be categorized according to the criteria shown in Fig. 2:

Current aquaculture farms can be classified according to four non-exclusive criteria. The first criterion, which is the physical medium, distinguishes between inland and marine farms. Inland farms are defined as those for freshwater systems in contrast with marine farms, which are open-sea installations. The second criterion used is their intensity degree of production, so that farms can be classified as extensive, semi-intensive and intensive installations. The difference among them is how the species grown are fed, ranging from the natural input used in extensive farms to the use of premixed feed typical of intensive installations. The third criterion is the species raised, which can be fish, molluscs, or algae. In some installations, as will be commented below, growing different species together is an option for water regeneration in aquaculture farms, since toxic byproducts for some species are nutrients for other species. The last criterion used is the geometry or shape of the containers where species are grown, so we can have ponds (single or multiple ponds) that are usually made from plastic or concrete, or land-based ponds where ponds are excavated in the land and the water circulation among them is controlled. There is another subtype of land-based ponds where the soil is irrigated, which exist in freshwater farms or inland saline systems located in estuaries or even intertidal spaces.

Nowadays, the aquaculture in Spain uses several types of inland and marine layouts (Acuicultura de España, OESA, 2025), which are depicted in Fig. 3. Single and multiple ponds are normally used for growing different freshwater fish species, whereas cages are used for marine fish. Ponds can be used isolated and/or as part

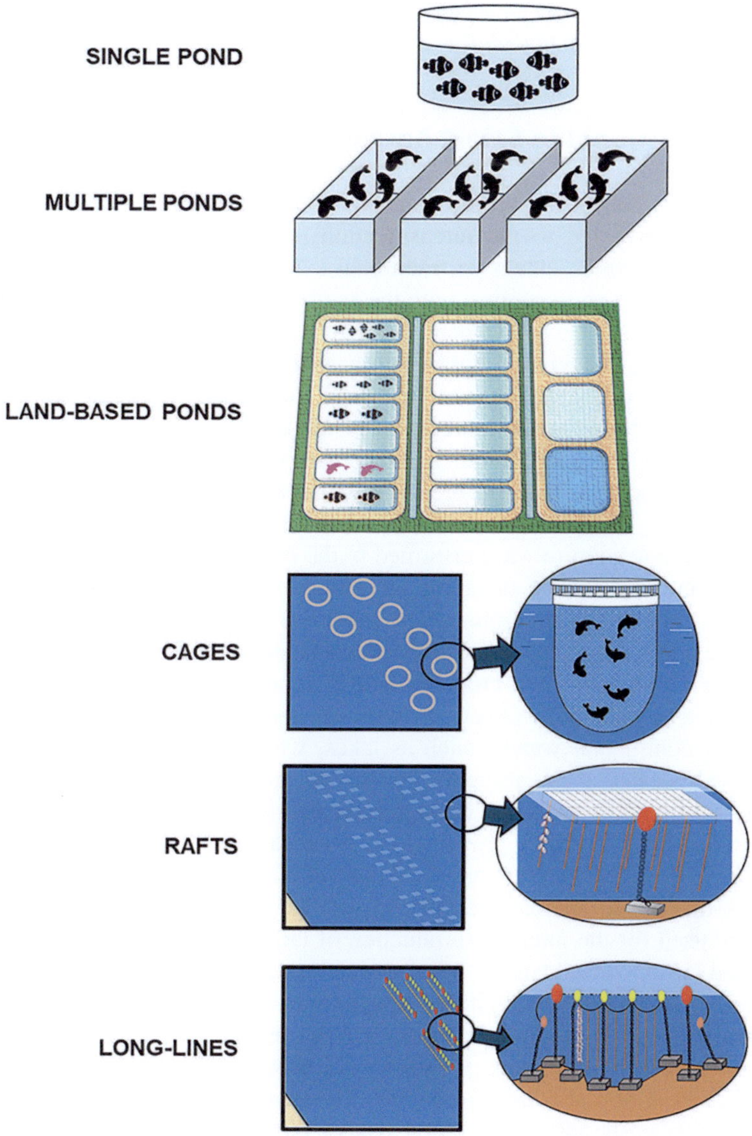

Fig. 3 Common layouts of aquaculture systems used in Spain (adapted from Acuicultura de España, OESA 2025)

of a recirculation system, which enhances water regeneration and decreases the toxicity of fish metabolites. Land-based installations can be useful to raise both fish and mollusc species, which often encompass digging to fabricate pool-like ponds, whereas intercoastal areas are used for growing molluscs. Rafts and long lines are floating devices especially designed for molluscs and algae. Ropes or wires are normally used, although net baskets can also be used at the initial stages. These floating platforms are often near the coast and can be seen from there.

From Conventional to Smart Aquaculture

The increasingly demanding requirements for food, especially for fish, crustaceans, and molluscs, have led to the intensification of aquaculture, with some disadvantages over the often-extensive traditional systems, such as the more efficient production of intensive in contrast with the environmental issues posed by intensive farming, such as effluent water pond management (Edwards 2015). These issues largely depend on the type of aquaculture system, which has evolved from static inland ponds to open sea cages. The term 'traditional aquaculture' refers to using locally available input sources. In contrast, agro-industrial pellets as feed characterize 'modern aquaculture' together with other measures, such as genetic improvement, hormone-induced breeding, therapeutic drugs, vaccines, and other chemicals.

Traditional aquaculture was performed in the past together with other activities, such as agriculture, fishery, or taking advantage of peri-urban wastewater to feed aquaculture farms. These combinations gave rise to different inland aquaculture farms. Rice farms producing a single crop have been modified to fabricate fishponds using dikes so that rice/fish crops integrate, although in many cases rice fields are converted to fishponds. Other conventional aquaculture farms were developed in cages in continental waters, such as lakes, where many eutrophication issues were solved because these nutrients are used to produce fish. These systems have been mostly used for extensive or semi-intensive aquaculture systems, except for trash fish that are intensively grown within these cages. In modern aquaculture, the excess of pelleted feed leads to a lack of control of cage density and eutrophication problems that result in massive fish deaths. Recirculating aquaculture systems (RAS) are used for the intensive production of large amounts of fish within small-size closed tanks. Closed tanks minimize water evaporation issues facilitated by the large surface area of open fishponds, especially in dry areas where water is scarce. Water recirculation is used to remove toxic fish waste from the tank effluent before returning the water to the tank. However, this process generates large amounts of sludge that must be removed periodically. The complexity of operations results in high operating costs and an increased risk of failure for these systems. A recent trend is the use of biofloc fish farms, which rely on microorganisms to provide nutrients while simultaneously remediating the water (Goswami et al. 2022; Yu et al. 2023). Bacterial, algal, or protozoal flocculates within a suspended organic matter water matrix are added to improve waste treatment and to prevent disease outbreaks, without changing the water from the tank. Thus, it can be regarded as an environmentally friendly option since more than 41% of the world's population lives in drought-affected areas (Yu et al. 2023). Nitrates and nitrites are removed by nitrifying bacteria and algae-based photoautotrophic that use nitrification and denitrification through aerobic and anaerobic processes, respectively.

The second big category of aquaculture farms is coastal farms, where cages, rafts, and long lines are the formats used. Cage culture, as mentioned before for inland farms, suffers from the eutrophication issue arising from the toxic waste of fish that diffuses through the cage and is released into the environment. Some alternatives have been to move to more open places, with more space to dilute the effluents, where the costs and risks considerably increase (dilution of feed, variable winds, large waves). These farms are mainly used for producing molluscs and seaweed.

The most recent trends in aquaculture rely on the combination of different strategies to take advantage of the strengths and minimize the weaknesses of the different aquaculture configurations, emphasizing inland production using polyculture or multitrophic options rather than open sea farm options, as global production data and the number of publications suggest (Zhang et al. 2022).

Thus, managing the production of plants and different animal species in aquaculture is connected to the concept of biosecurity, as it happens with agriculture or livestock husbandry (Aly and Fathi 2024). From the point of view of biosecurity management, three different areas of aquaculture can be distinguished: biological, physical, and operational aquaculture (Fig. 4). Biological aquaculture involves using biological means to avoid disease, such as the introduction of disease-resistant species, as well as the enhancement of immunity by probiotics together with the use of immunostimulants. These biological measures are essential for achieving sustainable aquaculture. Additionally, vaccines are preferred over antimicrobial drug treatments, such as antibiotics, as they help reduce the risk of antimicrobial resistance (Reverter et al. 2020). Physical aquaculture is related to the physical barriers that prevent the entry of predators or exotic species, which can lead to fewer specimens because of their predation or the infections or diseases they may spread. Some actions taken are using appropriate fences or other barriers, protection against storms or floods, and isolating different cohorts or batches to separate healthy from

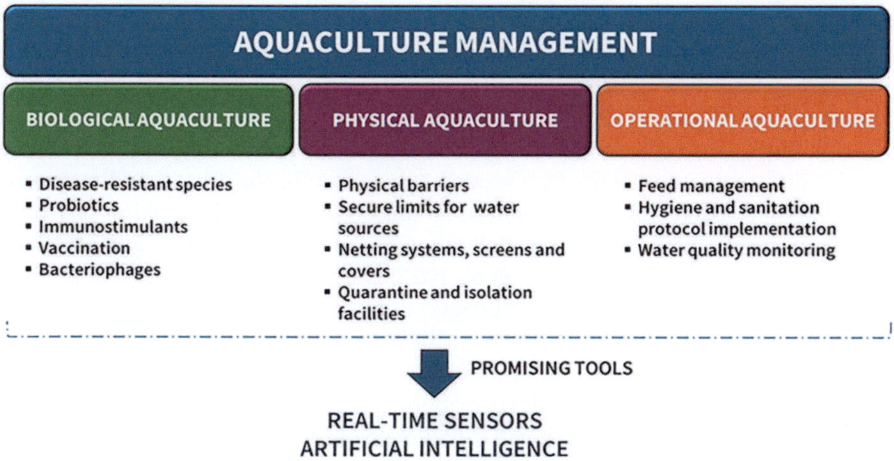

Fig. 4 Areas of aquaculture management (adapted from Reverter et al. (2020))

ill specimens, among other measures. Operational aquaculture involves feeding management, water quality monitoring (WQM), and implementation of hygiene and sanitation protocols. These operations contribute to animal welfare and biosecurity. Contaminant-free environments, together with suitable feed for the species farmed, as well as water quality control, are critical to keeping animals healthy.

The preservation of the environment, animal welfare, and biosecurity, as well as the need to control feeding systems and water quality in real-time, have led to advances in automation, encompassing artificial intelligence (AI).

Smart Aquaculture: Bringing AI and Real-Time Sensing Together

Smart aquaculture, also known as precision aquaculture, combines advanced technology, automated systems, and data analytics following OODA (Observe, Orient, Decide, Act) principles (Biazi and Marques 2023). The observation process is mediated by the data generated by a set of underwater sensing devices and in-farm sensors. The orient step is replaced by interpretation using AI experience-based reasoning, which allows structuring the knowledge to provide relevant information for a correct decision-making process in the decision step. AI tools and information technologies supplement the interpretation step with prediction models. Finally, the act process can be automated with mechanized devices for specific operations within the smart aquaculture farm.

According to these facts, AI plays an important role as a prediction and control tool in aquaculture systems, where computer systems replace human intelligence to perform some tasks by emulating human intelligence functions, such as perception, learning, reasoning, problem-solving ability, and language (Mandal and Ghosh 2024). The meaning of 'Smart Things' is wider than that of the term 'Internet of Things' (IoT), since the first ones provide intelligent and responsive actions, unlike the latter that are more restricted to data handling, collection, and sharing. These terms are sometimes used indistinctly, so the nomenclature of this rapidly increasing field is somewhat ambiguous in existing literature.

Smart aquaculture schemes usually integrate an AI subset to develop brain-like logical structures that can learn without previously programmed routines. AI systems have a key role in the development of smart monitoring systems, where fit with different sensor units (the second subset) are useful for several purposes, encompassing fish feeding, fish counting, monitoring fish welfare, and/or physical-chemical sensors for WQM (Hu et al. 2022; Chen et al. 2022; Sriram and Sai Shibu 2021; Goswami et al. 2022; Zuriati et al. 2022; Darmalim et al. 2021; Abdurohman et al. 2022; Kodali and Sabu 2022; Lozano et al. 2022; Chang et al. 2022; Rao et al. 2022; Yamabe et al. 2021; Rossi et al. 2022; Damiano et al. 2020; Li et al. 2021; Curley et al. 2021; Li et al. 2022). Several AI tools and methodologies are used to support farmers with data analysis, pattern recognition, and predictive analysis. Machine learning, deep learning, natural language processing, and computer vision are the most frequently used AI technologies in aquaculture, individually or as dual AI systems.

Intelligent Feeding Systems

Deep learning is the AI largely used in aquaculture in different fish farms, many of which use computer-vision systems to monitor feeding systems in intensive aquaculture (Son and Jeong 2024; Hu et al. 2022; Zuriati et al. 2022; Chen et al. 2022). These systems allow monitoring of the amount of feed remaining in ponds or tanks to avoid misuse of pelleted feed and water pollution owing to the nutrient excess. However, water turbidity, the small pellet size, intense fish activity, and fish and feed overlap may preclude an adequate analysis of the images taken. A new approach that monitors the waves of the fish-eating feed by using deep learning technology has been developed (Hu et al. 2022). This methodology, which can incorporate the information provided by sensors for WQM, is suitable to automatically monitor the feeding process because measurements can be taken at the upper layers of the water mass. The image sensing module uses a white light image sensor (an ultraviolet and anti-salt camera) with a full HD 1080p resolution, with a high sensitivity CMOS detector. The camera is fitted with 18 IR LEDs that can penetrate fog and get to a 20-m distance. Regarding the deep neural network used, an R(2 + 1)D architecture is fitted with an Adam optimizer. The signal processing module is designed to control the feeding process automatically and remotely using a 3G/4G wireless mobile network. Feed distribution portions within fish farms have been automated using a two-module system based on the use of a convolutional neural network (CNN) model to monitor fish growth and a second module for prediction of portions, so a time-series data is generated and processed using a gated recurrent unit (GRU) neural network. The first module takes images of fish using an underwater camera. It processes these images to train the CNN neural network to send data on fish growth to the second module, which processes them together with the time-series data collected to adjust the feed portions.

Homemade seawater aquariums have also been automated to make fish feeding easier for owners who want to remotely control the conditions of the aquarium without investing too much time in the aquarium conditioning and keeping (Chen et al. 2022). This system has been developed to monitor *Hippocampus* seahorses as a fish breeding model since it is a very delicate species to be raised. The system is composed of two webcam image sensors and five sensors for WQM. The system also contains actuators that allow taking the frozen feed with a robotic arm and adding it to a first container to de-ice the feed before its addition to the aquarium to prevent changes in water temperature within the aquarium. The image recording system takes images from seahorses to build a recognition system of external symptoms of seahorse disease using a You-Only-Look-at-Once version 3 (YOLOv3) object detector incorporating different algorithms based on a CNN and further residual neural network, together with a feature pyramid network (FPN) to improve the ability to detect small objects. Seahorses can be cured if the changes in the aquarium conditions are adequately controlled. This completely automated system, in addition to automatically measuring water quality and adding the feed to the aquarium tank, is capable of performing automatically different actions: (1)

measuring the water level by using an ultrasound sensor, (2) water purification by a reverse osmosis filter, (3) a heater to keep appropriate water temperature, (4) an oxygen pump to increase the water oxygen content, (5) pumping and draining motors to lower and raise the water level, (6) changing the water by using a container to store the water for the change, (7) UV water cleaning with a germicidal lamp, and (8) real-time underwater monitor to take images of seahorses inside the aquarium.

A decision tree regression algorithm has also been used to provide an IoT aquarium with decision capabilities for automatic feeding (Abdurohman et al. 2022). The algorithm allows the actuators to control in response to the information sent by water quality sensors through the IoT module. The conditions of the model are limited to static environments at small aquariums, which restricts the practical usefulness of this smart system. Amazon Web Service (AWS) platform has also been used to gather data from WQM sensors and to automatically control the aquarium remotely (Kodali and Sabu 2022). Fish feeders and aquarium lights are the two systems that have been checked for suitability of the AWS platform.

Automatic feeding systems have also been used in aquaponics by automating the supply of fish excrement to supplement plant feeding systems. For instance, IoT has been used to build a platform to implement an automated aquaponic system with real-time remote monitoring to keep the optimum conditions for integrating fish and plant farming (Sriram and Sai Shibu 2021). Temperature and pH sensors together with a water level sensor, a hydro water pump, and an auto feeder are controlled using an ESP8266-NodeMCU module with a Wi-Fi connection to transmit the control signals. The hardware cost (sensors, microcontroller, pump, time module, and servomotor) is about 45 USD. No AI techniques are incorporated to study the system behaviour.

Intelligent Systems for Detecting Fish Activity and Welfare

As mentioned before in this chapter, ensuring an adequate number of specimens in cages are essential since an excess of individuals can lead to a high mortality rate owing to disease outbreaks because of poor water quality. Also, animals in crowded cages are subjected to stressful conditions that affect their development and quality (Li et al. 2022). Machine vision has been used at the aquarium level to detect sea cucumber individuals infected with *Vibrio alginolyticus* (Xu et al. 2023). A behaviour analysis based on Detection and Tracking-You Only Look Once version 5 (DT-YOLOv5) and automated frame-matching coordinates for multiple objects was developed to achieve a qualitative and quantitative analysis of the infected population based on the trajectories of multiple objects. Infected sea cucumbers have different volumes of exercise in addition to shaking head and skin ulceration compared to healthy specimens.

Large-volume aquaculture systems, such as offshore cages, incorporate fish populations of heterogeneous sizes. Image processing has also been reported to develop biofloc farms incorporating RAS (Goswami et al. 2022). These environments often suffer from low-ambient light or turbid environments at underwater conditions that make fish detection difficult using only image sensors, so other detectors such as sonar (Rao et al. 2022; Chang et al. 2022) or accelerometers (Lozano et al. 2022). A two-mode system has been used to develop an underwater sensor for objects combining vision and acoustic detection systems (Chang et al. 2022). RGB image cameras can hardly estimate fish metrics, unlike sonar camera systems. Thus, the integration of the information of both methods is possible when they are working in the same coordinate frame to obtain 3D complementary information. CNNs can estimate the length and weight distribution of specimens in the tank, removing the overlaps that can be present in a crowded tank or cage. Gaussian mixture models are used to estimate the length of fish distribution. A challenge in using sonar images in very large volumes is the narrow frame of sonar imaging detectors. This issue can be solved using segmentation networks for estimating the size distribution and number of specimens in a certain group inside the considered segment. This bimodal system was checked in closed tanks and offshore cages where vision is challenged by the low ambient light, thus showing that different neural networks and Gaussian models helped to overcome the specific problems of vision and acoustic sensors within the different aquaculture systems assessed. Bimodal systems have also been used to predict relevant events and behaviour in crustacean farms (Li et al. 2021). This review paper discusses the advantages and limitations of acoustic, vision, and electrosensors, highlighting the application of each type of sensor. Electrosensor devices implanted in fish individuals, such as accelerometers and gyroscopes, measure metabolic activity, respiratory frequency, and fish swimming movement (Lozano et al. 2022). The data are processed using microcontrollers that provide information on individuals' IDs and the measured variables. This system is somehow invasive, so its use in large-volume systems with many specimens seems unfeasible.

Smart WQM Sensors

WQM is essential to ensure an optimum yield and capacity in aquaculture farms. Intelligent or precision aquaculture relies on real-time monitoring of water's physical-chemical properties through sensors controlled by an Arduino card and connected to a Wi-Fi device, which sends the data obtained and the results to the cloud and store them, where human or computer operators analyse them and make decisions to keep the aquaculture farm working under optimum conditions. The wide variety of sensors for WQM leads to different configurations depending on the sensor used and the connection capabilities of the location where they will be installed. In all instances, these sensors are a solid alternative to conventional laboratory or in situ methods, especially in difficult-to-reach areas and for long-term measurements, where machine-to-machine communication happens without the need for human intervention. The main requirements for these type of systems have been defined (López-Ramírez and Aragón-Zavala 2023) as quality of service (QoS) parameters: (1) to be energetically efficient, (2) data are effectively transferred, (3) communications have a wide coverage, (4) data are transferred at real-time, (5) costs are decreased, and (6) systems are flexible, so that new sensors can be incorporated to the sensing architecture. Thus, the critical points of these smart WQM sensor developments are tied to meeting these requirements, in addition to developing appropriate sensing units. Since WQM supports all the processes within smart aquaculture farms, the specific hardware (sensors) and information technologies used will be analysed in more detail in the next section of this chapter.

Intelligent Sensors for WQM

WQM is the core function of aquaculture because it provides the basis for the growth of healthy individuals within the aquaculture farm. All intelligent systems described in the literature for small to medium-sized aquaculture farms incorporate a sensor unit to control the standard parameters defined for water quality. Some European projects, such as AGILE, were able to introduce successful tools for WQM sensing with enhancements, such as a real-time monitoring interface and alert systems to help farmers control production without constant attention (Lafont et al. 2019). Regression analysis was used to develop dissolved oxygen sensing systems that could predict the behaviour of the system at least 24 h before (Hsu et al. 2020; Manoharan et al. 2020; Chatziantoniou et al. 2022).

Recent smart sensors for aquaculture are composed of six subsystems or layers: physical (sensors and microcontroller); network layer (Wi-Fi network, Internet and/or LoRa), cloud layer (integrated by user profile, historical data, and real-time sensor data), service layer (sensors or actuators mapping, storage and data management, and service applications), machine learning layer, and application layer (web and mobile App, data dashboard, and data report and notification) (Ahmed et al. 2024). This architecture has been used to improve the performance of a shrimp farm using machine learning analysis through six different classification approaches: logistic regression (LR), decision tree (DT), random forest (RF), multilayer perceptron (MLP), Adaboost, and eXtreme Gradient Boosting (XGB) to predict the production at three different levels: minimum, medium, and maximum with a 94% of success rate. Afterward, the GridSearchCV application is applied to determine the best parameters for each model. The system performance was checked among farmers to gather their opinions about the functionalities of the system, with positive aspects (ease of use, the alert system allows easy optimization) and challenges (initial costs of the system, unreliable Internet connectivity). The use of deep learning techniques has been described as having limitations, such as data overfitting and data interpretation. A method involving the use of Dilated Spatial-temporal Convolution Neural Network (DSTCNN) has been reported to overcome these difficulties in water ponds used in aquaculture (Arepalli and Jairam Naik 2024). This model can report the changes in water quality over time and space in more than one place at the same time. With the proposed model, a 99.28% overall accuracy and a 99.27% precision are obtained in predicting the mortality and growth of fish.

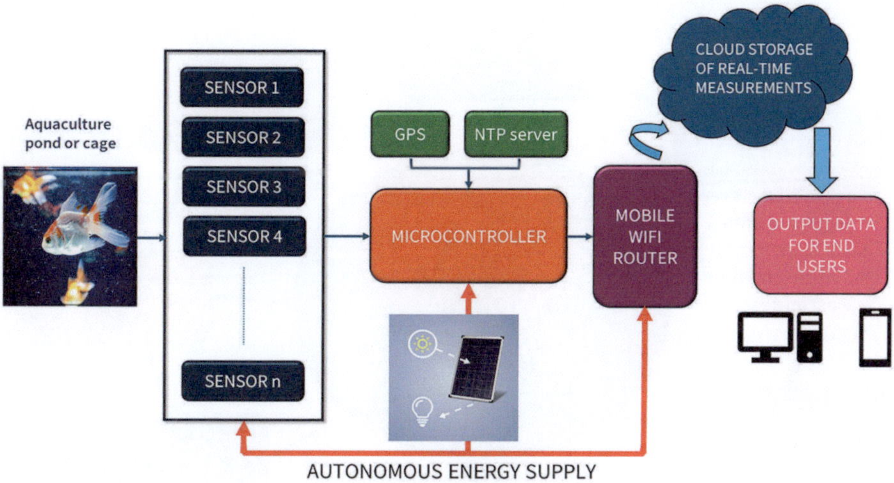

Fig. 5 General scheme of IoT sensors for WQM

The introduction of AI to make predictions from gathered data from WQM sensors is quite recent; most configurations already published rely on the use of IoT systems, where the data measured by WQM sensors are gathered and stored in the cloud using the Internet supplied by wire or a Wi-Fi system. The typical scheme of IoT sensors is depicted in Fig. 5:

These IoT sensing systems rely on the use of physical sensors connected to a microcontroller. These sensors are usually located in adequate places in the aquaculture system to ensure continuous real-time monitoring of water parameters. Once data are collected, they are transmitted by a wireless network to a server that allows their cloud storage and proper handling to deliver the information to end users through web or mobile applications. The main features of each component of IoT sensors are described below.

Regarding sensors, Fig. 6 shows the main real-time sensors described in the literature for this purpose. The most frequent sensors are pH, temperature, conductivity, and dissolved oxygen, although total dissolved solids and nutrient concentration are also important to monitor potential eutrophication issues due to excess feed in the medium. Sensors for pollutants are important to monitor the influents on aquaculture farms and to control the effluents to be delivered to natural water resources. Some effluents could deliver antimicrobials or other products from therapeutic monitoring of individuals in a cage or pond. For marine aquaculture, these pollutant sensors may allow for checking the presence of specific discharges that could affect the specimens' growth, or the food safety of the products finally developed in the aquatic environment.

There are optimum values for the main variables: pH, temperature, conductivity, salinity, and dissolved oxygen since they regulate important processes in aquaculture. For instance, pH greatly influences the solubility of nutrients and

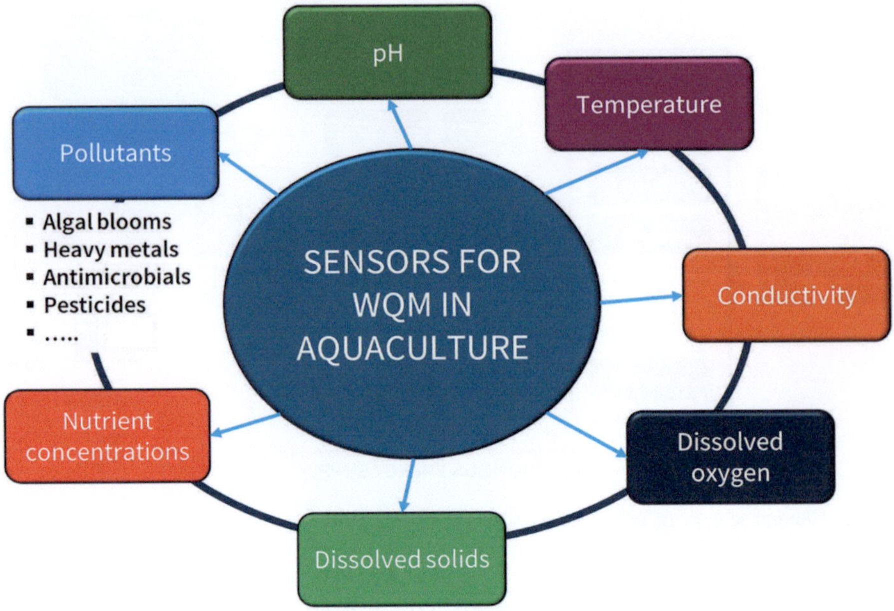

Fig. 6 Main types of sensors for WQM in aquaculture

harmful substances. Plant and animal metabolic rates, reproductive functions, and growth rates also depend on pH values. Values out of the optimum values will negatively affect aquaculture processes. Temperature affects the gas solubility so that at high temperatures, oxygen levels decrease, thus leading to hypoxia and stress and providing an adequate medium for the proliferation of algal blooms. The literature on IoT sensor systems drastically increased from 2019 until now, where the developments range from basic IoT systems without application to IoT systems applied to real aquaculture systems. As can be seen, pH, water temperature, humidity and temperature, dissolved oxygen sensors, electrical conductivity, and total dissolved solids are the most frequently determined parameters. Most sensors at this stage are commercial sensors, unlike those for pollutant detection, which is an emerging field with homemade applications under development. The information in Table 2 does not aim to be fully exhaustive because, in some articles, the information on sensors is missing.

Conductivity sensors are useful and can be modified to assess the salinity of a system, which has a definitive influence on plant and fish osmoregulation, especially in shrimp and fish aquaculture. Other parameters analysed are the amount of dissolved solids that influence light penetration and availability and, thus, photosynthetic processes and feeding patterns. Nutrient concentrations, such as nitrates and nitrites, influence eutrophication processes, so they must be kept to a minimum to ensure that undesirable effects are not present in the aquatic environment. For pollutant determination, real-time sensors are still in their infancy,

Table 1 Reference values for the most common WQM parameters

WQM parameter	Reference value
Temperature	25–32 °C
pH	6.5–8.5
Turbidity	0–25 NTU
Dissolved oxygen (DO)	>5 mg/L
Total dissolved solids (TDS)	400 mg/L
Total suspended solids (TSS)	<80 mg/L
Electrical conductivity (EC)	150–500 μS/cm
Hardness	>15 mg/L
Alkalinity	50–300 mg/L
Nitrite (NO_2^-)	0.2 mg/L
Nitrate (NO_3^-)	0–100 mg/L
Total ammonia nitrogen	0–0.2 mg/L

and their connection to provide IoT sensors useful in aquaculture is scarce, as will be commented below (Zakaria et al. 2021; Kanwal et al. 2024; Prathibanandhi et al. 2024). Table 1 shows a list of the reference values for the common WQM parameters encountered in the literature (Kanwal et al. 2024):

Most of the IoT approaches described use commercial sensors, summarized in Table 2. These sensors are mostly analog, so analog-to-digital converters must be used to convert analog signals into electrical signals.

The pH sensors mainly consist of pH electrodes and ISFET detectors. They work as those employed in pH meters and the precautions must be the same as for these instruments. The water temperature probe DS18B20 has been reported in many articles, probably owing to its operation simplicity, which involves dipping the metal end into water. Temperature changes modify electrical resistance value, with negative temperature coefficient (NTC) being the most common measurement mode. The DTH11 atmospheric humidity and temperature digital sensor has also been used to monitor environmental conditions in aquaculture systems. The variety of dissolved oxygen sensors seems to be higher, with analogic sensors as probes with dynamic ranges in the range of mg/L O_2. These probes are adequate for fresh and marine waters and are based on the diffusion of oxygen through a polyethylene membrane. Electrical conductivity detectors are mostly based on a probe that measures the amount of electricity that flows through a solution after applying an electrical current. Total dissolved solid detectors also measure the conductivity of the ions coming from dissolved solids measured in ppm, according to manufacturers.

Several IoT systems were developed some years ago to demonstrate the feasibility of sensor integration, but their application to aquaculture systems was not assayed (Rosaline and Sathyalakshimi 2019; Teja et al. 2020; Adiono et al. 2021; Tsai et al. 2021; Nayoun et al. 2024; Iniyan Arasu et al. 2024; Baena-Navarro et al. 2025). In these systems, sets of three to five sensors were used to build monitoring systems, with dissolved oxygen sensors as a common element. Some

Table 2 Commercial sensors described in the literature for WQM in aquaculture

Sensor		Manufacturer	Working range	References
pH sensors				
	4502C+E-201C Analog	VOKTTA GROVE – Dyimore	0–14 (±0.1 pH) 0–14 (±0.15 pH) 0–14 (±0.15 pH) 0–14 (±0.15 pH)	Tendolkar et al. (2021) Khaoula et al. (2021) Huan et al. (2020) Kanwal et al. (2024)
	SEN161-gravity Analog	DFROBOT	0–14 (±0.1 pH)	Nakajima et al. (2022)
	ENV-50-pH Analog	Atlas Scientific	0–14 (±0.002 pH)	Quevy et al. (2023)
	Oxyguard Analog Dissolved O_2 Temperature	Oxyguard	0–14 −5–45 °C	Jeon et al. (2021)
	SeapHOx ISFET sensor multiparameter	Sea bird scientific	6.5–9 (±0.05 pH) 120% surface Saturation O_2 −5 to 45 °C	Bresnahan et al. (2020)

Water temperature sensors				
	DS18B20	Maxim integrated	−55 to 125 °C	Tsai et al. (2021) Singh et al. (2022) Adiono et al. (2021) Nakajima et al. (2022) Tendolkar et al. (2021) Khaoula et al. (2021) Boonsong (2021) Chen et al. (2022) Goswami et al. (2022) Abdurohman et al. (2022)
	MAX6675 for k-thermocouple	Robotland Analog	0–800 (±0.25 °C) 0–1024 (2 °C)	–
Humidity and temperature sensors				
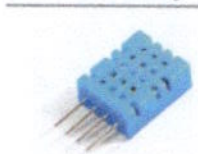	DGT11 digital	–	0–50 (±2 °C) 20–90 (±5% RH)	Kodali and Sabu (2022) Adiono et al. (2021) Teja et al. (2020)
Dissolved oxygen sensors				
	SEN0237 Gravity series Analog	DFROBOT	0–20 mg/L	Nakajima et al. (2022)

(continued)

Table 2 (continued)

Sensor		Manufacturer	Working range	References
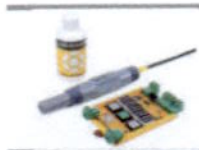	KIT 106D	Atlas Scientific	0–32 (±0.1 mg/L)	Boonsong (2021)
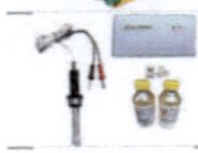	LJ 831	Lab Junction	0–20 mg/L	Tendolkar et al. (2021)
	DO-600	Jieotwice	0–20 mg/L	Huan et al. (2020)
Electrical conductivity sensors				
	SEN0244 Gravity series	DFROBOT	0–2000 µS/cm	Adiono et al. (2021) Nakajima et al. (2022)
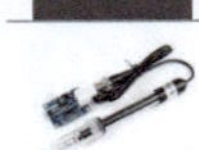	EC sensor	GROVE	0–2000 µS/cm	Khaoula et al. (2021)
Total dissolved solid sensors				
	TDS	GROVE	0–1000 ppm	Khaoula et al. (2021)

interesting applications were developed when the technology was more mature. These applications encompassed inland aquaculture made in ponds (Wiranto et al. 2015; Huan et al. 2020; Singh et al. 2022), where the WQM monitoring of shrimp farms was performed (Wardhany et al. 2021) or marine aquaculture. The usefulness of different sensor batteries was also checked in RAS aquaculture (Tendolkar et al. 2021; Jeon et al. 2021). These approaches have been applied in farms that grow different species, such as mud crabs (Boonsong 2021; Nakajima et al. 2022), and shellfish (Bresnahan et al. 2020). Other parameters, such as carbon dioxide (Mendes et al. 2019) and nitrite (Akhter et al. 2021), have also been assessed in aquaculture systems. Nanomaterials, such as multiwalled carbon nanotubes, have been used to develop an interdigitated electrode for nitrate detection in the range of 0.01–30 mg/L, which was prepared by a mould-based solution casting method. This sensor was integrated into a smart system, where data were transmitted using an IoT system, and machine learning was used to train the microcontroller system to measure temperature and nitrate concentration in water. The results obtained for different water samples were compared to those provided by a reference method with 95% accuracy.

Pollutant determination using smart sensors is a field still under development. A recent publication has reviewed some of the approaches of smart optical biosensors and devices in aquaculture (Simone Soares et al. 2024), where the scarce development of sensors for determining hormones related to fish stress, such as cortisol, is acknowledged. The same happens to bacteria, water pollutants, such as antimicrobial or other emerging pollutants, and micro- and nano-plastics. The paper presents some point-of-care methods, but no advances in smart sensors for these pollutants have been presented so far. Another review paper highlights the role of image analysis and vision techniques and highlights molecular biology methods as confirmatory methods (Bohara et al. 2024).

The decreased use of antimicrobials in aquaculture was recommended in July 2019 (Aquaculture Advisory Council 2019), where the need for vaccine development was emphasized. However, as recognized in the letter to DG Sante from the Finfish Working Group, this is still far from reality, and producers still need to use antimicrobials (Aquaculture Advisory Council 2022). Thus, according to the current strategic guidelines, the control of the supply chain to ensure traceability is essential, as well as the availability of analytical methods to monitor these treatments. The use of IoT remote sensing platforms is essential for real-time WQM systems. A sensor array based on molecularly imprinted polymers and machine learning has been developed to determine fluoroquinolone antibiotics (Wang et al. 2022). An electronic tongue has been developed by using gold-nanoparticle decorated multiwall carbon nanotubes. The method was built and validated using artificial neural networks, but it was applied to pharmaceutical and human urine samples. Another electrochemical method uses a glassy carbon electrode modified with electrochemically reduced graphene oxide to determine ofloxacin and norfloxacin (Wang et al. 2023). This electrochemical sensor was applied to aquaculture wastewater with satisfactory results. A gold-nanoparticle immune strip has been developed to simultaneously determine 83 antibiotics by using generic antibodies for 5 antibiotic classes (Lei et al. 2023). However, the couplings of these latter methods to smart systems have not been reported so far.

Algal blooms also pose a risk in aquaculture systems, so it is desirable to have sensing systems that can prevent their growth. Recently, a multiplex sensor named Tri-Lux has been used in marine aquaculture farms in South Africa to monitor key parameters, such as chlorophyll a, phycoerythrin, and turbidity (Ajmal et al. 2024). This three-parameter sensor was combined with machine learning approaches to forecast phytoplankton activity in aquaculture farms.

These sensors are controlled by digital microcontrollers, where the instructions given are automated and include analog-to-digital converters to change analog signals to digital ones. Only a few commercial sensors are digital, which involves transferring the signal directly to the cloud. Microcontrollers in WQM WSN must meet some requirements(López-Ramírez and Aragón-Zavala 2023): (1) low power consumption, (2) processing and storage capacity to process and store collected data before sending them, (3) connectivity to sensing units and (4) to wireless communication units, (5) control capacity over the sensing hardware, and (6) security capabilities for data protection. Arduino 1 has been the microcontroller type mostly used so far (Hsu et al. 2020; Wardhany et al. 2021; Tendolkar et al. 2021; Nakajima et al. 2022; Kanwal et al. 2024; Arepalli and Jairam Naik 2024; Prathibanandhi et al. 2024). In some instances, data were sent by Bluetooth to a Raspberry Pi device that worked as a server (Bhawiyuga et al. 2019; Rosaline and Sathyalakshimi 2019; Hsu et al. 2020; Nakajima et al. 2022). The microcontroller unit is sometimes completed with MCU ESP8266 to give rise to a microcontroller node (Boonsong 2021; Khaoula et al. 2021; Wardhany et al. 2021; Uddin et al. 2022; Arepalli and Jairam Naik 2024). Recently, ESP8266 has been replaced by ESP32 MCU (Teja et al. 2020; Tsai et al. 2021; Quevy et al. 2023; Kanwal et al. 2024; Prathibanandhi et al. 2024). Another type of microcontroller, namely, STM32, has been used to connect the sensors directly to the cloud (Huan et al. 2020; Ahmed et al. 2024). This latter microcontroller type has been reported to be energetically more efficient than Arduino or Raspberry Pi microcontrollers (López-Ramírez and Aragón-Zavala 2023). It also has good memory capacity and compatibility with wireless standards. There are some other specific microcontrollers, such as ATmega, SAMD, and RENESAS, that have shown good reliability and performance for WQM in hostile environments. Arduino and Raspberry Pi microcontrollers also have higher costs and sizes than these specific microcontrollers. Nevertheless, most articles published so far in the aquaculture field do not report the extensive use of these specific microcontrollers. Typical memory capacities of the commercial microcontrollers are 2kB of RAM, 32kB of flash, and 1kB of Electrically Erasable Programmable Read-Only (EEPROM) memory; the capacity thus depends on the number of nodes and the data volume of the applications considered. The transmission happens through a wireless network to yield the so-called *Wireless Sensing Network* (WSN) (Biazi and Marques 2023), where different sensors work dispersed in several places of a location (the aquaculture farm in this context) to exchange their information at real time so that their data are sent through the wireless network to be stored in the cloud.

The sensor calibration is an important aspect to ensure data quality of smart sensors (Kinar and Brinkmann 2022; Naloufi et al. 2024; Bhati et al. 2024; Shete et al. 2024; Bakar et al. 2025). Although WSNs present many advantages, such as cost-effectiveness and real-time monitoring, calibration remains an issue that impacts the accuracy of sensor measurements (Naloufi et al. 2024; Bakar et al. 2025). An extensive study on the long-term stability of WSNs for WQM in urban rivers (Naloufi et al. 2024) showed that low-cost IoT sensors for dissolved oxygen, conductivity, temperature, and pH worked similarly to other laboratory-based methods. Changes in the temperature for dissolved oxygen and pH lead to anomalous values because the internal corrections of the probes were not enough, so temperature compensation was required. A cleaning step and weekly calibration were needed to avoid signal drift. The results obtained for the turbidity sensor showed that the sensor response is affected by the sample matrix composition, showing poorer performance in freshwater systems because the sensor used in the study was developed for use in mariculture systems. Sensor biofouling and debris deposition are some factors that can prevent sensor performance. Precision and stability are two properties usually checked to assess the suitability of the WSN to the analysis of urban river waters (Kinar and Brinkmann 2022; Naloufi et al. 2024; Bhati et al. 2024; Bakar et al. 2025) and for aquaculture systems (Shete et al. 2024). Regarding cost-effectiveness, a WSN sensor module may cost less than €400, while, e.g., the price of some laboratory devices, such as Hydrolab multiprobes, is about €3000.

Another important factor is WSN communications. Figure 7 shows the three communication tasks in IoT systems, which can be extended to the smart sensors used for WQM in aquaculture. As aforementioned in this chapter, IoT communication modules for WSN in WQM must be chosen regarding QoS parameters (López-Ramírez and Aragón-Zavala 2023). Energy consumption is a critical factor because these sensors operate in an autonomous regime where the electricity supply relies on solar cells or batteries. Those energy sources must be replaced by operators, so their durability is a factor that considerably influences sensor maintenance routines and the costs of the overall WQM process. Similarly, the communication coverage and bandwidth must be chosen depending on the aquaculture farm location to achieve fast and reliable data transfer. The WSN choice will depend on the type of farm, the water physical-chemical properties, and the analyte to be monitored. For chemical measurements that are usually complex, controlling accuracy and precision through calibration and stability studies is essential for the correct monitoring of the aquaculture process.

Low Power Wide Area Networks (LPWANs) are powerful tools for remote sensing. Long Range Wide Area Network (LoRaWAN) is widely used in the first step to transfer data from the sensor module to either the gateway or directly to the cloud. It is a wireless transmission technology, like Wi-Fi, Bluetooth, etc. (Lafont et al. 2019; Bhawiyuga et al. 2019; Guarda et al. 2021; Quevy et al. 2023; Biazi and Marques 2023). Other technologies, such as Zigbee and Bluetooth are similar to LoRaWAN but end up being low range, so LoRaWAN is preferred because of the long distances that feature them and the low energy consumption.

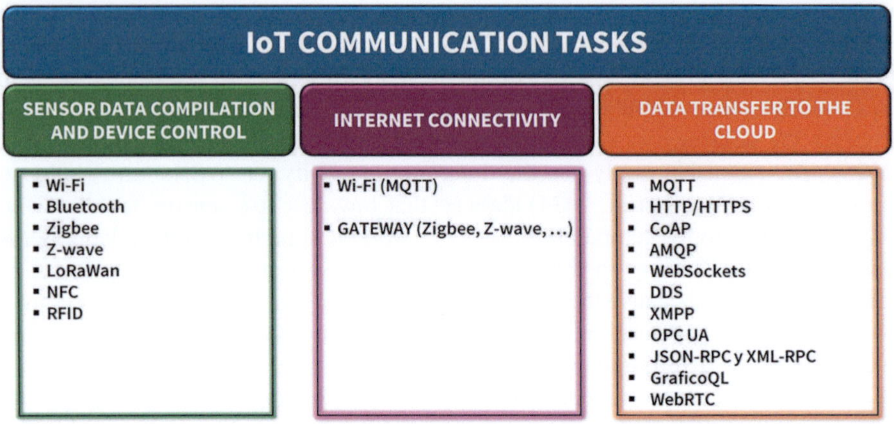

Fig. 7 Information technologies and protocols used in IoT

Message Queue Telemetry Transport (MQTT) is the Internet protocol chosen to transmit the data to the cloud, regardless of whether they come directly from the microcontroller or the gateway. MQTT is a communication protocol working above TCP and is used to manage and/or publish messages because of its reliability and good performance features, such as low power consumption and minimum bandwidth (Kim et al. 2018; Bhawiyuga et al. 2019; Angani et al. 2019a; Singh et al. 2022).

5G Wi-Fi connections have been used to preserve data transmission and optimize 5G IoT in aquaculture systems (Kim et al. 2018; Angani et al. 2019a, b; Liu et al. 2022). The latest application reinforces the usefulness of 5G connections in marine aquaculture by comparing Zigbee's performance to LoRaWAN's. This study has shown that the joint use of 5G connectivity with LoRaWAN provides cost-effective and reliable data transmission systems.

Conclusions

Smart sensors for WQM have come true in recent years, with a significant number of articles published to describe the control of physical-chemical variables relevant to animal and plant growth. WSNs for aquaculture have shown they are cost-effective, since data can be obtained continuously in real time with minimal presence of specialized staff at the aquaculture farm, which has considerably reduced logistics expenses. In contrast, data integrity related to signal stability and transmission must be ensured. Also, this field is still in its infancy regarding the availability of smart sensors for pollutants. There is a need for chemical and (bio)chemicals that can withstand the operational conditions within aquaculture farms and provide reliable measurements. Other aspects, such as the conversion of analog signals into digital and data transmission, are now in a relatively mature stage.

Acknowledgement This research was financially supported by the project "Plan Complementario de I + D + i en el área de Ciencias Marinas" funded by the European Union within the framework of the Recovery, Transformation and Resilience Plan—NextGenerationEU and by the Regional Government of Andalucia, through the Consejería de Universidad, Investigación e Innovación de la Administración de la Junta de Andalucía. (Grant No. PCM_00121). The views and opinions expressed are solely those of the author(s) and do not necessarily reflect those of the European Union or the European Commission. Neither the European Union nor the European Commission is responsible for them.

References

Abdurohman M, Putrada AG, Deris MM (2022) A robust internet of things-based aquarium control system using decision tree regression algorithm. IEEE Access 10:56937–56951. https://doi.org/10.1109/ACCESS.2022.3177225

Acuicultura de España (2025) https://acuiculturadeespana.es/acuicultura-espana/tipos-de-instalaciones-acuicolas/. Accessed 11 April 2025

Adiono T, Toha AM, Pamungkas S et al (2021) Internet of things for marine aquaculture. In: 2021 international symposium on electronics and smart devices (ISESD). IEEE, Bandung, pp 1–5

Ahmed F, Bijoy MHI, Hemal HR, Noori SRH (2024) Smart aquaculture analytics: enhancing shrimp farming in Bangladesh through real-time IoT monitoring and predictive machine learning analysis. Heliyon 10:e37330. https://doi.org/10.1016/j.heliyon.2024.e37330

Ajmal T, Mohammed F, Goodchild MS et al (2024) Mitigating the impact of harmful algal blooms on aquaculture using technological interventions: case study on a south African farm. Sustainability 16:3650. https://doi.org/10.3390/su16093650

Akhter F, Siddiquei HR, Alahi MEE, Mukhopadhyay SC (2021) An IoT-enabled portable sensing system with MWCNTs/PDMS sensor for nitrate detection in water. Measurement 178:109424. https://doi.org/10.1016/j.measurement.2021.109424

Aly SM, Fathi M (2024) Advancing aquaculture biosecurity: a scientometric analysis and future outlook for disease prevention and environmental sustainability. Aquac Int 32:8763–8789. https://doi.org/10.1007/s10499-024-01589-y

Angani A, Lee C-B, Lee S-M, Shin KJ (2019a) Realization of eel fish farm with artificial intelligence part 3: 5G based mobile remote control. In: 2019 IEEE international conference on architecture, construction, environment and hydraulics (ICACEH). IEEE, Xiamen, pp 101–104

Angani A, Oh SM, Kim ES, Shin KJ (2019b) Realization of eel fish farm with artificial intelligence part 2: IoT based flow control using MQTT. In: 2019 IEEE international conference on architecture, construction, environment and hydraulics (ICACEH). IEEE, Xiamen, pp 97–100

Aquaculture Advisory Council (2019) Promotion on the use of fish vaccines replacing antimicrobials. https://aac-europe.org/en/publication/promotion-of-the-use-of-fish-vaccines-replacing-the-use-of-antimicrobials/. Accessed: 1 Feb 2025

Aquaculture Advisory Council (2022) Correspondence: subject: new EU legal framework for the oral distribution of veterinary medicines in the aquaculture sector. https://aac-europe.org/en/publication/aac-letter-to-the-ec-on-the-new-eu-legal-framework-for-the-oral-distribution-of-veterinary-medicines-in-the-aquaculture-sector/. Accessed: 1 Feb 2025

Arepalli PG, Jairam Naik K (2024) An IoT based smart water quality assessment framework for aqua-ponds management using dilated spatial-temporal convolution neural network (DSTCNN). Aquac Eng 104:102373. https://doi.org/10.1016/j.aquaeng.2023.102373

Baena-Navarro R, Carriazo-Regino Y, Torres-Hoyos F, Pinedo-López J (2025) Intelligent prediction and continuous monitoring of water quality in aquaculture: integration of machine learning and internet of things for sustainable management. Water 17:82. https://doi.org/10.3390/w17010082

Bakar AAA, Bakar ZA, Yusoff ZM et al (2025) IoT-based real-time water quality monitoring and sensor calibration for enhanced accuracy and reliability. Int J Interact Mobile Technol 19:155–170. https://doi.org/10.3991/ijim.v19i01.51101

Bhati A, Hiran KK, Vyas AK et al (2024) Low-cost artificial intelligence internet of things-based water quality monitoring for rural areas. Internet Things 27:101255. https://doi.org/10.1016/j.iot.2024.101255

Bhawiyuga A, Amron K, Primanandha R et al (2019) LoRa-MQTT gateway device for supporting sensor-to-cloud data transmission in smart aquaculture IoT application. In: 2019 international conference on sustainable information engineering and technology (SIET). IEEE, Lombok, pp 187–190

Biazi V, Marques C (2023) Industry 4.0-based smart systems in aquaculture: a comprehensive review. Aquac Eng 103:102360. https://doi.org/10.1016/j.aquaeng.2023.102360

Birkeland MJ, Petersen JK, Timmermann K et al (2024) Identification and application of quality elements, indicators, and criteria for assessment of impact on habitats in marine Natura 2000 areas. J Environ Manag 369:122247. https://doi.org/10.1016/j.jenvman.2024.122247

Bohara K, Joshi P, Acharya KP, Ramena G (2024) Emerging technologies revolutionising disease diagnosis and monitoring in aquatic animal health. Rev Aquac 16:836–854. https://doi.org/10.1111/raq.12870

Boonsong W (2021) Embedded wireless dissolved oxygen monitoring based on internet of things platform. J Comm 16:363–368. https://doi.org/10.12720/jcm.16.9.363-368

Bresnahan PJ, Wirth T, Martz T et al (2020) Equipping smart coasts with marine water quality IoT sensors. Results Eng 5:100087. https://doi.org/10.1016/j.rineng.2019.100087

Chang CC, Ubina NA, Cheng SC et al (2022) A two-mode underwater smart sensor object for precision aquaculture based on AIoT technology. Sensors 22:7603. https://doi.org/10.3390/s22197603

Chatziantoniou A, Charalampis Spondylidis S, Stavrakidis-Zachou O et al (2022) Dissolved oxygen estimation in aquaculture sites using remote sensing and machine learning. Remote Sens Appl Soc Environ 28:100865. https://doi.org/10.1016/j.rsase.2022.100865

Chen LB, Liu YH, Huang XR et al (2022) Design and implementation of a smart seawater aquarium system based on artificial intelligence of things technology. IEEE Sensors J 22:19908–19918. https://doi.org/10.1109/JSEN.2022.3200958

Communication from the Commission to the European Parliament, the Council, the European Economic and Social Committee and the Committee of the Regions (2021) Strategic guidelines for a more sustainable and competitive EU aquaculture for the period 2021 to 2030

Consejería de Agricultura, Pesca, Agua y Desarrollo Rural (2022). Nueva Estrategia de la acuicultura marina en Andalucía (2021–2030)

Curley EAM, Valyrakis M, Thomas R et al (2021) Smart sensors to predict entrainment of freshwater mussels: a new tool in freshwater habitat assessment. Sci Total Environ 787:147586. https://doi.org/10.1016/j.scitotenv.2021.147586

Damiano E, Bibbiani C, Fronte B, Di Lieto A (2020) Smart and cheap scale for estimating live-fish biomass in offshore aquaculture. In: 2020 IEEE international workshop on metrology for agriculture and forestry (MetroAgriFor). IEEE, Trento, pp 160–164

Darmalim F, Hidayat AA, Cenggoro TW et al (2021) An integrated system of mobile application and IoT solution for pond monitoring. IOP Conf Ser Earth Environ Sci 794:012106. https://doi.org/10.1088/1755-1315/794/1/012106

Edwards P (2015) Aquaculture environment interactions: past, present and likely future trends. Aquaculture 447:2–14. https://doi.org/10.1016/j.aquaculture.2015.02.001

FAO Strategic Framework (2021) In strategic framework. https://www.fao.org/strategic-framework/en. Accessed 10 Feb 2025

Farcy P, Durand D, Charria G et al (2019) Toward a European coastal observing network to provide better answers to science and to societal challenges; the JERICO research infrastructure. Front Mar Sci 6:529. https://doi.org/10.3389/fmars.2019.00529

Goswami N, Sufian SA, Khandakar MS et al (2022) Design and development of smart system for biofloc fish farming in Bangladesh. In: 2022 7th international conference on communication and electronics systems (ICCES). IEEE, Coimbatore, pp 1424–1432

Guarda T, Portela F, Santos MF (2021) Advanced research in technologies, information, innovation and sustainability: first international conference. In: ARTIIS 2021, La Libertad, Ecuador, November 25–27, 2021, proceedings. Springer, Cham

Hsu WC, Chao PY, Wang CS et al (2020) Application of regression analysis to achieve a smart monitoring system for aquaculture. Information 11:387. https://doi.org/10.3390/info11080387

Hu WC, Chen LB, Huang BK, Lin HM (2022) A computer vision-based intelligent fish feeding system using deep learning techniques for aquaculture. IEEE Sensors J 22:7185–7194. https://doi.org/10.1109/JSEN.2022.3151777

Huan J, Li H, Wu F, Cao W (2020) Design of water quality monitoring system for aquaculture ponds based on NB-IoT. Aquac Eng 90:102088. https://doi.org/10.1016/j.aquaeng.2020.102088

Iniyan Arasu M, Ran SS, Thiyagarajan K, Appathurai A (2024) AQUASENSE: aquaculture water quality monitoring framework using autonomous sensors. Aquac Int 32:9119–9135. https://doi.org/10.1007/s10499-024-01606-0

Jeon JH, Lee NE, Lee YH et al (2021) Implementation of DAQ and monitoring system for a smart fish farm using circulation filtration system. J Inf Process Syst 17:1179–1190. https://doi.org/10.3745/JIPS.04.0226

Kanwal S, Abdullah M, Kumar S et al (2024) An optimal internet of things-driven intelligent decision-making system for real-time fishpond water quality monitoring and species survival. Sensors 24:7842. https://doi.org/10.3390/s24237842

Khaoula T, Abdelouahid RA, Ezzahoui I, Marzak A (2021) Architecture design of monitoring and controlling of IoT-based aquaponics system powered by solar energy. Procedia Comput Sci 191:493–498. https://doi.org/10.1016/j.procs.2021.07.063

Kim Y, Lee N, Kim B, Shin K (2018) Realization of IoT based fish farm control using mobile app. In: 2018 international symposium on computer, consumer and control (IS3C). IEEE, Taichung, pp 189–192

Kinar NJ, Brinkmann M (2022) Development of a sensor and measurement platform for water quality observations: design, sensor integration, 3D printing, and open-source hardware. Environ Monit Assess 194:207. https://doi.org/10.1007/s10661-022-09825-9

Kodali RK, Sabu AC (2022) Aqua monitoring system using AWS. In: 2022 international conference on computer communication and informatics (ICCCI). IEEE, Coimbatore, pp 1–5

Lafont M, Dupont S, Cousin P et al (2019) Back to the future: IoT to improve aquaculture: real-time monitoring and algorithmic prediction of water parameters for aquaculture needs. In: 2019 global IoT summit (GIoTS). IEEE, Aarhus, pp 1–6

Lei X, Xu X, Liu L et al (2023) Gold-nanoparticle-based multiplex immuno-strip biosensor for simultaneous determination of 83 antibiotics. Nano Res 16:1259–1268. https://doi.org/10.1007/s12274-022-4762-z

Li D, Liu C, Song Z, Wang G (2021) Automatic monitoring of relevant behaviors for crustacean production in aquaculture: a review. Animals 11:2709. https://doi.org/10.3390/ani11092709

Li D, Wang G, Du L et al (2022) Recent advances in intelligent recognition methods for fish stress behavior. Aquac Eng 96:102222. https://doi.org/10.1016/j.aquaeng.2021.102222

Liu T, Liu J, Wang J, Xu J (2022) Optimization of the intelligent sensing model for environmental information in aquaculture waters based on the 5G smart sensor network. J Sens 2022:1–12. https://doi.org/10.1155/2022/6409046

López-Ramírez GA, Aragón-Zavala A (2023) Wireless sensor networks for water quality monitoring: a comprehensive review. IEEE Access 11:95120–95142. https://doi.org/10.1109/ACCESS.2023.3308905

Lozano M, Cabruja E, Perez-Sanchez J et al (2022) AEFishBIT: a smart system to monitor fish activity and welfare. In: 2022 smart systems integration (SSI). IEEE, Grenoble, pp 1–4

Mandal A, Ghosh AR (2024) Role of artificial intelligence (AI) in fish growth and health status monitoring: a review on sustainable aquaculture. Aquac Int 32:2791–2820. https://doi.org/10.1007/s10499-023-01297-z

Manoharan H, Teekaraman Y, Kshirsagar PR et al (2020) Examining the effect of aquaculture using sensor-based technology with machine learning algorithm. Aquac Res 51:4748–4758. https://doi.org/10.1111/are.14821

Mendes JP, Coelho L, Kovacs B et al (2019) Dissolved carbon dioxide sensing platform for freshwater and saline water applications: characterization and validation in aquaculture environments. Sensors 19:5513. https://doi.org/10.3390/s19245513

Nakajima S, Sumiya D, Morii M et al (2022) IoT-based experimental aquarium environment for observing crabs. In: 2022 IEEE international conference on internet of things and intelligence systems (IoTaIS). IEEE, Bali, pp 317–321

Naloufi M, Abreu T, Souihi S et al (2024) Long-term stability of low-cost IoT system for monitoring water quality in urban rivers. Water 16:1708. https://doi.org/10.3390/w16121708

Nayoun MNI, Hossain SA, Rezaul KM et al (2024) Internet of things-driven precision in fish farming: a deep dive into automated temperature, oxygen, and pH regulation. Computers 13:267. https://doi.org/10.3390/computers13100267

Observatorio Español de Acuicultura (OESA) (2025) Unidad Didáctica 1. La acuicultura. Accessed 11 April 2025

Prathibanandhi K et al (2024) Smart systems for sustainable aquaculture: a focus on water quality. In: 2024 international conference on power, energy, control and transmission systems (ICPECTS). IEEE, Chennai, pp 1–4

Quevy Q, Lamrini M, Chkouri M et al (2023) Open sensing system for long term, low cost water quality monitoring. IEEE Open J Ind Electron Soc 4:27–41. https://doi.org/10.1109/OJIES.2022.3233919

Rao BB, Keerthana J, Raghavendra CG et al (2022) An overview on detemining fish population using image and acoustic approaches. In: 2022 IEEE international conference on data science and information system (ICDSIS). IEEE, Hassan, pp 1–6

Reverter M, Sarter S, Caruso D et al (2020) Aquaculture at the crossroads of global warming and antimicrobial resistance. Nat Commun 11:1870. https://doi.org/10.1038/s41467-020-15735-6

Rosaline N, Sathyalakshimi S (2019) IoT based aquaculture monitoring and control system. J Phys Conf Ser 1362:012071. https://doi.org/10.1088/1742-6596/1362/1/012071

Rossi L, Bibbiani C, Fronte B et al (2022) Validation campaign of a smart dynamic scale for measuring live-fish biomass in aquaculture. In: 2022 IEEE workshop on metrology for agriculture and forestry (MetroAgriFor). IEEE, Perugia, pp 111–115

Shete RP, Bongale AM, Dharrao D (2024) IoT-enabled effective real-time water quality monitoring method for aquaculture. MethodsX 13:102906. https://doi.org/10.1016/j.mex.2024.102906

Simone Soares M, Singh R, Kumar S et al (2024) The role of smart optical biosensors and devices on predictive analytics for the future of aquaculture systems. Opt Laser Technol 177:111049. https://doi.org/10.1016/j.optlastec.2024.111049

Singh M, Sahoo KS, Nayyar A (2022) Sustainable IoT solution for freshwater aquaculture management. IEEE Sensors J 22:16563–16572. https://doi.org/10.1109/JSEN.2022.3188639

Son S, Jeong Y (2024) An automated fish-feeding system based on CNN and GRU neural networks. Sustainability 16:3675. https://doi.org/10.3390/su16093675

Sriram GY, Sai Shibu NB (2021) Design and implementation of automated aquaponic system with real-time remote monitoring. In: 2021 advanced communication technologies and signal processing (ACTS). IEEE, Rourkela, pp 1–6

Teja KBR, Monika M, Chandravathi C, Kodali P (2020) Smart monitoring system for pond management and automation in aquaculture. In: 2020 international conference on communication and signal processing (ICCSP). IEEE, Chennai, pp 204–208

Tendolkar A, Choraria A et al (2021) Estuarine-star (smart estuarine aquaculture management with sensor network). In: 2021 IEEE international conference on systems, man, and cybernetics (SMC). IEEE, Melbourne, pp 1593–1598

Tsai HL, Lin JY, Lyu WH (2021) Design and evaluation of wireless multi-sensor IoT system for monitoring water quality of freshwater aquaculture. In: 2021 international automatic control conference (CACS). IEEE, Chiayi, pp 1–6

Uddin MA, Kumar Dey U, Akter M (2022) Proposing a cloud and edge computing based decision supportive consolidated farming system by sensing various effective parameters using IoT. In: 2022 IEEE international IOT, electronics and mechatronics conference (IEMTRONICS). IEEE, Toronto, pp 1–6

Wang M, Cetó X, Del Valle M (2022) A sensor array based on molecularly imprinted polymers and machine learning for the analysis of fluoroquinolone antibiotics. ACS Sens 7:3318–3325. https://doi.org/10.1021/acssensors.2c01260

Wang Y, Cui Y, Wang H et al (2023) Ofloxacin and norfloxacin simultaneous detection by ERGO/GCE and its application in medicine and aquaculture wastewater. Res Chem Intermed 49:741–761. https://doi.org/10.1007/s11164-022-04892-6

Wardhany VA, Yuliandoko H, Subono et al (2021) Smart system and monitoring of vanammei shrimp ponds. Int J Adv Sci Eng Inf Technol 11:1366. https://doi.org/10.18517/ijaseit.11.4.8557

Wiranto G, Maulana YY, Hermida IDP et al (2015) Integrated online water quality monitoring. In: 2015 international conference on smart sensors and application (ICSSA). IEEE, Kuala Lumpur, pp 111–115

Xu W, Wang P, Jiang L et al (2023) Intelligent recognition and behavior tracking of sea cucumber infected with vibrio alginolyticus based on machine vision. Aquac Eng 103:102368. https://doi.org/10.1016/j.aquaeng.2023.102368

Yamabe S, Fukae K, Imai T et al (2021) Multi-sensor platform for smart aquaculture. In: 2021 IEEE 10th global conference on consumer electronics (GCCE). IEEE, Kyoto, pp 356–357

Yu YB, Lee JH, Choi JH et al (2023) The application and future of biofloc technology (BFT) in aquaculture industry: a review. J Environ Manag 342:118237. https://doi.org/10.1016/j.jenvman.2023.118237

Zakaria S, Ahmad Jafri MAA, Engku R, Ariff EA et al (2021) Nature driven IOT based automation of aquaponic system. In: Intelligent manufacturing and artificial intelligence. IEEE, Kyoto, pp 555–562

Zhang W, Belton B, Edwards P et al (2022) Aquaculture will continue to depend more on land than sea. Nature 603:E2–E4. https://doi.org/10.1038/s41586-021-04331-3

Zuriati Z, Supriyatna AR, Arifin O (2022) Design and development of feeding automation system and water quality monitoring on freshwater fish cultivation. IOP Conf Ser Earth Environ Sci 1012:012077. https://doi.org/10.1088/1755-1315/1012/1/012077

Drought Impacts on Hydrological Ecosystem Services: Indicators and Methodological Processes

Nadia Falah, Jaime Solís-Guzmán, Cristina Torrecillas, and Madelyn Marrero

Contents

Abstract

This study identifies existing gaps in the sustainable management of drought by applying it to derived Hydrological Ecosystem Services (HES). The proposal integrates physical and socioeconomic indicators, assessing the impact of socioeconomic factors such as quality of life, economic development, communication networks, and demographic changes, tailored to regional characteristics to evaluate the impact of these factors on environmental change in the HES. Finally,

N. Falah (✉) · J. Solís-Guzmán · C. Torrecillas · M. Marrero
University of Seville, Seville, Spain
e-mail: Nadfal@alum.us.es

D. Gutiérrez et al. (eds.), *Smart Water Quality Monitoring*,
https://doi.org/10.1007/978-3-032-03049-8_8

the present work proposes a strategy to generate a comprehensive modeling for introducing practical indicators to process the evaluation of them for future research. The selection of drought analysis indices depends on available data and regional requirements, necessitating the integration of diverse data and advanced modeling. The research methodology is divided into three parts: theoretical, analytical, and developmental. The analytical part involves evaluating possible collected data, emphasizing statistical techniques and spatial monitoring tools. The developmental part focuses on scenario analysis and the evaluation of risks, mitigation, and adaptation strategies.

Keywords

Hydrological ecosystem services · Drought indicators · Multi-criteria decision making · Socioeconomic vulnerability · Sustainable drought management

Introduction

Drought as a Natural Disaster

Drought is a hidden and complex natural disaster defined as a prolonged period of deficient precipitation. It has significant environmental, social, and economic consequences. Unlike sudden disasters such as earthquakes or floods, drought develops gradually, making its short-term impacts less apparent. However, its long-term consequences are extensive and profound. Based on literature review, there are different types of droughts; therefore, causes and effects of each drought can be distinct. In general, there are seven main types classified based on the duration of the event of droughts (in the case that flash drought count as a one type); see Fig. 1, with unique impacts on environment and ecosystem that explain, as a result, different types of frequent indicators for each type of drought can be found (Alawsi et al. 2022; West et al. 2019; Zellou et al. 2023). The type of drought described in Fig. 1 is linked to distinct environmental and socioeconomic effects. For instance, meteorological drought is characterized by rainfall deficits, often resulting in temperature anomalies and increased climate variability. Agricultural drought, on the other hand, affects soil moisture levels, leading to reduced crop yields and stressed food systems. Hydrological drought causes significant depletion of water levels in rivers, reservoirs, and lakes, which impacts aquatic ecosystems and water availability for human consumption. Furthermore, groundwater drought reduces underground water reserves, potentially causing land subsidence and affecting long-term water sustainability (Ndayiragije and Li 2022; Zhang et al. 2013). Ecological drought disrupts natural ecosystems, leading to loss of forests, wetlands, and wildlife habitats, which further accelerates biodiversity loss. Lastly, socioeconomic drought represents the cumulative effects on social stability, economic systems, and human well-being, often exacerbating inequalities and threatening livelihoods (Laín 2008; Nandgude et al. 2023). Addressing these diverse impacts requires a

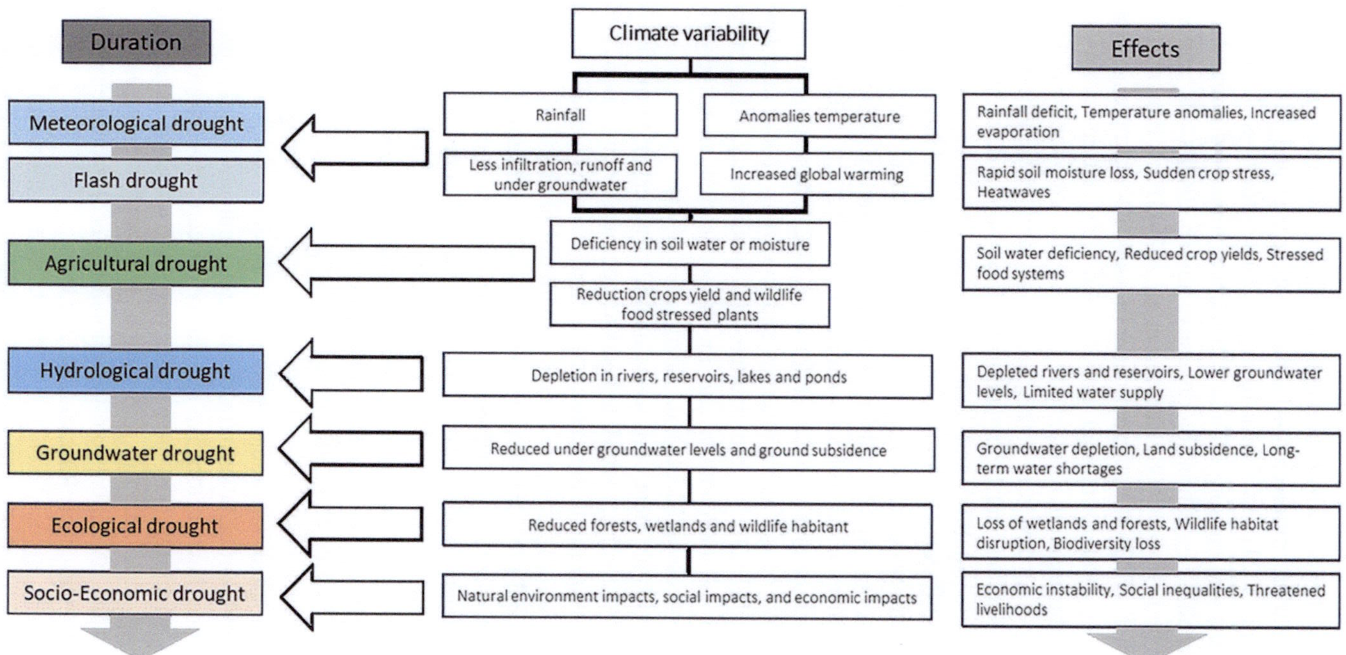

Fig. 1 Distinct types of main droughts linked to the duration of the event and their causes and effects based on (Ndayiragije and Li 2022)

comprehensive and multidimensional approach to better understand and mitigate the consequences of each type of drought.

These effects go beyond water shortages, impacting entire ecosystems, including soil health, vegetation, biodiversity, human livelihoods, and economic stability (Di Gregorio et al. 2024). As shown in Fig. 1, these effects go beyond reducing under water levels and causing ground subsidence, impacting entire ecosystems, including soil health, vegetation, biodiversity, human livelihoods, and economic stability (Burke and Brown 2008; Hariram et al. 2023). Each type of drought has distinct effects that contribute to these broader impacts. For example, meteorological droughts primarily lead to rainfall deficits and temperature anomalies, while agricultural droughts result in reduced soil moisture and crop failures. Hydrological droughts deplete water levels in rivers, reservoirs, and lakes, whereas ecological droughts disrupt habitats and biodiversity. Socioeconomic droughts extend these impacts to economic instability and social inequalities.

Drought is a multifaceted natural hazard with profound and far-reaching implications for ecosystems, water resources, agriculture, and human livelihoods. It serves as a critical driver of environmental and societal vulnerability, significantly altering the functions and sustainability of ecosystems (Alawsi et al. 2022; Hanadé Houmma et al. 2023; Ndayiragije and Li 2022; Zellou et al. 2023). Moreover, drought places significant stress on groundwater reserves and lowers river and lake levels. This situation directly threatens natural habitats and biodiversity. Also, reduced soil moisture leads to soil degradation and increased erosion, negatively impacting natural recovery processes (Mehta n.d.; Zhang et al. 2013).

Drought is one of the most destructive natural disasters in terms of scope and long-term effects. According to the Food and Agriculture Organization (FAO), more than 80% of all agricultural losses caused by natural disasters globally are attributed to drought. These losses include reduced agricultural productivity, increased food prices, and additional pressure on water resources (Canet-Martí et al. 2021; Tapia et al. 2021). The increasing frequency and severity of global drought events, exacerbated by climate change, have highlighted the limitations in existing drought management frameworks. Rising global temperatures, irregular precipitation patterns, and higher evaporation rates, as warned by the Intergovernmental Panel on Climate Change (IPCC), are intensifying these events, resulting in longer and more severe droughts (Gössling et al. 2023).

The urgency for robust and effective management strategies that integrate ecological, hydrological, and socioeconomic dimensions has grown significantly. Current gaps include deficiencies in methodologies, indicators, and approaches that effectively integrate these factors. An integrative concept, Hydrological Ecosystem Services (HES), which refers to the benefits derived from water-related ecosystem functions such as water supply, purification, and habitat preservation, remains underutilized despite its critical role in maintaining ecosystem health and human well-being (Bryła et al. 2021; Raposo et al. 2023).

Arid and semiarid regions, such as the Middle East, Africa, and parts of Central Asia, have been particularly affected. According to reports, coastal and island regions are also increasingly vulnerable due to rising soil salinity and reduced freshwater availability (Platform for Sustainable Cities and Bank 2018; World Bank 2015, 2021). Drought is not limited to environmental impacts; it also has far-reaching consequences for human societies. Reduced agricultural production leads to unemployment and migration, posing economic and social challenges for rural communities. Additionally, increased competition for limited water resources can create political and social tensions. Numerous instances of this have been reported in Africa and the Middle East, where drought has led to population displacement and heightened conflicts (Wilhite 2002). Despite efforts by organizations such as the Global Water Partnership (GWP) and the World Meteorological Organization (WMO) to enhance drought management plans, critical gaps persist in integrating ecological services concepts and creating comprehensive, actionable indicators (Gabiña et al. 2019; Meteorological Organization and Water Partnership 2016). Addressing these gaps requires the development of multifaceted indicators that integrate ecological and hydrological data alongside socioeconomic parameters (Burke and Brown 2008; Di Gregorio et al. 2024; Hariram et al. 2023; Li and Zhong 2024). These indicators can improve the predictive and monitoring capabilities for drought impacts while supporting sustainable water resource management and ecosystem conservation.

To further understand the wide-ranging and multifaceted impacts of drought, it is essential to delve into its specific consequences on water resources, biodiversity, food security, and socioeconomic systems, as highlighted below. This integrated approach ensures that policies are adaptive, resilient, and capable of mitigating the complex impacts of drought.

The main drought effects can be identified in the following:

1. Reduction in surface and groundwater resources.

   Drought significantly depletes surface water sources such as rivers, lakes, and reservoirs, while also overburdening groundwater aquifers. Prolonged dry periods result in lower water tables, reduced streamflow, and shrinking lakes, leading to severe water scarcity for communities and ecosystems. In this way:

   (a) In Southern Europe, water availability has dropped by 30% over the last decade due to recurring droughts and over-extraction of groundwater resources (Raposo et al. 2023).
   (b) California, USA, experienced its driest 22-year period in over 1200 years, exacerbating reliance on diminishing groundwater reserves and triggering a water crisis for both agriculture and urban centers (Chang and Bonnette 2016; Ekstrom et al. 2018; Meilinger and Monstadt 2021).
   (c) Globally, the United Nations Environment Programme (UNEP) estimates that by 2030, nearly 700 million people will be displaced by severe water scarcity if current trends persist (UNEP 2022).

   HES play a critical role in mitigating these impacts. For example, wetlands act as natural water storage systems, regulating flows and maintaining water quality during drought periods. However, the degradation of these ecosystems due to prolonged drought further intensifies water scarcity (Bryła et al. 2021).

2. Degradation of natural habitats and biodiversity loss.

Drought dries out wetlands, reduces the extent of aquatic habitats, and threatens species dependent on water availability. Ecosystem degradation disrupts food chains, weakens biodiversity resilience, and accelerates species extinction, as examples:

(a) In Africa, over 40% of wetlands have been lost in some regions due to persistent drought, adversely impacting aquatic bird populations and freshwater fish species (Zellou et al. 2023).
(b) In Australia, prolonged droughts have decimated water-dependent ecosystems like the Murray-Darling Basin, causing a sharp decline in native fish populations and widespread vegetation loss (Verdon-Kidd et al. 2023).
(c) The Ramsar Convention on Wetlands (2022) highlights that nearly 87% of the world's wetlands have been degraded or destroyed since 1700, with drought identified as a leading cause.

The loss of biodiversity also diminishes critical HES, such as pollination, soil stabilization, and habitat preservation. Restoring ecosystems and protecting biodiversity are essential steps in building resilience against drought and sustaining ecological services (Hanadé Houmma et al. 2023).

3. Food security challenges.

Drought directly threatens global food security by reducing water availability for irrigation and impacting agricultural productivity. This leads to lower crop yields, increased food prices, and higher risks of famine, particularly in developing regions. In India, droughts over the past decade have led to a 20% decrease in wheat production, affecting both domestic food supply and global markets. In sub-Saharan Africa, drought-induced crop failures are a leading cause of hunger, with maize yields falling by as much as 30% during prolonged dry spells. According to the Intergovernmental Panel on Climate Change (IPCC), drought will continue to exacerbate food insecurity globally, with agricultural production projected to decline by up to 25% in arid regions by 2050 (Burke and Brown 2008; Di Gregorio et al. 2024; Hariram et al. 2023; Li and Zhong 2024; Rank et al. 2024; Williams et al. 2024).

HES, such as natural water storage and regulation provided by ecosystems, are vital for maintaining irrigation systems and agricultural productivity. However, these services are severely compromised during drought, highlighting the need for sustainable agricultural practices and ecosystem restoration (Raposo et al. 2023).

4. Social and economic consequences.

In addition to environmental and food security impacts, drought has significant social and economic ramifications, including migration, loss of livelihoods, and heightened water conflicts:

(a) In Central America, recurrent droughts, such as those linked to the El Niño phenomenon, have forced thousands of farmers to migrate due to crop failures and economic instability.
(b) The Middle East has seen intensified political tensions over shared water resources, such as the Tigris and Euphrates rivers, where drought has exacerbated disputes among Turkey, Syria, and Iraq.

(c) Economic losses from drought are staggering. In the United States, drought-related impacts on agriculture, energy, and water supply cost an average of $9 billion annually (Altemus Cullen 2023; Bulama et al. 2021; Paneque n.d.; Williams et al. 2024).

The integration of HES into socioeconomic planning can mitigate some of these challenges. For instance, restoring wetlands and forests can reduce migration pressures by improving water availability and agricultural productivity. Furthermore, international cooperation on transboundary water management is essential to prevent conflicts in drought-prone regions (Hanadé Houmma et al. 2023).

Toward Sustainable Drought Management

To effectively address the escalating impacts of drought, it is essential to move beyond traditional approaches that focus solely on immediate physical consequences. Drought's multidimensional nature—affecting ecosystems, agriculture, water resources, and human livelihoods—necessitates innovative and sustainable management strategies. The increasing severity and frequency of drought, driven by climate change, demands solutions that account not only for environmental impacts but also for the social and economic dimensions of this growing crisis. Conventional drought management strategies, which primarily rely on rainfall predictions and water storage, often fail to address the complexities introduced by climate variability and ecosystem degradation (Altemus Cullen 2023; Moreno et al. 2022; Torelló-Sentelles and Franzke 2022). In this context, the concept of HES emerges as a promising framework for rethinking drought management. HES highlights the vital services provided by ecosystems, including natural water purification, flow regulation, and habitat preservation—services that are critical for building resilience to drought. By prioritizing these ecosystem functions, we can not only mitigate the impacts of drought but also enhance biodiversity, restore degraded ecosystems, and support the livelihoods of vulnerable communities.

This research will explore the evolving approaches to drought management, emphasizing the role of HES and key aspects of sustainability, by achieving objectives in Table 1.

By adopting this multidimensional perspective, including piloting methodologies in vulnerable regions, we can create more effective and resilient systems to combat the growing threat of drought, safeguarding ecosystems and communities for the future.

Research Background

Drought is one of the most pervasive and complex natural disasters, with far-reaching implications for ecosystems, water resources, agriculture, and human livelihoods. Its impacts extend beyond water scarcity to include biodiversity loss,

Table 1 Aims and key aspects, study objectives, and its description

Objective	Description
Analyzing the limitations of conventional drought management practices	– Identify gaps and inefficiencies in traditional approaches, such as reliance on single-dimensional indicators like precipitation or streamflow – Evaluate how these methods fail to capture the interconnected impacts on ecosystems, hydrology, and socioeconomic systems – Highlight the constraints of existing frameworks in addressing the increasing frequency and severity of drought events due to climate change
Examining the role of ecosystems and hydrological services in mitigating drought impacts	– Explore how ecosystem-based approaches contribute to drought resilience through natural processes such as water purification, flow regulation, and habitat preservation – Investigate the vulnerabilities of HES under drought conditions and their implications for water regulation and biodiversity – Assess the role of HES in supporting sustainable livelihoods and reducing socioeconomic disparities in drought-prone areas
Identifying gaps in current frameworks and proposing strategies for sustainable drought management	– Develop a comprehensive understanding of weaknesses in existing policies and tools for drought prediction, monitoring, and mitigation – Propose integrative strategies that incorporate multi-criteria indicators, encompassing climate, hydrology, vegetation, land use, and socioeconomic factors – Prioritize the inclusion of advanced technologies, such as remote sensing, machine learning, and big data analytics, to enhance the predictive capabilities of drought management systems
Developing actionable indicators and practical frameworks to integrate HES into drought response systems	– Design practical, region-specific indicators that enable real-time monitoring and comprehensive risk assessment – Establish frameworks for incorporating HES into existing drought management strategies, ensuring alignment with policy and decision-making processes – Pilot and refine these methodologies in vulnerable regions, demonstrating their scalability and effectiveness in diverse environmental and socioeconomic contexts

soil degradation, food insecurity, and social vulnerabilities. As climate change intensifies the frequency and severity of droughts, traditional management frameworks—often reliant on single metrics like precipitation—have proven inadequate in addressing the multidimensional challenges posed by drought. The evolution of drought management has witnessed a gradual shift from simplistic water-saving measures to integrated, multidimensional frameworks that recognize the interconnectedness of ecological, hydrological, and socioeconomic systems. Emerging concepts like HES now emphasize the critical role ecosystems play in mitigating drought impacts, including natural water purification, flow regulation, and habitat preservation. These services are essential for enhancing resilience to climate extremes and supporting sustainable livelihoods.

Despite advancements in drought assessment methodologies, significant gaps remain in integrating physical, ecological, and social dimensions. Early approaches primarily relied on meteorological data to evaluate drought severity, focusing on precipitation and temperature anomalies. While these metrics laid the foundation for drought assessment, they failed to address broader ecological and societal impacts, limiting their scope in modern contexts. Recent research highlights the importance of holistic approaches that incorporate diverse indicators, such as biodiversity metrics, socioeconomic parameters, and advanced hydrological modelling (Di Gregorio et al. 2024; Hidalgo-Triana et al. 2023; Torelló-Sentelles and Franzke 2022).

By the early 2000s, the increasing recognition of drought's cascading effects spurred the development of frameworks that bridged these gaps. Integrated approaches, like those promoted by the GWP and the WMO, combined physical, ecological, and social dimensions to create more comprehensive management strategies. Innovative research, such as that by Bryła et al. (2021) and Ha et al. (2023), has further refined drought monitoring tools, emphasizing the integration of biodiversity and water quality metrics and the development of actionable indicators tailored to regional and local contexts.

This section explores the historical progression of drought management approaches, emphasizing the transition from traditional single-metric tools to advanced multidimensional frameworks. It also highlights the increasing importance of HES in addressing the complex impacts of drought and fostering sustainable resilience strategies. Through this analysis, the necessity for integrated, adaptive, and ecosystem-based approaches becomes evident, paving the way for more effective drought mitigation and management strategies.

Evolution of Drought Management Approaches

Over the decades, drought management has undergone a profound transformation, evolving from basic water conservation techniques to sophisticated, multidimensional strategies that address the intricate interplay between ecological, hydrological, and socioeconomic factors. This shift reflects a growing understanding of drought's wide-ranging impacts, which extend beyond water scarcity to encompass biodiversity loss, soil degradation, and social vulnerabilities (Altemus Cullen 2023; Hidalgo-Triana et al. 2023; Torelló-Sentelles and Franzke 2022; Verdon-Kidd et al. 2023). Recognizing these complexities, researchers and policymakers have developed diverse frameworks that integrate physical, ecological, and human dimensions to create more comprehensive and adaptive solutions. The following sections trace the evolution of these approaches, detailing key methodologies, their strengths, and limitations, as they transitioned from reliance on meteorological data to holistic, integrated systems (Moreno et al. 2022; Torelló-Sentelles and Franzke 2022).

Early Approaches: Focus on Physical Indicators

In the early twentieth century, drought management primarily relied on meteorological data to quantify drought severity. Key indicators such as the Standardized Precipitation Index (SPI) and the Palmer Drought Severity Index (PDSI) became foundational tools for assessing drought conditions. These metrics were effective in short-term forecasting, as they focused on the immediate deficits in precipitation and anomalies in temperature. The SPI offered a standardized method for measuring rainfall deficits, which allowed for comparative analyses across different regions, proving valuable for early warning systems and response efforts. The PDSI, on the other hand, incorporated both precipitation and temperature anomalies, providing a more nuanced understanding of drought conditions by factoring in how temperature variations exacerbate the effects of rainfall shortages (Canet-Martí et al. 2021; Di Gregorio et al. 2024; Rank et al. 2024; Williams et al. 2024).

However, while these indicators were instrumental in understanding short-term drought dynamics, they had significant limitations. The SPI, being rainfall-centric, largely ignored the effects of drought on groundwater reserves and ecological systems, which are crucial for long-term water sustainability. Similarly, the PDSI struggled to capture the prolonged impacts of drought on ecosystems and lacked global adaptability, as it was heavily dependent on specific climatic data, making it less useful in areas with varying climate conditions (Di Gregorio et al. 2024; Hidalgo-Triana et al. 2023).

Emerging research has highlighted the need to complement these traditional meteorological metrics with indicators that address the broader ecological and socioeconomic dimensions of drought. For instance, Ha et al. (2023) proposed integrating in-situ measurements with hydrological models, which would better capture the multifaceted impacts of drought on water quality and ecosystem services. Such an integrated approach is essential to fully understand drought's effects and ensure that management strategies consider not only immediate water deficits but also long-term environmental and social resilience.

Emergence of Hydrological Approaches

By the mid-twentieth century, the increasing recognition of drought's impact on water resources led to the development of hydrological indicators that offered a more comprehensive approach to understanding water scarcity. These tools expanded beyond traditional precipitation measurements to assess both surface and groundwater availability, which are essential components of a region's overall water security. For example, the Streamflow Drought Index (SDI) was developed to focus specifically on reductions in river flows, offering critical insights for regions that rely heavily on surface water for their water supply. This index helped to quantify the severity and duration of drought conditions by monitoring streamflow patterns, which are directly impacted by changing precipitation and evaporation rates. The SDI proved particularly valuable in regions where river systems are the primary source of water for both agricultural and municipal needs, enabling better drought preparedness and response strategies (Di Gregorio et al. 2024). Similarly, groundwater resource indicators were introduced to monitor changes in aquifer levels, a vital metric for areas that are dependent on groundwater supplies. These indicators track the fluctuations in groundwater levels over time, helping to assess the sustainability of aquifers during periods of reduced recharge caused by drought. In regions where groundwater is the primary source of drinking water or irrigation, understanding these changes is crucial for managing water resources effectively, preventing over-extraction, and ensuring long-term availability (Altemus Cullen 2023; Williams et al. 2024). Together, these hydrological indicators provided a more integrated framework for drought assessment, enabling policymakers and water resource managers to take a more holistic approach to drought mitigation and planning. They offered critical insights not only into the availability of surface water but also into the hidden, often less-visible resource of groundwater, ensuring a more comprehensive understanding of drought's impacts on both types of water sources (Carvalho et al. 2022; Chang and Bonnette 2016).

The strengths of these hydrological approaches included a better understanding of water depletion during droughts and the ability to provide actionable insights for water resource planning and allocation. However, they also had limitations. They relied heavily on historical data, which reduced their ability to predict future drought conditions in the face of a changing climate. Additionally, they often ignored water quality and ecological services, which are vital for maintaining ecosystem health and biodiversity. These gaps underscored the need for more comprehensive tools that integrate both water quantity and quality in drought assessments.

Shift Toward Socioeconomic Approaches

In the 1990s, the global focus on drought management broadened to include socioeconomic dimensions. Recognizing that drought impacts extend beyond water shortages, new metrics were developed to assess how drought affects livelihoods, food security, and migration patterns. For example, socioeconomic drought indicators (SEDI) were designed to measure the effects of drought on agricultural productivity, labor markets, and access to water. These indicators provided a more comprehensive understanding of drought by linking water scarcity to economic and social outcomes, helping to assess the broader impact of drought on communities. Similarly, livelihood vulnerability indicators (LVI) were introduced to highlight the effects of drought on both rural and urban households, with a particular focus on food and water insecurity. These indicators shed light on how different socioeconomic groups are affected by drought, emphasizing the vulnerability of marginalized populations who may have limited access to resources (Hariram et al. 2023).

While these socioeconomic indicators improved the understanding of drought's impact on human systems, they were not without their drawbacks. They helped identify vulnerable populations and regions, enabling targeted interventions and support, and brought attention to the social and economic dimensions of drought beyond physical water scarcity. However, they were often disconnected from ecological and physical indicators, limiting their ability to provide a holistic view of drought impacts. Additionally, the lack of reliable data, especially in developing regions, hindered the effectiveness of assessments and interventions, making it difficult to address the full scope of drought's socioeconomic consequences.

Transition to Multidimensional Approaches

By the early 2000s, the increasing frequency and intensity of drought events—driven by climate change—underscored the need for integrated frameworks that combined physical, ecological, and social dimensions. These multidimensional approaches aimed to capture the complexity of drought impacts and improve management strategies. For example, Integrated Drought Management Programs (IDMP), spearheaded by organizations such as the WMO and GWP, focused on sustainable water management, ecosystem restoration, and reducing social vulnerabilities. These programs sought to address drought impacts in a comprehensive manner by balancing environmental sustainability with social and economic needs (GWP 2020). Similarly, Composite Drought Impact Indicators (CDII) combined data from multiple domains—physical (e.g., precipitation, temperature), ecological (e.g., biodiversity loss), and social (e.g., migration, food security)—to provide a more holistic perspective on drought impacts. This integration helped to understand the interconnected nature of drought's effects across different sectors (Bryła et al. 2021). Furthermore, innovative research on indicators has focused on enhancing drought monitoring frameworks by combining ecological, hydrological, and socioeconomic data. For instance, Bryła et al. (2021) emphasized the importance of linking biodiversity and water quality metrics to better assess the broader impacts of drought. Ha et al. (2023) further advanced this by developing 17 indicators that combined hydrological modelling with socioeconomic parameters, offering actionable insights for enhancing drought resilience.

These multidimensional frameworks have provided significant strengths. They offer a holistic view of drought impacts by integrating various dimensions into actionable insights. Such approaches enable better preparedness and adaptive management, with the ability to implement early warning systems and conduct comprehensive risk assessments. Moreover, they bridge the gap between traditional physical metrics and the complex realities of ecosystem and societal drought vulnerabilities, ensuring a more well-rounded approach to drought management.

However, these frameworks also face challenges. The high complexity of data collection and analysis, coupled with the need for advanced infrastructure and technical expertise, limits their accessibility in many regions. Financial and technical constraints, particularly in low-income areas, hinder the widespread adoption of these integrated frameworks. Additionally, a lack of sectoral coordination—often seen in the fragmented collaboration between agriculture, water management, and environmental policy sectors—can undermine the effectiveness of these frameworks, making it difficult to achieve comprehensive drought management outcomes. Table 2 summarizes the strengths and weaknesses of the three approaches.

Table 2 Strengths and weaknesses of the three approaches, early, emergence, and socioeconomic, and together in a multidimensional approach to drought mitigation and management

Approaches	Strengths	Weaknesses
Early approaches: focus on physical indicators	– SPI: Provided a clear, standardized method to measure rainfall deficits, allowing for comparative analyses across regions – PDSI: Incorporated both precipitation and temperature anomalies, offering more nuanced insights into drought conditions	– SPI was rainfall-centric and ignored groundwater and ecological impacts – PDSI struggled with capturing long-term effects on ecosystems and lacked global adaptability due to its reliance on specific climatic data
Emergence of hydrological approaches	– Enhanced understanding of water resource depletion during drought periods – Provided actionable insights for water resource planning and allocation	– Relied heavily on historical data, which limited predictive capabilities in a changing climate – Neglected water quality and ecological services, which are critical for ecosystem health and biodiversity
Shift toward socioeconomic approaches	– Helped identify vulnerable populations and regions, enabling targeted interventions – Brought attention to the social and economic dimensions of drought, beyond physical impacts	– Often disconnected from ecological and physical indicators, limiting their ability to provide holistic insights – Lack of reliable data, particularly in developing regions, hindered effective assessments and interventions
Transition to multidimensional approaches	– Provided a holistic view of drought impacts, integrating diverse dimensions into actionable insights – Enabled better preparedness and adaptive management through early warning systems and comprehensive risk assessments – Bridged the gap between traditional physical metrics and the complex realities of ecosystem and societal drought vulnerabilities	– High complexity in data collection and analysis: The need for advanced infrastructure and technical expertise limits accessibility in many regions – Financial and technical constraints: Limited resources in low-income areas hinder the widespread adoption of integrated frameworks – Lack of sectoral coordination: Weak collaboration across agriculture, water management, and environmental policy sectors often reduces the effectiveness of these frameworks

The Need for a Holistic Framework

Drought's wide-ranging impacts require management approaches that integrate multiple dimensions—physical, ecological, and social—into a cohesive strategy. Traditional methods, focused on single metrics like precipitation or streamflow, are insufficient to address the challenges posed by climate change and ecosystem degradation. This has led to the rise of innovative concepts like HES, which emphasize the critical role of ecosystems in drought resilience.

The Role of Ecosystem Services in Drought Management

Building on the need for a holistic framework discussed earlier, ecosystem services play a central role in drought management by integrating the ecological, social, and physical dimensions of environmental processes. This provides the foundation for innovative frameworks like HES, emphasizing the interconnectedness between human well-being and ecosystem functionality. Ecosystem services are the benefits humans derive from their natural functions. These services arise from complex biological and physical interactions, sustaining life, economic development, and societal well-being. Ecosystem services encompass the provision of natural resources, regulation of environmental processes, and opportunities for cultural and recreational activities (Millennium Ecosystem Assessment 2005). As outlined in the Millennium Ecosystem Assessment (2005), ecosystem services can be classified into four main categories, provisioning, regulating, cultural, and supporting services, all of which are directly or indirectly affected by drought, as identified in Fig. 2.

Drought significantly disrupts all categories of ecosystem services, magnifying environmental, social, and economic vulnerabilities:

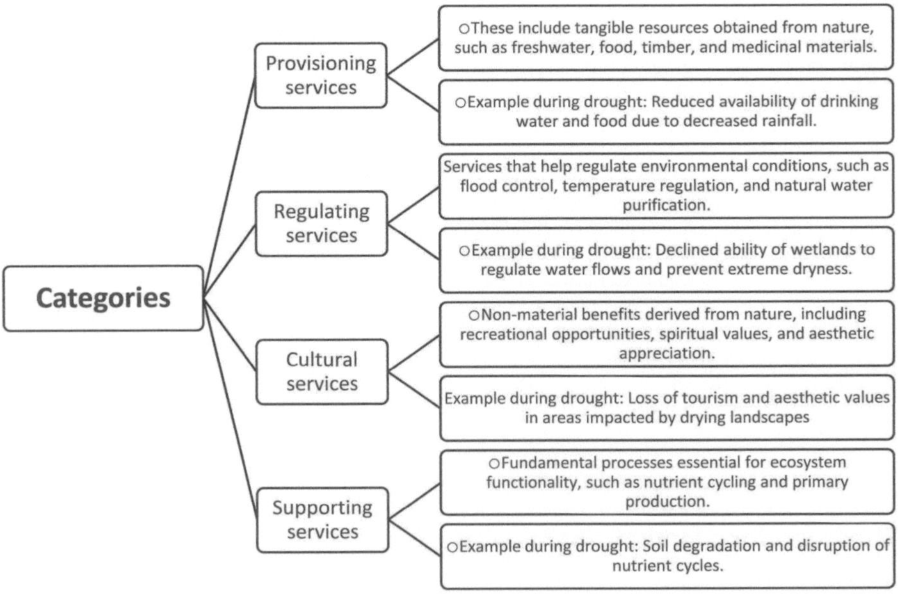

Fig. 2 Categories of ecosystem services and its impact in HES

- Provisioning services: Decreased water and food availability due to lower agricultural yields and reduced freshwater supplies.
- Regulating services: Impaired natural purification processes, increased water pollution, and reduced capacity of wetlands to buffer extreme drought effects.
- Cultural services: Decline in recreational opportunities and loss of cultural and aesthetic values in drought-affected regions.
- Supporting services: Soil degradation, reduced nutrient cycling, and decreased productivity of ecosystems.

Integrating ecosystem services into drought management frameworks, particularly through HES, enables a more comprehensive understanding of the long-term impacts of drought. By focusing on the value of ecosystems and their services, decision-makers can implement strategies that:

- Enhance ecosystem restoration and resilience.
- Support sustainable livelihoods in vulnerable communities.
- Foster adaptive and sustainable resource management practices.

This integration bridges the gap between ecological health and human well-being, aligning with the multidimensional framework outlined in section "Introduction". This foundation leads directly to the specific subset of ecosystem services known as HES, which focuses on water-related benefits and is critical for addressing drought.

The introduction of ecosystem services into drought management marked a paradigm shift. This concept expanded the focus from water availability alone to the broader benefits ecosystems provide, which are critical for mitigating the impacts of drought. See Fig. 3 for a summary of its integration, why is it introduced, who implemented it and its applications.

Similarities and Differences Between Ecosystem Services and HES

Ecosystem services and HES are both essential concepts for understanding the benefits that humans derive from nature. While they share several similarities in their role in environmental sustainability, they differ in scope, impact, and management applications. This comparison highlights how these two frameworks can complement each other in addressing the challenges posed by drought. In this way, the similarities and differences are:

Similarities:

1. Both ecosystem services and HES refer to benefits that humans derive from nature.
2. Both include direct and indirect impacts of drought.
3. Both contribute to natural resource management and environmental sustainability.

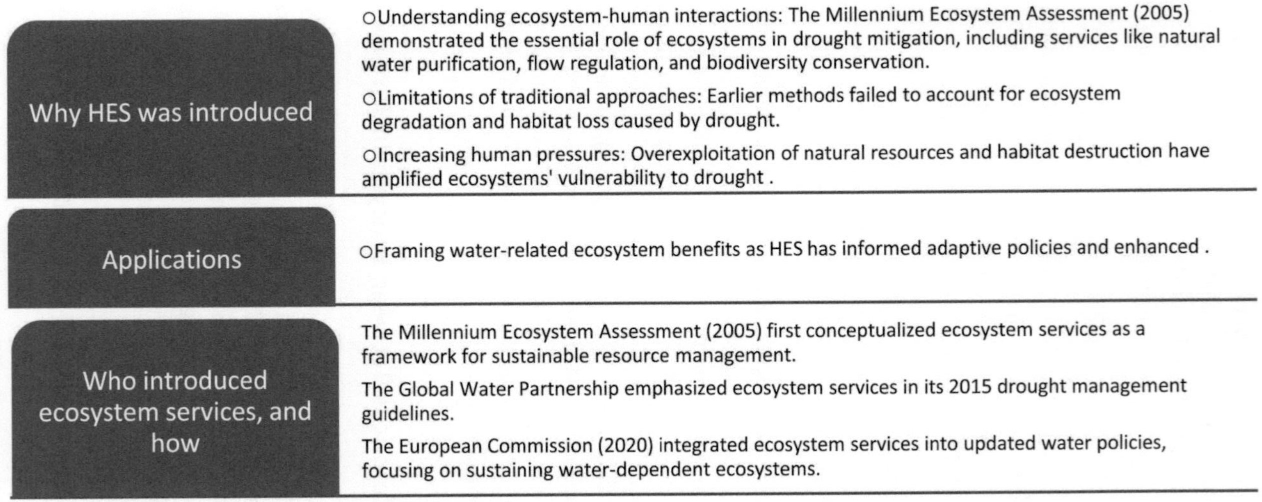

Fig. 3 The introduction of HES in drought management based on Altemus Cullen (2023), Bulama et al. (2021), Burke and Brown (2008), Di Gregorio et al. (2024), Hanadé Houmma et al. (2023), Hidalgo-Triana et al. (2023), Li and Zhong (2024), Millennium Ecosystem Assessment (2005), Torelló-Sentelles and Franzke (2022)

Differences:

1. Scope:
   (a) Ecosystem services encompass a broader range of services covering all ecosystems.
   (b) HES is specifically focused on water-related services and hydrological processes.
2. Impact of drought:
   (a) Drought affects ecosystem services more broadly, impacting soil, vegetation, and nonaquatic habitats.
   (b) Drought impacts on HES are specifically concentrated on water resources and related services.
3. Management goals:
   (a) Ecosystem services serve as a framework for overall environmental policy and sustainability.
   (b) HES is more directly applied to water resource management and drought mitigation.

The integration of ecosystem services and HES provides a comprehensive framework for managing natural resources in the face of drought. While ecosystem services address a wide range of ecosystem functions, HES focuses specifically on water resources and hydrological processes, making it particularly crucial during drought conditions. Combining these two concepts allows policymakers and natural resource managers to adopt more effective, sustainable, and adaptive strategies for mitigating drought impacts.

HES: A Focused Approach

HES is a subset of ecosystem services that emphasizes the critical role of water in sustaining ecosystems and human well-being. It focuses on the flow and quality of water as fundamental to ecosystem health, especially in the context of drought. The following key services of HES highlight their critical role in drought management:

- Key services in HES:
  - Natural water purification.
  - Regulation of water flows.
  - Preservation of water-dependent habitats.
- Impact of drought on HES:
  - Reduced water quality and quantity disrupt natural water purification processes (Nedkov et al. 2022).
  - Loss of biodiversity and habitat diminishes ecosystem stability and resilience (Raposo et al. 2023).
- Why HES matters in drought management:
  - Studies by Brauman et al. (2007) and Lüke and Hack (2017) highlight HES as a critical element in mitigating both environmental and socioeconomic impacts of drought.

– Composite indicators like the Composite Water Stress Index integrate HES data to provide a holistic view of drought impacts (Raposo et al. 2023).

HES focuses specifically on the benefits humans derive from water resources and hydrological processes. These services encompass a range of critical functions, such as the provision of drinking water, natural water purification, regulation of water flows, and preservation of water-dependent habitats (Kisser et al. 2024; Lüke and Hack 2017; Verdon-Kidd et al. 2023). HES are essential for maintaining the health of ecosystems, especially in water-scarce areas. HES are essential for maintaining the health of ecosystems, especially in water-scarce areas.

HES can be categorized as follows:

1. Water provision:
   (a) The supply of water for essential human needs, including domestic, agricultural, and industrial use.
   (b) Example during Drought: Reduced availability of drinking water, especially in arid and semiarid regions, due to lowered precipitation and reduced groundwater recharge.
2. Natural water purification:
   (a) Natural processes such as soil filtration and wetlands that remove pollutants and maintain water quality.
   (b) Example during Drought: Wetlands and soil systems lose their capacity to filter water effectively as water levels drop, resulting in poorer water quality.
3. Flow regulation:
   (a) Management of water flows, including flood control and mitigation of water scarcity.
   (b) Example during drought: Reduced River flows and declining groundwater levels, which increase the risk of extreme dryness and water shortages.
4. Aquatic habitat preservation:
   (a) Protection of water-dependent ecosystems that sustain biodiversity and provide critical services.
   (b) Example during drought: Shrinking habitats for aquatic species due to lower water levels, threatening biodiversity and ecosystem stability.

HES are vital for managing water resources, especially during droughts, when their role becomes even more critical:

- Drinking water supply: As freshwater resources decline during drought, managing HES becomes crucial to ensure a sustainable and adequate water supply for human consumption and agriculture. Flow regulation of wetlands and river systems, key components of HES, are essential in mitigating both flooding and water scarcity. However, during drought, these systems are under stress, which can exacerbate drought effects.

- Habitat preservation: Wetlands and other aquatic habitats support diverse species. Drought severely threatens these ecosystems, reducing biodiversity and the availability of vital ecosystem services (Nedkov et al. 2022).

Strengths and Challenges of HES Framework

The HES framework offers significant benefits, particularly in the areas of ecosystem restoration and drought resilience. By focusing on the restoration of ecosystems, the framework supports biodiversity conservation, which is crucial for maintaining healthy environments that can better withstand the impacts of drought. Additionally, the HES framework enhances resilience by leveraging nature-based solutions, allowing ecosystems to function in ways that mitigate the effects of drought, such as through water retention and soil protection. This approach not only benefits the environment but also supports sustainable livelihoods for vulnerable communities, ensuring that local populations can continue to rely on ecosystem services like water, food, and clean air, even in the face of environmental stresses.

The HES framework improves drought management by providing a more comprehensive and multidimensional approach to understanding drought impacts. By offering comprehensive assessments, HES enables a broader understanding of the long-term effects of drought on both ecosystems and communities, capturing not just immediate water shortages but also the ripple effects on biodiversity, food security, and livelihoods. In addition, enhanced resilience is a key feature of the HES framework. By promoting the restoration of natural buffers like wetlands and forests, HES helps mitigate drought impacts. These ecosystems serve as natural sponges, storing water and helping to regulate the climate, making areas more resilient to drought over time.

Furthermore, by shifting the focus from merely managing water quantity to the sustainability of ecosystems, the HES framework fosters sustainable policymaking. This approach encourages adaptive management strategies that can be modified as conditions change, ensuring that drought management is both proactive and responsive to new challenges (Kisser et al. 2024; Lüke and Hack 2017; Verdon-Kidd et al. 2023).

However, the implementation of the HES framework also faces several challenges. One of the primary obstacles is the substantial investment required for data collection, monitoring, and capacity-building. This investment is crucial for understanding and tracking the health of ecosystems and their ability to provide services during drought. Additionally, there is often limited integration of the HES framework with existing policies and governance structures. This lack of integration makes it challenging to incorporate ecosystem services into broader decision-making processes. Furthermore, the complexity of quantifying ecosystem services presents another challenge. These services are often difficult to measure and incorporate into policy frameworks, making it challenging for decision-makers to fully recognize their value when planning for drought resilience and sustainability. These challenges need to be addressed to fully realize the potential of the HES framework.

In conclusion, the evolution from traditional drought management approaches to those that incorporate ecosystem services, and the HES framework represents a significant advancement. These integrated frameworks offer a multidimensional perspective that combines physical, ecological, and social considerations, ultimately supporting more effective policy decisions and promoting long-term resilience to drought.

Methodology

The multidisciplinary approach adopted in this research involves a structured and systematic multistep process, designed specifically for addressing the complexity inherent in drought management. Initially, the research begins with an extensive review and selection of relevant criteria and indicators derived from current scientific literature, best practices, and case studies. To ensure the practicality, effectiveness, and applicability of the chosen indicators within sustainable drought management, a comprehensive Strengths, Weaknesses, Opportunities, and Threats (SWOT) analysis is conducted. This strategic assessment helps to critically evaluate the suitability and robustness of each indicator, highlighting areas for innovation and refinement. Subsequently, the selected indicators are systematically assessed and prioritized through advanced Multi-Criteria Decision Making (MCDM) techniques. These methodologies provide an analytical and structured decision framework specifically tailored to scenarios characterized by multiple, often conflicting criteria. In this context, modern MCDM approaches, including but not limited to the Analytic Hierarchy Process (AHP), Analytic Network Process (ANP), Technique for Order of Preference by Similarity to Ideal Solution (TOPSIS), and the VIKOR method, are suggested. Recent advancements also recommend integrating methods such as Fuzzy MCDM, which account for uncertainties and ambiguities frequently encountered in drought-related decisions. Specifically, the MCDM process involves three fundamental stages: (1) identification and clear definition of relevant criteria and alternatives, including critical hydrological, environmental, climatic, vegetation-related, and socioeconomic factors; (2) rigorous weighting and prioritization of the criteria based on expert judgment, stakeholder inputs, and data-driven techniques; and (3) systematic evaluation, ranking, and selection of the most optimal and practical indicators aligned with the integration of HES. Owing to their structured, transparent, and integrative nature, these advanced MCDM and complementary methodologies collectively empower decision-makers and practitioners, allowing them to navigate uncertainties, evaluate diverse alternatives, and make robust, informed, and adaptive decisions for effective and sustainable drought management. Predicting flash droughts, which are characterized by rapid onset and intensification of drought conditions, can be challenging but is crucial for water resource management, agriculture, and other sectors. The choice of index depends on the specific objectives of the analysis, available data, and regional needs. It is essential to integrate data from multiple sources and apply advanced modelling and forecasting techniques. Collaboration among meteorologists, climatologists, hydrologists, and agronomists is critical to improving the accuracy of flash drought predictions. While no single indicator can definitively predict flash droughts, a combination of multiple indicators can be used to assess the risk. In the final phase, mitigation and adaptation strategies are designed, ensuring that the outcomes align with the initial objectives and effectively inform decision-making processes. The structure of the methodology is outlined in Fig. 4.

However, previous research indicates various challenges, including limitations in data availability, complexity in calculation methods, and shortcomings in integrating socioeconomic factors. By following the steps, MCDM can be effectively utilized to evaluate ecosystem services. This approach enables stakeholders to consider trade-offs and make decisions that promote the sustainable management of ecosystems.

Results

Drought indicators provide essential tools for assessing and managing droughts by capturing their diverse dimensions and impacts. These indicators are broadly categorized into climate, soil texture, hydrology, vegetation, land use and land cover, and socioeconomic factors. Effective drought management necessitates the integration of these categories to comprehensively address the multifaceted nature of droughts.

Indicators for the Integration of HES in Drought Management

To formulate a sustainable drought management strategy utilizing MCDM techniques as it is mentioned in section "methodology", and integrating HES principles, the following indicators have been identified and group in climate, soil texture, hydrology, vegetation, land use, and socioeconomic impacts. MCDM methods are essential tools for assessing and managing drought, as drought is a complex, multi-dimensional phenomenon requiring simultaneous consideration of various aspects, including climatic, hydrological, environmental, vegetation, socioeconomic, and land-use factors. The structured and flexible nature of these methods allows experts to make optimal and targeted decisions by effectively evaluating multiple, and sometimes conflicting, criteria. In this study, aligned with the objectives of this chapter, the application of MCDM was carried out in two primary stages:

Stage 1 (Identification and Selection of Indicators and Criteria): In this stage, indicators relevant to drought management and Hydrological Ecosystem Services (HES) were identified and categorized through a comprehensive literature review and SWOT analysis (assessing Strengths, Weaknesses, Opportunities, and Threats). These indicators cover a broad range of factors, including climatic (e.g., SPI and PDSI), hydrological (e.g., SDI), soil properties (e.g., soil moisture), vegetation (e.g., NDVI), socioeconomic (e.g., migration rates and community vulnerability), and land-use changes.

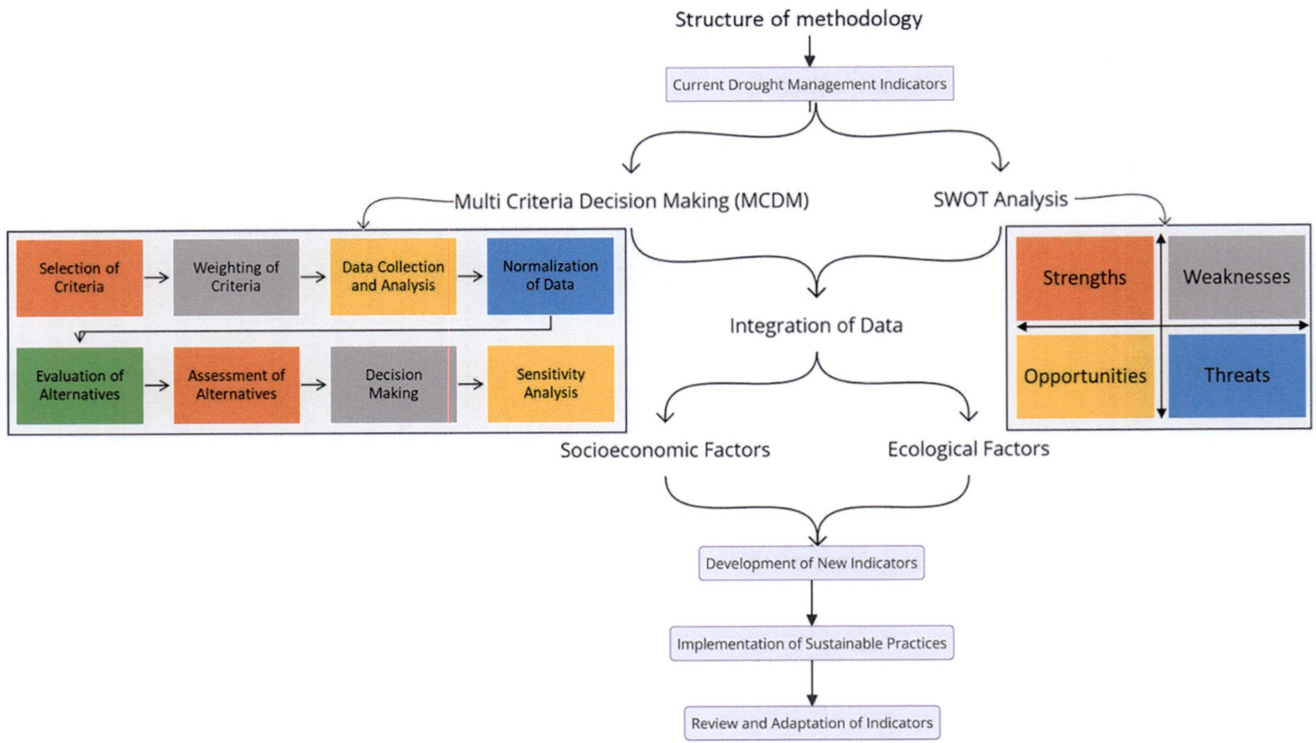

Fig. 4 Structure of the methodology for identifying indicators

Stage 2 (Weighting and Analysis of Indicators in Future Studies): It is recommended that future studies build upon this work by employing methods such as the Analytic Hierarchy Process (AHP), Analytic Network Process (ANP), and ranking techniques like VIKOR and TOPSIS. This will enable researchers to weight the selected indicators, rank areas based on drought vulnerability and develop comprehensive management scenarios and adaptive strategies. The detailed operational steps and procedures for implementing this MCDM approach are outlined in the subsequent sections of this chapter. See Table 3, the first stage of MCDM or the selection of criteria.

The selection of drought indicators in this study followed a systematic, multistep methodological approach to ensure comprehensiveness, relevance, and scientific robustness. Initially, an extensive list of drought-related indicators was compiled based on previous research experience and an in-depth review of the literature, focusing specifically on indicators frequently cited and validated across numerous studies in drought management. This initial compilation enabled the identification of commonly applied indices known for their reliability and wide acceptance in scientific communities dealing with drought assessment and monitoring. Subsequently, a critical gap analysis was conducted to identify which dimensions of drought assessment had received inadequate attention or lacked representation among these frequently used indicators. This gap analysis involved comparing the initial comprehensive list against the various dimensions of drought impact, such as climate, soil, hydrology, vegetation, land use, and particularly socioeconomic factors, to pinpoint any overlooked or underrepresented categories in the existing body of research. Following the identification of these gaps, an additional targeted literature review was performed, emphasizing the less-attended dimensions—most notably socioeconomic impacts. This stage aimed to discover and integrate the most influential and relevant indicators capable of addressing these specific gaps. Indicators such as population demographics, vulnerability groups, economic trends, social networks, and human development factors were thus explicitly integrated, ensuring a more holistic and balanced representation of drought impacts. In the final phase, all selected indicators were systematically classified into comprehensive categories consistent with MCDM methodologies, resulting in the refined, multidimensional framework presented in this research. This structured categorization not only provided clarity and coherence but also facilitated the effective application of MCDM techniques, enabling a robust evaluation, prioritization, and eventual integration of these indicators into drought management strategies. Through this rigorous and methodologically justified approach—starting from an evidence-based initial indicator selection, progressing through critical gap identification, targeted complementary literature reviews, and culminating in a comprehensive multi-criteria categorization—the resulting indicators set forth in this research provide a scientifically sound foundation for holistic drought assessment and effective decision-making.

Drought indicators, spanning climate, soil, hydrology, vegetation, land use, general drought indicators (mixed indicators) and socioeconomic domains (with

Table 3 Indicators for the integration of HES

Indicator type	Examples	Purpose	References
Climate indicators	Standardized precipitation index (SPI), palmer drought severity index (PDSI), evapotranspiration anomalies	Analyze weather patterns and atmospheric data to assess meteorological drought	Hidalgo-Triana et al. (2023), Verdon-Kidd et al. (2023), Williams et al. (2024)
Soil texture and moisture indicators	Soil moisture content, soil organic matter index, hydraulic conductivity	Evaluate soil properties critical for agricultural drought analysis	Bryła et al. (2021), Mehta (n.d.)
Hydrological indicators	Streamflow drought index (SDI), groundwater depletion metrics, reservoir and storage capacities	Assess water availability in surface and subsurface reservoirs	Brauman et al. (2007), Lüke and Hack (2017), Raposo et al. (2023), Wu et al. (2022)
Vegetation indicators	Normalized difference vegetation index (NDVI), leaf area index (LAI), vegetation condition index (VCI)	Measure the effects of drought on plant health and land cover	Hanadé Houmma et al. (2023), Meteorological Organization and Water Partnership (2016), Zellou et al. (2023)
Land use and land cover indicators	Wetland loss index, deforestation rates, urbanization metrics	Track changes in land use that influence and amplify drought risks	Bryła et al. (2021), Raposo et al. (2023), Zellou et al. (2023)
Socioeconomic indicators	Water stress index, economic loss indicators, human migration metrics	Capture broader human and economic dimensions of drought	Alawsi et al. (2022), Meteorological Organization and Water Partnership (2016), Tapia et al. (2021), Torelló-Sentelles and Franzke (2022)
Integrated approaches	Cross-domain collaboration, early warning systems, dynamic adaptation	Provide holistic understanding of drought impacts and enhance response strategies by integrating climate, hydrology, and socioeconomic data	Ha et al. (2023), Hanadé Houmma et al. (2023)

less attention), are indispensable tools for managing the impacts of droughts. These indicators not only provide valuable insights into the occurrence and severity of droughts but also inform decision-making for resource management. The selection of a suitable indicator for analyzing droughts hinges on several factors, including the goals of the analysis, the availability of data, and the specific regional requirements. No single indicator can conclusively predict droughts due to the complexity and diversity of the phenomenon; therefore, the integration of multiple indicators is essential for a comprehensive drought risk assessment.

The chosen indicators can be broadly classified into seven main categories based on the objectives and focus of the analysis:

1. Climate (e.g., precipitation, temperature).
2. Soil and texture (e.g., soil moisture content).
3. Hydrology (e.g., streamflow, groundwater levels).
4. Vegetation (e.g., vegetation health, NDVI).
5. Land use (e.g., land cover types, agricultural practices).
6. Drought mixed indicators.
7. Socioeconomic (e.g., population, vulnerability, economic activities, etc.)

These categories are crucial for generating drought hazard maps, which are derived from the interaction of hazard, exposure, and vulnerability. Specifically, the drought hazard indicator incorporates factors related to climate, soil conditions, hydrology, and land use, among others. For the socioeconomic criteria, the focus is on understanding the current conditions and trends within the region, including factors such as population density, vulnerable groups, and the economic impact of droughts. These data are instrumental in assessing both the immediate and long-term effects of drought on human societies and economies. Data such as population and vulnerable groups, economic conditions, and trends, help understand how the broader community and economy respond to drought conditions.

In regions with limited data, simpler indicators, such as the Standardized Precipitation Index (SPI), might be preferred due to their straightforward nature and ease of calculation (Torelló-Sentelles and Franzke 2022; Wu et al. 2022). In contrast, more complex indicators like the Standardized Precipitation Evapotranspiration Index (SPEI) (Alawsi et al. 2022; Nandgude et al. 2023; West et al. 2019) or the Palmer Drought Severity Index (PDSI) (Ha et al. 2023; Hanadé Houmma et al. 2023; Raposo et al. 2023) are more suited to regions with well-established monitoring networks and more comprehensive data availability. These indicators are often used in areas where continuous monitoring and a deeper understanding of climate, soil, and hydrological factors are essential.

To better understand drought impacts and enhance prediction accuracy, composite indices that combine multiple indicators are often proposed. These indices are designed to offer a more precise and reliable understanding of drought conditions, as they account for the interactions between different environmental and socioeconomic factors. The use of composite indices allows for a more nuanced interpretation of drought risks, considering the diverse factors that contribute to drought severity. The indicators presented in Table 4 offer a comprehensive summary of those commonly employed in literature for predicting and assessing droughts. These indices have been widely applied in studies related to drought, hydrology, and sustainability, demonstrating their productivity and reliability in identifying key trends and risk areas.

While most indicators primarily focus on meteorological aspects and physical parameters, they may not fully capture the broad spectrum of drought impacts on ecosystems, socioeconomic factors, and water resources. However, they provide valuable insights into the spatial and temporal dynamics of drought. A comprehensive list of commonly used indicators, categorized into meteorological, hydrological, and agricultural drought types as a most frequent types of droughts, is evaluated based on MCDM. Also, each indicator has its strengths and weaknesses, opportunities and treats based on SWOT and providing a robust framework for selecting the most appropriate indicator for a given application or region (See Table 5).

Evaluating and Predicting Drought

The comprehensive assessment of drought risk using multi-criteria analysis provides valuable insights for drought management and decision-making. Also, a general methodology and some commonly used software tools for evaluating and predicting drought risk in case studies have been identified. The steps can be combined and summarized in the following:

1. Define the study area and objectives.
    (a) Clearly define the geographic area under investigation.
    (b) Specify the study's objectives, such as assessing drought risk for agriculture, water resources, historical climate data, land cover classification or other sectors.
2. Data collection and compilation.
    (a) Explain the sources of data, including field measurements, satellite data, climate models, or other relevant sources.
    (b) Identify the satellite platforms such as Landsat, Sentinel, MODIS, Gravity Recovery and Climate Experiment (GRACE), Soil Moisture Active Passive (SMAP), ALOS PALSAR, and others. Choose the one or mixed with the appropriate spatial and temporal resolution by applying fusion method.
    (c) Gather historical weather data, soil data, hydrological data, vegetation health and land use, land cover, Texture and Change Detection (CORINE), and any relevant socioeconomic data that might be impacted by drought for real-time monitoring and time-series analysis.
3. Drought indicators selection.
    (a) Choose appropriate drought indicators for the study (based on Table 4. Common indicators include the Standardized Precipitation Index (SPI), Palmer Drought Severity Index (PDSI), and Standardized Precipitation

Table 4 Indicators, definitions and categories commonly employed (Alawsi et al. 2022; Altemus Cullen 2023; Aznarez et al. 2014; Di Gregorio et al. 2024; Falah et al. 2024; Ha et al. 2023; Hanadé Houmma et al. 2023; Li and Zhong 2024; Moreno et al. 2022; Nandgude et al. 2023; Raposo et al. 2023; Torelló-Sentelles and Franzke 2022; West et al. 2019; Wu et al. 2022)

Index		Definition	Categories
NDVI	Normalized difference vegetation index	Assesses vegetation health and density	Vegetation
SAVI	Soil-adjusted vegetation index	Includes a soil adjustment factor to account for soil brightness	
EVI	Enhanced vegetation index	Corrects for atmospheric influences and provides accurate assessments of vegetation	
VHCI	Vegetation health condition index	Combines measures of temperature and vegetation health to evaluate drought severity	
NDBI	Normalized difference built-up index	Detects built-up areas, urbanization, and impervious surfaces	Build-up area
LULCC	Land use and land cover change	Tracks LULC changes over time, clarifies how human activities are altering landscapes	
LST	Land surface temperature	Understanding urban heat islands, monitoring climate change impacts, and studying LULCC	Climate and whether
ET	Evapotranspiration rates	Measures the combined processes of water evaporation from the soil surface and transpiration by plants	
SPI	Standardized precipitation index	Assesses precipitation anomalies, identify droughts and wet periods	
SPI-6	Six-month standardized precipitation index	A version of SPI that considers a longer time scale (6 months) to assess drought conditions	
SPEI	Standardized precipitation-evapotranspiration index	Combines precipitation and temperature data, considering both water input (precipitation) and water demand (evapotranspiration)	
SML	Soil moisture levels or NDMI	Estimates soil moisture levels, based on the difference between actual and potential evapotranspiration	Soil and texture
BSI	Bare soil index	Detects bare soil areas, which can be valuable in agriculture and land cover	

(continued)

Table 4 (continued)

Index		Definition	Categories
SI	Salinity index	Estimates soil salinity levels, which is crucial for agriculture and land management	
NDMI	Normalized difference moisture index	Measures the wetness or moisture content in an image	
NDWI	Normalized difference water index	Quantifies water content and is helpful in monitoring changes in water bodies	Water source
WST	Water surface temperature	Monitors the temperature of water bodies helps in understanding the thermal dynamics	
WI	Water resources	Status of water reservoirs, rivers, and groundwater levels, assess water availability during drought periods	
RDI	Rainfall deficit index	Quantifies the cumulative rainfall deficit over a specific period and long-term trends in precipitation	
SDI	Streamflow drought index	Assesses drought conditions using streamflow data, is relevant for water resource	Drought factors
KBDI	Keetch-Byram drought index	Evaluates the fire potential and soil moisture conditions, for wildfire risk assessment	
WR	Wildfire risk	Identifies areas at high risk of wildfires due to dry conditions	
NDDI	Normalized difference drought index	Captures changes in vegetation associated with drought stress same as NDVI	
PDSI	Palmer drought severity index	A comprehensive drought index that considers both precipitation and evapotranspiration. It's used to assess long-term drought conditions	
DSI	Drought severity index	Combines vegetation health, soil moisture, and temperature for evaluate drought severity	

(continued)

Table 4 (continued)

Index		Definition	Categories
	Population Socity Economy Civilization	Population demographic trends, vulnerable groups Social health, habitat education conditions, potential social impacts, social capital Economic capacity, developmental levels in economical contexts, economic roles in region Food security and health, cost and benefits savings Nodes and communication networks link (including urban and rural networks), the region's specific morphological characteristics and structure	Socioeconomic

Evapotranspiration Index (SPEI). These indicators provide key insights into drought risk across multiple sectors, including agriculture and water resources.

4. Data preprocessing.
   (a) Clean and quality-check the collected data.
   (b) Normalize the indicators.
   (c) Aggregate data to the desired spatial and temporal scales for analysis.
   (d) Preprocess the satellite imagery to correct atmospheric effects, geometric distortions, and other artifacts.
5. Collect methods for analyzing.
   (a) Multi-criteria analysis: Some common multi-criteria methods include Weighted Sum Method, Analytic Hierarchy Process (AHP), Analytic Network Process ANP and FANP, Technique for Order Preference by Similarity to Ideal Solution (TOPSIS), and Electre Method (Elimination and Choice Expressing Reality) for ranking alternatives by comparing their performance and other innovation methods.
   (b) Spatial and geospatial analysis: Apply ArcGIS and Geographic Information System (QGIS) to perform geospatial analysis and use spatial data to create visualize maps for different indexes. Also, use Google Earth Engine and ENVI for analyzing satellite images and data, from environmental monitoring to land-use classification, time-series, and cloud-based analyses, and integrate data from multiple sources and develop algorithms.
   (c) Machine learning models: support vector machines, and deep learning neural networks, to monitor and predict drought conditions.
6. Drought monitoring and analysis.
   (a) Analyze historical drought events to understand their frequency, duration, and intensity into the likelihood of flash droughts in the study area.

Table 5 SWOT analysis of main frequent drought indicators

Frequented used indexes	Strengths	Weaknesses	Opportunities	Threats
Standardized precipitation index (SPI)	– Uses precipitation data alone, which is readily available – Easy to calculate and interpret	– Does not consider other factors like temperature or soil moisture – Relies on accurate and long-term precipitation data	– Can be integrated with other indexes for a more comprehensive assessment	– Limited effectiveness in regions where precipitation is not the primary driver of drought
Palmer drought severity index (PDSI)	– Considers both precipitation and temperature – Provides a historical perspective on drought severity	– Complex calculation requiring multiple inputs – Susceptible to errors due to incomplete or unreliable data	– Can be used for long-term trend analysis and planning	– May not capture short-term drought events effectively
Standardized precipitation-evapotranspiration index (SPEI)	– Incorporates both precipitation and potential evapotranspiration, providing a better understanding of moisture availability – Accounts for temperature variability	– Requires data on potential evapotranspiration, which may not be available in all regions – Calculation complexity may hinder widespread adoption	– Suitable for assessing agricultural drought and water resource management	– Limited applicability in regions with insufficient meteorological data
Surface water supply index (SWSI)	– Integrates various hydrological variables such as streamflow, reservoir levels, and snowpack – Provides a direct assessment of water availability	– Relies on data from gauging stations, which may be sparse in some areas – Limited effectiveness in regions where surface water is not the primary water	– Useful for water resource management and drought response planning	– Susceptible to errors and biases in data collection and measurement

Streamflow drought index (SDI)	– SDI provides specialized insights into streamflow conditions, making it vital for assessing water availability and managing resources during droughts	– It focuses solely on surface water flows, neglecting other important drought indicators like soil moisture or groundwater levels	– Enhancements in hydrological data collection and analysis methods could improve the accuracy and functionality of SDI	– Increased unpredictability in weather patterns due to climate change may affect the reliability of SDI in predicting water conditions
Vegetation health index (VHI)	– Utilizes satellite data to assess vegetation health, which can indicate drought stress – Provides spatially explicit information	– Dependent on satellite imagery availability and quality – May not capture drought impacts on non-vegetated areas	– Useful for monitoring agricultural drought and ecosystem health	– Limited effectiveness in heavily cloud-covered regions or areas with complex terrain
Soil moisture index (SMI)	– Directly measures moisture content in the soil, which is crucial for agriculture – Provides information on short-term drought conditions	– Requires soil moisture data from reliable sources, which may not be available in all regions – Limited spatial coverage of soil moisture monitoring networks	– Valuable for agricultural planning and drought early warning systems	– Challenges in integrating different soil moisture measurement methods and scales

(b) Drought indices are typically calculated using statistical software like R, Python, or specialized drought monitoring tools.
(c) Identify any patterns or trends in past drought occurrences to identify any changes in drought patterns and conditions over time and to classify drought categories (e.g., mild, moderate, severe, extreme of flash) based on indicator values.
(d) Meteorological data analysis: Tools like Climate Data Operators (CDO) and Climate Data Analysis Tools (CDAT) can be used to process and analyze meteorological data.

7. Hydrological management analysis.
   (a) Hydrological models: indicators same as the movement of water through the landscape, including rainfall, runoff, evapotranspiration, and groundwater recharge Soil and Water Assessment Tool (SWAT), the Variable Infiltration Capacity (VIC) and Software like Hydrology Studio, HEC-HMS (Hydrologic Engineering Centre's Hydrologic Modelling System), and MODFLOW.
   (b) Water resources modelling: evaluate Infrastructure and resilience such as the availability of water storage facilities, using software such as WEAP (Water Evaluation and Planning System) or MIKE SHE (Simulating the Hydrology of Catchments) for water resources planning and management.
8. Climate and weather analysis.
   (a) Regularly monitoring weather conditions and assessing long-term climate data to identify climate patterns and teleconnections that may influence drought in the region. This can include information on El Niño/La Niña events and other climate drivers.
   (b) Climate data analysis: using global and regional climate models (GCMs and RCMs) and Software such as R, Python.
   (c) Climate modelling: Climate Model Intercomparison Project (CMIP) models or Weather Research and Forecasting (WRF) model can be used for creating climate scenarios.
9. Statistical analysis and modelling.
   (a) Evaluate socioeconomic data such as population and vulnerability, particularly vulnerable groups, economic conditions and trends and the potential human impact.
   (b) Use statistical methods and modelling techniques to establish relationships between drought indicators, climate variables, and historical drought events.
   (c) Estimate the likelihood and severity of future droughts based on current conditions. Regression analysis, and time series analysis.
   (d) Statistical software like R, SPSS, SAS, and STATA can be employed for statistical analysis of data related to drought and water research.
10. Scenario analysis.
   (a) Develop different scenarios for future climate conditions, considering climate change projections.

   (b) Assess the impact of these scenarios on future drought risk by prediction models and evaluate their performance against actual conditions.
   (c) Evaluate how changes in the index parameters (e.g., timescales for SPI) impact the results and whether different parameters might better capture local conditions.
   (d) Compare the performance of drought index with other relevant indices, such as SPI, PDSI, or SPEI, if applicable.
11. Risk assessment and mapping.
   (a) Quantify the risk of drought for different sectors (e.g., agriculture, water resources) based on the developed models.
   (b) Create risk maps that highlight areas of high and low drought risk within the study area based on the current conditions and potential risks for predicting future.
12. Early warning systems.
   (a) Develop or enhance early warning systems for drought.
   (b) Specify trigger levels for initiating response actions.
13. Mitigation and adaptation strategies.
   (a) Propose strategies and measures to mitigate the impact of drought and adapt to changing conditions, based on the risk assessment.
14. Validation and calibration.
   (a) Validate the models and results using historical drought events and observed data.
   (b) Calibrate the models if necessary to improve accuracy. Assess how well the drought index aligns with observed impacts and the impact on agriculture, such as crop yield reductions, water shortages, or economic losses during drought events.
   (c) Use historical drought event records, observations, or other recognized drought indicators to validate the performance of the index.
15. Reporting and communication.
   (a) Spatial maps of drought risk, ranking of areas by risk level, and any trends or insights.
16. Continuous monitoring and policy implications.
   (a) How can the results be used for drought preparedness, mitigation, or response? What are the implications for local communities, agriculture, and water resource management?
17. Feedback and improvement.
   (a) Observed impacts on agriculture, water resources, or other sectors affected by drought.

The proposed methodological steps were summarized in Fig. 3.

Discussion

Primary Indicators

The latest research highlights the limitations of traditional drought indicators and underscores the need for multidimensional drought indicators that incorporate various aspects. This research suggests the development of primary indicators that integrate additional criteria, including socioeconomic factors, to offer a comprehensive evaluation of drought impacts. As outlined in Fig. 5, the SWOT analysis applied to drought management serves as a foundational step for introducing a range of indicators.

In this study, it is recommended that drought indicators should encompass six primary categories of criteria: climate, soil texture, hydrology, vegetation, land use and land cover, and socioeconomic data that based on different types of droughts, the weight and impact of each indicator will be changed, but in general all of them must be mentioned for evaluation the real conditions (Aznarez et al. 2014; Ha et al. 2023). For instance, hydrological indicators might include measurements of groundwater recharge rates, reservoir levels, and river flow regimes, and in type of the groundwater drought, these items have a stronger role. Effective collaboration among these six criteria is crucial for enhancing the accuracy of drought predictions. In addition, a key element in this research is identifying and analyzing socioeconomic factors in the context of HES, which adds a significant

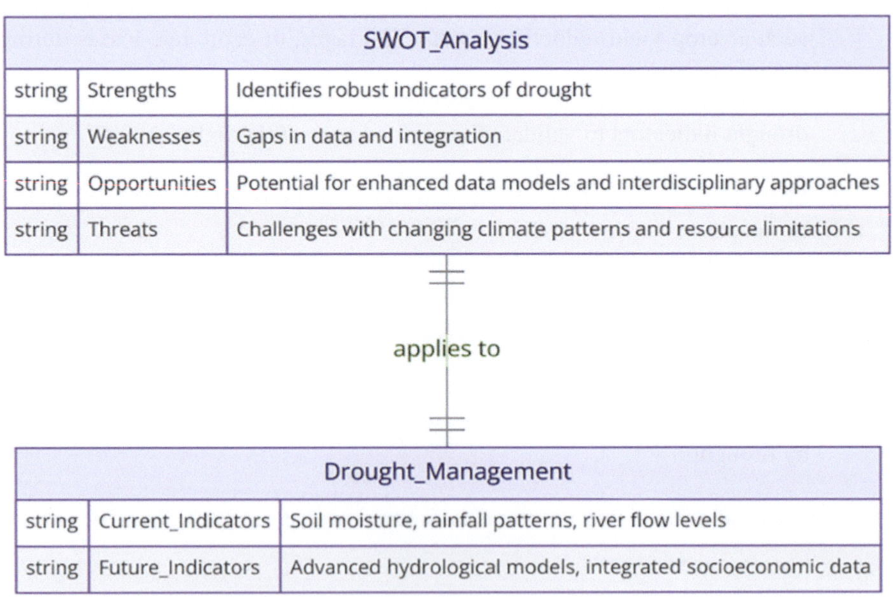

Fig. 5 Structure of SWOT analysis and the indicators identified

innovative layer to drought assessment strategies that must be mentioned in all types of droughts.

A comprehensive suite of indicators across various drought dimensions ensures a comprehensive approach to drought management, enhancing capabilities to predict, prepare for, and respond to drought conditions sustainably. Therefore, list practical and multifaceted indicators previously used in numerous studies, alongside novel ones introduced based on a SWOT analysis of traditional indicators in Tables 4 and 5 based on six main criteria, and specify criteria as a "Drought indicator," which its indicators are multifaceted and comprehensive items and will be used in all types of droughts. This list targets two main pillars: a drought risk map incorporating both the drought indicators and HES. Secondly indicators specifically tailored to the socioeconomic factors of the study area. This list, influenced by current trends and conditions in case studies, will focus on economic and social activities, evaluating socioeconomic data.

It is evident that no single indicator can independently provide a comprehensive forecast of drought conditions. A combination of multiple factors and indicators is essential for an accurate assessment of drought risk. Utilizing a diverse set of indicators enhances precision and minimizes errors in drought assessment. This underscores the importance of selecting indicators that match the specific objectives of the analysis. These mixed indicators that, alongside socioeconomic factors, can effectively evaluate the conditions of HES and provide a more professional analysis of drought:

1. Multi-Indicator Drought Indicator (MIDI): The MIDI aggregates several key indicators and may be updated less frequently, depending on the availability of underlying data such as precipitation, temperature, and soil moisture readings. This indicator offers an integrated view of drought conditions across various time scales, making it particularly valuable.
2. Standardized Precipitation-Evapotranspiration Index (SPEI): The SPEI combines precipitation and temperature data, considering both water input (precipitation) and water demand (evapotranspiration).
3. Vegetation Condition Indicator (VCI): This indicator assesses how current vegetation conditions deviate from historical norms, providing vital information on the severity of agricultural droughts that impact plant growth.
4. Palmer Drought Severity Indicator (PDSI): The PDSI is widely recognized for its comprehensive approach to measuring drought severity, which includes the effects of temperature, thereby making it highly pertinent in the context of climate change.
5. Integrated Water Resource Management (IWRM) Indicators: These could include metrics on the efficiency of water use, equity in water distribution, and the sustainability of water resources. IWRM focuses on the coordinated development and management of water, land, and related resources to optimize economic and social welfare.
6. Drought Severity Index (DSI): the DSI combines vegetation health, soil moisture, and temperature for evaluate drought severity.

7. Combined Drought Indicator (CDI): The CDI integrates diverse data sources, including precipitation, soil moisture, streamflow, and vegetation metrics, to deliver a holistic assessment of ecological drought conditions across different ecosystems.
8. And mix indicator related to socioeconomic parameters, it must be included:
   (a) Population (population demographic trends, vulnerable groups).
   (b) Society (social health, habitat education conditions, potential social impacts, social capital).
   (c) Economy (economic capacity, developmental levels in economical contexts, economic roles in context, food security and health, cost and benefits savings).
   (d) Civilization (nodes and communication networks link, including urban and rural networks, and the region's specific morphological characteristics and structure).

These indicators are instrumental in enhancing the analytical capabilities for evaluating drought impacts, supporting effective decision-making in drought management.

Current Challenges

Gaps and Challenges in Current Drought Management

Despite advances in understanding and managing drought, critical gaps persist in existing frameworks, indicators, and policies. These gaps undermine the ability to comprehensively address the environmental, social, and economic dimensions of drought, particularly as climate change intensifies the frequency and severity of drought events. A detailed examination of these challenges reveals the necessity of adopting an integrated, multidimensional approach to enhance resilience and sustainability (Ha et al. 2023). For instance, the absence of integrated drought indicators has led to inefficient water resource management and heightened socioeconomic vulnerabilities in regions like sub-Saharan Africa.

Lack of Integrated Indicators

One of the most significant barriers to effective drought management is the lack of holistic indicators that can capture the multidimensional nature of drought. Existing indicators, while essential for specific domains, often fall short in providing a comprehensive view of drought's interconnected impacts on water resources, ecosystems, and societies. Key challenges include:

- Fragmentation of metrics:
  - Commonly used indicators such as the Standardized Precipitation Index (SPI) and the Palmer Drought Severity Index (PDSI) focus primarily on climatic variables like precipitation and temperature. These tools are effective for meteorological drought assessments but fail to account for groundwater

depletion, vegetation stress, or ecosystem disruptions (Li and Zhong 2024; Tramblay et al. 2020; Williams et al. 2024).
  - Hydrological indicators, such as the Streamflow Drought Index (SDI), emphasize water availability in rivers but neglect downstream ecological and socioeconomic impacts (Lüke and Hack 2017; van Loon and Van Wassenhove 2020; Wu et al. 2022).
- Insufficient cross-domain integration:
  - Ecological disconnect: Indicators rarely integrate biodiversity loss, wetland degradation, or disruption of ES, despite their critical role in drought resilience (Zellou et al. 2023).
  - Social disconnect: Metrics tracking human displacement, economic losses, and food insecurity often remain isolated from ecological and hydrological data, resulting in siloed assessments (Ha et al. 2023).
  - Incompatibility across scales: Global and regional drought indicators often fail to address local variations in drought severity and impact, reducing their applicability in context-specific management strategies.
- Overreliance on historical data:
  - Traditional indicators like SPI and PDSI are built on historical climatic averages, making them less reliable in the context of increasing climate variability and extremes (Gabiña et al. 2019; West et al. 2019).
  - Lack of predictive models that integrate climate forecasts with socioeconomic and ecological data limits the ability to anticipate and mitigate future drought risks.
- Underutilization of advanced technologies:
  - Remote sensing limitations: While satellite tools such as the Normalized Difference Vegetation Index (NDVI) and Soil Moisture Active Passive (SMAP) provide valuable data, their integration with ground-based socioeconomic metrics is minimal.
  - Missed opportunities with AI: Artificial intelligence and machine learning have significant potential to analyze complex drought datasets but remain underused, particularly in developing regions.
- Limited consideration of emerging metrics:
  - Indicators incorporating ecosystem services and HES are still in developmental stages and not widely implemented in drought management frameworks. Metrics assessing natural water purification, flow regulation, and biodiversity resilience could enhance understanding of drought's long-term impacts (Brauman et al. 2007; Lüke and Hack 2017).

The need for integrated frameworks to address these challenges requires cohesive frameworks that:

1. Combine climatic, hydrological, ecological, and social metrics: Integrated tools such as the Composite Drought Index (CDI) or Composite Drought Impact Indicators (CDII) can provide a holistic view of drought impacts.

2. Leverage advanced technologies: Use satellite imagery, big data analytics, and machine learning models to create predictive and multidimensional drought monitoring systems.
3. Develop region-specific indicators: Tailor metrics to local conditions, such as soil salinity in arid regions or reservoir capacity in urban areas.
4. Incorporate emerging metrics: Expand the use of ecosystem services and HES indicators to measure ecosystem contributions to drought resilience.

The fragmented and narrow focus of current drought indicators prevents policymakers from fully understanding and mitigating the multi-faceted impacts of drought. By adopting integrated, advanced, and ecosystem-based metrics, drought management frameworks can evolve to address the growing challenges posed by climate change, ensuring resilience for both ecosystems and human societies.

Weak Integration of Ecological, Hydrological, and Social Data

Drought is inherently a multidimensional phenomenon, affecting ecological, hydrological, and social systems in interconnected ways:

- Ecological impacts: Biodiversity loss, habitat degradation, soil erosion, and the disruption of natural water purification and food chain stability (Zellou et al. 2023).
- Hydrological impacts: Severe depletion of surface and groundwater resources, reduced streamflow, and shrinking reservoirs exacerbate water scarcity.
- Social impacts: Forced migration, food insecurity, and rising poverty disproportionately affect vulnerable populations, amplifying inequalities (Ha et al. 2023).

However, current management frameworks often fail to integrate data across these domains:

- Ecological oversights: Drought's cascading effects on biodiversity, ES, and wetlands are rarely factored into water management plans (Zellou et al. 2023).
- Hydrological disconnection: Most drought monitoring systems rely heavily on precipitation and streamflow data, neglecting broader impacts on ecosystems and water quality.
- Siloed social metrics: Data on human displacement, economic losses, and social vulnerabilities often exist in isolation from ecological and hydrological analyses (Ha et al. 2023).

This lack of integration results in fragmented and ineffective responses, preventing decision-makers from developing strategies that balance ecological preservation, sustainable water resource management, and social resilience.

Limited Policies and Reactive Approaches

Despite the growing severity of droughts driven by climate change, the dominance of reactive, short-term policies continues to hinder effective drought management, highlighting the urgent need for proactive and sustainable approaches:

- Focus on crisis management: Most policies emphasize post-drought relief efforts, such as emergency water rationing or financial compensation, rather than addressing the root causes of vulnerability through long-term planning (GWP 2020).
- Neglect of preventative measures: Proactive strategies, such as investing in ecosystem restoration, improving agricultural resilience, and fostering sustainable water use practices, are often underfunded or absent from national policies.
- Policy fragmentation: Many regions lack coherent frameworks that align local, national, and international drought management efforts. For example, shared water resources, such as transboundary rivers, often become sources of conflict during droughts due to uncoordinated policies.
- Limited integration of HES: Few policies incorporate these frameworks, missing opportunities for comprehensive and sustainable drought management (Meteorological Organization and Water Partnership 2016).

These limitations underscore the need for a shift from reactive to proactive, ecosystem-based policies that prioritize resilience and sustainability.

Insufficient Data and Limited Use of Advanced Technologies

The availability and utilization of accurate, real-time data remain significant challenges in drought management, particularly in developing regions:

- Data gaps in vulnerable regions: Areas like sub-Saharan Africa and parts of South Asia often lack reliable data on water resources, soil health, and ecosystem conditions, limiting the accuracy of drought assessments (Raposo et al. 2023).
- Underutilization of technology: Advanced tools such as satellite-based remote sensing, machine learning algorithms, and IoT-based water monitoring are not widely adopted, even in regions with sufficient resources (Bryła et al. 2021).
- Challenges in real-time monitoring: While global initiatives like the European Drought Observatory (EDO) provide valuable data, their scope often excludes localized social and ecological impacts, leaving gaps in actionable intelligence.

Enhanced investment in data collection infrastructure, along with the application of cutting-edge technologies, is essential to improving drought forecasting and response strategies.

Challenges in Integrating HES

While the concept of ecosystem services and HES has gained traction, their integration into drought management remains limited:

- Lack of awareness and adoption: Policymakers often undervalue the role of ecosystems in mitigating drought impacts, such as wetlands' capacity to regulate water flows and purify water naturally (Brauman et al. 2007).
- Operational limitations: Current drought frameworks struggle to quantify and incorporate the benefits of HES, such as groundwater recharge, biodiversity preservation, and soil stabilization (Hanadé Houmma et al. 2023).
- Underdeveloped indicators: While there is progress in developing HES-specific metrics, these tools are not yet widely implemented or standardized (Bryła et al. 2021).

Greater emphasis on nature-based solutions, coupled with actionable HES indicators, can significantly enhance drought resilience.

Case Studies Highlighting Gaps and Challenges
Global case studies demonstrate the widespread impact of these gaps and challenges:

1. Southern Europe: Recurring droughts in Spain and Italy have exposed the limitations of relying solely on hydrological metrics, as social and ecological impacts are rarely integrated into decision-making (Zellou et al. 2023).
2. Sub-Saharan Africa: In regions like Kenya and Ethiopia, severe data shortages and underdeveloped monitoring infrastructure hinder the ability to manage recurring agricultural droughts effectively (Gabiña et al. 2019).
3. South Asia: India and Pakistan face compounded challenges of over-reliance on irrigation infrastructure and neglect of ecological restoration, leading to exacerbated groundwater depletion and ecosystem stress (Ha et al. 2023).

These examples highlight the urgent need for multidimensional, region-specific approaches to drought management.

Need for Comprehensive Frameworks
Addressing the gaps in current drought management requires a paradigm shift towards integrated and proactive strategies:

- Development of holistic indicators: Create metrics that combine climatic, hydrological, ecological, and social dimensions to provide a complete picture of drought impacts.
- Adoption of ecosystem-based approaches: Integrate HES into drought frameworks to leverage nature-based solutions such as wetland restoration and sustainable agriculture.
- Technological innovation: Invest in satellite imagery, big data analytics, and IoT systems to enable real-time drought monitoring and forecasting.
- Sustainable and inclusive policies: Develop policies that balance environmental conservation, water resource management, and socioeconomic development, with a focus on supporting vulnerable communities.

The gaps and challenges in current drought management highlight the need for a comprehensive, multidimensional approach that integrates physical, ecological, and social dimensions. By adopting advanced technologies, ecosystem-based solutions, and holistic indicators, drought management frameworks can become more effective, sustainable, and resilient. This evolution is essential to address the increasing frequency and severity of droughts in a changing global climate.

One of the principal challenges in applying the HES concept in drought-prone areas is establishing a set of multifaceted indicators, incorporating both distinct types of droughts and HES (as it mentions in previous part, socioeconomic indicators are essential for evaluation sustainability of HES). The Millennium Ecosystem Assessment and recent studies, such as those by Hanadé Houmma et al. (2023) and Alawsi et al. (2022), highlight the need for indicators in drought management. Ha et al. (2023) describes a set of 17 indicators using in situ measurements, hydrological modelling, GIS, and remote sensing techniques. This need for integration is supported by research focusing on quantifying the droughts impacts on services like water supply, quality, and biodiversity (Gabiña et al. 2019; Nedkov et al. 2022; Raposo et al. 2023). Innovative case studies, such as the California Water Action Plan, provide insights into effective integrated water resource management (Bryła et al. 2021; Chang and Bonnette 2016). This approach highlights the need to integrate advanced drought detection technologies and modelling tools for incorporating HES into drought risk analyses. In early 2022, the integration of hydrological modelling emerged as a priority in water-related HES research, as highlighted by the severe drought reported by the Global Drought Observatory (GDO) of the Copernicus Emergency Management Services (CEMS) (Bryła et al. 2021; Tramblay et al. 2020).

However, there is still a weakness in defining comprehensive and all-encompassing indicators (especially in socioeconomic criteria and link between these criteria and others). In addition, it is necessary to pay attention to two important points: (1) integration of ecological and hydrological data and practical indicator with focus not only on physical factor but also on seriocomic criteria), and (2) integration of advanced predictive modelling and innovative monitoring technologies. In recent studies, the integration of ecological and hydrological data has become essential for better management of HES, especially in growing drought challenges. Researchers, recognizing this need, have been actively quantifying the impact of droughts on essential services like water supply, quality, and biodiversity. This integrated approach is pivotal for informed decision-making and effective resource management, ensuring that actions are both sustainable and adaptive to changing conditions (Alawsi et al. 2022; Ha et al. 2023; Lüke and Hack 2017). Raposo et al. (2023) have specifically focused on quantifying the impact of droughts on those services. Their research highlights the growing recognition of the necessity to integrate ecological and hydrological data for more effective management of HES. They noted how droughts affect various HES, including water purification, flood control, and habitat functions, thereby enhancing our understanding of HES impacts and aiding informed decision-making and effective resource management (Nedkov et al. 2022). Also, innovative case studies, such as the implementation of the California Water Action Plan and similar initiatives, provide valuable insights into effective integrated water resource management in drought-prone areas. These studies offer actionable recommendations for policymakers and stakeholders, focusing on sustainable and adaptive management strategies to mitigate drought impacts (Zellou et al. 2023). Still, there is a gap in the integration of ecological and hydrological data that needs to be solved as a coherent and unified methodology for achieving this kind of approach. Also, there are some gaps in prediction and monitoring modelling that can cover all the characteristics of indicators and parameters. The challenges identified include addressing data gaps in drought-prone regions, improving the accuracy of predictions under changing climate scenarios, and ensuring equitable water resource management (Torelló-Sentelles and Franzke 2022). Recent literature indicates a gap in the ability to predict and rapidly respond to drought impacts on HES. For example, studies by Gabiña et al. (2019) highlight the challenges in accurately predicting drought patterns due to climate change, and West et al. (2019) emphasize the lack of real-time monitoring systems capable of assessing the immediate effects of drought on water quality and availability. Therefore, there is a need to contribute significantly to the sustainability and resilience of water resources under changing climate conditions. The research of Ha et al. (2023) underscores the need to contribute to this field by developing more accurate and nuanced predictive models that incorporate climate variability, land-use changes, and socioeconomic factors (Torelló-Sentelles and Franzke 2022; Wu et al. 2022). Integrating advanced machine learning algorithms and big data analytics has led to more sophisticated models for drought prediction, incorporating climate variability, land-use changes, and socioeconomic factors for improved accuracy (Ha et al. 2023; Hanadé Houmma et al. 2023; West et al. 2019).

These developments, notably enhanced post-2021, reflect an ongoing evolution in the approach to environmental forecasting (Alawsi et al. 2022). The importance of remote sensing and Internet of Things (IoT) technologies in real-time drought monitoring has been emphasized in recent research, underlining their critical role in providing timely and accurate data for drought management (Hanadé Houmma et al. 2023). These technologies offer detailed insights into soil moisture levels, groundwater reserves, and vegetation health, key indicators of drought impact (Aznarez et al. 2014; Nedkov et al. 2022), providing data for early warning systems, thus enhancing the capacity for rapid response to drought conditions. The development of advanced predictive modelling and innovative monitoring technologies in a sustainable and practical way has significantly enhanced our ability to predict, detect, and respond to drought impacts on HES and fill the methodology gap. This enhanced capability will not only improve the understanding of the immediate drought effects on HES but also aid in the development of more effective and adaptive management strategies and policies. Addressing drought challenges amidst climate change requires a multidisciplinary approach, integrating research, predictive models, technologies, and policy frameworks. This strategy aims to develop multifaceted indicators for drought and HES, crucial for minimizing droughts' effects and ensuring sustainable water management, and filling the gaps related to the integration approach and methodology that has been identified in previous studies and research. This enhances the resilience of water resources, safeguarding the well-being of ecosystems and human societies in an era of increasing climate variability.

Conclusion

Effective management of drought requires a comprehensive approach that integrates ecological, hydrological, and socioeconomic dimensions. It is critical to employ primary indicators and robust frameworks for the precise monitoring of drought conditions, assessment of impacts on HES, and the deployment of adaptive management strategies. This article contributes significantly to the advancement of sustainable drought management by emphasizing the importance of integrated indicator systems in enhancing resilience to droughts.

Drought management, as a multidimensional challenge, requires innovative and integrated frameworks that combine ecological, hydrological, and socioeconomic dimensions. This chapter reveals the limitations of traditional frameworks, often reliant on climate and hydrological indicators, in addressing the complex and interconnected impacts of drought. By focusing on HES, this study emphasizes the necessity of composite, multi-criteria indicators that account for physical (e.g., water quality and availability), social (e.g., migration and economic impacts), and environmental (e.g., biodiversity and land use) dimensions. Key Recommendations for Sustainable Drought Management:

1. Composite and data-driven indicators: Leveraging advanced indices like the CDI to integrate climatic, hydrological, and socioeconomic data for a holistic understanding of drought impacts.
2. Modern technologies integration: Utilizing tools such as AI, ML, remote sensing, and GIS to enhance drought prediction and analysis accuracy.
3. Ecosystem services utilization: Restoring wetlands, protecting natural resources, and strengthening ecosystem resilience to mitigate drought impacts.
4. Localized strategy design: Tailoring strategies to specific regional contexts, considering demographic trends, infrastructure, and socioeconomic vulnerabilities.

The results of this study underscore the importance of multidimensional approaches to drought management. Composite indices, like CDI, provide a more comprehensive framework by integrating diverse criteria such as climate variability, hydrological conditions, vegetation health, and socioeconomic factors. These indices are essential for developing risk maps, guiding resource allocation, and formulating targeted interventions such as adopting drought-resistant crops, initiating water conservation efforts, and planning for agricultural adaptation.

The adoption of modern technologies is another key finding. By combining geospatial analysis, remote sensing, and AI-driven models, policymakers can significantly improve early warning systems, enabling timely and precise responses to drought risks. Furthermore, emphasizing the role of HES in drought management highlights the critical need to restore ecosystems and leverage their natural capacity to regulate water flow, purify water, and sustain biodiversity.

Future efforts should prioritize refining indicators through continuous feedback from implemented strategies and emerging datasets. Long-term studies assessing the efficacy of MCDM models across different regions and climatic conditions will provide valuable insights into the adaptability and robustness of these frameworks. Such studies will also aid in localizing indicators to match specific socioeconomic and ecological realities, ensuring that policies remain dynamic and context-sensitive (Zarrineh and Azari 2014).

By embracing these multidimensional and adaptive strategies, communities and ecosystems can enhance their resilience to drought, effectively mitigating its impacts while adapting to climate variability.

References

Alawsi MA, Zubaidi SL, Al-Bdairi NSS, Al-Ansari N, Hashim K (2022) Drought forecasting: a review and assessment of the hybrid techniques and data pre-processing. Hydrology 9(7):115. https://doi.org/10.3390/HYDROLOGY9070115

Altemus Cullen K (2023) A review of applications of remote sensing for drought studies in the Andes region. J Hydrol Reg Stud 49:101483. https://doi.org/10.1016/j.ejrh.2023.101483

Aznarez C, Jimeno-Sáez P, López-Ballesteros A, Pacheco JP, Senent-Aparicio J, Rio D, Michaelides S (2014) Analysing the impact of climate change on hydrological ecosystem services in Laguna del sauce (Uruguay) using the SWAT model and remote sensing data. Remote Sens 13(10):2014. https://doi.org/10.3390/rs13102014

Brauman KA, Daily GC, Duarte TK, Mooney HA (2007) The nature and value of ecosystem services: an overview highlighting hydrologic services. Annu Rev Environ Resour 32:67–98. https://doi.org/10.1146/ANNUREV.ENERGY.32.031306.102758

Bryła M, Walczykiewicz T, Skonieczna M, Żelazny M (2021) Selected issues of adaptive water management on the example of the Białka River basin. Water 13(24):3540. https://doi.org/10.3390/W13243540

Bulama L, Lawrence EU, Umar AA (2021) The effects of wetland droughts on a reserve in the African continent. Asian J Geogr Res 4:22–30. https://doi.org/10.9734/ajgr/2021/v4i230128

Burke EJ, Brown SJ (2008) Evaluating uncertainties in the projection of future drought. J Hydrometeorol 9(2):292–299. https://doi.org/10.1175/2007JHM929.1

Canet-Martí A, Pineda-Martos R, Junge R, Bohn K, Paço TA, Delgado C, Alenčikienė G, Skar SLG, Baganz GFM (2021) Nature-based solutions for agriculture in circular cities: challenges, gaps, and opportunities. Water 13(18):2565. https://doi.org/10.3390/W13182565

Carvalho PN, Finger DC, Masi F, Cipolletta G, Oral HV, Tóth A, Regelsberger M, Exposito A (2022) Nature-based solutions addressing the water-energy-food nexus: review of theoretical concepts and urban case studies. J Clean Prod 338:130652. https://doi.org/10.1016/J.JCLEPRO.2022.130652

Chang H, Bonnette MR (2016) Climate change and water-related ecosystem services: impacts of drought in California, USA. Ecosyst Health Sustain 2(12):e01254. https://doi.org/10.1002/EHS2.1254

Di Gregorio S, Lupiano V, Forno F, Calidonna CR, Catelan P, Chidichimo F (2024) A transdisciplinary analysis of the 2009 catastrophe in giampilieri superiore by land use evolution over time. Earth Syst Environ. https://doi.org/10.1007/s41748-024-00454-5

Ekstrom JA, Klasic MR, Fencl A, Lubell M, Baker E, Einterz F, Brown EG (2018) Drought management and climate adaptation of small, self-sufficient drinking water systems in California. California's Fourth Climate Change Assessment. www.climateassessment.ca.gov

Falah N, Solis-Guzman J, Falah N (2024) Thermal footprint of the urbanization process: analyzing the heat effects of the urbanization index (UI) on the local climate zone (LCZ) and land surface temperature (LST) over two decades in Seville. Land 13(11):1877. https://doi.org/10.3390/land13111877

Gabiña D, Garrido A, Garrote L, Hajispyrou S, Ibáñez JC, Iglesias A, Keravnou-Papailiou E, Lapeña A, Lebdi F, López-Francos A, Louati MH, Mathlouthi M, Mellouli HJ, Moneo M, Ouassou A, Pangalou D, Pashardes P, Quiroga S, Rossi G, Ziyad A (2019) Drought management guidelines European Commission-Europe Aid co-operation office Euro-Mediterranean regional programme for local water management (MEDA water) Mediterranean drought preparedness and mitigation planning (MEDROPLAN)

Gössling S, Neger C, Steiger R, Bell R (2023) Weather, climate change, and transport: a review. Nat Hazards 118(2):1341–1360. https://doi.org/10.1007/s11069-023-06054-2

GWP (2020) Integrated Drought Management Programme in Central and Eastern Europe Activity 1.2 Final Report Review of the current status of the implementation of Drought Management plans and measures within RBMP according to WFD Integrated Drought Management Programme in CEE

Ha LT, Bastiaanssen WGM, Simons GWH, Poortinga A (2023) A new framework of 17 hydrological ecosystem services (HESS17) for supporting river basin planning and environmental monitoring. Sustainability 15(7):6182. https://doi.org/10.3390/SU15076182

Hanadé Houmma I, Gadal S, El Mansouri L, Garba M, Gbetkom PG, Mamane Barkawi MB, Hadria R (2023) A new multivariate agricultural drought composite index based on random forest algorithm and remote sensing data developed for Sahelian agrosystems. Geomat Nat Haz Risk 14(1):2223384. https://doi.org/10.1080/19475705.2023.2223384

Hariram NP, Mekha KB, Suganthan V, Sudhakar K (2023) Sustainalism: an integrated socio-economic-environmental model to address sustainable development and sustainability. Sustainability 15(13):10682. https://doi.org/10.3390/su151310682

Hidalgo-Triana N, Solakis A, Casimiro-Soriguer F, Choe H, Navarro T, Pérez-Latorre AV, Thorne JH (2023) The high climate vulnerability of western Mediterranean forests. Sci Total Environ 895:164983. https://doi.org/10.1016/j.scitotenv.2023.164983

Kisser J, Wirth M, De Gusseme B, Van Eekert M, Zeeman G, Schoenborn A, Vinnerås B, Finger DC, Repinc SK, Bulc TG, Bani A, Pavlova D, Staicu LC, Atasoy M, Cetecioglu Z, Kokko M, Haznedaroglu BZ, Hansen J, Istenič D, Beesley L (2024) A review of nature-based solutions for resource recovery in cities. In: Towards circular cities: nature based solutions for creating a resourceful circular city. IWA Publishing, London. https://doi.org/10.2166/9781789064360_CH5

Laín MM (2008) Drought and climate change impacts on water resources: management alternatives

Li T, Zhong S (2024) Advances in optical and thermal remote sensing of vegetative drought and phenology. Remote Sens 16(22):4209. https://doi.org/10.3390/rs16224209

Lüke A, Hack J (2017) Modelling hydrological ecosystem services-a state of the art model comparison. Hydrol Earth Syst Sci Discuss. https://doi.org/10.5194/hess-2017-436

Mehta A (n.d.) Applications of remote sensing to soil moisture and evapotranspiration speakers: Erika Podest. www.nasa.gov

Meilinger V, Monstadt J (2021) Articles-from the sanitary city to the circular city? Technopolitics of wastewater restructuring in Los Angeles, California. Int J Urban Reg Res. https://doi.org/10.1111/1468-2427.13014

Meteorological Organization and Water Partnership (2016) Integrated drought management programme handbook of drought indicators and indices. www.droughtmanagement.info

Millennium Ecosystem Assessment (2005) Millennium ecosystem assessment. Island Press, Washington. www.islandpress.org

Moreno M, Bertolín C, Ortiz P, Ortiz R (2022) Satellite product to map drought and extreme precipitation trend in Andalusia, Spain: a novel method to assess heritage landscapes at risk. Int J Appl Earth Obs Geoinf 110:102810. https://doi.org/10.1016/j.jag.2022.102810

Nandgude N, Singh TP, Nandgude S, Tiwari M (2023) Drought prediction: a comprehensive review of different drought prediction models and adopted technologies. Sustainability 15(15):11684. https://doi.org/10.3390/SU151511684

Ndayiragije JM, Li F (2022) Effectiveness of drought indices in the assessment of different types of droughts, managing and mitigating their effects. Climate 10(9):125. https://doi.org/10.3390/cli10090125

Nedkov S, Campagne S, Borisova B, Krpec P, Prodanova H, Kokkoris IP, Hristova D, Le Clech S, Santos-Martin F, Burkhard B, Bekri ES, Stoycheva V, Bruzón AG, Dimopoulos P (2022) Modeling water regulation ecosystem services: a review in the context of ecosystem accounting. Ecosyst Serv 56:101458. https://doi.org/10.1016/J.ECOSER.2022.101458

Paneque P (n.d.) Resistances to address water scarcity and drought in Andalusia. https://www.researchgate.net/publication/265253485

Platform for Sustainable Cities and Bank (2018) Urban sustainability framework

Rank PH, Patel K, Patel R, Damji Rank H, Paghdal AM, Patel KC, Patel DJ, Gohil GD, Patel RJ, Rank HD (2024) Leveraging AI and IOT based technologies for sustainable water management in reservoir command area Agri-India today leveraging AI and IOT based technologies for sustainable water management in reservoir command area. https://www.researchgate.net/publication/378744797

Raposo VMB, Costa VAF, Rodrigues AF (2023) A review of recent developments on drought characterization, propagation, and influential factors. Sci Total Environ 898:165550. https://doi.org/10.1016/J.SCITOTENV.2023.165550

Tapia C, Randall L, Wang S, Aguiar Borges L (2021) Monitoring the contribution of urban agriculture to urban sustainability: an indicator-based framework. Sustain Cities Soc 74:103130. https://doi.org/10.1016/J.SCS.2021.103130

Torelló-Sentelles H, Franzke CLE (2022) Drought impact links to meteorological drought indicators and predictability in Spain. Hydrol Earth Syst Sci 26(7):1821–1844. https://doi.org/10.5194/hess-26-1821-2022

Tramblay Y, Koutroulis A, Samaniego L, Vicente-Serrano SM, Volaire F, Boone A, Le Page M, Llasat MC, Albergel C, Burak S, Cailleret M, Kalin KC, Davi H, Dupuy JL, Greve P, Grillakis M, Hanich L, Jarlan L, Martin-StPaul N, Polcher J (2020) Challenges for drought assessment in the Mediterranean region under future climate scenarios. Earth-Sci Rev 210:103348. https://doi.org/10.1016/J.EARSCIREV.2020.103348

UNEP (2022) Annual report 2020

van Loon P, Van Wassenhove LN (2020) Transition to the circular economy: the story of four case companies. Int J Prod Res 58(11):3415–3422. https://doi.org/10.1080/00207543.2020.1748907

Verdon-Kidd DC, Sandi SG, Metcalfe AG, Kidd LJ (2023) Challenges of classifying and mapping perennial freshwater systems within highly variable climate zones: a case study in the Murray Darling basin. Austr Sci Total Environ 905:167260. https://doi.org/10.1016/j.scitotenv.2023.167260

West H, Quinn N, Horswell M (2019) Remote sensing for drought monitoring & impact assessment: progress, past challenges and future opportunities. Remote Sens Environ 232:111291. https://doi.org/10.1016/J.RSE.2019.111291

Wilhite DA (2002) Combating drought through preparedness. Nat Res Forum 26:36. http://digitalcommons.unl.edu/droughtfacpub/36

Williams EL, Abatzoglou JT, Hegewisch KC, Williams AP (2024) Anthropogenic climate change has reduced drought recovery probabilities across the western US. Commun Earth Environ 5(1):546. https://doi.org/10.1038/s43247-024-01640-z

World Bank (2015) A World Bank Group flagship report

World Bank (2021) Urban development overview. https://www.worldbank.org/en/topic/urbandevelopment/overview

Wu H, Su X, Singh VP, Zhang T (2022) Predicting hydrological drought with Bayesian model averaging ensemble vine copula (BMAViC) model. Water Resour Res 58(11):e2022WR033146. https://doi.org/10.1029/2022WR033146

Zarrineh N, Azari M (2014) Integrated water resources management in Iran: environmental, socio-economic and political review of drought in Lake Urmia. Int J Water Resour Environ Eng 6(1):40–48. https://doi.org/10.5897/ijwree2012.0380

Zellou B, El Moçayd N, El Bergou H (2023) Review article: towards improved drought prediction in the Mediterranean region - modeling approaches and future directions. Nat Hazards Earth Syst Sci 23(11):3543–3583. https://doi.org/10.5194/NHESS-23-3543-2023

Zhang Y, Peng C, Li W, Fang X, Zhang T, Zhu Q, Chen H, Zhao P (2013) Monitoring and estimating drought-induced impacts on forest structure, growth, function, and ecosystem services using remote-sensing data: recent progress and future challenges. Environ Rev 21(2):103–115. https://doi.org/10.1139/ER-2013-0006

Miniaturised (Bio)sensors in Aquatic Environmental Monitoring

Vanesa Román-Pizarro, Ángela Écija-Arenas, and Juan Manuel Fernández-Romero

Contents

Abstract

Water security is a critical global concern due to its impact on future generations. This concept is multidisciplinary, combining not only water availability but also its quality, encompassing chemical, physical, and biological characteristics to ensure suitability and safety. Among these aspects, water quality is the central focus of water security, particularly considering intensified human activities that pose significant risks of contamination. Different practices in this area are increasing nowadays, which may compromise water quality. To address these challenges, innovative technologies such as microfluidics are gaining attention for accelerating analytical processes, thereby reducing the timeline of test procedures and enhancing monitoring efficiency. Microfluidic devices have arisen as powerful tools to detect environmental pollutants. These devices are

V. Román-Pizarro (✉) · Á. Écija-Arenas · J. M. Fernández-Romero
Departamento de Química Analítica, Instituto Química para la Energía y el Medioambiente (IQUEMA), Universidad de Córdoba, Campus de Rabanales, Edificio Anexo "Marie Curie", Córdoba, Spain
e-mail: q52ropiv@uco.es; q92ecara@uco.es; qa1feroj@uco.es

D. Gutiérrez et al. (eds.), *Smart Water Quality Monitoring*,
https://doi.org/10.1007/978-3-032-03049-8_9

portable, easy to use, cost-effective, and have rapid response capabilities. Their application extends across various fields, including monitoring of environmental matrices (water, air, and soil), as well as agricultural monitoring. This chapter explores the development and implementation of microfluidic devices for detecting and determining prominent contaminants such as pesticides, heavy metals, microorganisms, etc. A range of detection methods, including electrochemical, colourimetric, and fluorescent, is used in combination with these devices to implement analytical platforms for monitoring water quality.

Keywords

Water quality monitoring · Microfluidics paper-based analytical devices · Environmental water samples · Pesticides · heavy metals · Microorganism resistance · Smartphone-based detection method

Introduction

In recent decades, human activities have significantly transformed nearly every aspect of the world's water systems. The rapid pace of industrialisation, urban expansion, and population growth creates impacts that have led to an increase in significant environmental pollution, affecting and putting both human health and the ecosystem at serious risk. The global challenge of ensuring access to safe drinking water remains a pressing issue, with nearly one-third of the worldwide population lacking reliable access to clean water. There is alarming data from the World Health Organisation (WHO) revealing that around 2 billion people depend on contaminated water sources, contributing to the statistic of nearly a million deaths per year. Contaminants in water, such as heavy metals, pesticides, or pathogens, have the ability to permeate through water systems via multiple pathways, posing significant long-term threats to public health and causing widespread disruptions and negative consequences in ecosystems (Chaudhry and Malik 2017).

Conventional methods for environmental pollutants monitoring, such as ultraviolet-visible (UV-Vis) spectroscopy, fluorescence spectroscopy, atomic absorption spectroscopy (AAS), mass spectrometry (MS), high-performance liquid chromatography (HPLC), or inductively coupled plasma optical emission spectroscopy (ICP-OES), offer high sensitivity and precision for detecting analytes at trace levels (Broekaert 1982; Kanu 2021; Mukherjee et al. 2021; Murphy and Riley 1962; Sekar et al. 2021). However, despite their analytical strength, they face several drawbacks, including high equipment costs, lengthy processes, and the requirement of skilled operators (Dhar and Lee 2018), which hinder their applicability for point-of-need testing and testing platforms, particularly in low-resource areas. As a result, there is an increasing interest in developing rapid, user-friendly, and sustainable detection technologies to address these limitations.

The microfluidic field has emerged as a promising solution for the addressed challenges. This technology has the ability to enable faster and more efficient analysis for water pollutant detection, utilising miniaturised and automated complex analysis to incorporate multiple chemical processes. Microfluidics is well-suited for environmental applications in water contaminant analysis due to its ease of operation, adaptability in handling water samples, minimal reagent consumption, and connection with a wide range of detection techniques (Saez et al. 2021). Since water is a natural fluid, its physical properties, like low viscosity and surface tension, facilitate smooth flow and adequate mixing within the microchannels, ensuring accurate sample processing (Chaudhry and Malik 2017). Moreover, other capabilities of microfluidic devices, including portability and automation, make them ideal for on-site water monitoring, especially in remote or resource-constrained areas where access to infrastructure to ensure safe drinking water is limited (Araci and Brisk 2014; Yang et al. 2024).

Newly, microfluidic paper-based analytical devices (μPADs) have garnered significant consideration due to their low cost, portability, ease of manufacture, usability, storage, transport, and minimal requirement for reagents and samples. Different examples are illustrated in Fig. 1. The advantages of paper over traditional device materials include its availability, low cost, high porosity, flexibility, and compatibility, making it a suitable substrate material for microfluidic assays (Almeida et al. 2018; Devadhasan and Kim 2018). The fabrication of μPADs requires the use of hydrophobic barriers on the substrate to define the region of fluid flow (Bilen et al. 2019). The hydrophilic nature of paper induces lateral flow action, facilitating fluid transport within its porous structure (Sharifi et al. 2020). Figure 1A shows the assembly process to obtain hydrophobic μPADS.

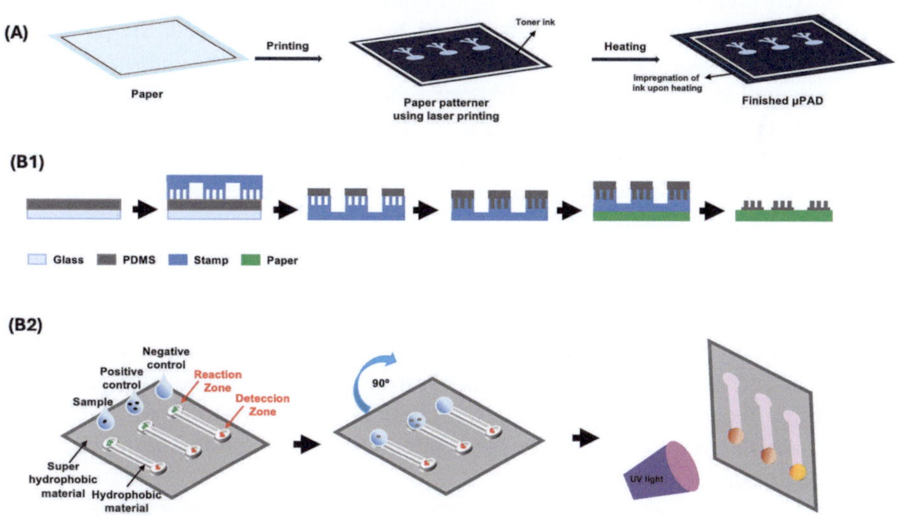

Fig. 1 Schemes of the μPADs fabrication process. (**A**) Hydrophobic μPADs assembly. (**B1**) Superhydrophobic coating of the SD-μPADs assembly process to obtain PDMS coating on the paper surface. (**B2**) Reaction process to determine lead (II) ions using SD-μPADs. (Adapted from Sun et al. (2018))

Table 1 Comparative summary of commonly used detection techniques in environmental sensing applications

Detection technique	Advantages	Disadvantages	References
Electrochemical	– Plain system – Miniaturisation – Cost-effectiveness – Real detection – Simple equipment	– Redox elements are required – Time-consuming – Sensitive to the surrounding environment	Luka et al. (2015)
Fluorometric	– Outstanding sensitivity – Outstanding selectivity – Harmless – On-site measurement	– Time-consuming – Temperature-sensitive – Low precision	Elsayed et al. (2021)
Colourimetric	– Quick – Inexpensive – Portable – Small sample size	– Large sets of RGB values – Imaging reproducibility stability	Sriram et al. (2017)

The basis of conventional PADS relies on capillary forces to drive liquid. Besides, suspending droplet μPADS (SD-μPADS) combine the use of capillary force with gravity force to drive liquid and transport the droplets between two reservoirs. Fig. 1B shows the use of SD-μPADs to determine Pb2+ with G-quadruplex-based luminescence detection (G-quadruplex formation under heat and mixing it with iridium (III) complex) (Sun et al. 2018). Figure 1.B1 shows the assembly process of a superhydrophobic coating pattern with hydrophobic channels SD-μPADs to obtain a PDMS coating paper surface. The process for creating the platform involves using a negative Teflon stamp to imprint onto a PDMS-coated glass slide. The measurement process is observed in Fig. 1.B2. The water samples were added to the reaction zone, and after 3 min, the chip was tilted by 90°. The droplet rolled into the detection zone, where it reacted with an iridium(III) complex and was then illuminated with ultraviolet light.

Microfluidic systems must be coupled with detection techniques to enable sensor and biosensor applications. Various signal transduction strategies have been widely employed in environmental sensing using microfluidic devices, including electrochemistry, fluorometry, and colourimetry, among others. In Table 1, a summary of the key benefits and drawbacks of commonly used detection techniques is presented, highlighting their primary benefits and limitations. These methods form the basis of many sensing strategies that are later integrated into microfluidic platforms to achieve (bio)sensor applications.

As microfluidic systems continue to evolve, their integration with biosensing technologies has opened new avenues for real-time and on-site monitoring of environmental contaminants. These hybrid platforms combine the advantages of microfluidics with the high specificity and sensitivity offered by biosensors. The

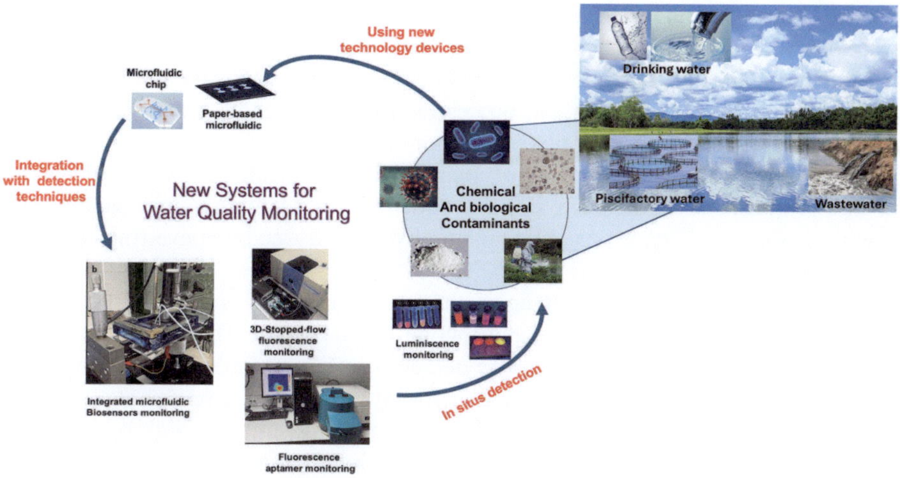

Fig. 2 Microfluidic-based application integrated with optical detection techniques for monitoring common contaminants in water (adapted from the AlMashrea et al. (2024))

synergy broadens the scope of applications in water quality monitoring. The core of each biosensor is the transducer. This component is an element that converts the binding event between the target analyte and the biorecognition element into a measurable signal, which includes electrochemical, optical, temperature, electric, magnetic, and mechanical signals. Biosensors can be categorised based on the type of signals the transducer produces, with electrochemical and optical biosensors being the most widely studied (Abbasi et al. 2024). Recent developments in water quality monitoring combine microfluidic platforms with a range of detection strategies. Figure 2 illustrates how microfluidic-based applications are integrated with optical detection techniques for monitoring common contaminants in various water sources, including drinking water, aquaculture systems, and wastewater treatment systems.

Distinct generations can trace the progress of biosensors, each bringing significant advancements. The first biosensor generation utilised natural receptors, such as enzymes, antibodies, and whole cells, as biorecognition elements, to interact with target analytes to produce a signal. In the second generation, the focus shifted to improving the transducer by incorporating nanomaterials, which enhanced sensitivity and signal amplification. Miniaturisation is also a focus for this generation, making them suitable for point-of-care (POC) and portable applications. The third generation is marked by the ability to detect multiple analytes simultaneously. Multiple detections established direct signal transfer between transducers and analytes. Moreover, the integration of multifunctional platforms has become the future direction to develop methods capable of handling complex samples. Additionally, integrating artificial intelligence (AI) technology with advanced data analysis and mining capabilities enhanced the interpretability and automation of biosensors.

The fourth generation represented a convergence of several cutting-edge technologies. They integrated nanotechnology, the Internet of Things (IoT), and biomimetic materials. This integration resulted in instantaneous measurements with real-time monitoring of biological processes. Devices became more wearable and implantable, with lab-on-chip platforms representing this generation. Smartphones have facilitated the development of platforms that enable the visual detection of colourimetric or fluorescent signals, allowing for accurate signal capture and visual monitoring. Signals could be quantitatively analysed by converting colour or fluorescence images into RGB values.

The fifth generation is built upon the combination of the on-off-on signal-switching analysis methods and smartphone technologies. This combination has gained popularity for real-time analyte detection using portable sensing platforms. This approach holds significant potential for environmental assessments, ensuring water safety. This generation led to the concept of hospital-on-a-chip, expecting to refine existing technologies. Looking ahead, the sixth generation is expected to incorporate 6G wireless networks and holographic technologies into microfluidic systems. With this knowledge, various innovations are becoming feasible, such as surgery-on-a-chip and the Internet of Medical Things (IoMT), which are expanding applications in telemedicine (Aryal et al. 2024).

Recently, a new generation based on the IoT is coming, which consists of a network of interconnected operating systems, industrial machinery, smart devices, the environment, and individuals assigned unique identifier (UID) numbers, with the ability to transfer data over a network without human intervention. IoT-integrated biosensors leverage IoT technology to enable continuous, real-time monitoring of biological and biochemical parameters (Abbasi et al. 2024). With the support of artificial intelligence, detection systems can analyse large volumes of experimental data, identifying patterns and correlations difficult for humans to detect. Machine learning algorithms can adapt and improve over time, enhancing the accuracy of detection processes and reducing the need for manual intervention. Among the biosensors, lab-on-a-chip (LoC) provides a compact solution for in situ, real-time sensing and mixture analysis in water environments.

Biosensors play a crucial role as one of the most critical sensing components within the IoT in multiple applications, such as environmental monitoring (Banerjee et al. 2024; Himeur et al. 2022). AI has emerged as a transformative technological science that enhances the functionality of these systems. Machine learning (ML) is a key branch of AI that enables computer systems to learn from data, enhancing their performance by identifying patterns and improving over time without requiring explicit programming. ML involves predictions and decisions based on data, so it is an advanced form of statistical analysis. ML consists of the development of neural networks and algorithms. For example, architectures such as convolutional neural networks (CNNs) and recurrent neural networks (RNNs) are essential for image processing (Cossu et al. 2021). Despite its potential and significant advantages, AI also has some limitations. The development of algorithms requires large, high-quality datasets to train models. Microfluidic chip platforms can overcome some AI limitations, offering cost-effective, scalable, and automated options. When integrated with AI algorithms, microfluidic platforms can make more efficient data acquisition and enable robust analysis of waterborne contaminants. This synergy represents a powerful approach for next-generation water quality monitoring by combining rapid, on-chip sensing with advanced data interpretation capabilities.

The application of AI in water quality monitoring offers significant advantages for environmental analysis due to its capability for real-time monitoring. The fate of the aquatic environment has become a serious global concern. Due to high salinity, approximately 97% of the Earth's water is suitable for direct drinking. Conversely, the water available for drinking is heavily impacted by population growth, urbanisation, and industrialisation in many countries. The primary source of water pollution is the unregulated and unauthorised disposal of organic and inorganic waste from industrial and domestic sources into surface water. These point sources introduce harmful substances into surface water, including heavy metals, pesticides, and pathogens (Ibrahim et al. 2020). Heavy metals such as arsenic (As), cadmium (Cd), copper (Cu), and lead (Pb), among others, naturally occur in the environment. They are essential for human health when present in permissible concentrations. However, levels of heavy metals exceeding safe limits can have a severe impact on public health due to their mutagenic or carcinogenic properties. Therefore, periodic monitoring of water quality is essential. Modern techniques for monitoring water quality include atomic absorption spectroscopy (AAS), electrochemical methods, electrothermal atomic absorption spectrometry (ETAAS), inductively coupled plasma mass spectrometry (ICP-MS), and photocatalytic degradation, among others. Whilst these techniques are selective and sensitive, they require skilled labour and expensive equipment (Butcher 2010; Rai et al. 2022). Although these analytical methods can provide high selectivity, sensitivity, and low detection limits for identifying trace amounts of metals, they are time-consuming and necessitate complex, costly instruments and trained specialists for operation. Moreover, these methods are not efficiently utilised as portable analytical tools for on-site detection of metal ions. Thus, researchers have developed simple, portable, and low-cost sensors for the rapid, sensitive, and selective detection of heavy metals, suitable for on-site assays and point-of-care tests (Devadhasan and Kim 2018).

Most importantly, these methods are ineffective in the field due to their lack of portability and the complexity of their infrastructure. Consequently, there is a need for a highly sensitive and selective, portable, cost-effective, user-friendly, reliable, and field-effective alternative. Microfluidic devices have emerged as an ideal solution, meeting all these criteria for point-of-care applications (Gale et al. 2018; Santana et al. 2020). Although microfluidics initially relied on silicon-based processes, the development of various alternative microfabrication methods including inkjet printing, 3D printing, injection moulding, and xerography has provided numerous options for microfluidic devices that utilise novel materials such as paper, polymers, and thermoplastics (Akyazi et al. 2018; Niculescu et al. 2021). These techniques have also enabled the integration of innovative sensing and detection units, which are essential for maintaining water quality.

Heavy metals are naturally present in water. They have densities at least five times greater than water. Some of them, such as arsenic (As), cadmium (Cd), and lead (Pb), are naturally present in the environment and play essential roles in physical functions. However, they should be maintained within safe concentration limits. Going beyond these thresholds can pose severe health risks due to their mutagenic or carcinogenic properties (Balali-Mood et al. 2021; Tchounwou et al. 2012). For this, developing methods to determine them in environmental matrices is needed. In recent years, smartphones have been utilised as sensors to detect heavy metals through integration with rapid test strips, sensor chips, and handheld detectors (Yetisen et al. 2013). The use of smartphones for this purpose implies an improvement over traditional configurations, which often require complex prism-based optics or specialised laboratory infrastructure. Modern smartphone-based platforms utilise nanomaterials, such as metal nanoparticles, to achieve enhanced sensitivity and portability, thereby enabling real-time and user-friendly environmental monitoring.

Over the years, microfluidic-based systems have been explored for use in water quality monitoring systems, featuring various detection mechanisms. The most popular methods include electrochemical, photoelectrochemical, and fluorescence detection. Other detection mechanisms include colourimetric, chemiluminescence, absorbance-based, and surface plasmon resonance. In this chapter, only colourimetric, fluorescence, and electrochemical detection systems are discussed, as they are highly popular in the microfluidics community. According to studies reviewed by Xu et al., exposure to environmental pollutants, including heavy metals, biogenic toxins, and industrial effluents, is associated with adverse health effects. These contaminants are estimated to contribute to approximately 22% of the global disease burden and 23% of deaths worldwide (Xu et al. 2018).

On the other hand, pesticides play a crucial role in modern agriculture in meeting global food production demands by controlling pests. Exposure to pesticides has been associated with both acute and chronic effects, including asthma, hormone disruption, neurological problems, cancer, and more. They can be categorised based on their chemical family or action mechanism. The most commonly used pesticides in agriculture belong to the organophosphate family, which are considered organic contaminants (Hassaan and El Nemr 2020). Due to their toxicological effects and hazardous nature, organic contaminants have various adverse health impacts. Sensors designed to detect volatile organic compounds (VOCs) have received considerable attention, but most sensor fabrication processes involve complex procedures. In recent years, smartphones have provided a promising digital platform for pesticide diagnostics. They enable rapid, simple, portable, and ubiquitously available sensing, making them a technology suitable for minimally trained users to perform chemical analyses in the field (Motalebizadeh et al. 2018).

Over the past 5 years, research has focused on the potential of microfluidic devices for detecting pesticides in water, exploring various innovative detection methods. Standard rapid pesticide detection techniques that utilise microfluidics include enzyme inhibition, immunoassays, molecular imprinting, and related approaches (Uka et al. 2021). Organophosphate pesticides have been investigated as analytes using a paper microfluidic device based on a lateral flow assay (LFA). In these spot-based assays, the disadvantages of LFA sensors, such as colour gradient

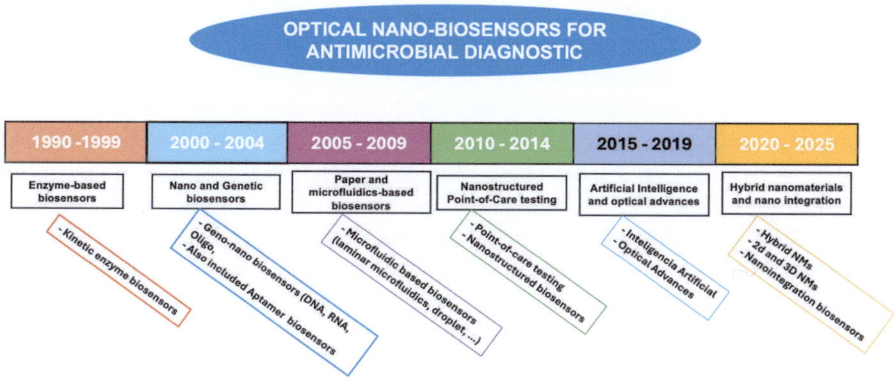

Fig. 3 Timeline of developing optical nano-biosensors for antimicrobial detection. (Adapted from Chen (2012))

development, were avoided; however, sensitive performance with detection limits (LoD) has yet to be achieved. Other drawbacks of paper-based devices include multistep operations, time consumption, and a small test zone area.

Conventional antimicrobial resistance diagnostics, such as antibiotic susceptibility testing, have drawbacks, including high costs, a time-consuming process, and challenges with on-site diagnostic analysis. Nano-enabled, AI-supported, innovative optical biosensors offer potential solutions by enabling rapid point-of-care detection of antimicrobial resistance (AMR) with sensitive, real-time, and portable capabilities. Figure 3 illustrates a timeline of the development of optical nano-biosensors for AMR from 1990 to 2025.

Standard microbial water-testing methods enable the detection of viable bacteria, which distinguishes their ability to produce progeny that poses a risk to public health (Bagheri et al. 2015). They may be underestimated by methods detecting non-viable bacteria (Giménez-Gómez et al. 2019). However, standard water-testing methods are time-consuming and necessitate a substantial laboratory setup. Pathogenic microorganisms include viruses, bacteria, and other microscopic organisms that can cause disease in humans and animals. Therefore, developing effective detection techniques for monitoring waterborne pathogens is crucial in safeguarding public health and safety. Traditional methods for detecting microorganisms using FPR include electrochemistry, spectroscopy, fluorescence, and surface plasmon resonance (SPR).

Detection Techniques

Over the last 5 years, various detection techniques have been used for water monitoring using microfluidic-based systems. The most widely used methods are fluorescence, colourimetric, and electrochemical detection. Other methods include chemiluminescence, SERS, and surface plasmon resonance (SPR). Here, we focus on electrochemical, fluorescence, chemiluminescence, and colourimetric methods, the most prevalent techniques today.

Electrochemical analysis encompasses techniques such as potentiometry, conductometry, and voltammetry, each of which is dependent on specific electrical characteristics (Li et al. 2018). These offer low detection limits, which are in the picomole range, representing an essential benefit over widely used techniques such as colourimetric or fluorescence methods, which have yet to achieve comparably low detection limits. In many cases, electrochemical detection offers benefits, and for this reason, electrochemical systems have been used to detect heavy metals.

Fluorescence techniques used for environmental detection include designing and synthesising several fluorophores which incorporate recognition to sense environmental analytes effectively. A method for assessing water quality through fluorescence detection in microfluidic systems involves mixing the sample with a fluorescent probe for the analyte, passing it through the device, illuminating it to induce fluorescence, and subsequently measuring the intensity of the emitted light. Over the past 5 years, there has been a significant increase in the development of colourimetric platforms for detecting environmental analytes. Most researchers have focused on designing paper-based microfluidic colourimetric sensors, recognising them as cost-effective, user-friendly alternatives for analytical detection.

Electrochemical Biosensors

Electrochemical analysis is an effective detection technique that relies on the interaction between an analyte and the electrode surface. In the context of water quality monitoring, contaminants are identified through variations in the electrical response caused by the presence of pollutants. Due to its straightforward operation, portability, low cost, excellent sensitivity, and selectivity, electrochemical detection has been extensively investigated for use in on-chip and real-time monitoring systems. Advances in microfabrication have enabled the miniaturisation of electrodes, facilitating the integration of three-electrode electrochemical detection systems within microfluidic platforms. Among the most employed techniques are amperometry, voltammetry, conductometry, and potentiometry (Mohan et al. 2022).

The integration of electrochemical techniques into microfluidic systems has enabled the development of electrochemical microfluidic devices, resulting in the creation of micro total analysis systems (μTAS). Employing potentiometry with microfluidic paper-based solution sampling can alleviate some of the limitations of the electrochemical techniques. Potentiometry is a rapid, low-cost, and straightforward electrochemical technique that does not necessitate complicated sample pre-treatment before measurements (Lisak et al. 2015). Recently, inorganic modifications of microfluidic paper-based solution substrates have been found to excel in eliminating the super-Nernstian response of heavy metal-sensitive ion-selective electrodes (ISEs). However, during measurements, ion desorption from the inorganic salt-modified microfluidic paper-based solution sampling substrates may increase.

Modern techniques, such as electronic and electrochemical analyses, have been developed to address the limitations of traditional methods. Nevertheless, these approaches face challenges such as instability, low compatibility, low selectivity, and complicated on-site sampling. Thus, optical methods represent the best option for detecting toxic ions. They are sensitive, cost-effective, fast, straightforward, efficient, and selective.

Paper-based platforms have gained recognition for developing analytical devices. Since their conception, microfluidic paper-based analytical devices (μPADs) and electrochemical paper-based analytical devices (ePADs) have attracted significant attention due to their versatility and applicability in different fields, including metal analysis. Several reports use μPADs for colourimetric or ePADS for electrochemical detection of select heavy metals, including iron, copper, zinc, mercury, nickel, lead, and cadmium (Kokkinos et al. 2018).

Moreover, many electrochemical sensors still rely on conventional discrete electrodes, lacking the integration and miniaturisation necessary for in situ detection. Consequently, their limitations restrict their applicability in real-time heavy metal detection and the establishment of multiple sensor nodes for water quality assessment. In contrast, microfluidic paper-based sampling solutions platforms, integrated with electrodes for heavy metal detection, utilise the natural affinity of negatively charged cellulose material to adsorb metal ions, enhancing preconcentration and detection efficiency. A miniaturised Arduino-based microcontroller was programmed to generate pulses, read capacitance signals, and continuously record data for practical applications. Santangelo et al. investigated the potential of epitaxial graphene on silicon carbide for detecting heavy metals, such as Cd(II) and Pb(II), in water. The sensor allows the study of changes in conductivity when Pb and/or Cd ions interact with the graphene surface (Santangelo et al. 2019). Gimenez Gomez et al. investigated an innovative technique for the automated determination of As(III) in water. The method uses a modular microfluidic system with an electrochemical sensor integrated inside. The sensor demonstrated excellent performance in spike recovery analysis, representing a promising alternative for quantifying As(III) (Giménez-Gómez et al. 2019).

Ma et al. developed a portable microfluidic electrochemical sensing platform designed for the rapid detection of Pb (II). Using this metal ion as a model analyte, they proposed a microfluidic electrochemical sensing chip compatible with a smartphone-based electrochemical sensor for quick and efficient detection (Ma et al. 2023). Wang et al. bonded a PDMS material to silicon to form an electrochemical detection chamber, thereby simplifying the testing process (Wang et al. 2016). Laschi et al. developed an automatically portable system based on square-wave anodic stripping voltammetry (SWASV), an online sensor for detecting heavy metals (Laschi et al. 2024). The sensors' surfaces are appropriately modified and coupled in a flow cell. A software tool was introduced to handle the potentiometric remote control. Prasad et al. presented a highly sensitive aptamer-based nano biosensor for the efficient electrochemical detection of arsenic in water, employing

Table 2 Microfluidic devices for heavy metal and pesticide electrochemical detection in water

Pollutants	Device or material	Detection limit	References
Cu(II)	MEMS using multi-parameter integrated in a portable system	2.3 ppb	Wu et al. (2020)
Pb(II)	Glass microfluidic system integrated with Ag/AgCl reference electrode	0.13 ppb	Dai et al. (2021)
Paraoxon	Origami multi paper	2 ppb	Arduini et al. (2019)
Parathion	Microfluidic paper-based chip with Fe_3O_4 nanozyme-supported carbon quantum dots and silver terephthalate MOFs	1.1×10^{-11} M	Li et al. (2021)
Glyphosate	Glass microsensor using MIP-based	–	Uka et al. (2021)

glassy carbon electrodes modified with tungsten trioxide nanomaterial (Sciuto et al. 2023).

The determination of pesticides in water quality assessment remains a challenging task due to the complexity of the extraction and detection procedures. Despite the advantages offered by the integration of wireless sensor networks, the Internet of Things, and machine learning. Arduini et al. reported an origami, paper-based electrochemical biosensor for the simultaneous detection of paraoxon and the pesticides 2,4-D and atrazine (Arduini et al. 2019). Their approach also employed enzymatic inhibition alongside a portable potentiostat. This method showcased the biosensor's ability to detect three classes of pesticides. Table 2 summarises the microfluidic electrochemical devices for detecting heavy metals and pesticides in water.

Detection of pathogenic adenovirus for critical applications requires laborious and expensive diagnostic tools for virus detection. However, user-friendly, cost-effective biosensors, along with computational simulations, are used to assess the conformational stability of the selected epitope. Li et al. developed an electromagnetic swimmer system (SES) to monitor ions and SARS-CoV-2 in water (Li et al. 2024). They constructed an experimental confined pipe filled with tap water. A stationary SES with integrated NFC sensing modules was placed at one end of the pipe. This system utilises soft electronic technologies to integrate thin multifunctional bio/chemical sensors and a wireless data acquisition module. Altintas et al. developed a microfluidic device equipped with an Au biochip for detecting *Escherichia coli* in water samples (Altintas et al. 2018).

Optical Biosensors

Biosensors that employ an optical transduction mechanism are known as optical biosensors. In these systems, the optical transducer converts the binding interaction between the target analyte and the biorecognition element into optical signals, such as light emission or spectral changes. Optical biosensors offer several advantages, including high sensitivity, specificity, reliability, versatility, and the capability for real-time detection. They have been extensively researched as highly sensitive and user-friendly devices, particularly in lab-on-chip applications. Moreover, optical biosensors are significant in point-of-care (POC) and point-of-need (PON) applications, including on-site pollution detection for environmental monitoring. These biosensors can be broadly classified into three main types: colourimetric, luminescent, and refractive index-based, each possessing distinct mechanisms for signal generation. Colourimetric biosensors detect variations in light absorption or scattering, often involving coupling with chromogenic indicators or nanoparticle agglutination that produce visible colour changes in the presence of a target analyte. Luminescent biosensors measure the light emitted from analyte-specific biochemical reactions, whereby excited molecules emit light as they return to their ground state. Refractive index-based biosensors detect changes in the refractive index of samples using plasmonic, resonance, or interference methods (Aryal et al. 2024).

The field of optical biosensors is evolving rapidly, with each new generation bringing notable advancements in both technology and functionality. The integration of quantum sensing and machine learning is poised to revolutionise biosensing further, enhancing sensitivity, specificity, and usability. In particular, the application of machine learning (ML) to enhance chemiluminescence (CL) and electrochemiluminescence (ECL) biosensors has attracted considerable attention. Additionally, ML offers significant potential for extracting meaningful insights from images acquired via smartphone cameras. One study employed a portable ECL system based on molecularly imprinted polymers, coupled with smartphone image capture and artificial neural networks (ANNs), to detect nitrofurazone, a banned carcinogenic compound. Similarly, another study utilised convolutional neural networks (CNNs) to analyse images from an ECL system designed to detect the harmful pesticide 2,4-Dichlorophenoxyacetic acid (2,4-DC) (Idros and Chu 2018). These examples highlight the versatility of this approach for detecting a wide range of analytes (Ibrahim et al. 2020).

Colourimetric Biosensors

Colourimetry biosensors integrated with microfluidic devices are primarily based on paper-based technology. They are widely recognised as some of the most suitable detection methods. Figure 4 illustrates microfluidic devices with colourimetric detection that detail the steps involved in detecting contaminants in wastewater. These biosensors depend on a colour change when the targeted analyte is present. This change can be due to the product of a chemical reaction between the analyte and a reagent. Colourimetric biosensors are preferred for their straightforwardness, affordability, and minimal equipment requirements for reading results. These biosensors provide quick and simple responses and do not need an operator or advanced equipment. However, their low sensitivity poses limitations. Colourimetric analysis can be developed using container-free methods. They are alternatives to strip containers and provide reproducible and cost-effective measurements. The

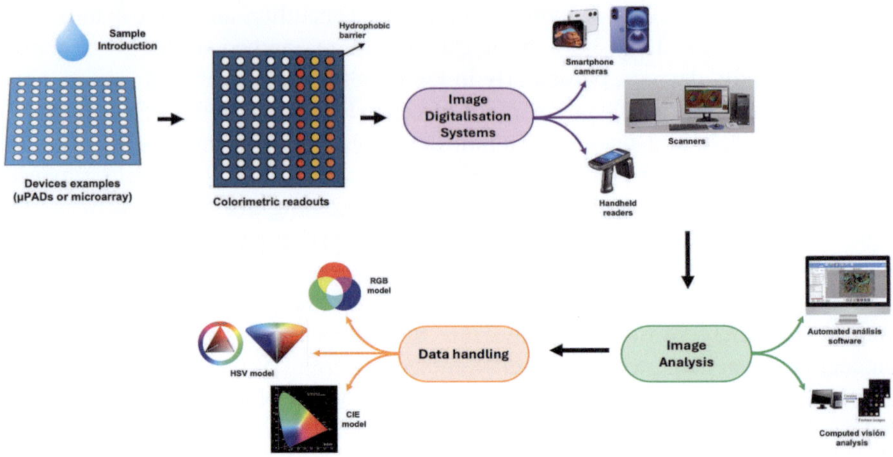

Fig. 4 Stages to proceed with colourimetric detection in the analysis of contaminated water samples (adapted from Rai et al. (2022))

difference between container-free and container-based affects only how the analyte area is captured from the image.

Smartphone spectroscopy involves taking pictures with the camera and examining them with proper applications. The enhanced sensitivity and resolution of smartphones in capturing the light intensity and colour allow for precise discrimination of signal variations between samples. A type of smartphone-based optical sensor is the micro-imaging sensor, which can be used for visualising and identifying biological components. The smartphone-assisted colourimetric biosensor was an intelligent and portable sensor for the operational process. The concentrations of all water contaminants could be output directly from the mobile app. Smartphone applications, equipped with built-in functions, enable the efficient analysis and digital processing of changes in light intensity and colour. To use images in colourimetric sensing, a portable platform can be designed to put the smartphone camera on top and make a support device. Uniform images can be captured this way. If a detection box is used to block light from outside the sensing area, better results are obtained. White LEDs are used to illuminate the sample area within the detection box, thereby creating a reproducible ambient environment. Utilising integers, a new RGB colour reader app can be employed to convert RGB values into variations in colour intensity. Table 3 summarises a microfluidics colourimetric device for pollutants in water.

Accurate quantification of colour from images of the analyte region remains a challenge in colourimetric analysis. The visual characteristics of this area, especially in highly saturated images, can significantly influence the measurement results. Bright images can be quantified in greyscale, RGB, HSV, and other colour spaces. Greyscale is consistently used to represent colour intensity and brightness instead

Table 3 Some microfluidic devices for the colourimetric detection of contaminants in water

Contaminants	Device or material	Detection limit	References
Cu(s)	3D origami paper-based combined with a PVC membrane	1.7 mg mL^{-1}	Sharifi et al. (2020)
Fe(II), Ni(II)Cr(VI), Cu(II)	3D microfluidic paper combined with a smartphone	0.2, 0.3, 0.1, 0.003 mg mL^{-1}	Lace et al. (2019)
Ni(II), Cr(VI), Hg (II)	Functionalised paper-based	0.24, 0.18, 0.19 mg mL^{-1}	Devadhasan and Kim et al. (2018)
As(II), Hg(II)	Portable smartphone based PDMS microfluidic	–	Almeida et al. (2018)
Chlorpyrifos	2D paper digital microfluidic	8.60 mg mL^{-1}	Uka et al. (2021)
Carbaryl, carbosulfan, furathiocarb	Paper microfluidic device	0.4, 0.24 and0.46 mg mL^{-1}	Beshana et al. (2022)
Profenol methomyl	Paper microfluidic device coupled to a smartphone	55 and 34 mg mL^{-1}	Wu et al. (2022)
E. coli	Paper devices with micro-pupil holes and single view angle	10^3 CFU	Rodoplu et al. (2022), Wang et al. (2016)
E. coli	3D-printed chip	2 CFU	Rodoplu et al. (2022)
E. coli	Paper based in quantitative picture of the cell response to SARS-CoV-2	10^3 CFU	Snyder et al. (2020)

of colour channels, whilst RGB and HSV channels can be unstable under complex lighting conditions (Pebdeni et al. 2022).

Colourimetric assays for detecting heavy metals in water are well-established methods. These assays are standardised and automated for laboratory use; however, the challenge lies in collecting, preserving, and processing the samples. Many of these heavy methods are unsuitable for typical water sampling locations. This issue can be resolved through miniaturisation, which can be achieved using microfluidic techniques. This method is called lab-on-a-chip, where laboratory procedures are miniaturised.

Wang et al. developed an innovative sensor capable of simultaneously detecting multiple heavy metal ions. This compact and user-friendly device incorporates automatic sample handling, on-chip chemical reactions, gravitational-magnetic separation, and a distance-based readout system (Wang et al. 2022). The chemosensory system comprises a 3D-printed layer of PDMS and Norland Optical Adhesive 63. Magnetic and polystyrene microparticles bind to the DNAzyme, forming a stable structure in the absence of target metal ions. The trapping distance of the polystyrene microparticles can be utilised to quantify the metal ions visually.

Many microfluidic devices and paper-based analytical devices (PADs) have been reported for detecting Hg(II) in various types of water, including drinking water, pond water, river water, and industrial wastewater. Pourreza et al. developed PADs to monitor Hg (II) concentration using curcumin nanoparticles (CURNs). The paper-based probe was created using the wax dipping technique, incorporating CURNs as the sensing reagent. The device reaches a limit of detection of 0.17 $\mu g\ mL^{-1}$ (Lace and Cleary 2021; Pourreza et al. 2016). Meelapsom et al. developed a chromatic analysis to monitor the colour change of μPADs based on a simple RGB colour model for the sensitive and selective determination of Hg(II). The colour change was detected using a digital camera with a small sample volume. The device was fabricated by inkjet printing using alkyl ketene dimer doped with unmodified silver nanoparticles. The developed system enabled the quantification at low concentrations, with a detection limit of 0.001 $mg\ L^{-1}$ (Meelapsom et al. 2016).

Wu et al. presented a study using polystyrene-divinylbenzene microspheres for heavy metal ion detection. A colourimetric reactant, 1-(2-pyridylazo)-2-naphthol (PAN), was immobilised on wax-printed filter paper to establish a colourimetric Cu (II) ion detector (Q. Wu et al. 2019). Li et al. developed a straightforward double-layered 3D μPAD that employs a smartphone and a flat LED lamp to achieve multiplexed colourimetric detection of six metal ions: Fe (III), Ni (II), Cr (VI), Cu (II), Al (III), and Zn (II) simultaneously (Li et al. 2019). The 3D μPAD device, illustrated in Fig. 5, consists of two paper layers: the pretreatment layer on top, and the colourimetric detection layer on the bottom. Each channel designed on paper is directed to analyse one metal ion.

Kamnoet et al. proposed a new PAD device for the simultaneous detection of heavy metal ions, including Cu (II), Co (II), Ni (II), Hg (II), and Mn (II), using bathocuproine (Bc), dimethylglyoxime (DMG), dithizone (DTZ), and 4-(2-pyridylazo) resorcinol (PAR) as complexing agents designed a colourimetric

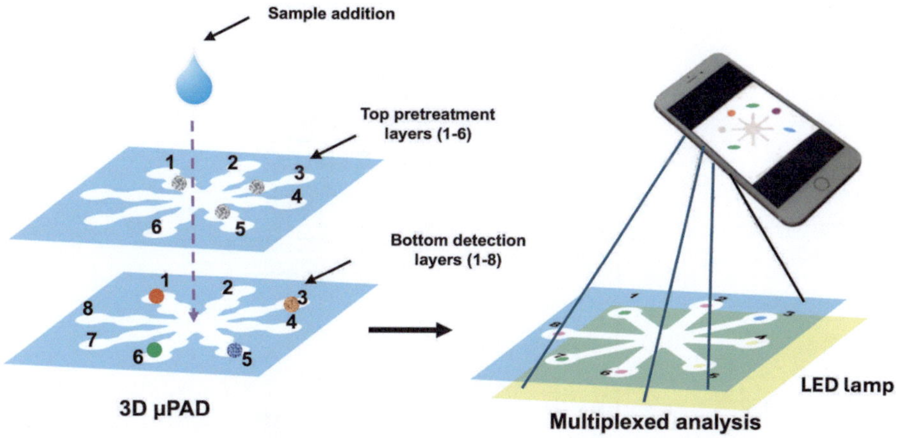

Fig. 5 Fabrication of the double-layered three-dimensional microfluidic paper-based Analytical device (3D μPAD) for multiplexed colourimetric detection of six metal ions. Pretreatment zones 1–6 for Fe (III), Ni (II), Cr (VI), Cu (II), Al (III), and Zn (II), respectively. Detection zones 1–8 for Fe (III), Ni (II), Cr (VI), Cu (II), Al (III), Zn (II), pH, and blank reference, respectively. (Adapted from Li et al. (2019))

technique (Kamnoet et al. 2021). The device has two pretreatment zones, where masking agents remove interfering ions, and the detection zone is prepared with optimised reagents for each metal, where the colourimetric reaction occurs. A scanner was used to obtain images.

Lace et al. developed a microfluidic system for detecting arsenic in water using leuco-malachite green (LMG) dye (Lace et al. 2019). This method was the first to integrate LMG into a microfluidic device for arsenic detection. Additionally, the colour complex displayed long-term stability, allowing the microfluidic analysis. Cao et al. developed a colourimetric analysis method using a smartphone to measure the concentration of copper ions in drinking water based on the blue value, eliminating the need for a strip container (Cao et al. 2019). Solmaz et al. developed an Android app for conducting single-strip colourimetric analysis without a strip container, utilising a least-squares support vector machine (Solmaz et al. 2018) and a random forest model based on average RGB values, as illustrated in Table 4 (Sharifi et al. 2020).

Colourimetric μPADS has been described as a method to determine organophosphate pesticides (OPs) based on the inhibition of acetylcholinesterase (AChE). This enzyme hydrolyses the acetylthiocholine iodide substrate into thiocholine base and acetic acid. Due to the inhibition of AChE, the colour intensity of the reaction product decreases with increasing pesticide concentration. Kim et al. demonstrated a colourimetric paper-based sensor for detecting chlorpyrifos, an organophosphate pesticide. Organophosphate pesticides inhibit AChE, producing a less intense colour. This colour enabled the pesticide to be quantified through image analysis (Kim et al. 2018).

Table 4 Some microfluidic devices for the fluorometric detection of contaminants in water

Contaminants	Device or material	Detection limit	References
Cu (II), Cd(II), Pb(II)	3D pinwheel-shaped paper-based microfluidic device using ion imprinting polymer aided by quantum dots	0.007–0.0015 mg mL^{-1}	Wang et al. (2021)
Hg(II)	Feedback controlled digital microfluidic droplet chip with fluorescent probe	0.07 mg mL^{-1}	Zhang et al. (2019)
Cu(II), Hg(II)	Filter paper-based device using candle wax channels with nitrogen-doped carbon dots	6.2 nmol L^{-1} and 2.304 nmol L^{-1}	Patir and Gogoi (2019)
Pb(II)	Hydrophobic suspending-droplet paper-based device using G-quadruplex-based luminescence	–	Sun et al. (2018)
E. coli	PDMS microfluidic device using electroosmotic force with magnetic microbeads	–	Miller et al. (2019)
E. coli	PDMS droplet microfluidic with DNAzyme mixture	–	Rauf et al. (2022)
Norovirus	Paper-based device with antibody-conjugated, fluorescent submicron particles with smartphone-based microscope	1 genome copy/μL	Chung et al. (2019)

Among the biorecognition molecules used, aptamers show promising results in the development of aptasensors. They have a similar action mechanism to antibodies, as they interact with the target selectively. Nevertheless, aptamers present higher stability in storage conditions. The use of smartphone-based aptasensors is the development of a fully portable detection system. Ulloa-Gomez et al. developed a portable biosensor for rapid and label-free colourimetric detection using gold microparticles functionalised with specific aptamers on a microfluidic paper-based analytical device (Ulloa-Gomez et al. 2024; Wang et al. 2024).

Li et al. developed a reliable analytical method for determining ibuprofen in environmental samples based on portable paper strips and a 3D-printed miniaturised testing system (Li et al. 2019). This biosensor has a fast response, high sensitivity, and a visually detectable colour change. Chen et al. developed a hydrogel kit for on-site quantitative monitoring of pesticides. Using a smartphone as an assistive device, the colour change in the portable kit could be converted into digital information, thereby establishing a linear range for pesticide detection (Chen et al. 2024).

Nowadays, smartphones have emerged as valuable tools in diagnosis. Compared to conventional tools, smart sensors possess distinctive characteristics; for example, the device can be miniaturised significantly, enhancing portability and facilitating rapid analysis. Additionally, smartphones are widely available today and are less expensive than traditional laboratory equipment.

The risk of infection with the waterborne pathogen has increased in recent years. The combination of biosensors with smartphones enables real-time diagnosis, which is particularly helpful for doctors in identifying pathogens and initiating treatments promptly. Innovative optical biosensors, supported by nano-enabled artificial intelligence, present a potential solution for facilitating rapid, point-of-care detection of antimicrobial resistance with real-time, sensitive, and portable capabilities.

Sehit et al. combined computational approaches for epitope selection and determination of polymer composition to obtain highly sensitive and epitope-specific imprinted polymers (eIPs) for the detection of human adenovirus. The computationally constructed eIPs achieved approximately twice the performance of conventional eIPs with only one functional monomer, demonstrating that this strategy can lead to the design of smart MIPs (Sehit et al. 2024). Wang et al. developed a colourimetric sensing platform integrated with an in situ immunomagnetic separation function for the ultrasensitive detection of *E. coli*. The concentration of *E. coli* can be achieved with a smartphone in one step by performing a digital image colourimetric analysis (Wang et al. 2024).

Luminescent Biosensors

Luminescence-based biosensors are particularly useful in applications that require high sensitivity, thanks to their superior signal-to-noise ratio and minimal background interference. They play a significant role in clinical diagnostics and environmental monitoring. Luminescence can originate through different mechanisms, including fluorescence, chemiluminescence (CL), bioluminescence (BL), and electrochemiluminescence (ECL). Fluorescence-based biosensors operate by detecting the light emitted from molecules called fluorophores upon excitation by a specific wavelength. In contrast, the basis of chemiluminescence-based biosensors relies on capturing the light generated by the excitation of species through chemical reactions. Bioluminescence is a biologically mediated form of chemiluminescence that occurs naturally in living systems. Electrochemiluminescence, conversely, is a type of chemiluminescence in which the light is stimulated electrochemically by an electrical potential. Notably, CL, BL, and ECL biosensors offer significant advantages by eliminating the need for external light sources, reducing background noise and simplifying instrumentation.

In luminescence-based biosensors, optical detectors operate by converting emitted light into electrical signals. Traditional detectors for these biosensors can be a point detector, such as photomultiplier tubes (PMTs). However, they require delicate materials and high voltages that are expensive, making them not the best choice for POC applications. As an alternative, avalanche photodiodes (APDs) utilise doped semiconductor materials to convert photons into electrons, which are amplified by impact ionisation. Arrays of APDs can be configured into compact, cost-effective, sensitive, and high-speed image sensors. These properties make them highly suitable for POC and PON optical detection. Whilst conventional image sensors are typically used with lenses to control focal distance, field of view (FOV), and spatial resolution, APDs introduce a trade-off: improving one parameter, such as FOV, often compromises another, like spatial resolution. However, lens-free image sensors overcome this limitation, as the FOV is defined by the number and size of their pixels, and the spatial resolution is determined by pixel size, eliminating the need for optical focusing elements. In fluorescence imaging, an emission filter should be placed between the sample and the sensor to hinder excitation wavelengths, which complicates the implementation of lens-free, contact-based setups. For this reason, luminescence-based biosensors based in lens-free contact configurations commonly utilise CL, ECL, and BL as signal generation mechanisms, as these techniques do not require external light excitation.

Fluorescence is also a widely used optical technique for detecting water contamination. It is typically favoured in microfluidic devices due to their rapid, sensitive, real-time analysis, high stability, non-destructive nature, and ability to be employed for on-site or online monitoring. However, because this technique depends on external components, APDs become less portable, more complex, and more costly. This limitation could hinder their suitability for home use or self-diagnosis applications, reducing their commercial potential. The need and demand for diversification in society necessitate the evolution of sensors from sophisticated, slow, and large-scale equipment to more straightforward and compact alternatives. This tendency is driven by the need to improve on-site detection, which involves developing portable and precise devices, as well as advancements in smartphone-based sensing devices, including high-resolution cameras and robust data processing capabilities. These devices often operate as standalone platforms that transmit signals through USB ports, audio jacks, or cameras, significantly simplifying design complexity and reducing the costs of detection systems. Table 4 illustrates various microfluidic devices for the fluorometric detection of contaminants in water.

3D μPADs have gained significant attention thanks to their advancements in miniaturisation and integration, particularly digital microfluidics (DMF) (Zhou et al. 2020). DMF manipulates liquids in discrete micro-droplets through surface tension modulation induced by an electric field. Individual droplets reduce the need for sample preparation and provide greater hardware flexibility, minimising the risk of blockages typically associated with continuous liquid streams. Serving as a micromixer, an L-shaped droplet can enhance mass transfer, reduce reaction times, and improve both yield and selectivity of products. DMF offers advantages in mechanical simplicity, scalability, reconfigurability, and low energy consumption, highlighting its potential for portable systems.

Therefore, combining fluorescence sensors with smartphone cameras and image processing technology enables more accurate on-site water safety analyses. This framework includes integrated smartphone-based sensing platforms that combine 3D components, smartphones, and fluorescence probes. Harnessing the "on-off-on" luminescent mechanism and utilising RGB values analysis software on portable smartphone devices has led to rapid advancements in digital imaging. The combination of the "on-off-on" luminescent method and smartphone technology has gained popularity for real-time monitoring. However, the "on-off-on" luminescent principle, combined with the portable detection strategy, still faces restrictions and challenges for its application in water analysis.

Regarding the improvement of ML-aided bioluminescent detection, an investigation utilised a whole-cell sensing array comprising genetically modified *E. coli* strains, each with a bioluminescence-based reporter mechanism. Distinct bioluminescence patterns observed with this system enable the classification and detection of antibiotics. The patterns can be processed to train a multiclass decision forest model (Huang et al. 2022). Another study utilised a bioluminescent bacterial system to detect water contaminants. In this case, a multilayer perceptron model was trained on light emission kinetics in response to various pollutants, showcasing promising results for rapid detection (Park et al. 2023).

Colourimetry offers numerous benefits in qualitative analyses but falls short in quantitative situations. Quantitative studies on PADs demand more rigorous and carefully regulated measurements to achieve precise and accurate analyte levels. It is the most favourable detection system due to its remarkable instrumental benefits, such as portability and the direct detection of heavy metal presence in water, which is measurable to the naked eye (Abdollahiyan et al. 2021). Qin et al. designed and fabricated microchips to recognise metal ions utilising a single indicator through multiple fluorescent channels (Qin et al. 2019). Meanwhile, Leng et al. developed multidimensional sensing platforms for various heavy metals based on the colour or fluorescence responses of a single indicator (Leng et al. 2016). Sun et al. developed a suspending droplet mode μPADS coupled with a G-quadruplex-based luminescence switch for sensing lead (II) ions (Sun et al. 2018). The device uses tension and gravity to actuate droplets. Pennachio et al. developed a portable biosensor to detect mercury (II) in seawater with a fluorescent analysis. They used a new chimaera formed by Vmh2 and H3w peptides adhered to a polystyrene multi-well plate. The developed biosensor was combined with three machine-learning prediction models (multilayer perceptron, random forest, and XGBoost) (Pennacchio et al. 2022).

Detecting pesticides in water is one of the most essential measures to protect water quality and human health. The biosensor identifies pesticide residues in water with high sensitivity and selectivity, combining biological components with sensor technology. Zhang et al. developed a paper-based microfluidic chip that utilises ratiometric fluorescence imaging to determine 2,4-dichlorophenoxyacetic acid (Zhang et al. 2020).

AMR poses a critical global One Health concern due to unintentional and ongoing antibiotic exposure, as well as difficulties in accurately diagnosing infections. Gung et al. developed a paper-based microfluidic device for detecting norovirus in environmental samples using fluorescence. The norovirus was captured on nitrocellulose paper through electrostatic interactions, and then it was retained by specific antibodies conjugated with polystyrene particles. The fluorescence of these aggregates was observed via images taken by a smartphone-based fluorescence microscope (Gowda et al. 2022).

Other Biosensors

Surface Plasmon Resonance (SPR) Biosensors

Surface plasmon resonance (SPR) is an optical label-free detection technique in which the interaction of (bio)molecules with a thin metal surface on a prism changes the direction of the reflected light. This interaction shifts the surface plasmon resonance angle and changes the refractive index. These changes can be monitored in real time to obtain information about the binding (bio)molecules. The SPR signal is related to changes in the intensity of the angle of reflection of a thin gold film. A variety of SPR is named as localised SPR (LSPR). They use metallic NPs as the metal that reflects the light beam. These sensors are increasingly used as localised strong electromagnetic field particles appear on the surface of NPs.

In addition to new data collection and processing technologies, these technologies can also be utilised to design and model biosensor surfaces. In this scenario, the surface was designed and modelled using simulation models that employ solid elements based on the Finite Element Method (FEM), divided into tetrahedral meshing, to conduct numerical calculations on the geometries. The biosensor incorporates graphene, gold, and silver meta-surface resonators. The models assess parameters such as the distance between resonators, the shape and size of the resonators, and the thickness of the material layers. The model is optimised by iteratively evaluating various geometric configurations to enhance the performance of the sensor, and validation is carried out by analysing the simulated results.

Additionally, the authors employed new data collection and processing technologies, such as the polynomial regression (PR) model, to identify the identity and concentration of biomolecules in water samples. This machine learning method relies on the relationship between independent and dependent variables to predict outcomes. In this context, the R^2 metric is utilised to assess the effectiveness of the predictions. Once trained, the model can interpolate and/or extrapolate values to forecast changes for identifying biomolecules and their concentrations. The biosensor produces simulated output signals in the form of spectra or sensor response curves, which are subsequently used to determine the identity and concentration of the analytes of interest within the sample.

Following the principles of an SPR sensor that utilises the impact of two-dimensional materials on the Au chip for identifying pathogens in clinical water samples, a smartphone-based SPR platform was used to assess the accuracy and repeatability of measurements by detecting various concentrations of antibodies binding to a functionalised sensing element.

Ozkaya et al. developed a multichannel platform for SPR sensors capable of measuring antibiotic concentrations. SPR sensor technology distinguishes between antibiotic-resistant and susceptible bacteria in approximately 3–6 h. This technique has shown promising results as a real-time method for the early identification of bacteria that have developed antibiotic resistance (Ozkaya et al. 2019).

Surface-Enhanced Raman Spectroscopy (SERS)

The SERS, a supersensitive and non-destructive technique, is becoming one of the most promising for water analysis. Methods based on SERS allow the detection of low concentrations of adsorbed molecules on the surface of plasmonic substrates. Table 5 presents various microfluidic devices for contaminant detection based on SERS technique.

Lafuente et al. introduced an innovative method for pesticide detection by creating SERS active regions within a microfluidic channel. The technique relies on the adsorption of pesticide molecules onto the substrate, which induces a measurable change in SERS activity (Lafuente et al. 2020).

Emonds et al. write about developing a new quantitative method for determining pesticides using microfluidics combined with SERS (Emonds-Alt et al. 2022).

Table 5 SERS-based microfluidic devices to determine contaminants in water

Contaminants	Device or material	Detection limit	References
Cd (II)	3D glass channels with 2D periodic metal nanostructures fabricated by all-femtosecond laser	10 ppb	Bai et al. (2018)
Paraoxon	PDMS microchannels with polyoxometalate-decorated gold nanostructures fabricated in situ	0.1 μM	Lafuente et al. (2020)
Glyphosate	PFA coils with in situ silver nanoparticles synthesis	40 $\mu g\ mL^{-1}$	Emonds-Alt et al. (2022)
E. coli	Two PDMS perpendicular channels connected by a nanoporous membrane using silver nanoparticles to make the surface enhancement	–	Krafft et al. (2021)

Several authors used machine learning (ML) to identify four pesticides using silver nanoparticles. The Raman spectra were utilised in their original form without any pre-treatments. The authors developed an optimised learning model to classify multiple pesticides in environmental water. This model achieved nearly 100% accuracy in both classification and regression. Zhou et al. discussed the use of SERS to identify pathogenic microbes (Zhou et al. 2020). SERS can map and characterise bacteria without the need for additional labelling.

Conclusion

Numerous detection strategies have been implemented in microfluidic platforms to identify and determine water contaminants. They are based on electrochemical, fluorescent, and colourimetric methods. Amid water contaminants, metal ions, including heavy metals, pesticides, and biological contaminants, are among the most challenging environmental pollutants. Their persistence and severe health issues can lead to significant problems for both humans and animals. This chapter has explored the potential of different types of microfluidic devices as analytical tools, their functional capabilities, their ability to introduce innovative materials, and their practical applications in water quality monitoring. Among microfluidic platforms, colourimetric-based methods are the most effective for detecting heavy metal ions, whilst fluorescent-based methods are more suitable for biological contaminants.

Due to the advantages of using microfluidic platforms in this field, it is expected that these systems will play a crucial role in environmental analysis. These advantages rely on offering cost-effective, portable, and on-site alternatives, making them novel options to conventional laboratory-based detection methods. Despite the benefits of real-time, on-site, and rapid detection compared to traditional methods, they present some challenges in the data analysis step. In this context, artificial intelligence (AI) technology emerges as a promising solution to enhance the monitoring performance and reliability of microfluidic systems. By applying machine learning algorithms, AI can support the processing, classification, and interpretation of sensor outputs to offer valuable tools for analysing contaminants in water.

Despite the advances achieved in combining microfluidic platforms and AI, this arrangement must be optimised for accuracy and reliability in the obtained results. The complex data generated from a microfluidic chip requires large, high-quality datasets to train AI models, ensuring data quality. This training can be challenging in real-world environmental monitoring scenarios, as acquiring large amounts of labelled data can pose difficulties for real-world water quality analysis. However, addressing the issue through improved data acquisition strategies and innovative interdisciplinary approaches could be crucial for advancing microfluidic-AI systems as standard methods for monitoring contaminants in water in future determinations.

Perspective Towards Future Trends

Notable progress has been made in the development of microfluidic technologies for detecting water contaminants over the past two decades. Nevertheless, these systems offer considerable potential for further innovation in their development. For example, their benefits could be utilised in applications that require point-of-need analysis. The synergy of microfluidic devices and smartphones or tablets has proven to be a promising innovation, enhancing parameters such as sensitivity and selectivity, reducing analysis time whilst offering real-time results, and eliminating the need for expensive data transmission infrastructures. Advancements in wireless connectivity, combined with the expansion of IoT technology and infrastructure, will be key to overcoming challenges associated with autonomous and decentralised monitoring.

AI presents significant opportunities, but certain limitations should also be considered. However, the effectiveness of AI depends on the availability of large, high-quality datasets for training and working with this considerable amount of information, and insufficient data can affect the predictions. ML will be a powerful subset of AI, as it allows systems to analyse complex datasets, identify patterns and make data-driven predictions without explicit programming. The challenges can be mitigated using microfluidic chip technology, which provides cost-effective, scalable, and automated platforms.

Combining IoT technologies and AI with microfluidic platforms enables more efficient data acquisition and analysis. These synergistic approaches are expected to lead to more intelligent, portable, and efficient diagnostic platforms capable of providing real-time, on-site assessments across diverse environmental settings, converting them into powerful tools for analysing contaminants in water.

Funding and Acknowledgement This work was financially supported by the project "Plan Complementario de I+D+i en el área de Ciencias Marinas" (PCM_00121), funded by the European Union within the framework of the Recovery, Transformation, and Resilience Plan-NextGenerationEU and by the Regional Government of Andalusia. Additionally, the author AEA thanks the financial support of the Margarita Salas postdoctoral fellowship (UCOR02MS), funded by the European Union within the framework of the Recovery, Transformation, and Resilience Plan-NextGenerationEU and submodality 2.6 "UCOProyecta", from the Own Research Plan of the Universidad de Córdoba "Enrique Aguilar Benítez de Lugo" for 2024.

Competing Interest The authors declare that they have no known competing financial interests or personal relationships that could have influenced the work reported in this chapter.

References

Abbasi R, Hu X, Zhang A, Dummer I, Wachsmann-Hogiu S (2024) Optical image sensors for smart analytical chemiluminescence biosensors. Bioengineering 11(9):912. https://doi.org/10.3390/bioengineering11090912

Abdollahiyan P, Hasanzadeh M, Seidi F, Pashazadeh-Panahi P (2021) An innovative colorimetric platform for the low-cost and selective identification of Cu(II), Fe(III), and Hg(II) using GQDs-DPA supported amino acids by microfluidic paper-based (μPADs) device: multicolor plasmonic patterns. J Environ Chem Eng 9(5):106197. https://doi.org/10.1016/j.jece.2021.106197

Akyazi T, Basabe-Desmonts L, Benito-Lopez F (2018) Review on microfluidic paper-based analytical devices towards commercialisation. Anal Chim Acta 1001:1–17. https://doi.org/10.1016/j.aca.2017.11.010

AlMashrea BA, Almehdi AM, Damiati S (2024) Simple microfluidic devices for in situ detection of water contamination: a state-of-art review. Front Bioeng Biotechnol 12:1355768. https://doi.org/10.3389/fbioe.2024.1355768

Almeida MIGS, Jayawardane BM, Kolev SD, McKelvie ID (2018) Developments of microfluidic paper-based analytical devices (μPADs) for water analysis: a review. Talanta 177:176–190. https://doi.org/10.1016/j.talanta.2017.08.072

Altintas Z, Akgun M, Kokturk G, Uludag Y (2018) A fully automated microfluidic-based electrochemical sensor for real-time bacteria detection. Biosens Bioelectron 100:541–548. https://doi.org/10.1016/j.bios.2017.09.046

Araci IE, Brisk P (2014) Recent developments in microfluidic large scale integration. Curr Opin Biotechnol 25:60–68. https://doi.org/10.1016/j.copbio.2013.08.014

Arduini F, Cinti S, Caratelli V, Amendola L, Palleschi G, Moscone D (2019) Origami multiple paper-based electrochemical biosensors for pesticide detection. Biosens Bioelectron 126:346–354. https://doi.org/10.1016/j.bios.2018.10.014

Aryal P, Hefner C, Martinez B, Henry CS (2024) Microfluidics in environmental analysis: advancements, challenges, and future prospects for rapid and efficient monitoring. Lab Chip 24(5):1175–1206. https://doi.org/10.1039/d3lc00871a

Bagheri H, Afkhami A, Khoshsafar H, Rezaei M, Sabounchei SJ, Sarlakifar M (2015) Simultaneous electrochemical sensing of thallium, lead and mercury using a novel ionic liquid/graphene modified electrode. Anal Chim Acta 870:56–66. https://doi.org/10.1016/j.aca.2015.03.004

Bai S, Serien D, Hu A, Sugioka K (2018) 3D microfluidic surface-enhanced Raman spectroscopy (SERS) chips fabricated by all-femtosecond-laser-processing for real-time sensing of toxic substances. Adv Funct Mater 28(23):1706262. https://doi.org/10.1002/ADFM.201706262

Balali-Mood M, Naseri K, Tahergorabi Z, Khazdair MR, Sadeghi M (2021) Toxic mechanisms of five heavy metals: mercury, Lead, chromium, cadmium, and arsenic. Front Pharmacol 12. https://doi.org/10.3389/fphar.2021.643972

Banerjee D, Adhikary S, Bhattacharya S, Chakraborty A, Dutta S, Chatterjee S, Ganguly A, Nanda S, Rajak P (2024) Breaking boundaries: artificial intelligence for pesticide detection and eco-friendly degradation. Environ Res 241:117601. https://doi.org/10.1016/j.envres.2023.117601

Beshana S, Hussen A, Leta S, Kaneta T (2022) Microfluidic paper based analytical devices for the detection of carbamate pesticides. Bull Environ Contam Toxicol 109(2):344–351. https://doi.org/10.1007/s00128-022-03533-3

Bilen S, Bilen M, Turan V (2019) Relationships between cement dust emissions and soil properties. Pol J Environ Stud 28(5):3089–3098. https://doi.org/10.15244/pjoes/92521

Broekaert JAC (1982) The application of ICP-OES to industrial analytical problems. TrAC Trends Anal Chem 1(11):249–253. https://doi.org/10.1016/0165-9936(82)80080-0

Butcher DJ (2010) Advances in inductively coupled plasma optical emission spectrometry for environmental analysis. Instrum Sci Technol 38(6):458–469. https://doi.org/10.1080/10739149.2010.517884

Cao Y, Liu Y, Li F, Guo S, Shui Y, Xue H, Wang L (2019) Portable colorimetric detection of copper ion in drinking water via red beet pigment and smartphone. Microchem J 150:104176. https://doi.org/10.1016/j.microc.2019.104176

Chaudhry FN, Malik MF (2017) Factors affecting water pollution: a review. J Ecosyst Ecogr 7(1):1000225. https://doi.org/10.4172/2157-7625.1000225

Chen GQ (2012) Synthetic biology: research perspectives from China. ACS Synth Biol 1(7):251. https://doi.org/10.1021/sb300057c

Chen J, Hao M, Xin Y, Zhu R, Gu Z, Zhang L, Guo X (2024) A novel phosphotriesterase hybrid nanoflower-hydrogel sensor equipped with a smartphone detector for real-time on-site monitoring of organophosphorus pesticides. Int J Biol Macromol 276:133979. https://doi.org/10.1016/j.ijbiomac.2024.133979

Chung S, Breshears LE, Perea S, Morrison CM, Betancourt WQ, Reynolds KA, Yoon J-Y (2019) Smartphone-based paper microfluidic Particulometry of norovirus from environmental water samples at the single copy level. ACS Omega 4(6):11180–11188. https://doi.org/10.1021/acsomega.9b00772

Cossu A, Carta A, Lomonaco V, Bacciu D (2021) Continual learning for recurrent neural networks: an empirical evaluation. Neural Netw 143:607–627. https://doi.org/10.1016/j.neunet.2021.07.021

Dai J, Gao W, Yin J, Liang L, Zou J, Jin Q (2021) Microfluidic sensor integrated with nanochannel liquid conjunct Ag/AgCl reference electrode for trace Pb(II) measurement. Anal Chim Acta 1164:338511. https://doi.org/10.1016/j.aca.2021.338511

Devadhasan JP, Kim J (2018) A chemically functionalized paper-based microfluidic platform for multiplex heavy metal detection. Sens Actuators B Chem 273:18–24. https://doi.org/10.1016/j.snb.2018.06.005

Dhar BC, Lee NY (2018) Lab-on-a-chip technology for environmental monitoring of microorganisms. Biochip J 12(3):173–183. https://doi.org/10.1007/s13206-018-2301-5

Elsayed AA, Erfan M, Sabry YM, Dris R, Gaspéri J, Barbier J-S, Marty F, Bouanis F, Luo S, Nguyen BTT, Liu A-Q, Tassin B, Bourouina T (2021) A microfluidic chip enables fast analysis of water microplastics by optical spectroscopy. Sci Rep 11(1):10533. https://doi.org/10.1038/s41598-021-89960-4

Emonds-Alt G, Malherbe C, Kasemiire A, Avohou HT, Hubert P, Ziemons E, Monbaliu JCM, Eppe G (2022) Development and validation of an integrated microfluidic device with an in-line surface enhanced Raman spectroscopy (SERS) detection of glyphosate in drinking water. Talanta 249:123640. https://doi.org/10.1016/j.talanta.2022.123640

Gale BK, Jafek AR, Lambert CJ, Goenner BL, Moghimifam H, Nze UC, Kamarapu SK (2018) A review of current methods in microfluidic device fabrication and future commercialization prospects. Inventions 3(3):60. https://doi.org/10.3390/inventions3030060

Giménez-Gómez P, Baldi A, Ayora C, Fernández-Sánchez C (2019) Automated determination of as(III) in waters with an electrochemical sensor integrated into a modular microfluidic system. ACS Sens 4(12):3156–3165. https://doi.org/10.1021/acssensors.9b01286

Gowda HN, Kido H, Wu X, Shoval O, Lee A, Lorenzana A, Madou M, Hoffmann M, Jiang SC (2022) Development of a proof-of-concept microfluidic portable pathogen analysis system for water quality monitoring. Sci Total Environ 813:152556. https://doi.org/10.1016/j.scitotenv.2021.152556

Hassaan MA, El Nemr A (2020) Pesticides pollution: classifications, human health impact, extraction and treatment techniques. Egypt J Aquat Res 46(3):207–220. https://doi.org/10.1016/j.ejar.2020.08.007

Himeur Y, Rimal B, Tiwary A, Amira A (2022) Using artificial intelligence and data fusion for environmental monitoring: a review and future perspectives. Inf Fusion 86:44–75. https://doi.org/10.1016/j.inffus.2022.06.003

Huang WC, Wei CD, Belkin S, Hsieh TH, Cheng JY (2022) Machine-learning assisted antibiotic detection and categorization using a bacterial sensor array. Sensors Actuators B Chem 355. https://doi.org/10.1016/j.snb.2021.131257

Ibrahim S, Azab El-Liethy M, Abia ALK, Abdel-Gabbar M, Mahmoud Al Zanaty A, Mohamed Kamel M (2020) Design of a bioaugmented multistage biofilter for accelerated municipal wastewater treatment and deactivation of pathogenic microorganisms. Sci Total Environ 703. https://doi.org/10.1016/j.scitotenv.2019.134786

Idros N, Chu D (2018) Triple-indicator-based multidimensional colorimetric sensing platform for heavy metal ion detections. ACS Sens 3(9):1756–1764. https://doi.org/10.1021/acssensors.8b00490

Kamnoet P, Aeungmaitrepirom W, Menger RF, Henry CS (2021) Highly selective simultaneous determination of Cu(II), Co(II), Ni(II), Hg(II), and MN(II) in water samples using microfluidic paper-based analytical devices. Analyst 146(7):2229–2239. https://doi.org/10.1039/d0an02200d

Kanu AB (2021) Recent developments in sample preparation techniques combined with high-performance liquid chromatography: a critical review. J Chromatogr A 1654:462444. https://doi.org/10.1016/j.chroma.2021.462444

Kim HJ, Kim Y, Park SJ, Kwon C, Noh H (2018) Development of colorimetric paper sensor for pesticide detection using competitive-inhibiting reaction. Biochip J 12(4):326–331. https://doi.org/10.1007/s13206-018-2404-z

Kokkinos C, Economou A, Giokas D (2018) Paper-based device with a sputtered tin-film electrode for the voltammetric determination of Cd(II) and Zn(II). Sensors Actuators B Chem 260:223–226. https://doi.org/10.1016/j.snb.2017.12.182

Krafft B, Tycova A, Urban RD, Dusny C, Belder D (2021) Microfluidic device for concentration and SERS-based detection of bacteria in drinking water. Electrophoresis 42(1–2):86–94. https://doi.org/10.1002/elps.202000048

Lace A, Cleary J (2021) A review of microfluidic detection strategies for heavy metals in water. Chem 9(4):60. https://doi.org/10.3390/chemosensors9040060

Lace A, Ryan D, Bowkett M, Cleary J (2019) Arsenic detection in water using microfluidic detection systems based on the leucomalachite green method. Anal Methods 11(42):5431–5438. https://doi.org/10.1039/c9ay01580a

Lafuente M, Pellejero I, Clemente A, Urbiztondo MA, Mallada R, Reinoso S, Pina MP, Gandía LM (2020) In situ synthesis of SERS-active au@POM nanostructures in a microfluidic device for real-time detection of water pollutants. ACS Appl Mater Interfaces 12(32):36458–36467. https://doi.org/10.1021/acsami.0c06725

Laschi S, Sfragano PS, Tadini-Buoninsegni F, Guigues N, Palchetti I (2024) Development of a flow system for decentralized electrochemical analysis of heavy metals using screen-printed electrodes: the importance of sensor stability. Analyst 149(16):4239–4249. https://doi.org/10.1039/d4an00616j

Leng Y, Qian S, Wang Y, Lu C, Ji X, Lu Z, Lin H (2016) Single-indicator-based multidimensional sensing: detection and identification of heavy metal ions and understanding the foundations from experiment to simulation. Sci Rep 6. https://doi.org/10.1038/srep25354

Li S, Zhang C, Wang S, Liu Q, Feng H, Ma X, Guo J (2018) Electrochemical microfluidics techniques for heavy metal ion detection. Analyst 143(18):4230–4246. https://doi.org/10.1039/C8AN01067F

Li F, Hu Y, Li Z, Liu J, Guo L, He J (2019) Three-dimensional microfluidic paper-based device for multiplexed colorimetric detection of six metal ions combined with use of a smartphone. Anal Bioanal Chem 411(24):6497–6508. https://doi.org/10.1007/s00216-019-02032-5

Li S, Pang C, Ma X, Zhang Y, Xu Z, Li J, Zhang M, Wang M (2021) Microfluidic paper-based chip for parathion-methyl detection based on a double catalytic amplification strategy. Microchim Acta 188(12):438. https://doi.org/10.1007/s00604-021-05084-6

Li D, Zhou J, Zhao Z, Huang X, Li H, Qu, Ao, Zhou C, Yao K, Liu Y, Wu M, Su J, Shi R, Huang Y, Wang J, Zhang Z, Liu Y, Gao Z, Park W, Jia H, Yu X (2024) Battery-free, wireless, and electricity-driven soft swimmer for water quality and virus monitoring. Sci Adv 10:6301

Lisak G, Cui J, Bobacka J (2015) Paper-based microfluidic sampling for potentiometric determination of ions. Sens Actuators B Chem 207:933–939. https://doi.org/10.1016/j.snb.2014.07.044

Luka G, Ahmadi A, Najjaran H, Alocilja E, DeRosa M, Wolthers K, Malki A, Aziz H, Althani A, Hoorfar M (2015) Microfluidics integrated biosensors: a leading technology towards lab-on-a-chip and sensing applications. Sensors 15(12):30011–30031. https://doi.org/10.3390/s151229783

Ma S, Zhao W, Zhang Q, Zhang K, Liang C, Wang D, Liu X, Zhan X (2023) A portable microfluidic electrochemical sensing platform for rapid detection of hazardous metal Pb2+ based on thermocapillary convection using 3D ag-rGO-f-Ni(OH)2/NF as a signal amplifying element. J Hazard Mater 448:130923. https://doi.org/10.1016/j.jhazmat.2023.130923

Meelapsom R, Jarujamrus P, Amatatongchai M, Chairam S, Kulsing C, Shen W (2016) Chromatic analysis by monitoring unmodified silver nanoparticles reduction on double layer microfluidic paper-based analytical devices for selective and sensitive determination of mercury(II). Talanta 155:193–201. https://doi.org/10.1016/j.talanta.2016.04.037

Miller S, Weiss AA, Heineman WR, Banerjee RK (2019) Electroosmotic flow driven microfluidic device for bacteria isolation using magnetic microbeads. Sci Rep 9(1):14228. https://doi.org/10.1038/s41598-019-50713-z

Mohan JM, Amreen K, Javed A, Dubey SK, Goel S (2022) Emerging trends in miniaturized and microfluidic electrochemical sensing platforms. Curr Opin Electrochem 33:100930. https://doi.org/10.1016/j.coelec.2021.100930

Motalebizadeh A, Bagheri H, Asiaei S, Fekrat N, Afkhami A (2018) New portable smartphone-based PDMS microfluidic kit for the simultaneous colorimetric detection of arsenic and mercury. RSC Adv 8(48):27091–27100. https://doi.org/10.1039/c8ra04006k

Mukherjee S, Bhattacharyya S, Ghosh K, Pal S, Halder A, Naseri M, Mohammadniaei M, Sarkar S, Ghosh A, Sun Y, Bhattacharyya N (2021) Sensory development for heavy metal detection: a review on translation from conventional analysis to field-portable sensor. Trends Food Sci Technol 109:674–689. https://doi.org/10.1016/j.tifs.2021.01.062

Murphy J, Riley JP (1962) A modified single solution method for the determination of phosphate in natural waters. Anal Chim Acta 27:31–36. https://doi.org/10.1016/S0003-2670(00)88444-5

Niculescu A-G, Chircov C, Bîrcă AC, Grumezescu AM (2021) Fabrication and applications of microfluidic devices: a review. Int J Mol Sci 22(4):2011. https://doi.org/10.3390/ijms22042011

Ozkaya GU, Durak MZ, Akyar I, Karatuna O (2019) Antimicrobial susceptibility test for the determination of resistant and susceptible S. aureus and Enterococcus spp. using a multi-channel surface plasmon resonance device. Diagnostics 9(4):191. https://doi.org/10.3390/DIAGNOSTICS9040191

Park J, Yang JH, Jung J, Kwak IS, Choe JK, An J (2023) Comparative analysis of the capability of the extended biotic ligand model and machine learning approaches to predict arsenate toxicity. Chemosphere 344:140350. https://doi.org/10.1016/J.CHEMOSPHERE.2023.140350

Patir K, Gogoi SK (2019) Nitrogen-doped carbon dots as fluorescence ON–OFF–ON sensor for parallel detection of copper(II) and mercury(II) ions in solutions as well as in filter paper-based microfluidic device. Nanoscale Adv 1(2):592–601. https://doi.org/10.1039/C8NA00080H

Pebdeni AB, Roshani A, Mirsadoughi E, Behzadifar S, Hosseini M (2022) Recent advances in optical biosensors for specific detection of *E. coli* bacteria in food and water. Food Control 135:108822. https://doi.org/10.1016/j.foodcont.2022.108822

Pennacchio A, Giampaolo F, Piccialli F, Cuomo S, Notomista E, Spinelli M, Amoresano A, Piscitelli A, Giardina P (2022) A machine learning-enhanced biosensor for mercury detection based on an hydrophobin chimera. Biosens Bioelectron 196. https://doi.org/10.1016/j.bios.2021.113696

Pourreza N, Golmohammadi H, Rastegarzadeh S (2016) Highly selective and portable chemosensor for mercury determination in water samples using curcumin nanoparticles in a paper based analytical device. RSC Adv 6(73):69060–69066. https://doi.org/10.1039/C6RA08879A

Qin M, Li F, Song Y (2019) A photochromic sensor microchip for high-performance multiplex metal ion detection. Methods Mol Biol 2027:49–59. https://doi.org/10.1007/978-1-4939-9616-2_4

Rai PK, Islam M, Gupta A (2022) Microfluidic devices for the detection of contamination in water samples: a review. Sens Actuators A Phys 347:113926. https://doi.org/10.1016/j.sna.2022.113926

Rauf S, Tashkandi N, De Oliveira Filho JI, Oviedo-Osornio CI, Danish MS, Hong PY, Salama KN (2022) Digital E. coli counter: a microfluidics and computer vision-based DNAzyme method for the isolation and specific detection of E. coli from water samples. Biosensors 12(1):34. https://doi.org/10.3390/BIOS12010034/S1

Rodoplu D, Chang C, Kao C, Hsu C (2022) A micro-pupil device for point-of-care testing of viable Escherichia coli in tap water. Microchem J 178:107390. https://doi.org/10.1016/j.microc.2022.107390

Saez J, Catalan-Carrio R, Owens RM, Basabe-Desmonts L, Benito-Lopez F (2021) Microfluidics and materials for smart water monitoring: a review. Anal Chim Acta 1186:338392. https://doi.org/10.1016/j.aca.2021.338392

Santana HS, Silva JL, Aghel B, Ortega-Casanova J (2020) Review on microfluidic device applications for fluids separation and water treatment processes. SN Appl Sci 2(3):395. https://doi.org/10.1007/s42452-020-2176-7

Santangelo MF, Shtepliuk I, Filippini D, Puglisi D, Vagin M, Yakimova R, Eriksson J (2019) Epitaxial graphene sensors combined with 3D-printed microfluidic chip for heavy metals detection. Sensors 19(10):2393. https://doi.org/10.3390/s19102393

Sciuto S, Volpatti D, Esposito G, Pastorino P, Khalil SMI, Stoppani N, Esposito G, Prearo M, Gabetti A, Maganza A, Fariano L, Montemurro V, Bozzetta E, Acutis PL, Colussi S (2023) Near infrared spectroscopy as a novel non-invasive tool for the detection of lactococcosis in rainbow trout. Aquac Rep 33:101862. https://doi.org/10.1016/j.aqrep.2023.101862

Sehit E, Yao G, Battocchio G, Radfar R, Trimpert J, Mroginski MA, Süssmuth R, Altintas Z (2024) Computationally designed epitope-mediated imprinted polymers versus conventional epitope imprints for the detection of human adenovirus in water and human serum samples. ACS Sens 9(4):1831–1841. https://doi.org/10.1021/acssensors.3c02374

Sekar A, Yadav R, Basavaraj N (2021) Fluorescence quenching mechanism and the application of green carbon nanodots in the detection of heavy metal ions: a review. New J Chem 45(5):2326–2360. https://doi.org/10.1039/D0NJ04878J

Sharifi H, Tashkhourian J, Hemmateenejad B (2020) A 3D origami paper-based analytical device combined with PVC membrane for colorimetric assay of heavy metal ions: application to determination of Cu(II) in water samples. Anal Chim Acta 1126:114–123. https://doi.org/10.1016/j.aca.2020.06.006

Snyder SA, Boban M, Li C, VanEpps JS, Mehta G, Tuteja A (2020) Lysis and direct detection of coliforms on printed paper-based microfluidic devices. Lab Chip 20(23):4413–4419. https://doi.org/10.1039/D0LC00665C

Solmaz ME, Mutlu AY, Alankus G, Kılıç V, Bayram A, Horzum N (2018) Quantifying colorimetric tests using a smartphone app based on machine learning classifiers. Sens Actuators B Chem 255:1967–1973. https://doi.org/10.1016/j.snb.2017.08.220

Sriram G, Bhat MP, Patil P, Uthappa UT, Jung H-Y, Altalhi T, Kumeria T, Aminabhavi TM, Pai RK, Madhuprasad, Kurkuri MD (2017) Paper-based microfluidic analytical devices for colorimetric detection of toxic ions: a review. TrAC Trends Anal Chem 93:212–227. https://doi.org/10.1016/j.trac.2017.06.005

Sun H, Li W, Dong ZZ, Hu C, Leung CH, Ma DL, Ren K (2018) A suspending-droplet mode paper-based microfluidic platform for low-cost, rapid, and convenient detection of lead(II) ions in liquid solution. Biosens Bioelectron 99:361–367. https://doi.org/10.1016/j.bios.2017.07.073

Tchounwou PB, Yedjou CG, Patlolla AK, Sutton DJ (2012) Heavy metal toxicity and the environment. Exp Suppl 101:133–164. https://doi.org/10.1007/978-3-7643-8340-4_6

Uka B, Kieninger J, Urban GA, Weltin A (2021) Electrochemical microsensor for microfluidic glyphosate monitoring in water using MIP-based concentrators. ACS Sens 6(7):2738–2746. https://doi.org/10.1021/acssensors.1c00884

Ulloa-Gomez AM, Waimin JF, Yu YC, Lucas A, Stanciu LA (2024) A smartphone-integrated aptasensor for pesticide detection using gold-decorated microparticles. Microchim Acta 191(4):194. https://doi.org/10.1007/s00604-024-06255-x

Wang N, Kanhere E, Miao J, Triantafyllou MS (2016) Miniaturized chemical sensor with bio-inspired micropillar working electrode array for lead detection. Sensors Actuators B Chem 233:249–256. https://doi.org/10.1016/j.snb.2016.04.048

Wang M, Song Z, Jiang Y, Zhang X, Wang L, Zhao H, Cui Y, Gu F, Wang Y, Zheng G (2021) A three-dimensional pinwheel-shaped paper-based microfluidic analytical device for fluorescence detection of multiple heavy metals in coastal waters by rational device design. Anal Bioanal Chem 413(12):3299–3313. https://doi.org/10.1007/S00216-021-03269-9

Wang G, Li J, Wu S, Jiang T, Chen TH (2022) A fully integrated, ready-to-use distance-based chemosensor for visual quantification of multiple heavy metal ions. Anal Chem 94(46):15925–15929. https://doi.org/10.1021/ACS.ANALCHEM.2C04712

Wang C, Deng R, Li H, Liu Z, Niu X, Li X (2024) An integrated magnetic separation enzyme-linked colorimetric sensing platform for field detection of Escherichia coli O157: H7 in food. Microchim Acta 191(8). https://doi.org/10.1007/s00604-024-06497-9

Wu Q, He J, Meng H, Wang Y, Zhang Y, Li H, Feng L (2019) A paper-based microfluidic analytical device combined with home-made SPE column for the colorimetric determination of copper(II) ion. Talanta 204:518–524. https://doi.org/10.1016/j.talanta.2019.06.006

Wu Z, Wang J, Bian C, Tong J, Xia S (2020) A MEMS-based multi-parameter integrated chip and its portable system forwater quality detection. Micromachines 11(1):63. https://doi.org/10.3390/mi11010063

Wu H, Chen J, Yang Y, Yu W, Chen Y, Lin P, Liang K (2022) Smartphone-coupled three-layered paper-based microfluidic chips demonstrating stereoscopic capillary-driven fluid transport towards colorimetric detection of pesticides. Anal Bioanal Chem 414(5):1759–1772. https://doi.org/10.1007/s00216-021-03839-x

Xu X, Nie S, Ding H, Hou FF (2018) Environmental pollution and kidney diseases. Nat Rev Nephrol 14(5):313–324. https://doi.org/10.1038/nrneph.2018.11

Yang Z, Zhang J, Zhao J, Zhou W, Cheng Y, Xu Z, Wei P, Wang Z, Liang H, Li C (2024) A high-sensitivity lab-on-a-chip analyzer for online monitoring of nitrite and nitrate in seawater based on liquid waveguide capillary cells. Lab Chip 24(14):3528–3535. https://doi.org/10.1039/D4LC00248B

Yetisen AK, Akram MS, Lowe CR (2013) Paper-based microfluidic point-of-care diagnostic devices. Lab Chip 13(12):2210. https://doi.org/10.1039/c3lc50169h

Zhang Q, Zhang X, Zhang X, Jiang L, Yin J, Zhang P, Han S, Wang Y, Zheng G (2019) A feedback-controlling digital microfluidic fluorimetric sensor device for simple and rapid detection of mercury(II) in costal seawater. Mar Pollut Bull 144:20–27. https://doi.org/10.1016/j.marpolbul.2019.04.063

Zhang Z, Ma X, Li B, Zhao J, Qi J, Hao G, Jianhui R, Yang X (2020) Fluorescence detection of 2,4-dichlorophenoxyacetic acid by ratiometric fluorescence imaging on paper-based microfluidic chips. Analyst 145(3):963–974. https://doi.org/10.1039/c9an01798d

Zhou J, Li B, Qi A, Shi Y, Qi J, Xu H, Chen L (2020) ZnSe quantum dot based ion imprinting technology for fluorescence detecting cadmium and lead ions on a three-dimensional rotary paper-based microfluidic chip. Sens Actuators B Chem 305:127462. https://doi.org/10.1016/j.snb.2019.127462

Autonomous Surface Vehicle for Water Monitoring Using Artificial Intelligence Methodologies

Alejandro Casado Pérez, Alejandro Mendoza Barrionuevo, Samuel Yanes Luis, Dame Seck Diop, Luis Miguel Díaz, Manuel Perales-Esteve, Sergio L. Toral, and Daniel Gutiérrez Reina

Contents

A. C. Pérez (✉) · A. M. Barrionuevo · S. Y. Luis · D. S. Diop · L. M. Díaz · M. Perales-Esteve · S. L. Toral · D. G. Reina
Electronics Engineering Department, Universidad de Sevilla, Sevilla, Spain
e-mail: acasado4@us.es; amendoza1@us.es; syanes@us.es; dseck@us.es; luismidp@us.es; mperales@us.es; storal@us.es; dgutierrezreina@us.es

D. Gutiérrez et al. (eds.), *Smart Water Quality Monitoring*,
https://doi.org/10.1007/978-3-032-03049-8_10

Abstract

Autonomous surface vehicles are characterized by a high payload capability and battery autonomy being able to perform long-range missions. When equipped with sensors, they allow for a smart and adaptative deployment in water resources for continuous monitoring. This chapter presents a real implementation of an autonomous surface vehicle prototype equipped with high-quality sensors to gather environmental data from water resources. This chapter provides an extensive definition and description of the hardware and electrical configuration as well as the software present in the vehicle, proposing an architecture that provides a framework to test and deploy new algorithms. The presence of a powerful embedded computer promotes the use of artificial intelligence algorithms and enhanced sensing techniques, increasing the scope of action of the vehicle. In this way, the vehicle can detect and locate macro-plastics using deep visual models, represent data in a continuous space domain and perform autonomous exploration and actuation, promoting methodologies such as adaptative informative path planning policies, monitoring data processing or predictive algorithms. This chapter provides experimental results to validate the capabilities of the vehicle, providing real monitoring data and example cases of artificial intelligence methodologies.

Keywords

Environmental data · Autonomous surface vehicle · Robotic system architecture · Macro-plastic detection and localization · Data processing and visualization · Real-time water monitoring · Artificial Intelligence (AI)

Introduction

Traditionally, water quality monitoring is a methodical work that involves time- and tool-consuming tasks. Traditional methods require for an operator to reach the designated mass of water, take samples of different locations, seal and send them to a laboratory that must analyse them, record the measures and dispose of the samples (Roy 2019). The task becomes further complicated when taking into account that it must be repeated periodically and samples require for specific and expensive tools to take depending on the parameters to monitor. For example, physical parameters are best taken in place, biological and bacteriological values may vary in the process of delivery to the laboratory and radioactive or toxic samples require extra protective measures, aside from the risk suffered by the operator.

The appearance of unmanned vehicles be them aerial (unmanned aerial vehicle), surface (unmanned surface vehicle) or underwater (unmanned underwater vehicle) (Kwon et al. 2023) has had a huge impact in the field of water quality monitoring. These vehicles have the capacity to reach places inaccessible for operators, take and seal samples, providing fast and accurate descriptions of the location and time of the sample and recovering them reducing considerably the risk of the operator. Furthermore, vehicles can be equipped with specialized sensors, measuring water quality in place without the need for laboratories and tools that allows them to perform specialized tasks. Overall, autonomous surface vehicles (ASVs) stand out as low-cost vehicles with higher battery autonomy and payload capacity than the other unmanned counterparts, allowing for long-range missions and continuous monitoring and gathering more extensive data (Ferri et al. 2014).

Water quality sensors present in ASVs are classified mainly into three groups (Kwon et al. 2023): electrochemical sensors that are in contact with water taking direct measures of chemical properties; spectral sensors that do not require direct contact with water to measure and are subdivided in two groups—active sensors that emit artificial waves to the water like light detection and ranging (LiDAR) and sound navigation and ranging (SONAR), allowing to measure the depth of the water and the underwater structure and passive sensors such as red-green-blue (RGB) or infrared (IR) cameras—lastly, biological sensors that measure biological properties. Additionally, ASV can be equipped with pumps to collect water samples or robotic arms to take sediment and biological samples (Chang et al. 2021).

The increase of battery autonomy and computation power of these ASVs has allowed them to perform autonomously without the need for an operator, and more recently, to perform coordinated tasks as a fleet of vehicles (Nothacker 2024) for contamination detection like oil spills (Al Maawali et al. 2019) or algae blooms (Beckler et al. 2019) while dynamically avoiding obstacles using detection systems (Ren et al. 2021; Chen et al. 2018). In recent years, the use of learning methods is on the rise, allowing for vehicles to perform coordinated tasks and process gathered monitoring data providing valuable models about the water quality status of a water mass (Yanes-Luis 2024). The problem of coordinating a fleet translates directly to a path planning problem (Popović et al. 2024) where vehicles limited by their movement capabilities and battery autonomy need to adapt to an unknown environment and its dynamics. In (Yanes Luis et al. 2021) (DRL) is used to create a policy that guides the movements of a fleet of ASV to solve a coverage problem localizing contamination and algae focal points. The processing of monitoring data promotes the use of heterogeneous vehicles and sensors, (Ferri et al. 2014) providing valuable insights on the water quality of a water body. In this way, vehicles can have different movement and sensing capabilities, where a fleet consisting of multiple low-cost low-performing vehicles can provide water quality values (Nothacker 2024). In (Samaniego et al. 2021), water quality data was processed to model water quality distribution, providing an estimated water quality value on areas of water not measured. In (Casado-Pérez et al. 2025), the data gathered by a fleet of ASV was used to estimate the evolution of a contamination accident in a water body.

This chapter describes an ASV developed by the group 'Investigation of Cybernetics Applications to Information Technologies' (ACE-TI) at the University of Seville in collaboration with Universidad Loyola (see Fig. 1). The group has published several papers on water monitoring using learning methods. The research includes algorithms that allow the deployment of a fleet of ASVs capable of cooperation, where ASVs can be assigned multiple tasks simultaneously and adapt

Fig. 1 Photo of the catamaran-shaped autonomous surface vehicle developed by ACE-TI

to the changes of the environment in real time following adaptive informative path planning (AIPP) policies (Yanes Luis et al. 2023). The tasks include intensification promoting areas based on contamination level, or coverage; and prioritizing the monitoring of all the points present on the map. This problem has been further researched, involving heterogeneous ASV with different monitoring or movement capabilities (Barrionuevo et al. 2024). Furthermore, more advanced techniques combine local Gaussian processes and deep reinforced learning (DRL) to optimize monitoring policies (Yanes Luis et al. 2024).

The ASV framework described in this paper presents heuristics based on a collision-free consensus for the safe deployment of multiple ASVs. This framework allows the integration of new algorithms and a unified neural network for all agents. In this way, the proposed ASV provides a robust system for large-scale environmental monitoring without direct supervision, which is able to detect floating debris and obstacles present on the water surface using computer vision algorithms and represent the data in a continuous space domain.

This chapter is organized as follows. This chapter presents a description of an ASV, starting with the hardware present including all developed and commercial modules for water quality monitoring, navigation and computation, among others. It continues with the software architecture that handles each of the modules present in the ASV as well as the different AI techniques applied for object detection, adaptative informative path planning, and sensor processing. Additionally, real water quality measures collected by the ASV in a water environment will be presented, as well as the different tools developed for fleet management and data visualization. Lastly, future and ongoing developments will be discussed.

Autonomous Surface Vehicle Structure

The ASV is a catamaran-shaped boat. It is formed by two hollow fiberglass hulls from Tumi Robotics[1] separated 39 cm apart. The wide stance and space between hulls of a catamaran provide a higher stability and payload capacity than the mono-hull counterparts. Furthermore, having two hulls, each with an engine, allows for better manoeuvrability and equipment redundancy, which translates in more security and a more robust and precise control.

The hulls are connected by an aluminium structure consisting of two rails connected side to side attached by three screws on each side. On the stern rail, two more rails are mounted facing upwards and connected by another rail. These act as a tower for holding the camera and the antennas, providing a longer communication range minimizing by decreasing the reflection with water and a higher point of view for the steroscopic camera.

The materials and shape were selected with the objective of minimizing its weight to maintain the maximum payload while keeping the design simple and the cost low. The ASV presents a length of 129 cm with a draft of 8 cm, a beam of 98 cm, a beam centreline of 69 cm and a clearance of 79 cm. These measures were selected to allow the boat to be transported in a van. A blueprint showing the ASV dimensions can be seen in Fig. 2. The ASV weighs around 24.5 kg without batteries (for transport) and 29 kg with the whole payload, allowing for 20 kg of extra payload. On each hull, there are two waterproof apertures allowing to access the interior and place new payload, electronics, or tools.

Over the rail, with the centre of gravity in the centre of the ASV in order to minimize the risk of instability, a waterproof box containing all the electronic components is mounted. As this ASV started as a prototype and presents a continuous development, having a big space with easy accessibility proves to be a good design choice. Just under the box, there is a structure for mounting an electrochemical multisensor probe for quality monitoring, and on the stern side of the rail, there is another rail to place a SONAR. The mountings of the sensors have been 3D-printed using polylactic acid (PLA).

Hardware Components

The ASV is formed by several interconnected components. The actuators that are in charge of moving the ASV, the antennas in charge of communication and GPS positioning, and the water quality sensors are located outside of the electronic box. Inside the electronics box are placed all the electronic modules and electric connections, maintaining a waterproof environment. On the back side of the box, a red switch is present, working as an on/off switch of the power line of the ASV.

[1] https://tumirobotics.com/en/

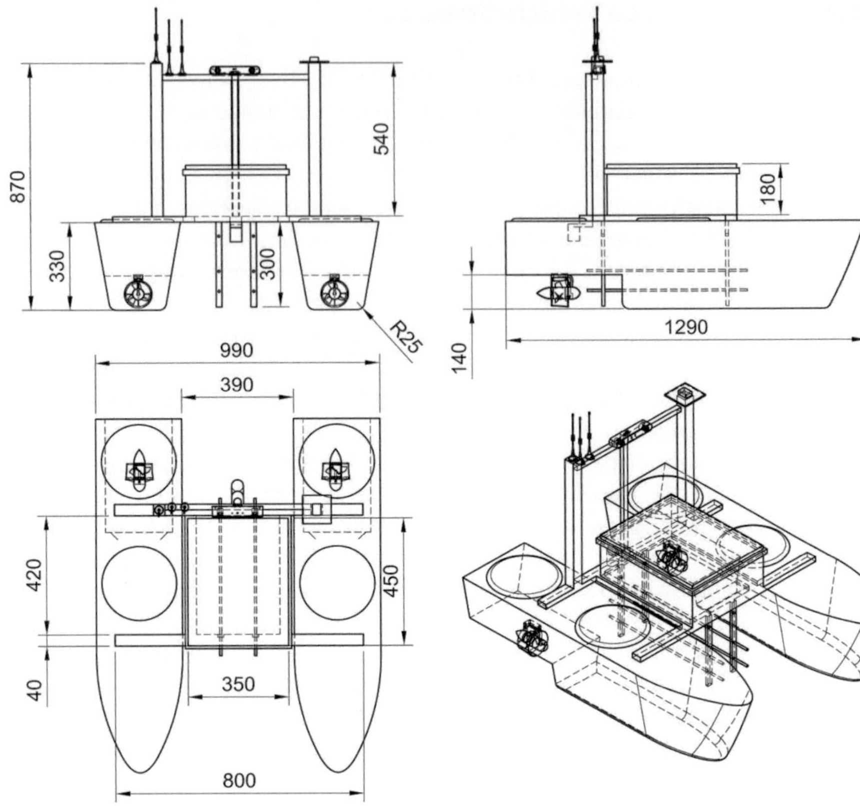

Fig. 2 ASV blueprint

Figure 3 illustrates the arrangement of the different components. Figure 4 provides a visual of all the modules placed within the ASV electronics box.

Propulsion System

The ASV is propelled by two BlueRobotics T200 (brushless direct current motors) thrusters (BlueRobotics 2024) (see Fig. 5), allowing for maximum speed of 4 knots (2 m/s). They are each controlled via pulse width modulation (PWM) by a BlueRobotics basic electronic speed control (ESC) directly powered by the battery with a peak current of 30 A.

The environment in which the ASV operates is often polluted and there are present lots of algae, sticks, and material that can damage thruster blades made of plastic. To protect it, a 3D-printed net of PLA has been placed to prevent the intrusion of big objects.

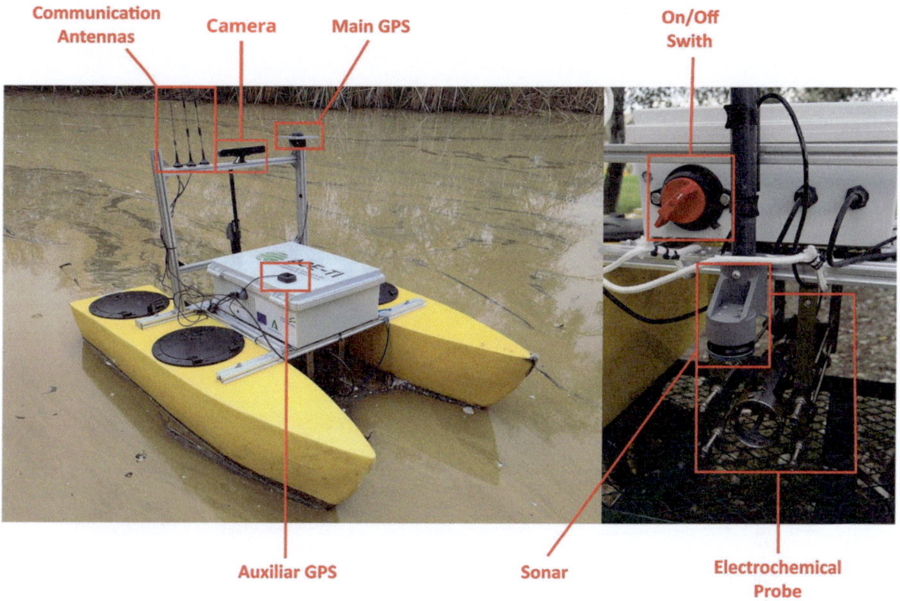

Fig. 3 Autonomous surface vehicle component depiction

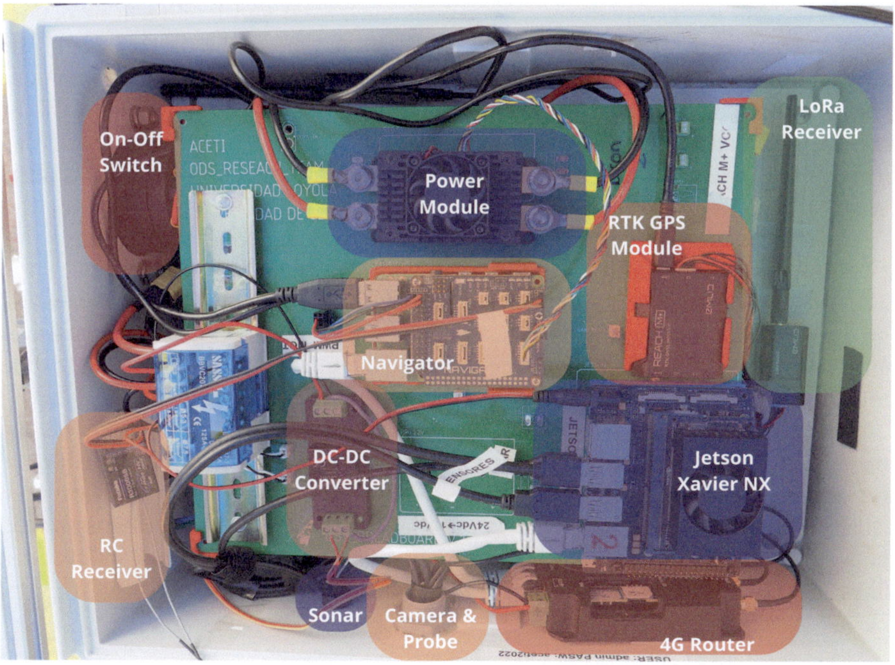

Fig. 4 Placement of the different hardware modules inside the electronics box

Fig. 5 BlueRobotics T200 thruster installed in the ASV and protective net

Controller Units

As a vehicle in need to perform autonomous movement and intelligent actuation, it is equipped with a central processing unit (CPU). Due to the complexity of the environment and the high number of parallel tasks that an intelligent autonomous vehicle must process, in most systems, control is divided in two or more CPUs, featuring a companion computer (Okudera et al. 2025). Thus, control is divided between low-level control, related to the physical world, and high-level control, in charge of the more abstract processes

The low-level controller unit is a Raspberry Pi 4 with a HAT Navigator.[2] Low-level tasks include the most basic ones, usually connected to the physical world tasks of the ASV. These include sensing (inputs from the world) like processing the data from a six-axis inertial measurement unit and a three-axis magnetometer and integrating GPS data to calculate via sensor fusion (Riseborough 2015) the orientation, position, and speed of the drone, providing an attitude and heading reference system (AHRS). On the other hand, related to the control and the movement of the drone (outputs to the world), it includes tasks like regulating the thrusters' power to guide the drone towards a goal position, maintain roll and pitch values within safe values or maintain a position rejecting disturbances. Furthermore, the Navigator HAT provides the Raspberry Pi 4 with several inputs and outputs that allow it to become a fully featured flight controller.

[2] https://bluerobotics.com/store/comm-control-power/control/navigator/

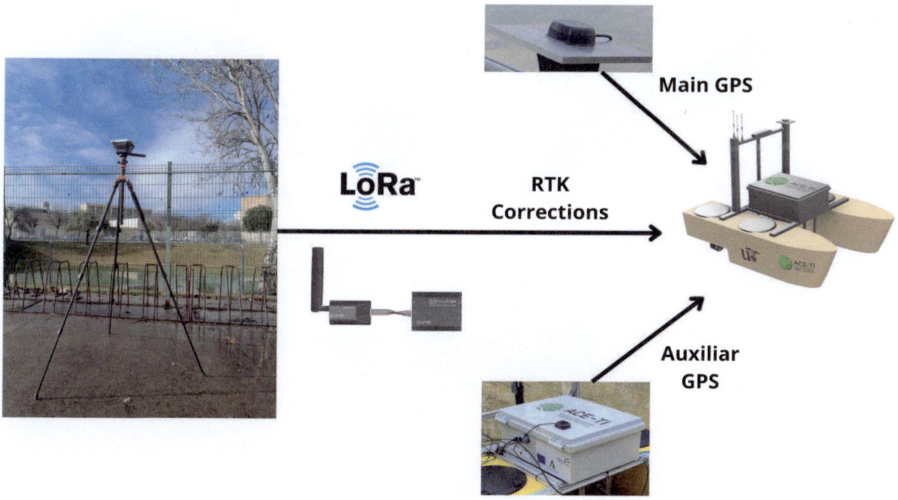

Fig. 6 GPS system placement and antennas

The high-level controller unit is a Jetson Xavier NX.[3] It is a small format computer with 16 GB of RAM and 21 tera operations per second (TOPs) designed for AI-driven applications. It serves as the main controller unit of the ASV, acting as the endpoint of the drone and being in charge of all the external communications. Its higher computation capability facilitates the tasks of water quality data processing, signal handling, and obstacle detection and avoidance.

GPS System

The ASV is equipped with two global positioning system (GPS) and global navigation satellite system (GNSS) antennas. The main antenna is a Tallysman multi-GNSS connected to an Emlid Reach M+ module. This module is used in conjunction with an Emlid Reach RS+ ground station working as a GPS/GNSS real-time kinematic (RTK) system allowing for a horizontal precision of 7 mm (emlid 2021). As RTK corrections are transmitted via LoRa, to achieve this precision, the ASV must be within an 8 km range of the ground station. A GPS/GNSS Navio2 auxiliary antenna is located on the top side of the box with a horizontal precision of 2.5 m (ublox 2015). This antenna is mainly used for position reference in case the main antenna has lost connection or few satellites are available.

After several tests, placing the main antenna over a ground metal plane as seen in Fig. 6 showed an increase of GPS signal of around 14 dBm on average, allowing to detect satellites even in unfavourable weather conditions.

[3] https://www.nvidia.com/es-es/autonomous-machines/embedded-systems/jetson-xavier-nx/

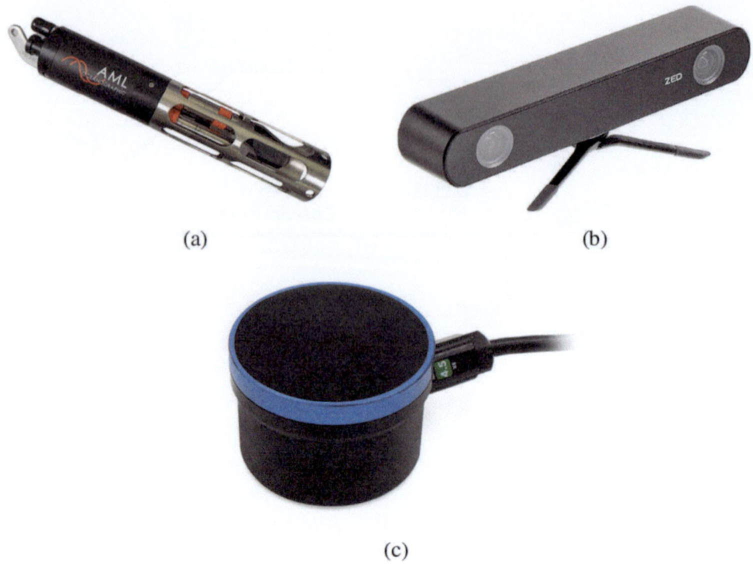

Fig. 7 The three sensors available in the prototype. (**a**) AML WQP probe, (**b**) ZED2i camera and (**c**) BlueRobotics Ping2 SONAR

Sensors

The high payload capacity of the ASV allows it to be equipped with several sensors and gather extensive and varied data about the environment.

Water quality parameters are measured via an AML-3 XC Oceanographic[4] electrochemical multisensor probe (see Fig. 7a). With a weight of 1.4 kg, this probe can be submerged up to 500 m and provide continuous measures of several water quality properties including pH, conductivity, dissolved oxygen, CDOM, chlorophyll A and B, crude and refined oils, fluorescein, phycocyanin, phycoerythrin, rhodamine, temperature, turbidity and tryptophan. The selected multisensor probe is equipped with three sensor ports providing up to three different water quality parameters at the same time, allowing for sensor interchangeability if the water body to measure requires a different water quality parameter. The probe is mounted under the electronic box, using a structure made of aluminium that positions the probe under the ASV's waterline while it protects the probe from the seabed or streambed.

A Stereolabs ZED2i stereo camera[5] (see Fig. 7b) is used as a spectral passive sensor above the water. As a stereoscopic camera, it is equipped with two identical cameras, replicating how animals perceive the environment it is able to provide depth perception of objects and background. It is equipped with polarized lenses

[4] https://amloceanographic.com/instruments/aml-3

[5] https://www.stereolabs.com/products/zed-2

reducing glare effects and offering a wide-angle field of view up to 120 degrees. This camera is prepared for computer vision, providing inertial data and characterizing objects captured by their size and position. It is located at a centred position at the top of the tower present in the ASV. This position provides protection from glares and takes advantage of the wide field of view present in the camera.

Lastly, a BlueRobotics Ping2 SONAR[6] is used as a spectral active sensor (see Fig. 7c). It is a single-beam echosounder providing depth data up to 100 meters with a 25-degree beam width. Designed taking bottom-tracking into account, it uses a frequency of 115 kHz and shows no interference with other market solutions. Additionally, it uses a specialized algorithm that provides a low noise deepness value with an error of up to 0.5% even in complicated situations with multiple echoes. It is located in a rail behind the electronics box allowing to lift and sink the sonar to a desired altitude depending on the water body to analyse.

Power Diagram

The ASV is powered by a set of two 4S LiPo batteries of 10,000 mAh each connected in parallel, providing a raw voltage that moves in the range of 14.8–16.8V depending on the remaining charge. This voltage needs to be regulated before powering all the ASV modules. The power consumption of the ASV is of $1.4W$ in standby mode without moving, reaching up to $430W$ of instantaneous power while it is moving at maximum speed during a mission. This allows for a lower bound autonomy of 2 hours when using it intensively with two batteries. The payload of the boat allows for more batteries to be connected in parallel, thus increasing the battery autonomy. It is important to take into account the health of the battery before a mission; batteries need to be stored in a storage voltage if they are not used; otherwise, the surplus of energy will promote chemical reactions that will lower the total charge available.

Figure 8 sums up the electric connections present in the ASV detailing all the power paths. The AML sensor is self-powered by an internal rechargeable battery and features a mechanical on/off switch for control. The 4G router includes an internal voltage regulator, being able to be connected directly to the batteries.

Communication Diagram

For a vehicle running autonomously, maintaining the communication between modules is critical for a correct operation. The ASV is equipped with several modules in charge of different communication channels, providing redundancy and different paths to communicate new requests and information to and from the drone. The connection between the different ASV modules is summed up in Fig. 9.

[6] https://bluerobotics.com/store/sonars/echosounders/ping-sonar-r2-rp/

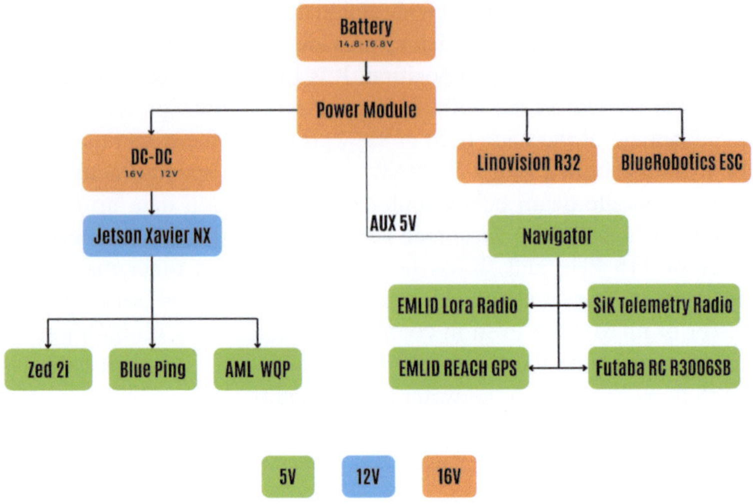

Fig. 8 Power scheme of the electronics box

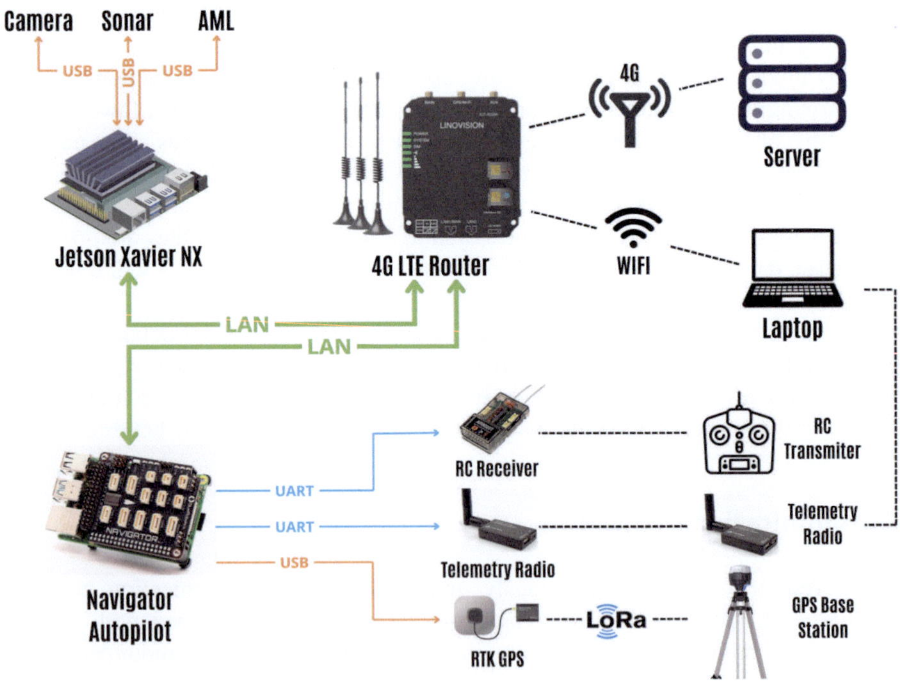

Fig. 9 ASV communication scheme

The main communication channel present in the drone is provided by a Linovision R32 industrial router featuring two LAN ports, Wi-Fi and 4G LTE connection via an IoT 4G SIM card. The LAN connection establishes a local network between the Jetson Xavier NX and the navigator autopilot, working as the main communication interface. However, robotic vehicles are characterized by vibrations that may cause the disconnection of wired cables. To reinforce this connection, Wi-Fi is used as a redundancy communication interface. The 4G LTE connection allows for external connection with the drone, providing an external channel for connection with the drone. This connection is managed by a remote server allowing to send and store relevant data as well as receive commands. As a second channel, Wi-Fi allows the connection from a close-by computer to monitor the ASV state and health as well as configure parameters or send commands online, providing flexibility and accessibility in the ASV operating environment. Additionally, as the bandwidth and quality of the Wi-Fi network created by the router are limited, a 915 MHz telemetry radio is connected to the Navigator autopilot, providing an auxiliary communication channel to monitor ASV state and send commands.

The communication with the sensors and the Reach M+ GPS module present in the ASV occurs via USB connection. However, the corrections provided by the Reach RS+ ground station are transmitted via LoRa,[7] a long-range low-power technology in the frequencies of 862–1020 MHz allowing for a range up to 8 km.

If all the communications fail, the ASV has a last lifeline consisting of an RC receiver R3006SB[8] working at 2.4G Hz. It is paired with an RC transmitter FUTABA 6K as a manual controller, allowing communications up to 1 km. The RC works as a safety device, allowing to halt autonomous operating mode and control manually the drone to recover it in states of fatal error or to prevent accidents.

Software Architecture

The architecture integrated in the ASV proposes a system formed by two hardware components: the mentioned Jetson Xavier NX, in charge of managing the high-level operations and communications, and the Navigator autopilot, in charge of the low-level control. Lastly, the management of the ASV occurs via a remote server in charge of storing and processing the monitoring data as well as issuing new commands.

Each of these processing units is running a dedicated software. However, in an ASV in continuous development where software and hardware are constantly changing, an environment where software modules are flexible and facilitate replicability, Reproducibility, and reusability is desired. To achieve this, all software runs in a dockerized environment. Docker[9] is a platform that allows to run applications

[7] https://store.emlid.com/products/reach-m-lora-radio?variant=47253675540785

[8] https://futabausa.com/wp-content/uploads/2018/09/R3006SB.pdf

[9] https://docs.docker.com/

in an isolated environment, abstracting the software from the hardware, so that the software runs independently of the host machine. This allows to integrate new software in any new processing unit in an easy way without any risk of incompatibilities. In this section, the software contained in each system will be discussed.

Navigator Autopilot

The Navigator HAT is in charge of the hardware abstraction layer, being the middleware between the peripheral sensors, the controller and the output devices (ESCs, servos, gimbals and other peripherals) present in the ASV. The software architecture running in the autopilot is depicted in Fig. 10. The processing unit is a Raspberry Pi 4 with the attached Navigator HAT that runs a modified image of Raspbian Bookworm called BlueOS[10] that integrates ArduPilot[11] autopilot firmware inside.

ArduPilot, being an open software present in the market for many years, provides users with a broad range of algorithm templates that allow to develop new algorithms without the need for an expert programmer. The software is structured in blocks containing routines executed periodically at different frequencies while allowing real-time interruptions. This block structure facilitates the integration of heterogeneous hardware and asynchronous communications between them, becoming a software that works on a wide variety of different hardware to control unmanned vehicles of all types.

The programming blocks are divided into three main groups: High-level libraries part of the hardware abstraction layer, perform the sensor fusion of the GPS and inertial data. They incorporate an extended Kalman filter (EKF) to integrate the attitude, heading and reference systems. Controller libraries, associated with the vehicle type, provide position, velocity and attitude control. High-level libraries incorporate packages that allow autonomous mission planning, object avoidance and post-mission analysis.

The flow of information between programming blocks occurs via MAVLink,[12] a standard message protocol. These messages can also be accessed externally connecting via IP through a user interface, also called ground control station (GCS), that allows to set up, configure, test and tune the ASV. ArduPilot also promotes the execution of different sets of programming blocks by the so-called vehicle modes. Thus, each vehicle mode allows for different sensor, control and high-level libraries to be executed, providing different vehicle behaviours. When powering the ASV, ArduPilot software has an internal flag that states the status 'disarmed', halting any

---

[10] https://blueos.cloud

[11] https://ardupilot.org/dev/index.html

[12] https://mavlink.io/

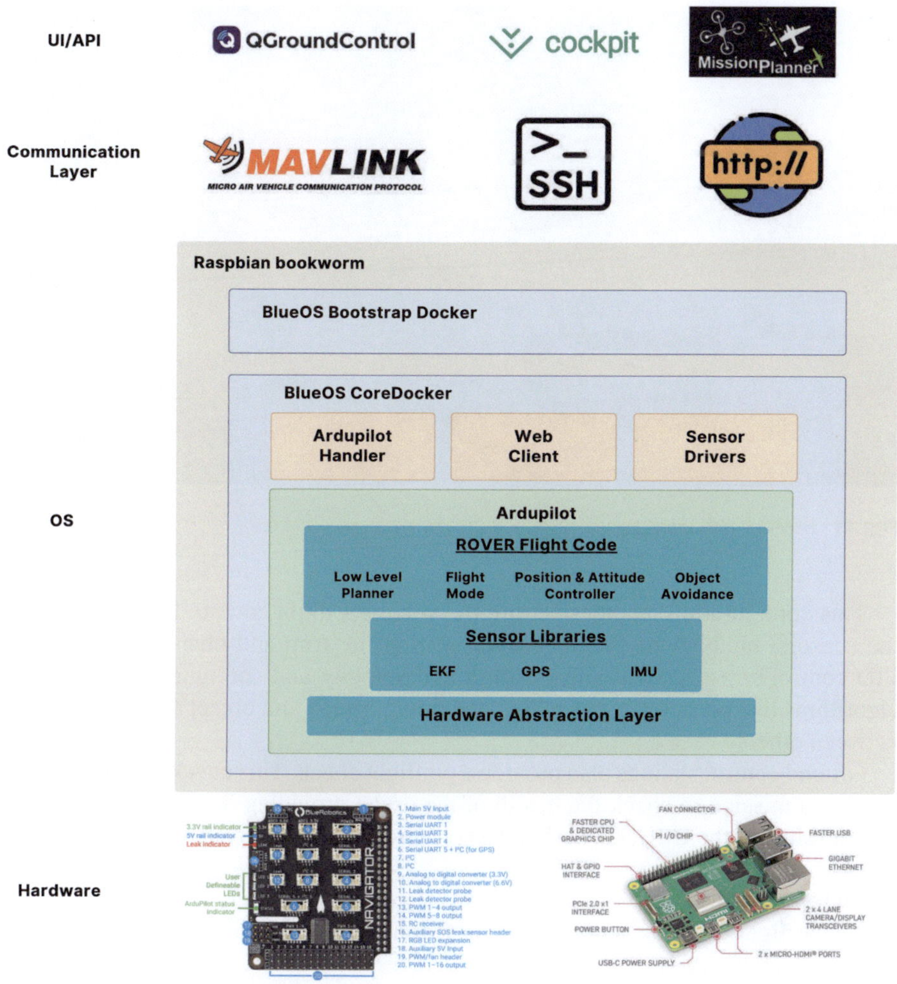

Fig. 10 Navigator software architecture

movement sent to the thrusters to avoid accidents while manually manoeuvring the boat.

ASV Configuration

Before going on a mission, the ASV needs to be set up and configured. This configuration is done via MAVLink messages that change the internal parameters of ArduPilot. To achieve this, a user can use an external ground control station like Mission Planner, providing an interface for the task, or make use of the web interface provided by BlueOS (see Fig. 11).

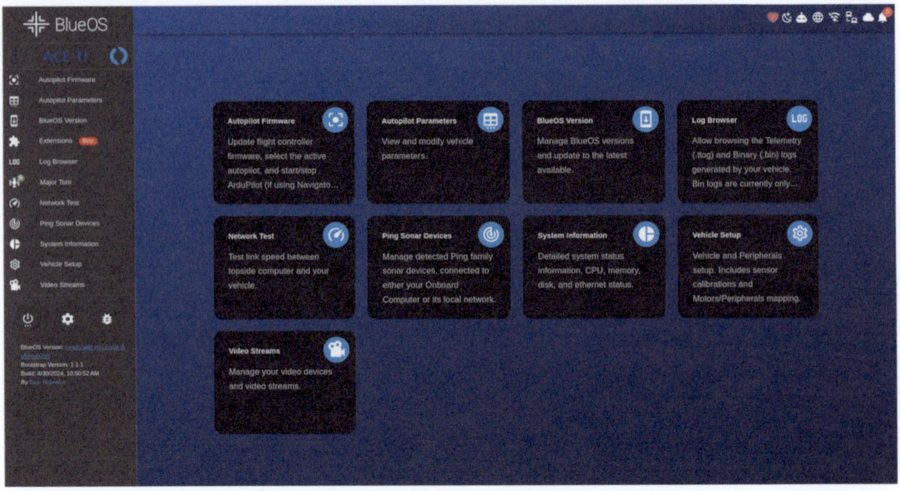

Fig. 11 BlueOS web interface

This configuration includes setting the vehicle's flight code to boat, calibrating the sensors, binding the RC controller, setting the trim and channels, tuning the PID controller associated with position and velocity and configuring behavioural algorithms like pivoting, loiter radius, maximum speed and object avoidance radius between others.

Ground control stations also provide a preflight check. This check is useful before starting a mission to check the health of the drone providing information about the health status of the IMU, GPS, thrusters, control and sensor systems.

Vehicle Modes

The boat flight code has a configuration featuring two thrusters and providing five different vehicle modes:

- **Manual Mode:** Allows the operator to control the ASV directly through an RC transmitter. In this vehicle mode, the inputs are connected directly to the movements of the vehicle (forward-backward, yaw).
- **Loiter Mode:** The vehicle executes no active path planning. The aim of this mode is to maintain position, staying idle if there are no currents or inertia and using the thrusters to recover the last position if it moves away at a distance greater than a predefined parameter.
- **Guided Mode:** The vehicle acts the same way as in loiter mode, but allows message reception from MAVLink to execute a linear route towards a requested waypoint.
- **Mission Mode:** The vehicle follows a preloaded path autonomously. This path is made of different waypoints. At each waypoint, a different action can be performed like wait or take a picture between others.

- **Follow Mode:** This mode requires constant position updates from a different MAVLink component, be it another vehicle, or an emulated component like a floating object. The vehicle will try to face and keep a minimum safe distance towards the provided component's position.

Jetson Xavier NX

The Jetson Xavier NX runs an Ubuntu Fossa (20.04 LTS) OS that implements a docker running Robot Operating System 2 (ROS2 Humble). ROS2 is an open-source middleware that promotes independent processes able to communicate and cooperate between them asynchronously, providing a flexible and real-time system for robotic applications. There is a huge community using ROS, providing a broad set of libraries and tools with common interfaces that ease the development of new robotic systems.

Each process is contained in a structure called node. When creating a node, the inputs, outputs and capabilities of the process are defined. This structure allows for interchangeability and scalability of different processes performing similar actions like the case of developing a new sensor library. The communication of these inputs and outputs of the nodes is constructed via Data Distribution Service (DDS) promoting automatic discovery between nodes and fulfilling the requirements of a real-time system. Nodes are able to communicate between them in three different ways: Topics, following a publisher/subscriber mechanism, are formed by one-way messages and are helpful when broadcasting data like the ASV health. Services, following a server/client mechanism, are useful when a process needs to call a function from another node and return a value. An example of service is the case of setting a new vehicle mode, answering with the success state of the mode change. Lastly, actions are message types characterized by a function call that needs to update the state of completion. An example of an action is path following, where the ASV needs to reach a final point while providing information about all the points traversed during the route. The developed ASV ROS2 system is formed by six different nodes. Figure 12 shows the nodes present in the ASV:

- **Path planner node:** This node implements an infrastructure based on Nav2,[13] a navigation framework broadly used in autonomous vehicles. It manages path planning using algorithms like the Dijkstra algorithm (Dijkstra 1959) providing a feasible route from the current ASV position to a goal position.
- **Sensors node:** Node in charge of the libraries managing the multiparametric probe and sonar sensors.
- **Perception node:** This node implements a module able to detect obstacles and macro-plastic using computer vision over the images provided by the Zed 2i camera.

[13] https://docs.nav2.org/

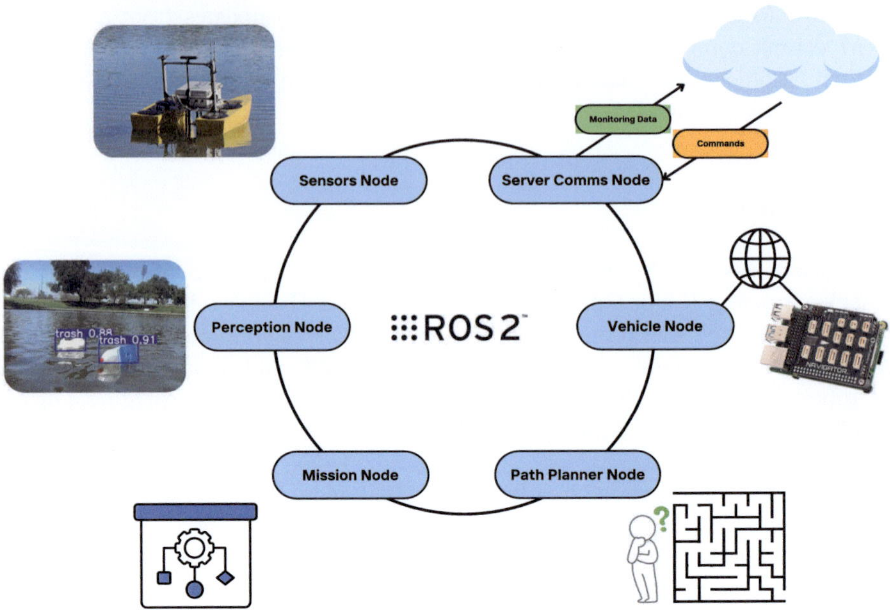

Fig. 12 ROS2 node structure

- **Server communications node:** In charge of handling the bidirectional communication with the remote server, providing the monitoring data and receiving new commands.
- **Vehicle node:** Utilizes a ROS library called MAVROS. MAVROS provides a transparent transformation between MAVLink messages and ROS topics and services, being in charge of handling the communication with the navigator autopilot.
- **Mission node:** Performs the role of ASV control. This node is in charge of maintaining the ASV health and executes the path to follow (when called from the server) by implementing a finite state machine (FSM):
  - Idle: Initial state from the drone, no actions taken. Vehicle is disarmed and deactivated. Only manual mode available.
  - Waiting Request: Vehicle activated and autonomous mode enabled, waiting for request from the server.
  - Processing Request: Transitory state where the status of the vehicle is updated and a path is created fulfilling the requests sent from the server.
  - Path Monitoring: In this state, the status of the drone and the environmental data is continuously checked, modifying the path to follow if an obstacle or point of interest is discovered or transitioning to 'Waiting Request' mode if the mission is completed.

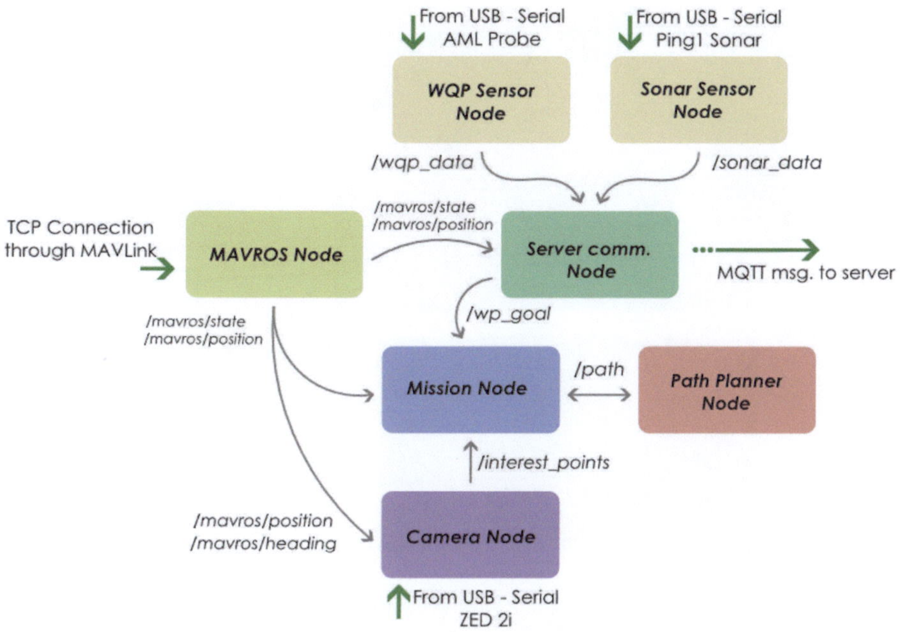

Fig. 13 ROS2 topics and services present in the ASV software

Figure 13 shows another view of the nodes present in ROS2 and the communication channels between them. This structure allows for flexibility and robustness, allowing the drone to continue operations even if a node fails and recover from errors. Furthermore, ROS2 provides a logging system that facilitates debugging.

Macro-plastic Detection and Localization

The ASV is equipped with a ZED2i stereo camera that provides the drone with both RGB and depth images simultaneously that, in the case of the ASV, is used for object detection. An instance of YOLOv5 (Kim et al. 2022), a popular vision AI module, has been fine-tuned to perform detection of macro-plastic, pollutants and other objects present on the water surface (see Fig. 14). The GPU-accelerated architecture of the Jetson Xavier NX allows for high-performance AI computation in real time with reaching over 30 frames per second. Figure 14 depicts the scheme for object detection and localization. The drone camera captures an RGB image, and YOLOv5 processes the image by detecting and providing the bounding box of each detected object. This bounding box is compared with the depth information contained in the object bounds, providing information about the object size and position with respect to the drone. Lastly, the position is postprocessed through a transformation matrix to translate from the ASV reference frame to global GPS coordinates.

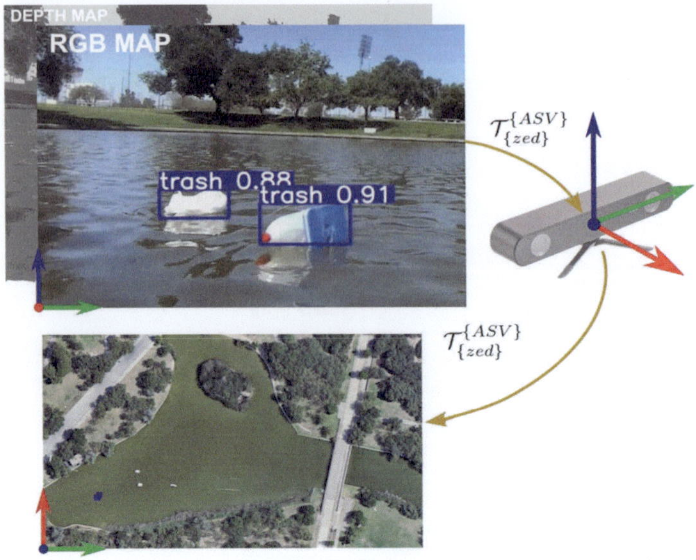

Fig. 14 RGB+depth map, using YOLOv5 detection scheme of macro-plastic in a real scenario. In blue are depicted the detected objects

In this way, as the ASV explores a water environment, an information map is filled with visual and spatial information about the ASV's surroundings. The macro-plastics detected become interest points for tracking and cleaning, while obstacles become forbidden areas to be avoided. This information contributes to the mitigation of water pollution and preservation of aquatic ecosystems.

Central Server

The last computation unit is a remote server working as a proxy-computation server for the ASV within the computation infrastructure of the University of Sevilla. It is equipped with multiple high-end GPUs and multiple cores, enabling them to run multiple programs at once and perform several tasks concurrently. In this scheme, the control of the ASV is centralized, requiring a server to execute a mission. However, the dockerized architecture allows the server to be executed in any computer, including the ASV in a critical case of communication loss. The main tasks the server needs to fulfil are maintaining the communications, handling monitoring data and executing heavy tasks like processing the monitoring data and providing a movement policy for the ASVs and alleviating the computational charge of the ASV.

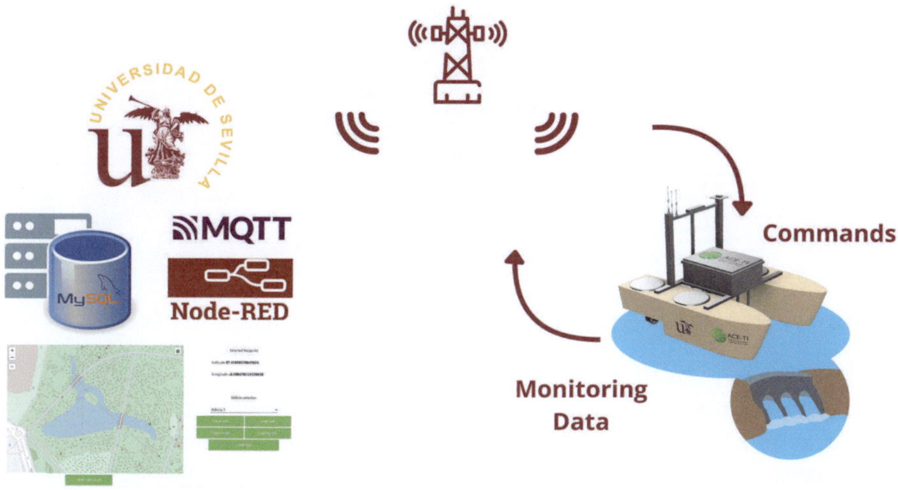

Fig. 15 Schematic of the communication between the server and the ASV

Server Communications

The communication with the ASV occurs via message queuing telemetry transport (MQTT). This protocol defines a publisher/subscriber architecture via communication channels decoupled from each other called topics. The main advantages of this protocol include a small size dataframe structure, providing a fast lightweight protocol stable in areas with narrow bandwidth, resulting in a low data consumption. The data received by the server is filtered, processed and stored using a robust and scalable relational database (mySQL[14]), providing a structured and organized access to monitoring data. This database is doubled, dividing the data in raw monitoring data (read only) and processed monitoring data, updated periodically.

To maintain the modularity of the software and scalability with a higher number of ASV, the messages transmitted present a standard structure, replicating the same messages used in MAVLink but adapted to the MQTT protocol. Each vehicle has assigned an identification number that is used to select the topics assigned to each vehicle. For each vehicle, there exist two topics. Take, for example, an ASV with assigned id $'1'$. There will be a topic in charge of the data transmission towards the vehicle *'asv1cmd/'* and other topics in charge or the monitoring data received from the vehicle *'asv1data/'*. Thus, messages are classified in commands to the drone (go to coordinate, load and execute mission, record video) and monitoring data from the drone (water quality values, position, ASV status) as seen in Fig. 15.

[14] https://www.mysql.com/

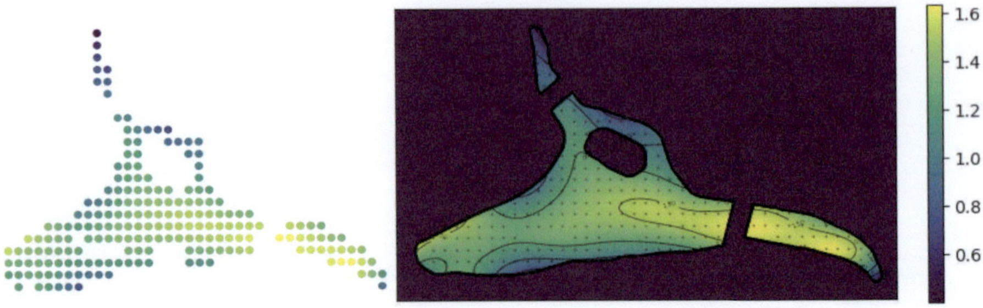

Fig. 16 Transformation from local discrete monitoring data to continuous data through GPs

Data Processing

The map provided by the ASV provides water quality measures at discrete coordinates, offering a representation that is hard to interpret. To provide a desirable overview, this value needs to be processed so that measures are interpolated and extrapolated to represent the monitoring data in a continuous spatial domain. This problem is hard and mathematically complex. Gaussian processes (GPs) allow to represent functions as random variables with Gaussian distributions (Rasmussen and Williams 2006), offering a solution to incorporate new water quality measures and make predictions based on them. The proposed solution receives the name of radial basis function (RBF), broadly used in environments where data experiments smooth and continuous variations like the case of water quality maps (Samaniego et al. 2021). The weight of a measure relies on a kernel defined by a length scale parameter that determines the distance between correlated samples. An example of this algorithm can be seen in Fig. 16.

Another use of the higher computation capability of the server is to provide a movement policy for the ASV from informative path planning algorithms (Yanes Luis et al. 2025), maintaining a unified neural network for all ASVs able to handle ASVs with different capabilities.

User Interfaces

This chapter has defined the physical capabilities and software present in the ASV, providing a broad range of tasks that the ASV can perform. However, by itself, the ASV requires an expert to manage and control the fleet and prepare a mission specifying all the tasks the ASV must achieve. To facilitate the use of the proposed ASV, it is necessary to develop a user interface that allows any operator to command the ASV, manage a mission, specify the tasks to fulfil and review the monitoring data gathered by the ASVs. Two different user interfaces have been designed: a web interface hosted in the remote server accessible via the Internet at a public URL, providing tools for supervision and management of the fleet of ASV and accessible from a mobile phone without the need of a computer, and a desktop

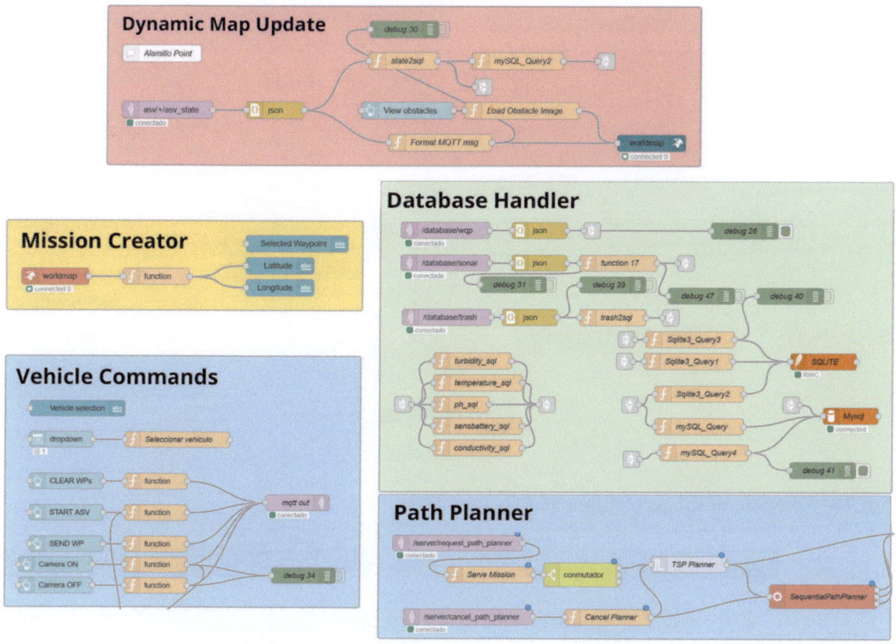

Fig. 17 Node-RED flow diagram

interface, designed to run in a laptop, that allows to access low-level configuration of the ASV and extract valuable and postprocessed monitoring data.

Web-Based Interface

The aim of the web user interface is to provide a simple straightforward interface that allows any operator to intuitively manage and control any ASV. It is constructed using Node-RED,[15] a framework running in Node.js that provides developers with several customizable modules and an easy interconnection of messages and events by visually wiring nodes that share information. These nodes are in charge of parsing the MQTT messages, managing the database and providing a user interface. More advanced nodes allow the execution of external scripts or bash commands. Figure 17 shows the main flow of the developed UI, containing more than 200 nodes and featuring five different tabs.

Any operator accessing the web page will find an interface like the one found in Fig. 18. On the front of the page appears an interactive map from the open-source

[15] https://nodered.org/

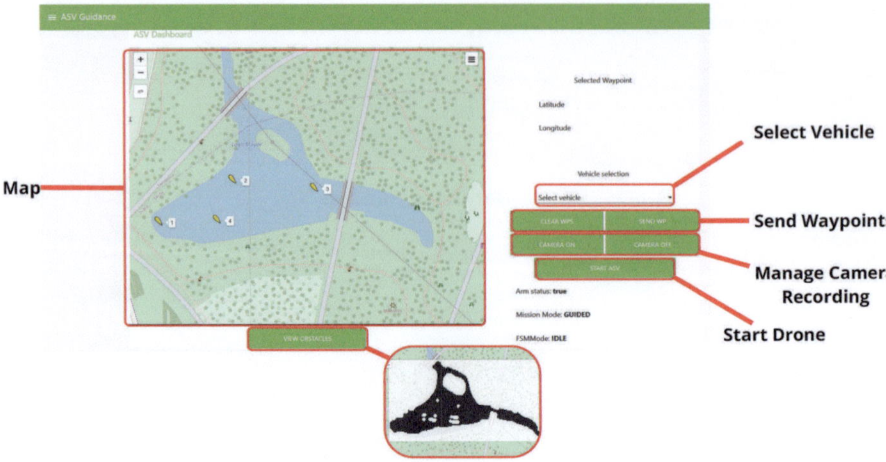

Fig. 18 Diagram of the web graphical user interface

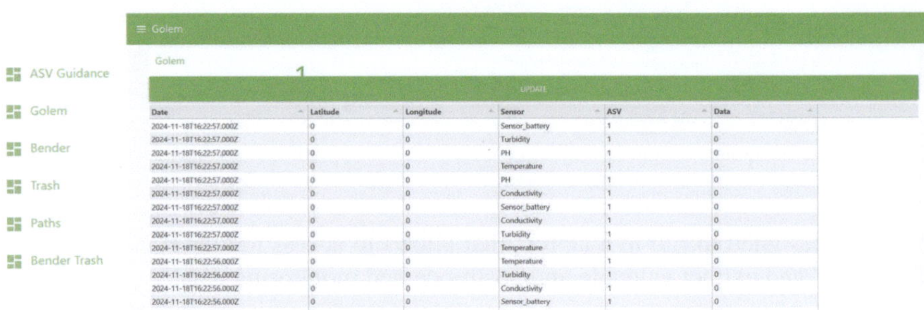

Fig. 19 Auxiliary tabs present in the web GUI

map OpenStreetMap.[16] By clicking the map, an operator can establish points of interest for the ASV to follow. Once the desired points are marked, a route is established. By the use of the menu located in the right side, the mission can be assigned to a specific ASV. The buttons located under the selector allow the operator to start the mission, pause it if an error occurs and configure whether to record image data from the mission. The three lines found in the top left corner allow the operator to access the auxiliary tabs, as seen in Fig. 19. The auxiliary tabs provide an extraction of the most recent data gathered by the ASVs and stored in the database.

[16] https://www.openstreetmap.org

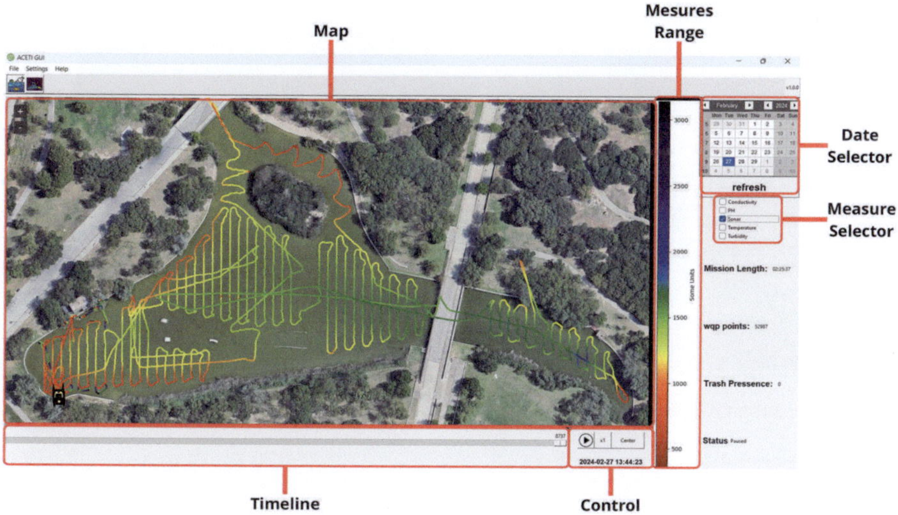

Fig. 20 Diagram of the desktop graphical user interface main tab

Desktop Interface

The desktop interface presents an extended version of the web interface providing the operator with more options to represent and interpretate the water quality data measured by the ASVs. It has been developed using Tkinter, a Python library designed to create graphical user interfaces. Later, this Python program has been encapsulated into a portable executable file, allowing to execute the interface from any computer without the need for installing any program. It features two tabs, handling all the communication with the server to represent graphically all the monitoring data in continuous and discrete maps. Figure 20 shows a visualization of the main tab. On the right side, there is a calendar where the dates with data available are shown in green. Under the calendar, there is a selector used to choose from conductivity, pH, deepness, temperature and turbidity. By selecting a date, the data is downloaded and the selected measure is displayed on the map. The representation of the different water quality values is done through a colour map. On the right side of the map is located a colour bar, indicating the value corresponding to the measure. The slider located below the map works as a timeline, enabling the user to relive the mission performed by the ASV with a second accuracy, showing measures and detected plastics by timestamp (see Fig. 21). Figure 22 shows the second tab available on the desktop interface. On this tab, the data collected by the vehicle is processed via Gaussian processes, providing a continuous value of the environmental data gathered by the ASV.

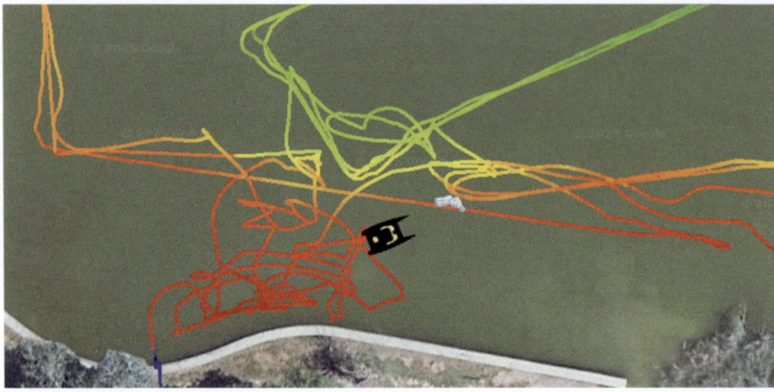

Fig. 21 Desktop GUI detail of macro-plastic detection

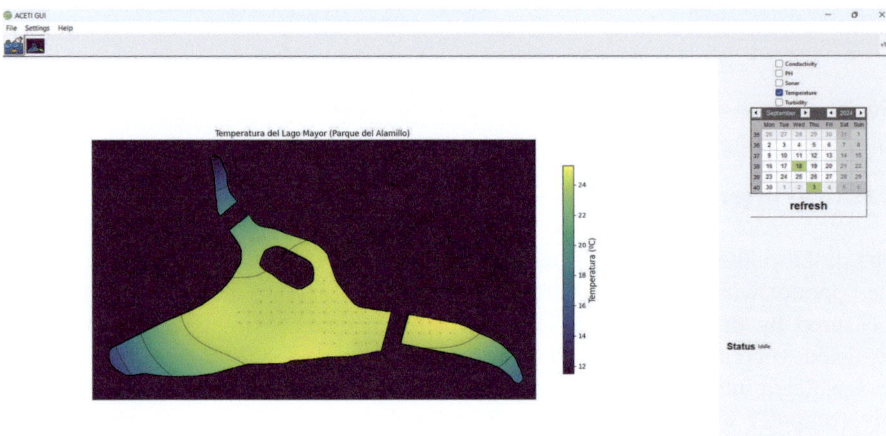

Fig. 22 Desktop GUI Gaussian process tab

Experimental Results

Several tests have been carried using the described ASV in an artificial lake present in Parque del Alamillo (Sevilla) presenting an area around 32,000 m^2. The tests gathered monitoring data of the temperature, pH, turbidity, conductivity and deepness of the lake.

The whole area was explored in a span of 100 minutes, detecting three different floating macro-plastics and gathering more than 37,000 water quality measures from the surface of the lake following a lawnmower path (Galceran and Carreras 2013). Figure 23 shows a comparison of the path assigned to the ASV during mission on the left and the resulting final path actually followed by the ASV on the right. From a technical point of view, the ASV was able to follow the proposed

Fig. 23 ASV expected and actual path during a mission in Parque del Alamillo

path thoughtfully without experimenting any crash with the water mass borders or any floating obstacles. The deployment point was the southwest point of the lake, location where a vehicle is shown. The point located outside the water bounds on the right side of the map is an outlayer due to a loss of GPS. Additionally, there is a line going from the last planned point to the deployment position in order to retrieve the drone after the mission.

Water Quality Environmental Results

Once data has been gathered by the ASV, it needs to be processed and studied by a competent entity. A good interpretation of the data shown is crucial for understanding the ecological status of the water body and identifying any potential environmental concerns. For this reason, it is important to make an emphasis in the representation of this data. As mentioned before, machine learning modelling techniques are commonly employed in spatial data analysis. The length scale parameter has been adjusted to 80 meters with approximate bounds between 55 and 110 meters.

Figure 24 shows the temperature map obtained after applying Gaussian processes. Being a small lake temperature should be close to the ambient temperature and uniform in the lake. However, analysing the values shown, there is a difference of 2 °C. This is due to the temporal evolution of the measures, as later into the mission, the measure is taken the higher the ambient temperature. pH values shown in Fig. 25 are closely linked to temperature, as pH experiments a variation with temperature. As this lake presents no water flows, the pH is the same through the whole lake (0.3 variation), showing a higher value in the first measures.

Figure 26 shows the conductivity. Conductivity is affected by temperature, presenting distribution similar to it. It can be said that conductivity is uniformly distributed in the lake. Figure 27 shows the turbidity map. Turbidity is variable

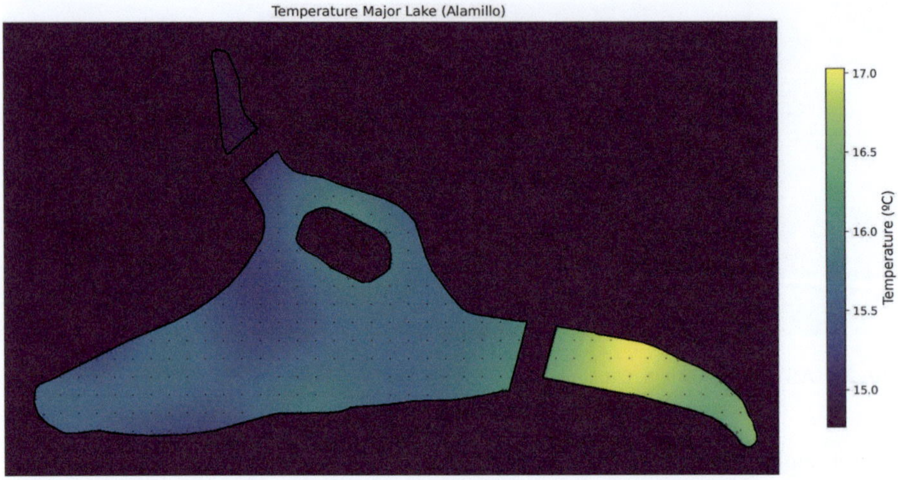

Fig. 24 Temperature estimation generated from the Lago Mayor monitoring mission

Fig. 25 PH estimation generated from the Lago Mayor monitoring mission

and heavily dependent on currents. The data show that there is an area with higher turbidity in the northern side of the lake.

Lastly, Fig. 28 shows the bathymetry of the map. While the bathymetry has no relationship with the water quality parameters, it provides important information about the water body shape and where may pollutants stay. The map shows how the depth of the lake is minimal on the southern edge and how it reaches a maximum in the middle right side of the lake.

Comparing the results shown in Figs. 24, 25, 26, 27, and 28 with the international standards (World Health Organization et al. 1963) and with the Spanish regulations

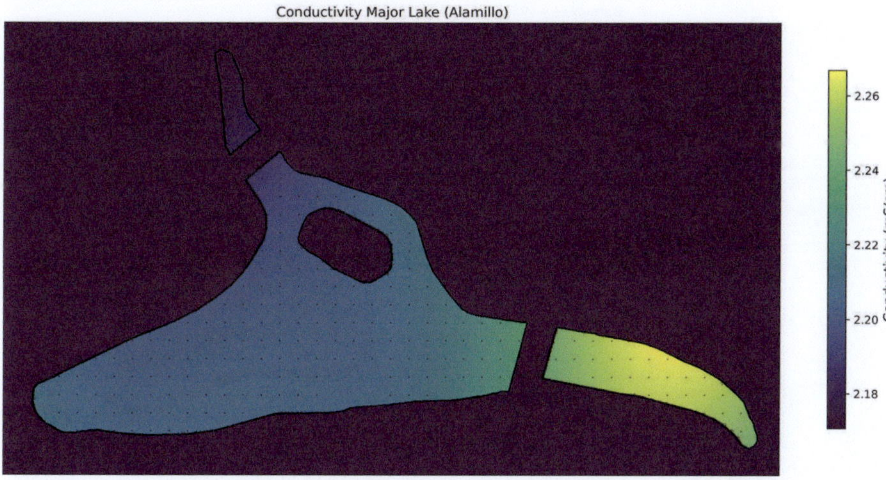

Fig. 26 Conductivity estimation generated from the Lago Mayor monitoring mission

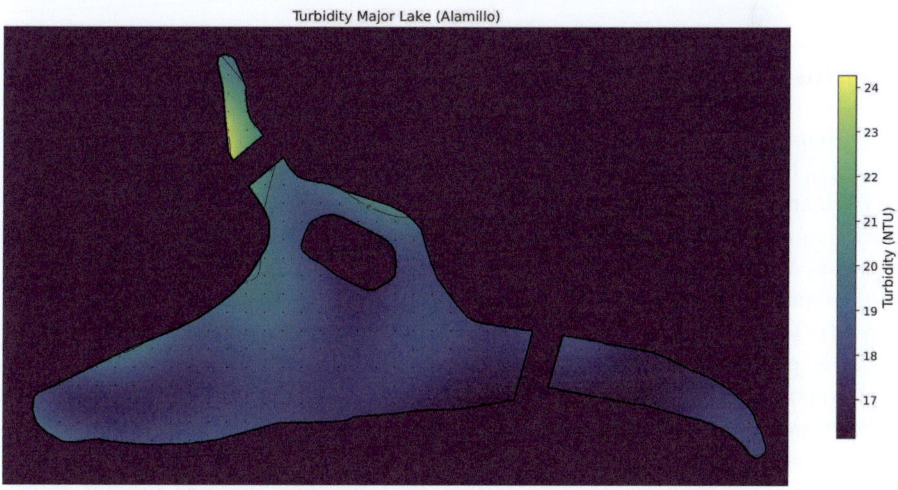

Fig. 27 Turbidity estimation generated from the Lago Mayor monitoring mission

(Ministerio de Sanidad, Consumo y Bienestar Social 2023) for drinking water, the water present in the lake presents healthy values without presence of contamination, allowing the presence of wildlife. However, the water is not suitable for human consumption, being the most undesirable parameter the high turbidity present in the lake.

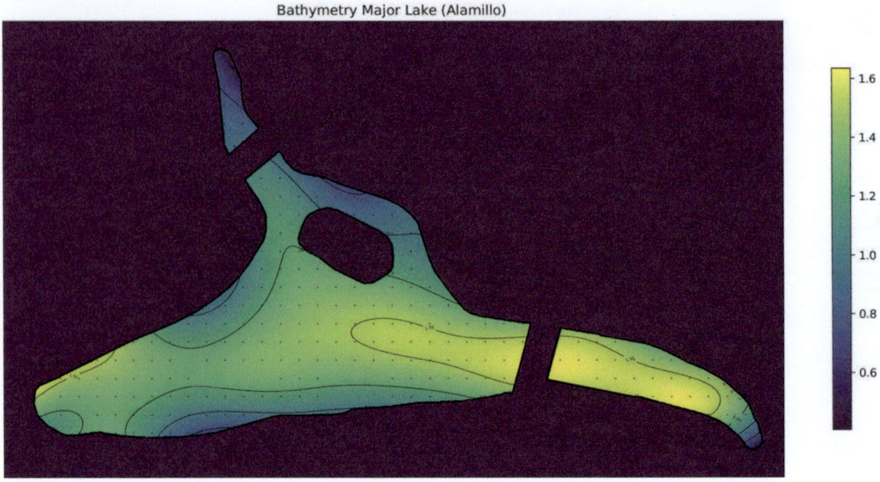

Fig. 28 Bathymetry estimation generated from the Lago Mayor monitoring mission

Conclusion and Future Works

This chapter introduces an innovative ASV for aquatic exploration and monitoring designed by the Department of Electronics of the University of Seville. It is capable of navigating through extensive water bodies while gathering real-time measures of a wide range of environmental data through the integration of onboard sensors such as sonar, camera and an electrochemical probe.

The monitoring data gathered by the ASV is provided using a user interface that provides tools for data interpretation and allow ASV management without the need for an expert. The whole system has been tested in a real environment, providing high importance environmental data in real time. Thus, the ASV contributes to the preservation of aquatic ecosystems and the detection of emerging contaminants, being a robust system with high autonomy.

Future lines of work include testing the investigation lines present in the department to guarantee a complete autonomous system able to make intelligent decisions and plan missions without the need for an operator. Furthermore, the information gathered by the ASV is relevant from an environmental point of view. The monitoring data will be accessible publicly so that it can be studied and analysed by competent entities.

The physical capabilities of the ASV will be further enhanced integrating a winch to provide water quality measures from submerging the electrochemical probe, making a profile of the water column and providing information about the water quality dynamics. In addition, making use of the payload capacity, the ASV will be equipped with a peristaltic pump and a series of water containers to collect geolocated water samples from areas of special interest.

Acknowledgments This work was partially funded by the following projects: TED2021-131326B-C21 of the Spanish Ministry of Science MCIN/AEI/10.13039/ 501100011033, and the European Union with 'NextGenerationEU/PRTR' funds, TED2021-131326A-C22 of the Spanish Ministry of Science MCIN/AEI/10.13039/ 501100011033 and the European Union with 'Next Generation EU/PRTR' funds, and the 'Junta de Andalucía: Consejería de Universidad, Investigación e Innovación' through the project 'Monitorization of Environmental Dangers with Unmanned Surface Agents: (MEDUSA)' under Grant PCM_00019, public funding. Activity: C23.I1.P03.S01.01 Andalucía Public subsidy for the development of the ≪INVESTIGO Programme ≫, financed by funds from the ≪Recovery and Resilience Mechanism ≫, Regional Government of Andalucía and the European Union. The authors have no conflicts of interest to declare that are relevant to the content of this chapter.

References

Al Maawali W, Al Naabi A, Al Yaruubi M, Saleem A, Al Maashri A (2019) Design and implementation of an unmanned surface vehicle for oil spill handling. In: 2019 1st international conference on unmanned vehicle systems-Oman (UVS), p 1–6

Barrionuevo AM, Yanes Luis S, Gutiérrez Reina D, Marín SLT (2024) Informative deep reinforcement path planning for heterogeneous autonomous surface vehicles in large water resources. IEEE Access 12:71835–71852

Beckler JS, Arutunian E, Moore T, Currier B, Milbrandt E, Duncan S (2019) Coastal harmful algae bloom monitoring via a sustainable, sail-powered mobile platform. Front Marine Sci 6:587

BlueRobotics (2024) T200 Thruster. https://bluerobotics.com/store/thrusters/t100-t200-thrusters/t200-thruster-r2-rp/. Accessed 04 Dec 2024

Casado-Pérez A, Yanes S, Toral SL, Perales-Esteve M, Gutiérrez-Reina D (2025) Variational autoencoder for the prediction of oil contamination temporal evolution in water environments. Sensors 25(6):1654

Chang H-C, Hsu Y-L, Hung S-S, Ou G-R, Wu J-R, Hsu C (2021) Autonomous water quality monitoring and water surface cleaning for unmanned surface vehicle. Sensors **21**(4):1102

Chen Y, Chen X, Zhu J, Lin F, Chen BM (2018) Development of an autonomous unmanned surface vehicle with object detection using deep learning. In: IECON 2018—44th annual conference of the IEEE industrial electronics society, pp 5636–5641

Dijkstra EW (1959) A note on two problems in connexion with graphs. Numerische mathematik 1(1):269–271

emlid (2021) Reach rs datasheet. emlid (28.4)

Ferri G, Manzi A, Fornai F, Ciuchi F, Laschi C (2014) The hydronet ASV, a small-sized autonomous catamaran for real-time monitoring of water quality: From design to missions at sea. IEEE J Ocean Eng **40**(3):710–726

Galceran E, Carreras M (2013) A survey on coverage path planning for robotics. Robot Auton Syst **61**(12):1258–1276

Kim JH, Kim N, Park YW, Won CS (2022) Object detection and classification based on yolo-v5 with improved maritime dataset. J Marine Sci Eng 10(3):377

Kwon DY, Kim J, Park S, Hong S (2023) Advancements of remote data acquisition and processing in unmanned vehicle technologies for water quality monitoring: an extensive review. Chemosphere 343:140198

Ministerio de Sanidad, Consumo y Bienestar Social (2023) Real decreto 3/2023, de 10 de enero, por el que se establecen los criterios técnico-sanitarios de la calidad del agua de consumo, su control y suministro. Boletín Oficial del Estado, 9 del 11/01/2023

Nothacker JS (2024) Sensor evaluation and fleet modeling of long-range low-cost autonomous surface vehicles. PhD thesis, Massachusetts Institute of Technology

Okudera R, Li Y, Matsubara Y, Takada H (2025) A lightweight, low-power and versatile companion computer for nano UAVs. In: Drones and unmanned systems, p 292

Popović, M Ott J, Rückin J, Kochenderfer MJ (2024) Learning-based methods for adaptive informative path planning. Robot Auton Syst 179:104727

Rasmussen CE, Williams CKI (2006) Gaussian processes for machine learning. Adaptive computation and machine learning. MIT Press, Cambridge. OCLC: ocm61285753

Ren J, Zhang J, Cui Y (2021) Autonomous obstacle avoidance algorithm for unmanned surface vehicles based on an improved velocity obstacle method. ISPRS Int J Geo-Inform 10(9):618

Riseborough P (2015) Application of data fusion to aerial robotics. In: Proceedings of embedded Linux conference

Roy R (2019) An introduction to water quality analysis. ESSENCE Int J Env Rehab Conserv **9**(1):94–100

Samaniego FP, Reina DG, Marín SLT, Arzamendia M, Gregor DO (2021) A Bayesian optimization approach for water resources monitoring through an autonomous surface vehicle: the ypacarai lake case study. IEEE Access 9:9163–9179

ublox (2015) Neo-m8 u-blox m8 concurrent GNSS modules. u-blox, (UBX-13003366 Rev. 10)

World Health Organization et al (1963) International standards for drinking-water. Journal-American Water Works Association

Yanes Luis S, Reina DG, Toral Marín SL (2021) A multiagent deep reinforcement learning approach for path planning in autonomous surface vehicles: the ypacaraí lake patrolling case. IEEE Access 9:17084–17099

Yanes Luis S, Shutin D, Gómez JM, Reina DG, Marín ST (2024) Deep reinforcement multi-agent learning framework for information gathering with local gaussian processes for water monitoring

Yanes Luis S, Basilico N, Antonazzi M, Gutiérrez-Reina D, Marín ST (2025) Variational model-based deep reinforcement learning for non-homogeneous patrolling aquatic environments with multiple unmanned surface vehicles. Expert Syst Appl 270:126483

Yanes-Luis S (2024) Deep reinforcement learning for adaptive monitarizacion and patrolling of water resources with unmanned surface vehicles. PhD thesis, Universidad de Sevilla

Yanes Luis S, Perales Esteve M, Gutiérrez Reina D, Marín ST (2023) Deep reinforcement learning applied to multi-agent informative path planning in environmental missions. In: Mobile robot: motion control and path planning. Springer, Berlin, pp 31–61

From Concept to Control: Development of an Advanced ASV Platform for Testing

Manuel Gantiva, Thalia Morel, Guillermo Bejarano, Pablo Millan, and Federico Peralta

Contents

M. Gantiva (✉) · T. Morel · G. Bejarano · P. Millan · F. Peralta
Universidad Loyola Andalucía, Seville, Spain
e-mail: megantiva@uloyola.es; tamorel@uloyola.es; gbejarano@uloyola.es; pmillan@uloyola.es; fdperalta@uloyola.es

D. Gutiérrez et al. (eds.), *Smart Water Quality Monitoring*,
https://doi.org/10.1007/978-3-032-03049-8_11

Abstract

The growing development and commercialisation of marine surface vehicles have driven significant advances in their design and integration. These vehicles, equipped with onboard computers, communication systems, and various sensors, offer flexibility for multiple applications. However, despite progress in motion control techniques, a considerable gap remains between validating algorithms in simulation and implementing them in real-world settings due to environmental uncertainties and inaccuracies in system modelling. This chapter addresses this issue by describing an experimental implementation of a path-following control strategy combined with an extended state observer. The control architecture is developed in Robot Operating System 2 (ROS 2) using open-source software and validated on the Yellowfish platform. Two cleaning tools designed for this vehicle are also introduced, expanding its practical applications. The experimental results confirm the robustness of the proposed strategy against external disturbances and validate the feasibility of transferring theoretical models into real-world implementations.

Introduction

The continuous development of mobile robots across various application areas has shown a consistent upward trend in recent years (Bai et al. 2022). Their integration into specific environments, such as marine and aquatic ones, has driven an increase in the variety of available models and simultaneously fostered a research focus on their control and automation. Surface vehicles stand out among these diverse categories due to their technical challenges, particularly given the wide range of their designs (Liu et al. 2016), as can be seen in Fig. 1. Key aspects of these vehicles include the design of the hull and the propulsion system, which significantly influence their dynamics and, consequently, the approaches required for their control and automation.

These vehicles have been deployed in various scenarios to fulfil a variety of purposes. Their applications include cleaning water bodies, monitoring environmental variables, and search and rescue operations, among others (Luo et al. 2024). Each function imposes specific requirements, requiring autonomous vehicles to adapt accordingly (Vitor et al. 2019). This challenge has led to the development of a wide array of control approaches, each tailored to specific objectives aimed at addressing or optimising different characteristics within the same type of vehicle.

Motion control is a fundamental aspect of autonomous vehicle performance and a technical challenge structured around different objectives depending on

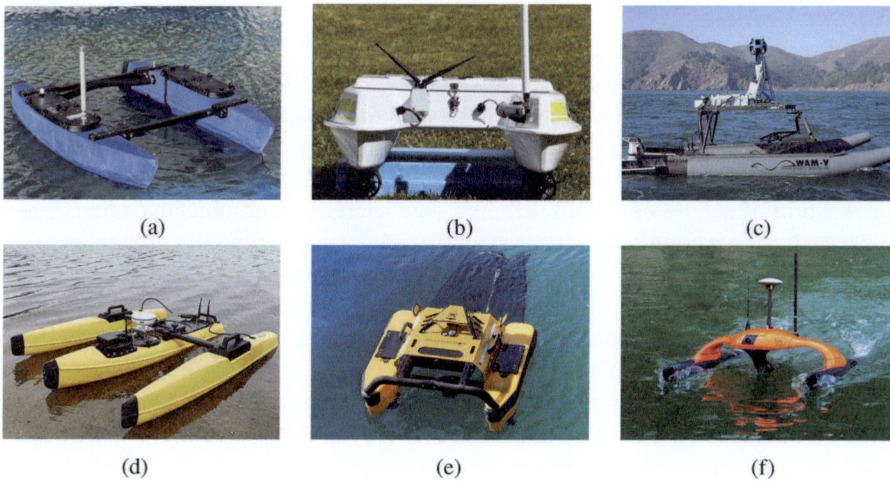

Fig. 1 Examples of commercial unmanned surface vehicles. (**a**) Blueboat (**b**) Duckling (**c**) WAM-V 16. (**d**) Z-Boat 1250 (**e**) Jellyfishbot (**f**) Sonobot-5

the application. For example, in an application where an environmental variable is monitored (Kathen et al. 2021), the vehicle initially travels from one point to another while maintaining specific parameters such as speed and orientation. Subsequently, it remains stationary at a designated point, compensating for external disturbances, such as wind or currents. Finally, it navigates to a docking area for recharging or maintenance. Each application stage demands a unique control approach, significantly increasing the complexity of developing an autonomous vehicle capable of efficiently meeting all operational requirements.

Notable approaches discussed in the literature include path following (Gu et al. 2023), which is particularly useful for tasks where precision in movement takes precedence over execution speed. In contrast, trajectory tracking places greater emphasis on temporal factors (Paliotta et al. 2019), adopting a more time-restrictive approach at the cost of reduced accuracy in tracking. In collaborative control, strategies such as flocking, inspired by swarm behaviour (Peng et al. 2021a), aim to maintain collective formations among vehicles in motion while avoiding collisions between agents or with external obstacles. Conversely, a more rigorous approach is formation control, which focuses on preserving a specific geometric configuration between vehicles while they are in motion (Peng et al. 2021b).

Although each approach uses distinct strategies, they share a common goal: achieving effective control of one or more vehicles. In this context, this introductory section aims to summarise and consolidate the development of control algorithms, regardless of the specific approach. The scope spans from the design of experimental platforms to a detailed description of the control architectures implemented, providing a comprehensive overview of the topic.

Surface Vehicles

Before delving into specific control strategies and objectives, it is essential to provide an overview of existing autonomous surface vehicle (ASV) or unmanned surface vehicle (USV) platforms. This will enable the reader to become familiar with the structure and layout of these systems. These platforms are presented as a collection of subsystems with different characteristics and objectives that, when integrated, form a comprehensive system capable of supporting the implementation of various control algorithms for different purposes (Zhang et al. 2021).

The diversity of ASVs is vast and is commonly classified based on criteria such as size, hull design, and power source (Qiao et al. 2023). Each category has specific attributes that make it suitable for different applications, depending on factors such as payload capacity (whether for materials or equipment), speed, or physical dimensions. These characteristics determine their suitability for particular scenarios and influence the control strategies that can be implemented (Bai et al. 2022).

To define the scope of this chapter, the focus will be on medium-sized or mini-surface vehicles. Although these are not designed to carry passengers, they have payload capacities and operate using electric power sources. However, the concepts and strategies discussed here can easily be extrapolated to larger or fossil fuel-powered vehicles.

At a high level, these platforms can be divided into two main components: software and hardware (Majid and Arshad 2016). Hardware encompasses all physical elements of the vehicle, including processors, hull, and actuators. Software, on the other hand, comprises the operating system, programmes, and algorithms required to ensure proper functionality of the platform (Tanakitkorn 2019). This initial division is primarily conceptual, distinguishing between the tangible and intangible elements of the system. However, for a more detailed analysis, it is essential to note that each component includes interdependent subsystems that actively interact to achieve the desired vehicle performance.

Hardware

Upon addressing the hardware first, six main subsystems can be identified, as illustrated in Fig. 2. The first of these is the hull or structure of the vessel, which forms the physical foundation of the vehicle and serves as the cornerstone of its design. The hull determines critical aspects of the dynamics of the vessel, and its design must carefully consider the specific application and environmental conditions in which it will operate. An appropriately designed hull ensures efficient hydrodynamics, enabling the required speeds to be achieved while minimising the impact of external factors and optimising energy consumption (Ghazali et al. 2024). Additionally, the hull defines the vehicle's payload capacity, an immutable characteristic once it is manufactured, emphasising the importance of thoroughly planning its capabilities and limitations during the design phase.

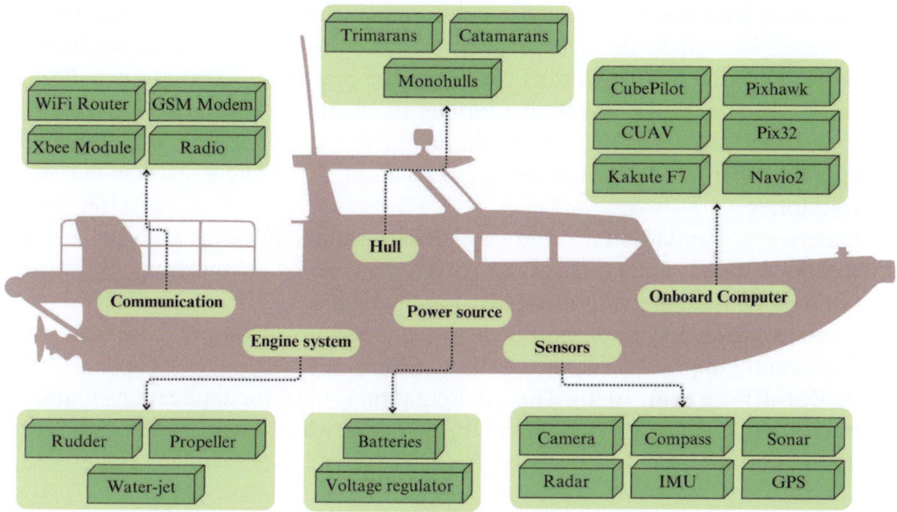

Fig. 2 Typical subsystems comprising the hardware of a surface vehicle

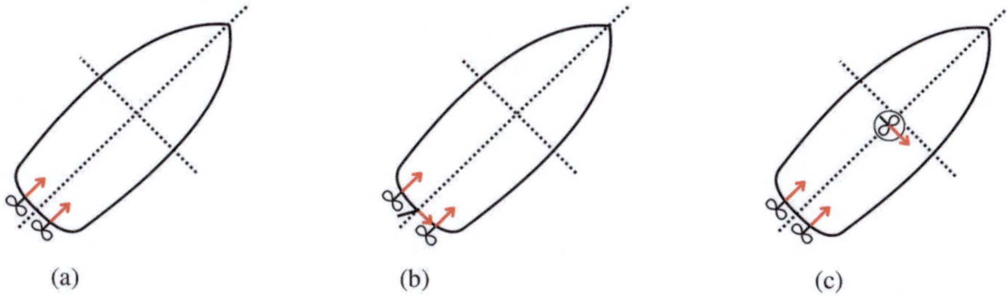

Fig. 3 Common actuator configurations in surface vehicles and the forces exerted by each. (**a**) Differential (**b**) Propeller-rudder (**c**) Transverse

The next subsystem is closely related to the vehicle's dynamics, and like the hull design, its components must be selected based on the specific application. The engine or propulsion system is crucial in defining the dynamic capabilities of the vessel, as the number and configuration of actuators directly influence its forward and turning skills (Fossen 2011). The most common configurations, illustrated in Fig. 3, include the following:

- **Differential:** This setup employs two or more thrusters located at the rear of the vehicle. The thrust force is proportional to the sum of the forces the thrusters generate. Conversely, the torque is proportional to the distance and the difference between these forces. This solution is versatile for basic manoeuvres and speed control (Li et al. 2019; Zhou et al. 2020; Ringback et al. 2021).

- **Propeller-Rudder**: In this setup, one or more thrusters work in tandem with a rudder located at the rear of the vehicle. The thrusters generate forward thrust, while the rudder controls the turning by producing a lateral force through angle variation, thus generating a turning moment (Arrichiello et al. 2011; Han et al. 2017; Zhang et al. 2019). Although this configuration has limitations regarding sharp turns, it provides greater stability for straight and low-curvature trajectories.
- **Transverse**: Similar to the differential configuration, this setup incorporates two or more thrusters at the rear of the vehicle and one or more azimuthal thrusters mounted beneath the hull along the longitudinal axis. The azimuthal thrusters generate lateral forces proportional to their distance from the transverse axis, significantly improving manoeuvrability (Wang et al. 2019b; Souissi et al. 2020; Del-Rio-Rivera et al. 2020). This configuration is ideal for applications requiring precise movements, such as in confined spaces where position and directional control are critical.

Each configuration offers specific advantages that should be carefully analysed based on the vehicle's operational requirements and environmental conditions. In addition, some variations include using servomotors to adjust the angle of thrusters (Sarda et al. 2016; Wu et al. 2025). This adaptability enables actuators to generate forces at specific angles as the dynamic conditions require, often providing greater control over the vehicle's degrees of freedom and enhancing its manoeuvrability and operational versatility.

The third hardware subsystem is the power supply, which primarily consists of electric batteries. Its design must balance weight and autonomy since excess batteries significantly increase the vehicle's weight, leading to higher energy consumption and reduced payload capacity (Eugene et al. 2019). Voltage regulators ensure a stable energy supply–essential devices that maintain a constant voltage level during battery discharge and protect the integrity of actuators and electronic components against fluctuations. This balance between energy capacity and efficiency is crucial for optimising the system's performance and reliability in operation.

The subsystems discussed so far are primarily related to vehicle dynamics and performance. In contrast, the following three subsystems are more closely linked to the software and provide a broader range of options, depending on the specific application. One such subsystem is the communication system, whose configuration largely depends on the operational environment (Tabish and Chaur-Luh 2024). For example, a Global System for Mobile Communications (GSM) modem can establish a stable Internet connection to facilitate information exchange with cloud servers. In contrast, Wireless Fidelity (Wi-Fi) modems are better suited to create reliable connections with ground stations. Modules like XBee are particularly advantageous for vehicle communication in a fleet, allowing for quick and reliable network configuration. The selection of communication system components should consider key factors such as bandwidth, transmission speed, and latency, ensuring they meet the application's specific requirements.

The sensors installed on an ASV can be classified into two groups depending on their purpose: navigation sensors and application-specific sensors. It is worth noting that information from a navigation sensor, such as position data, can also be used for application purposes; however, the distinction lies in the fact that navigation sensors are considered minimal and essential for motion control, whereas the second group of sensors is specifically oriented towards the functionality of the application.

On the one hand, navigation sensors include global positioning systems (GPSs), inertial measurement units (IMUs), and compasses, which collectively provide sufficient information to enable autonomous navigation of an ASV, as these devices focus primarily on the perception of the vehicle itself (El-Sheimy and Youssef 2020). On the other hand, sensors such as cameras and light detection and ranging (LiDAR) provide detailed environmental information, which is particularly useful for tasks such as obstacle detection and target tracking, thereby enhancing the vehicle's interaction with its operational environment (Leong and Ahmad 2024).

The variety of application-specific sensors is as broad as the possible applications themselves, and their selection depends on the specific requirements of each case. However, certain factors must be considered before choosing the sensors, such as their size and weight, as these directly impact the vehicle's payload capacity. Additionally, the increase in energy consumption associated with powering the sensors and the added weight they introduce must be considered. Standard application-specific sensors include sonars, water quality sensors (e.g., for measuring pH, electrical conductivity, and turbidity), and ultrasonic sensors, which are widely used for specific measurement and monitoring tasks (Thombre et al. 2022).

The final hardware subsystem is the onboard computer, which serves as the vehicle's central brain and plays a pivotal role in the control and automation of surface vehicles. Its primary function is to store and execute algorithms for control, navigation, and guidance, processing information received from sensors to make real-time decisions. This device must be selected carefully, considering cost, memory capacity, processing speed, and energy consumption.

One of the most common and accessible commercial solutions is flight controllers, electronic boards designed to control unmanned vehicles. These boards simplify system development by integrating various sensors and peripherals required for implementation, offering an efficient and optimised solution for a wide range of applications.

Software

The software of an autonomous vehicle can be structured into a layered architecture, where each layer groups and executes algorithms with a specific purpose, sharing information seamlessly with adjacent layers. This organisation not only enhances modularity but also simplifies the development, integration, and maintenance of the system, ensuring that each set of algorithms operates efficiently within its scope of responsibility.

Although no strict convention exists for defining and delimiting these layers, the following general structure aims to be application- and control-agnostic. This structure comprises four levels and five layers, as illustrated in Fig. 4. The top layer

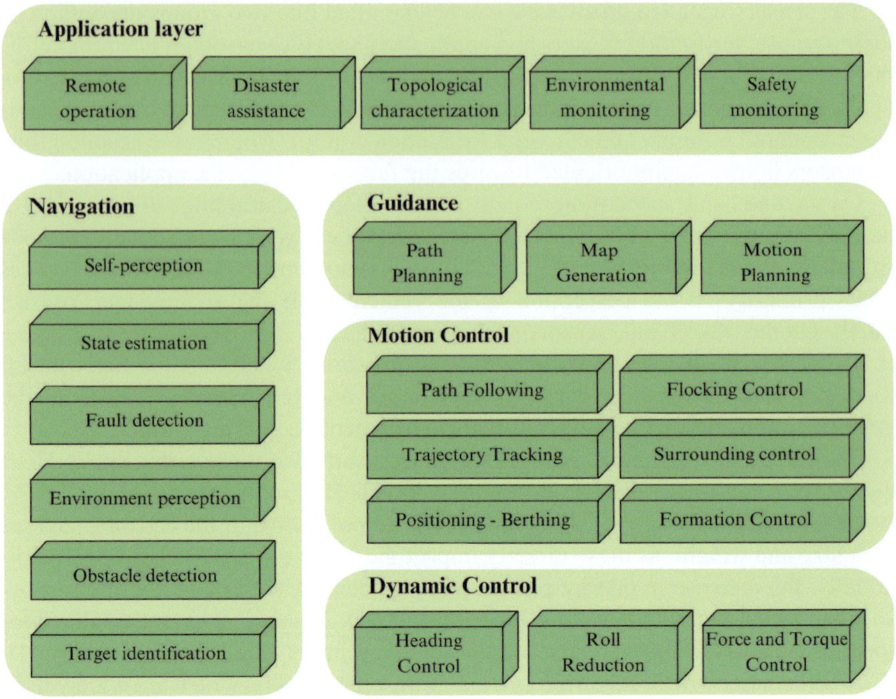

Fig. 4 Layered architecture of the software in a surface vehicle

is devoted to application-related algorithms and is the furthest from motion control. In many cases, the development and deployment of these vehicles focus primarily on this layer, assuming that the lower layers work correctly (Qiao et al. 2023).

The remaining three levels focus mainly on motion control, addressing tasks ranging from navigation to executing specific manoeuvres. Below is a description of the purpose of each layer and the information typically exchanged with its neighbours, highlighting the interdependence that ensures cohesion and performance across the entire system.

The Navigation Layer encompasses algorithms for reading and processing sensor data (Fossen 2011). It plays a cross-functional role across all three control levels, providing critical information to the Application Layer. The algorithms in this layer are divided into two main groups:

- **Self-Perception:** This group includes algorithms that use sensor data to determine the vehicle's current state, such as its position, velocity, acceleration, and other relevant parameters, as well as fault detection. Common examples include state observers and extended Kalman filters, which process and filter information to enhance system accuracy and reliability.

- **Environmental Perception:** This group focuses on algorithms that enable the vehicle to interpret its surroundings and respond to environmental changes. Tasks include obstacle detection and target identification using advanced image processing or point-cloud analysis techniques. These algorithms ensure the vehicle can interact safely and effectively with its operational environment.

The interaction between these groups enables the navigation layer to provide key information for higher-level motion control and decision-making.

Some authors combine the following three layers into a single layer. Still, this chapter separates them into three distinct layers with defined focuses to emphasise their interaction with different control objectives. The first of these layers, known as Guidance, bridges application algorithms and various control scenarios (Fossen 2011). This layer is crucial in ensuring applications work correctly by implementing diverse control objectives. Examples of algorithms found in this layer include the following:

- **Path Planning:** These widely studied algorithms, available in a variety of approaches, aim to generate safe paths that meet mission constraints and requirements. They utilise environmental information provided by the Navigation Layer, along with waypoints or specific instructions defined by application algorithms. Path planners can be classified into global planners, which design routes considering a complete view of the environment, and local planners, which focus on adjusting trajectories in immediate areas (Vagale et al. 2021). In addition, some planning algorithms are reactive, allowing them to dynamically re-plan routes in response to environmental changes or new application requirements, ensuring efficient and adaptive operation.
- **Map Generation:** Like path planning algorithms, these rely on information from the Navigation Layer and requirements defined by the Application Layer. However, their goal is to create maps of specific areas that the application can use for decision-making. These maps may represent one or more variables, depending on the available data and mission objectives (Anderson et al. 2022). Their value lies in providing a structured representation of the environment that facilitates the evaluation of conditions and planning specific actions during the vehicle's operation.
- **Motion Planning:** These algorithms bridge the decisions or instructions from the Application Layer and various control scenarios. Their primary goal is to use the information from generated paths or maps to plan the sequence of steps necessary to meet mission requirements (Chiang and Tapia 2018). These algorithms enable the transition between different control approaches, allowing for integrating and coordinating multiple objectives in a single operation. They make it possible to develop complex missions composed of various stages, ensuring smooth and adaptive execution to meet the specific requirements of each mission phase.

The next layer is Motion Control, also known as Kinematic Control, and it encompasses a set of algorithms designed to address various motion scenarios, each with its challenges and complexities. The number and type of algorithms incorporated into this layer determine the vehicle's autonomy level and ability to execute complex applications with high demands.

This layer distinguishes between two types of scenarios: individual and collective. Individual scenarios focus on controlling a single vehicle, while collective scenarios involve coordinating multiple agents. Although some algorithms for collective scenarios share principles with those for individual scenarios, they are classified separately due to the added complexity of managing and coordinating the behaviour of multiple vehicles simultaneously (Peng et al. 2021b). This level of control is essential for addressing advanced applications and tasks requiring cooperation among various agents. Examples of such scenarios include the following:

- **Path Following:** The goal is for the vehicle to follow a predefined path while adhering to forward velocity constraints without considering time constraints.
- **Trajectory Tracking:** This involves following a target moving along a defined trajectory at a specific and possibly time-varying velocity, meeting both time and position constraints.
- **Berthing:** One of the most complex control problems is positioning the vehicle at a specific location and orientation while respecting strict velocity constraints in the transient.
- **Flocking Control:** The objective is for a group of agents to move in a coordinated manner, emulating swarm behaviour. This approach ensures collision avoidance and the maintenance of safe distances between agents.
- **Formation Control:** The goal is for a group of agents to move in a coordinated manner while maintaining a constant geometric formation. In this case, the relative positions among agents are more strictly controlled than in flocking.
- **Surrounding Control:** This aims to have a group of agents surround or enclose a target in a coordinated manner, maintaining a specific geometric shape that can either remain static or rotate around the target.

The final layer is Dynamic Control, which is responsible for receiving commands generated by the kinematic control layer and processing them along with the vehicle's current positions and velocities to calculate the most appropriate control actions for following the established references. While there are control techniques that integrate dynamics and kinematics simultaneously, it is essential to emphasise that the vehicle's dynamics are more directly linked to its structural design and the placement of actuators, as mentioned in the previous section (Fossen 2011). For this reason, this software layer must be based on the vehicle's specific mathematical model, increasing its complexity level. Accurate modelling ensures effective dynamic control tailored to the vessel's unique characteristics.

Although various approaches exist within this control layer, they can be grouped into three categories:

- **Heading Control**: This approach maintains a constant forward velocity while adjusting the vehicle's orientation to reach the desired references. It is commonly used in scenarios where the trajectory primarily depends on the vehicle's heading.

- **Roll Reduction:** This approach aims to minimise the effects of roll (side-to-side oscillation) of the vessel, which is particularly critical in highly unstable environments, such as rough waters. Its effectiveness heavily depends on the vessel's design and environmental conditions.
- **Force and Torque Control:** This approach focuses on regulating the forces and torques generated by the vehicle's actuators to achieve the desired velocity references. This approach also enables lateral force control in vessels equipped with transversal thrusters, helping stabilise the vehicle and improve its manoeuvrability under challenging conditions.

Motivation

Over the past decade, there has been a significant increase in the number of manufacturers and startups dedicated to commercialising and producing ASVs designed to operate in a wide range of environments and meet diverse operational requirements. These vehicles typically include a variety of integrated hardware, such as sensors, onboard computers, and communication systems, and allow for adding extra sensors to adapt to specific user's needs (Tanakitkorn 2019).

By integrating key components from the initial design phase, ASVs significantly reduce development times, addressing essential aspects such as hull dynamics and actuator configuration, simplifying their implementation across various applications. Table 1 summarises the features of some of the latest models on the market, and below are some highlights of the most notable components of some vehicles, some of which are shown in Fig. 1 for better visualisation of their structural configurations and design differences:

- **BlueBoat:** This model is a catamaran notable for its hull design, significantly reducing the risk of entanglement with algae or other debris. It is equipped with a Raspberry Pi and a Navigator module, serving as its onboard computer. One of its most outstanding features is the use of open-source software, BlueOS, which facilitates the integration of various sensors, including side-scan sonars.
- **NX 100:** This robust catamaran features an all-in-one long-range controller that combines control, telemetry, and full high-definition (HD) video into a single device. It is equipped with high-capacity swappable batteries, each providing 3 hours of autonomy. The NX 100 is versatile and capable of integrating sensors such as water quality meters, thermal cameras, and LiDAR, making it suitable for a wide range of applications.
- **HyDrone:** Also a lightweight unmanned surface catamaran, this model is designed for hydrographic survey applications. It is equipped with real-time kinematic (RTK) GPS modules, remote radio control (RC) with a range of 1 km, and an AutoNav^{TM} module that enables motion control and autonomous mission execution.

Table 1 Description of key features in recent commercial ASVs

Model	Supplier	Release year	Hull type	Length (m)	Vehicle weight (kg)	Cruise speed (m/s)	Endurance[a] (h)
BlueBoat (Blue Robotics 2025)	Blue Robotics	2023	Catamaran	1.2	15	1.5	18
Apache 3 (CHCNAV 2025)	CHC Navigation	2021	Monohull	1	7	2	6
Sonobot 5 (EvoLogics 2025)	EvoLogics	2020	Catamaran	1.3	27	1	9
NX 100 (NX Innovation 2025)	Nxinnovation	2020	Catamaran	1	16	1–2	3
SL20 USV (OceanAlpha 2025)	OceanAlpha	2021	Monohull	1.05	17	1.5	6
HyDrone (Seafloor Systems 2025a)	Seafloor Systems	2013	Catamaran	1.16	9.8	1	6
Tridrone (Seafloor Systems 2025b)	Seafloor Systems	2022	Trimaran	1.2	10	1.5	6–8
Z-Boat 1250 (Teledyne Marine 2025)	Teledyne Oceanscience	2017	Trimaran	1.3	18	1–1.5	1–4

[a] Approximate battery endurance at constant cruising speed (in hours) specified by the manufacturer of each model

Alongside the development and commercialisation of these vehicles, research into motion control and the generation of new approaches has experienced significant growth in recent decades. This has resulted in a wide variety of methods designed to address the diverse challenges posed by these vehicles (Castano-Londono et al. 2024). On the one hand, progress has been made in applying classical control techniques such as backstepping and model predictive control (MPC), which provide robust and well-established solutions. On the other hand, there is growing interest in integrating machine learning-based techniques, such as radial basis functions (RBF) and deep reinforcement learning (DRL), which enable handling more complex problems with improved adaptability and accuracy. This combination of traditional and advanced methodologies has significantly driven the development of innovative solutions for controlling ASVs.

However, despite the increase in proposals and physical models, a significant gap remains between validating these algorithms through simulation and their experimental implementation on real surface vehicles (Gantiva Osorio et al. 2024b). This transition faces substantial challenges, such as managing uncontrolled environmental disturbances and addressing mathematical issues, including model uncertainties caused by equipment wear. These factors highlight the importance of developing robust approaches that can be validated in practical and dynamic scenarios.

Main Contribution

Based on the identified transitional gap, this chapter aims to provide an experimental perspective by implementing a path-following control strategy combined with an extended state observer on a real ASV (Gantiva Osorio et al. 2024a). Both techniques are programmed in Robot Operating System 2 (ROS 2) (Macenski et al. 2022) using a node-based architecture and have been experimentally validated on a Yellowfish surface vehicle, shown in Fig. 5, employing open-source software.

The remainder of the chapter is organised as follows: section "Yellowfish ASV" introduces the ASV Yellowfish platform, describing its hardware, including its physical structure and integrated sensors. Section "Description of Software" details the open-source software employed in the system, providing the reader with the necessary tools to understand its implementation. Section "Cleaning and Collection Tools" introduces two cleaning tools specifically designed for this model, illustrating examples of their practical applications and validating their potential. Section "Modular Control Architecture" addresses the control strategy and its implementation, starting with the mathematical foundations of the control method and proceeding to the details of its programming in ROS 2. Section "Experiments and Results" presents some relevant experimental results, evaluating the performance of the implemented system. Finally, section "Conclusions and Advice for Future Developments" summarises the main findings of the work and provides recommendations for future developments in this field.

Fig. 5 The Yellowfish ASV in operation in a lake

Yellowfish ASV

The Yellowfish ASV is an autonomous surface vehicle developed at Universidad Loyola Andalucía as part of the *ASV4Sustainability* research initiative (Universidad Loyola Andalucía 2025). This platform is designed to advance the experimental validation of motion control strategies while supporting environmental monitoring and sustainability-focused applications. The Yellowfish ASV integrates advanced hardware and software components, enabling autonomous navigation, real-time data collection, and adaptability for various missions.

The vehicle comprises six key hardware subsystems: hull and structure, propulsion system, power supply, communication system, sensor suite, and onboard computing unit. The following subsections provide a comprehensive overview of each subsystem, highlighting its role in the vehicle's overall performance.

Hull and Structure

The Yellowfish ASV follows a catamaran configuration. The structure consists of two identical fibreglass hulls, each measuring 128 × 25.5 × 30 cm, connected by a lightweight yet durable aluminium frame. The total vehicle dimensions are 128 × 98 × 30 cm, with an approximate weight of 24.5 kg.

A waterproof electronics box is mounted on the hulls, housing critical components and communication interfaces. The hulls are designed to withstand harsh conditions, making them ideal for long-duration missions. The electronic box is shown in Fig. 6, while Table 2 lists each component.

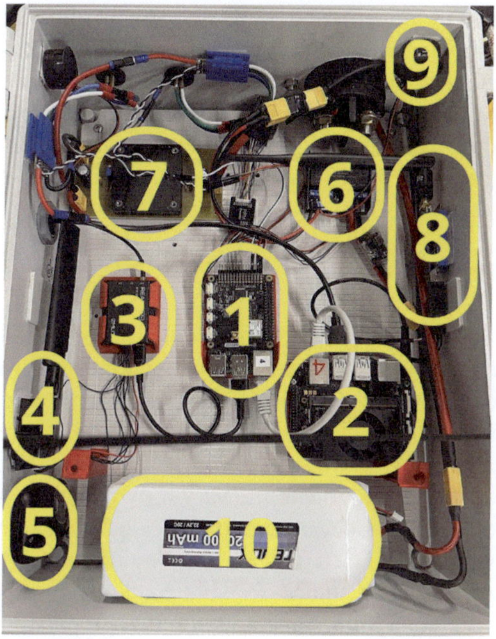

Fig. 6 The Yellowfish ASV and its hardware

Table 2 Hardware components of the Yellowfish ASV

Number in Fig. 6	Component
1	Raspberry Pi 4 Model B and Navio2
2	Jetson Xavier NX
3	GPS module
4	Long-range (LoRa) GPS antenna
5	Wi-Fi modem
6	12 V voltage regulator
7	16 V voltage regulator for thrusters
8	Xbee module
9	Digital voltmeter
10	22.2 V 20,000 mAh lithium-ion polymer (Li-Po) battery

Propulsion System

The Yellowfish ASV has two T200 thrusters from Blue Robotics, arranged in a differential thruster configuration at the stern. Each thruster operates on a 22.2V Li-Po battery, with a thrust capability ranging from -45.32 N to 57.40 N.

This setup enables precise manoeuvrability, allowing for sharp turns, dynamic positioning, and station-keeping. The propulsion system is integrated with pulse-width-modulation-based (PWM) electronic speed controllers (ESC) to regulate thrust levels accurately.

For information purposes, Fig. 7 illustrates the performance curves of each propeller and their corresponding thrust values for each ESC PWM setting. Multiple curves are provided for different voltage levels from the power source. It can also be observed that there is a dead zone between the forward and reverse movement of the propeller. The figure is derived from the propeller's datasheet; for further details, the reader is referred to (Blue Robotics 2025).

Power Source

The primary power source is a 22.2 V 20,000 mAh Li-Po battery, which supplies energy to the thrusters and the onboard electronics. To monitor and optimise energy consumption, a digital voltmeter is connected to the power system, providing real-time power level readings and detecting any fluctuations to improve operational reliability.

To accommodate the varying voltage requirements of the onboard components, the system integrates two dedicated voltage regulators. A 12 V regulator provides stable power to the Jetson Xavier NX onboard computer, ensuring consistent performance under variable load conditions. Additionally, a 16 V regulator is specifically implemented for the propulsion thrusters. Although the main LiPo battery provides a nominal voltage of 22.2 V (6S configuration), the thrusters operate within a recommended voltage range of 10–20 V, as depicted in Fig. 7. The chosen 16 V regulated supply guarantees operation within this optimal voltage range, ensuring consistent thrust performance throughout battery discharge (see Fig. 8).

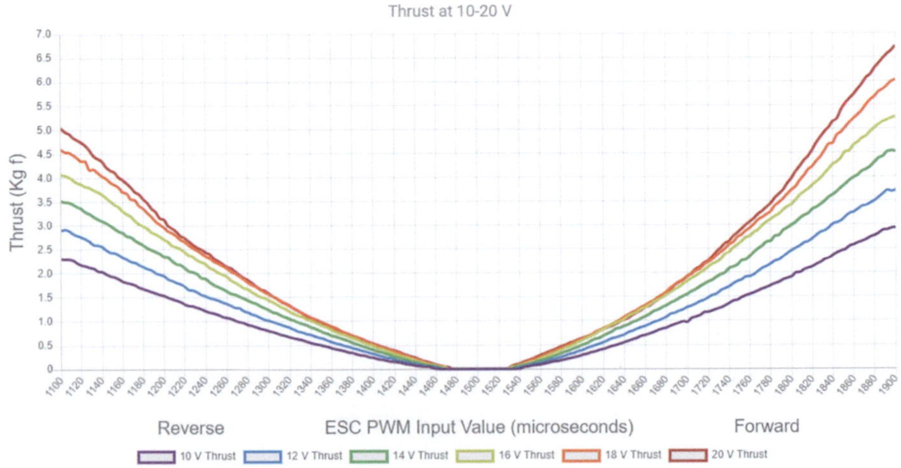

Fig. 7 Propeller PWM thrust datasheet curves

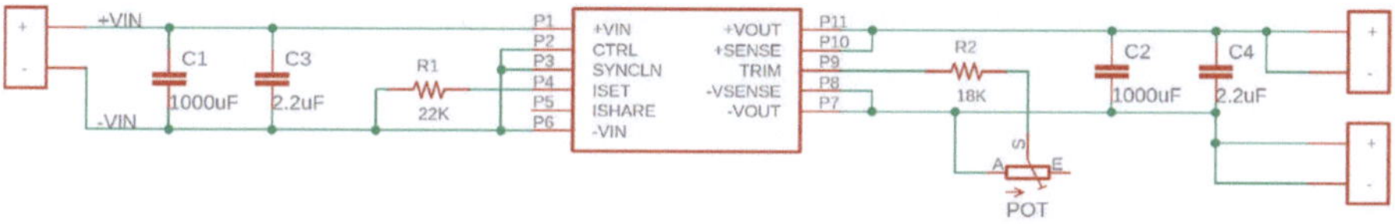

Fig. 8 Connection diagram of the 16 V regulator

This modular power system allows for efficient energy management and adaptability, optimising performance for both computational and propulsion requirements while maintaining operational safety.

Communication System

A robust communication system ensures uninterrupted data transmission and efficient mission execution. This system integrates multiple modules, each tailored to specific operational needs.

A Wi-Fi modem provides high-speed connectivity for short-range control and data transfer, enabling direct communication with ground control. The XBee module facilitates seamless inter-vehicle communication, which supports coordination between multiple ASVs. This functionality is particularly valuable for cooperative missions that require synchronised fleet operations.

Sensor Suite

High-precision navigation sensors ensure accurate and stable autonomous operation. The Reach RS+ RTK GPS provides centimetre-level accuracy, enhancing the precision of geolocation for navigation. In addition, the integrated LoRa GPS antenna improves positioning in low-bandwidth environments, enabling efficient long-range operations even under challenging conditions.

For attitude estimation and precise directional control, the Navio2 IMU and the compass continuously monitor orientation and movement, allowing real-time adjustments to maintain optimal course stability.

Onboard Computing Unit

A high-performance computing system manages processing and control, ensuring efficient execution of autonomous tasks. The core of this system includes the Raspberry Pi 4 Model B with Navio2, which handles mission planning, telemetry, and sensor integration. Complementing this, the Jetson Xavier NX provides advanced data processing capabilities that support the implementation of control algorithms for autonomous operations.

This architecture enables real-time navigation and adaptive control, making it a robust platform for marine research applications.

Description of Software

This section focuses on the open-source software used to implement the control strategy configured on the previously described hardware. It is important to note that, given the scope of this chapter, only essential concepts related to these programmes will be discussed without delving into technical details. Readers interested in developing and implementing their algorithms are encouraged to undertake a more in-depth study of each of these software solutions, as they offer a wide range of applications and have the potential to enrich future developments.

ArduPilot

ArduPilot (ArduPilot 2025) is a widely used open-source software for controlling autonomous vehicles, including drones, aircraft, ground vehicles, and watercraft. It is currently one of the leading projects in robotics and autonomous systems, along with PX4 Autopilot (PX4 2025), offering a flexible and efficient flight control and navigation solution. Some features that make it a preferred choice for autonomous vehicle development include the following:

- **Compatibility with various vehicles and configurations:** It supports a wide variety of platforms, such as quadcopters, aeroplanes, helicopters, rovers, submarines, and more.
- **Adaptability and customisation:** As open-source software, it allows advanced users to modify and customise the code to meet the specific needs of their projects.
- **Integration with a wide range of sensors:** It is compatible with essential sensors for autonomous navigation, such as GPS, magnetometers, accelerometers, barometers, and others, ensuring precise and robust control.
- **Execution on low-cost hardware:** It can run on affordable flight controllers, such as those based on Pixhawk, ArduPilot Mega (APM), and CubePilot.
- **Compatibility with complementary tools:** It works seamlessly with tools such as Mission Planner, which simplify mission planning, monitoring, and management, improving user experience and operational capabilities.

This software was configured on a Raspberry Pi 4 using the *Emlid Raspberry Pi OS Buster* (Emlid 2025), an operating system provided by the manufacturer of Navio2. Its architecture comprises three layers: the hardware abstraction layer, responsible for sensor readings and actuator control; the shared libraries layer, which integrates standard algorithms such as the extended Kalman filter (EKF); and the vehicle-specific code layer, which, for surface vehicles, is located in the rover

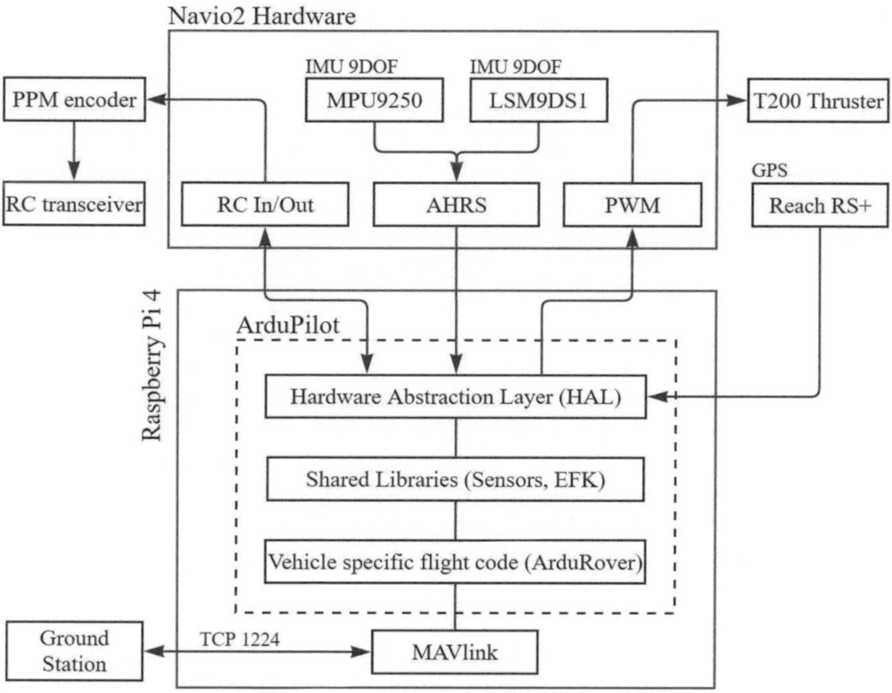

Fig. 9 ArduPilot architecture and its integration with hardware, where AHRS refers to Attitude and Heading Reference Systems, and PPM stands for pulse position modulation

section of ArduPilot, offering specialised functions for these applications. Figure 9 illustrates these layers and their interconnection with the previously described hardware.

Another notable feature of ArduPilot, illustrated in Fig. 9, is its compatibility with Micro Air Vehicle Link (MAVLink) (MAVLink 2025), a lightweight communication protocol designed for autonomous vehicle systems. MAVLink is widely used for data exchange between flight controllers, ground stations, and other components, offering advantages such as using a common standard, efficient communication, high versatility, and extensibility to add new commands. This protocol enables remote communication with a ground station configured via Transmission Control Protocol (TCP) on port 1224. However, as explained later, it can also be configured to use other links and ports depending on the system's specific needs.

Within the scope of this chapter, this software is primarily used for sensor readings and as a controller for forward speed and turning. Additionally, each sensor has been calibrated to ensure optimal performance.

ROS 2

ROS 2 is an open-source software platform designed for developing robotic applications (Macenski et al. 2022). As an evolution of ROS1, it was created to overcome its predecessor's limitations, particularly in scalability, support for distributed systems, and security (Rico 2022). Its main features include tools, libraries, and conventions that simplify robot development, simulation, control, and monitoring. It is important to note that ROS 2 is not an operating system but rather a framework composed of four main elements: a core framework, a set of tools, a plug-and-play architecture through plugins, and an active online community (Renard 2024). This integration has been fundamental in driving its adoption and establishing it as a cornerstone in robotics.

Readers interested in gaining a deeper understanding of ROS 2 are advised to acquire object-oriented programming (OOP) skills in Python or C++, depending on their preferred language, and to become familiar with using Linux through the command line. ROS 2 has established itself as a standard in robotics, and to facilitate comprehension for readers without advanced knowledge of the platform, some key concepts are briefly described below, along with the terminology used in subsequent sections:

- **Package:** An organisational unit that groups everything necessary for a specific functionality or application within a robotic system. It includes nodes, scripts, configurations, dependencies, and other related files.
- **Node:** The basic processing unit in ROS, representing an independent program or process that performs a specific task. Nodes communicate with each other to collaborate in executing an application.
- **Topic:** A communication channel based on the publish/subscribe paradigm, allowing a node to publish messages to a topic while one or more subscribed nodes receive the data.
- **Service:** A synchronous communication mechanism in which one node requests a specific action or information from another node and receives an immediate response. It operates under a request-response model.
- **Action:** An asynchronous communication mechanism designed for tasks requiring time or supervision. It consists of two services (one for starting and another for completing the action) and a topic for providing continuous feedback to the client on the task's status.
- **Interfaces:** Predefined data structures used to exchange information between nodes. They allow developers to define the data format communicated via topics, services, and actions.

The use and combination of these elements enable the development of complex architectures that implement advanced control strategies and support the creation of highly sophisticated applications. To aid in understanding the following sections, Fig. 10 presents the nomenclature used in ROS 2 node diagrams, which will be

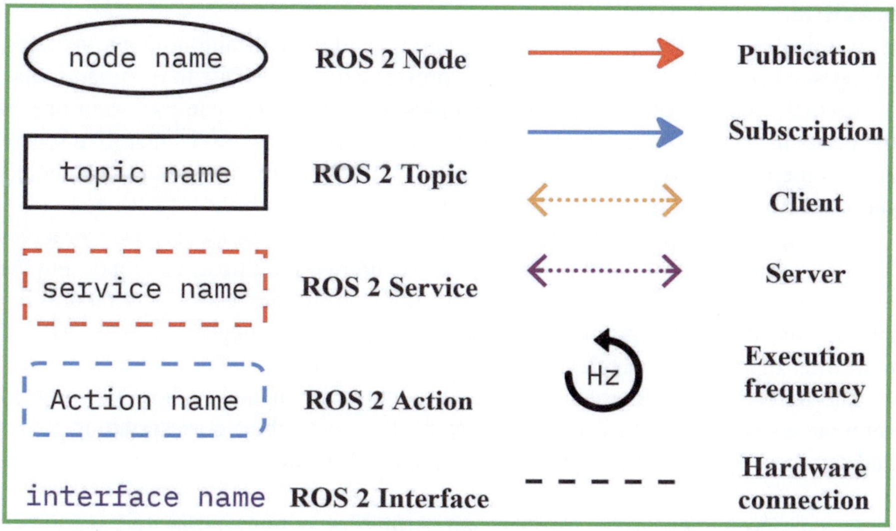

Fig. 10 Description of symbols used in ROS 2 graph diagrams

applied throughout. Additionally, it is clarified that the name of each component follows the `package/component_name` format, allowing for precise identification of the package to which each system element belongs.

Another advantage of using ROS 2 in robotics projects is the ability to reuse specialised packages that have been previously developed and validated, significantly reducing development times. In the implementation of the control strategy described in this chapter, specialised interface packages are primarily used, providing standardised data structures for handling various types of information. The packages utilised are as follows:

- **geometry_msgs:** It defines messages for representing geometries and spatial transformations in 2D and 3D, such as points, poses, vectors, and quaternions.
- **std_msgs:** It provides basic data types, such as Booleans, integers, and strings, to facilitate simple communication between nodes in ROS.
- **nav_msgs:** It includes messages specific to robotic navigation, such as occupancy maps, trajectories, and odometry data.

These packages simplify communication between nodes and ensure compatibility within more complex architectures.

Up to this point, the two main software tools used in the implementation described in this chapter have been outlined. The following section focuses on the MAVROS package, which serves as the primary bridge between ROS 2 concepts and using ArduPilot on the flight controller.

MAVROS

It is a package designed to facilitate communication with autonomous vehicles, acting as a bridge between the MAVLink protocol and ROS. Due to its broad scope, it provides a variety of commands and topics that developers can use according to the needs of their projects. In the context of this chapter, the topics that give sensor and telemetry data, as well as the services that enable interaction with the vehicle's state and configuration parameters, are described.

It is important to highlight that MAVROS enhances the proposed node structure by connecting different flight controllers with minimal modifications, rendering the hardware layer virtually transparent. This significantly increases the compatibility and flexibility of the proposed structure, simplifying integration with various platforms and accelerating development.

The MAVROS deployment executes a wide range of nodes; however, Fig. 11 summarises the nodes used in this chapter, along with their corresponding topics and services. Specifically, the following nodes will be utilised:

- **/mavros/sys:** This node manages all ArduPilot configuration and parameter management aspects.
- **/mavros/imu:** This node collects and processes information from the IMU.
- **/mavros/cmd:** This node receives and executes commands, such as mode changes and vehicle arming.
- **/mavros/rc:** This node collects and processes information from the remote controller.
- **/mavros/global_position:** This node collects and processes information the GPS provides.

These nodes facilitate the integration and operation of the components required to implement the control strategy, ensuring efficient communication between ROS 2 and ArduPilot. To provide context, the topics shown in Fig. 11 are also described below, highlighting those most relevant to the control strategy implemented in this chapter.

- **/mavros/state:** This topic publishes the current state of ArduPilot, which is useful for monitoring the operating mode of the vehicle and determining whether it is armed (ready to execute missions) or disarmed (on standby). The key information extracted from this message is the armed field, indicating whether the vehicle is in an armed or disarmed state.
- **/mavros/imu/data:** This topic includes inertial data of the vehicle, such as linear accelerations, angular velocities, and orientation in quaternions, provided by the IMU.
- **/mavros/global_position/compass_hdg:** This topic publishes the orientation relative to north, expressed in degrees, measured by the compass.
- **/mavros/global_position/global:** This topic provides the vehicle's global position in GPS coordinates (latitude, longitude, and altitude), along with the GPS status.

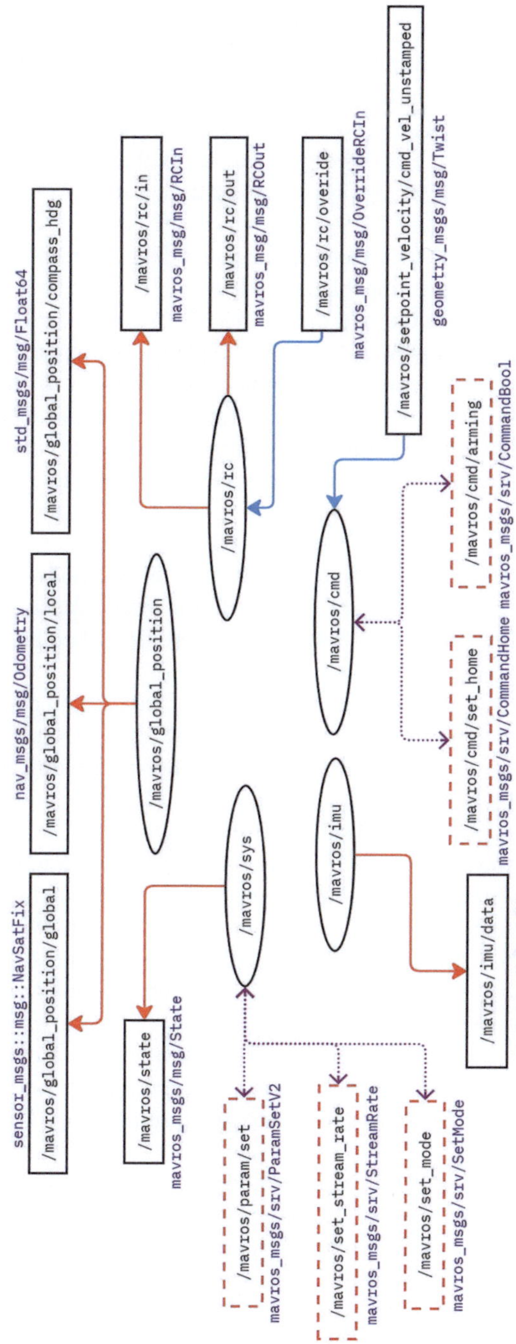

Fig. 11 Diagram of nodes used from the MAVROS package

- **/mavros/global_position/local:** This topic publishes the position in metres within an East, North, Up (ENU) coordinate system relative to the 'Home' reference point previously set in ArduPilot or MAVROS. Additionally, this topic includes the vehicle orientation expressed as a quaternion.
- **/mavros/rc/in:** This topic publishes the input values, expressed in milliseconds, of the radio control channels received from the transmitter on the vehicle.
- **/mavros/rc/out:** This topic publishes the output of the radio control channels, meaning the PWM commands, expressed in milliseconds, sent to each servomotor.
- **/mavros/rc/override:** This topic allows for overriding the radio control inputs, providing an external control mechanism. However, for this to be possible, it is necessary first to configure the parameter `SYSID_MYGCS` in ArduPilot to match the MAVROS system identifier and modify the PWM assignment settings.
- **/mavros/setpoint_velocity/cmd_vel_unstamped:** This topic allows sending velocity commands in the body reference frame within the ENU coordinate system to control the vehicle motion. It is important to note that for ArduPilot to execute these velocity commands correctly, the vehicle must have been previously configured in guided mode.

Similarly, the services enable interaction between ROS 2 nodes and ArduPilot, allowing functions such as parameter changes, flight mode selection, and vehicle arming or disarming. The services shown in Fig. 11 are described below:

- **/mavros/cmd/arming:** This service allows sending an arm and disarm command from the control architecture. The request consists of a Boolean value, which will be *True* when arming is desired and *False* when disarming is desired. The response consists of a Boolean value and an unsigned integer, which will be *True* and 1 if the command is executed successfully.
- **/mavros/cmd/set_home:** This service allows sending a command to set a new 'Home' reference point. The data required for the request include the latitude, longitude, and altitude, along with a Boolean value that should be *False* if the new 'Home' is to be different from the current position. The response consists of a Boolean value and an unsigned integer, which will be *True* and 1 if the command is executed successfully.
- **/mavros/set_stream_rate:** This service allows sending a command to set the publishing frequency of the topics. The data required for the request include the desired frequency and a Boolean value to enable transmission. This service does not include a response.
- **/mavros/param/set:** This is one of the most useful services in the MAVROS package, as it allows modifying any parameter of the running ArduPilot. The request for this service depends on the parameter, value, and type to be changed.
- **/mavros/set_mode:** This service allows sending a command to change the vehicle's operating mode. Currently, only two modes are used: `MANUAL` and `GUIDED`. The request for this service consists of a string with the mode name, and the response is a Boolean value confirming the action.

This information and descriptions give the reader the necessary tools to understand the following implementation sections. However, those interested in developing their own control strategies or leveraging other functionalities offered by the MAVROS package are encouraged to consult the official documentation and explore this tool in depth to adapt it effectively to their specific needs.

Docker

The installation of ROS 2 Humble requires some minimum requirements, including an Ubuntu 22.04-based operating system. This creates an initial incompatibility with the configuration of the Jetson Xavier NX, which uses the JetPack 5.1 operating system based on Ubuntu 20.04. While one solution could be to opt for ROS 2 Foxy, which is compatible with this Ubuntu version, a more robust and beneficial alternative is using Docker to create a container. This approach allows ROS 2 Humble to run in an isolated environment, overcoming the limitations of the base operating system and providing greater flexibility and portability. Figure 12 presents a diagram illustrating the configuration scheme.

Docker is a lightweight virtualisation platform that enables the creation, deployment, and execution of applications within containers. A container is a packaged unit that includes everything necessary for an application to run (Mouat 2016), such as code, libraries, and dependencies, ensuring consistent execution across any environment. In the robotics domain, Docker offers several key advantages:

- **Portability:** It ensures consistent development and deployment environments, regardless of the operating system or hardware used.

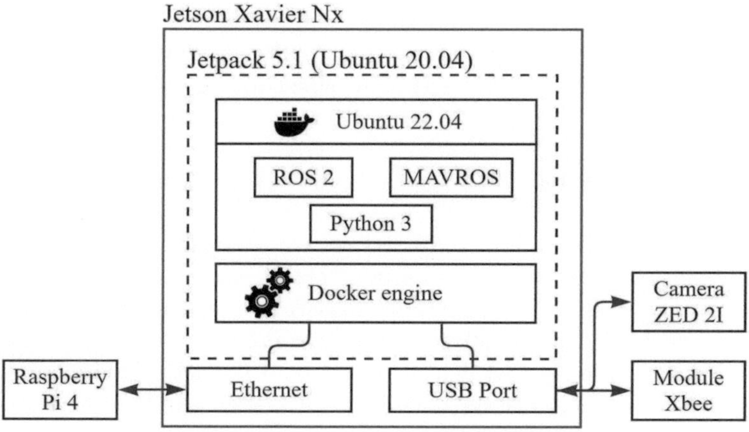

Fig. 12 Integration of JetPack 5.1 with ROS 2 and MAVROS in a Docker environment

- **Isolation:** It enables multiple applications or software versions to run on the same system, avoiding dependency conflicts.
- **Reduced Development Time:** It simplifies the configuration of robotic environments, eliminating issues associated with manual installations and configurations.
- **Scalability:** It is ideal for distributed systems, allowing efficient deployment and management of multiple containers across various devices.

The container running on the Jetson Xavier NX is based on the official Ubuntu 22.04 image, configured with ROS 2 Humble, and includes the necessary dependencies for the project, such as MAVROS and Python. This container was set up to allow Ethernet connectivity, facilitating communication with the Raspberry Pi, and access to Universal Serial Bus (USB) ports for using the XBee module. A dedicated folder was also created to store test data in ROSBAG files, enabling the recording and analysis of experimental data.

Cleaning and Collection Tools

Generally speaking, ASVs can move from waypoint to waypoint using their actuators, but their mission is not always to move but also to accomplish some high-level objective. In the specific case of this platform, the ASVs can, for example, be used as water-cleaning agents, focusing their motion to follow trajectories that can take advantage of collaborative planning to clean large water surfaces using cleaning and collection tools. These tools were selected to provide efficient waste collection. Considering that plastics and hydrocarbons are the two largest waste sources (Pattiaratchi et al. 2022) and that the relatively small vehicles may not be able to utilise the standard industrial tools appropriately, the cleaning tools were designed to collaboratively accomplish cleaning using two ASVs interconnected with the cleaning tool itself. In that sense, the ASVs are considered equipped with cleaning tools that can effectively accomplish the cleaning objective.

Net for Surface Waste

For surface macroplastics, a nylon-based net was designed for waste collection. Nylon presents some characteristics useful in plastic waste gathering and is already used in water environments. The design consists of a rectangular-shaped net with two pulling ropes on either smaller sides, as seen in Fig. 13. Additional descriptive characteristics include the fact that the top side of the net should admit buoyant foams to ensure that the net stays above surface level and that the bottom side includes ballast chains so that the net stays vertically extended.

Considering a first-principle design, the net will produce a drag force F related to the usual parameters as $F = \mathbb{C}_D \mathbb{A} \rho u^2/2$, where $\mathbb{C}_D$ is equal to the typical drag coefficient of a nylon-based net conservatively set to 1.5, $\mathbb{A}$ is the projected surface

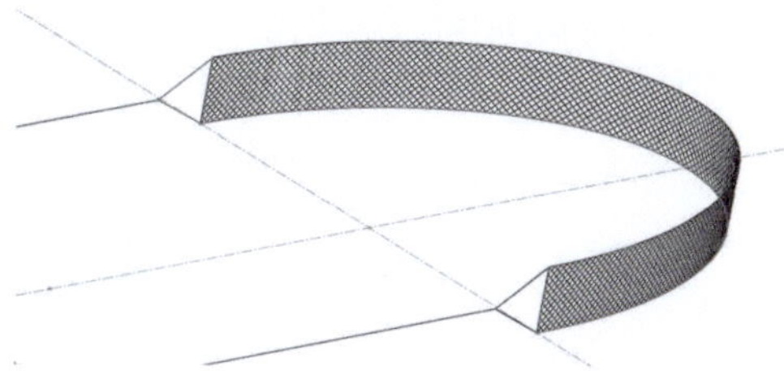

Fig. 13 Design of surface plastic waste collection net

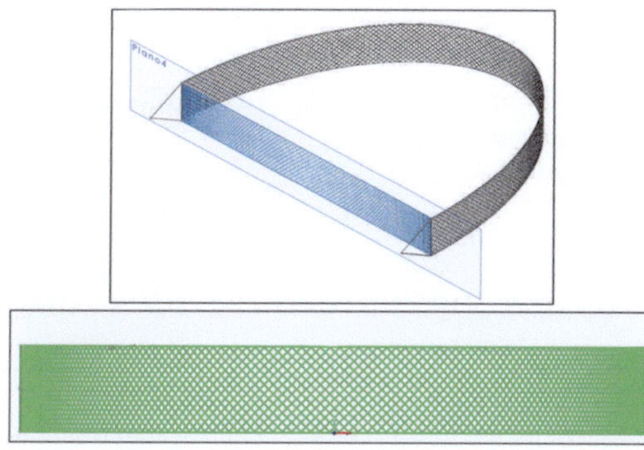

Fig. 14 Example of a projected area when considering a working extended net

Table 3 Maximum drag force considering different mesh sizes without loads

Mesh size (mm × mm)	Net thickness (mm)	Projected area (m^2)	Zero-load max. force (N)
15 × 15	0.2	0.37	142.08
25 × 25	0.3	0.25	96
30 × 30	0.35	0.19	72.96

of the net onto the forward movement orthogonal plane with u equal to the forward velocity, and ρ being the water density. Among the mentioned parameters, only the surface area $\mathbb{A}$ can be modified considering two different net design parameters: net thickness or diameter and mesh size. These two parameters yield the projected surface area. Figure 14 shows the projected area that will be subject to drag forces when towing the nets. It can be observed that with a smaller mesh size, a larger projected area will be obtained when considering similar net thicknesses.

Numerous bollard pull tests have been performed to obtain the nominal maximum force of the Yellowfish ASV, obtaining thrusts ranging from 0 up to a maximum of 80 N. Finally, since two different vehicles will be dragging the net, the total maximum force can be considered twice the maximum value, and from there, three different nets have been designed considering the available net thickness according to different vendors. Table 3 shows the maximum drag force of unloaded nets, i.e. no loads, considering different mesh sizes. Since the three nets produced drag forces within the limits, they were bought for testing. Figure 15 shows one of the nets being utilised by two ASVs, where the coordination of the ASVs is crucial since they must manage to divide the workload efficiently.

Fig. 15 One of the nets being dragged by two ASVs

Fig. 16 Hydrocarbon absorption barrier being carried by two ASVs

With these results, the preferred net is the one with mesh size 25 mm × 25 mm due to the fact that it is a balanced trade-off between small load force and greater waste collection capabilities. Nonetheless, experimental tests were done confirming that a larger drag force was obtained when using smaller mesh-size nets.

Absorption Barriers for Hydrocarbons

On the other side of the water waste spectrum, hydrocarbon waste consists of liquid hydrocarbons that are again subdivided according to their density. Lighter hydrocarbons, such as oil spills, are found on the water surface and can be collected with absorption barriers specifically designed to absorb the mentioned hydrocarbon. Considering the designed nominal drag force of the plastic waste net, two absorption barriers with similar drag forces were purchased. The first barrier considered a pessimistic scenario, with a fully loaded barrier considered for the weight, which was about 165 litres, according to the vendor. This particular barrier consisted of five 5-metre interconnected long barriers with diameters of 20 cm. The second barrier was chosen for smaller spillage of, at most, 96 litres of absorption and a diameter of 15 cm. Figure 16 shows one of the absorption barriers at use. Notably, the absorption barriers are significantly lighter than the nets, and their total surface in contact with water is much smaller, making less hydrodynamic efforts and easier manoeuvrers to tackle the hydrocarbon collection problem.

Modular Control Architecture

This section describes the motion control architecture implemented in the Yellowfish ASV for path following, including the main related layers: Navigation, Kinematic Control, and Dynamic Control. Firstly, details about the dynamic modelling of ASVs and the main variables involved are given. Then, some specific algorithms in each of the mentioned layers are detailed. In particular, in the Navigation Layer, an extended state observer (ESO) (Bejarano et al. 2022) is implemented to estimate the current state variables of the ASV, such as its linear position and orientation, linear and angular velocities, and the so-called lumped generalised disturbances. In the Kinematic Control layer, a surge-guided line-of-sight (LOS) controller (Wang et al. 2019a) is implemented to generate the references of the forward velocity and heading for the low-level or dynamic controller. Finally, the PID speed controller included in ArduPilot is applied in the Dynamic Control layer. Some details about the implementation of the state observer and the kinematic controller in ROS 2 are also given, as well as the required coordination and communication tasks.

The notation described in Table 4 is followed throughout the whole chapter.

Dynamic Modelling of ASVs

The kinematics and kinetics of an ASV moving in a horizontal plane are modelled, according to (Fossen 2011), as in (1):

$$\begin{cases} \dot{\eta} = \mathcal{R}(\psi)\nu, \\ \mathcal{M}\dot{\nu} = -\mathcal{C}(\nu)\nu - \mathcal{D}(\nu)\nu - g(\eta) + \mathcal{T}_w + \mathcal{T}, \end{cases} \tag{1}$$

where $\eta = [x, y, \psi]^T$ is the position and orientation vector on the plane expressed in the Earth-fixed inertial frame $\{n\}$ shown in Fig. 17; $\nu = [u, v, r]^T$ is the vector of the surge, sway (linear), and yaw (angular) velocities, respectively, expressed in the body-fixed frame $\{b\}$; $\mathcal{T} = [F_u, 0, \tau_r]^T$ refers to the force and torque vector, representing the available control actions, also expressed in $\{b\}$; and $\mathcal{T}_w = \left[F_{w,u}, F_{w,v}, \tau_{w,r}\right]^T$ refers to the environmental forces and torque due to wind, waves, and currents.

$\mathcal{M}$ is the inertia matrix, $\mathcal{C}(\nu)$ represents the Coriolis and centrifugal forces matrix, $\mathcal{D}(\nu)$ is the damping matrix, and $g(\eta)$ describes the gravitational and buoyancy forces. The inertia matrix, under manoeuvring assumptions, such as near-constant positive surge and calm waters, and with homogeneous mass distribution and X_b-Z_b-plane symmetry, can be assumed as symmetric and non-diagonal (Fos-

Table 4 Notation summary

Latin symbols		Greek symbols	
$\mathbb{A}$	Area (m^2)	β	Sideslip angle (rad)
$\mathcal{A}$	Linear acceleration ($m\ s^{-2}$)	γ	Path angle (rad)
A	State matrix in a state-space model	Δ	Look-ahead distance (m)
a	Identified scalar parameter (p.u.)	δ	Normalised PWM signal (p.u.)
B	Input matrix in a state-space model	η	Position and orientation vector
b	Identified scalar parameter (p.u.)	ν	Linear and angular velocity vector
$\mathbb{C}$	Scalar coefficient (p.u.)	ρ	Density ($kg\ m^{-3}$)
C	Coriolis matrix	$\boldsymbol{\sigma}$	Lumped generalised
C	Output matrix in a state-space model		disturbance vector
c	Identified scalar parameter (p.u.)	σ	Lumped generalised
$\mathcal{D}$	Damping matrix		disturbance ($m\ s^{-2}$ or $rad\ s^{-2}$)
D	PID derivative gain	τ	Torque (N m)
d	Identified scalar parameter (p.u.)	ψ	Heading angle (rad)
FF	PID feedforward gain	ω	Path variable (p.u.)
F	Force (N)		
g	Gravitational and buoyancy vector	Subscripts	
I	Identity matrix	$\{b\}$	related to the body frame
IG	Input gain ($m\ s^{-2}$ or $rad\ s^{-2}$)	D	related to drag
I	PID integral gain	d	desired
k	Control parameter (s^{-1})	e	error
L	Observer gain	$\{F\}$	related to the Frenet-Serret frame
$\mathcal{M}$	Inertia matrix	LOS	line-of-sight
m	Term of the inertia matrix	L	left
$\mathcal{N}$	Measurement noise vector	max	maximum
n	Measurement noise signal (m or rad)	min	minimum
O	Origin of a given frame	$\{n\}$	related to the inertial frame
$\mathbb{P}$	Propeller PWM signal	P	related to the desired path
$\mathcal{P}$	Matrix solution of the LMI problem	$path$	related to the two-dimensional path
P	PID proportional gain	p	related to linear variables
$\mathcal{R}$	Rotation matrix	R	right
r	Yaw velocity ($rad\ s^{-1}$)	ref	reference
$\mathcal{T}$	Generalised torque vector	r	related to yaw
T	Auxiliary observer matrix	tar	related to target
U	Total linear velocity ($m\ s^{-1}$)	u	related to surge
u	Surge velocity ($m\ s^{-1}$)	v	related to sway
v	Sway velocity ($m\ s^{-1}$)	w	environmental
$\mathcal{W}$	Matrix solution of the LMI problem	ψ	related to rotational variables
$\mathcal{X}$	State vector in a state-space model		
X	X-axis of a given frame		
x	North position (m)		
$\mathcal{Y}$	Output vector in a state-space model		
Y	Y-axis of a given frame		
y	East position (m)		

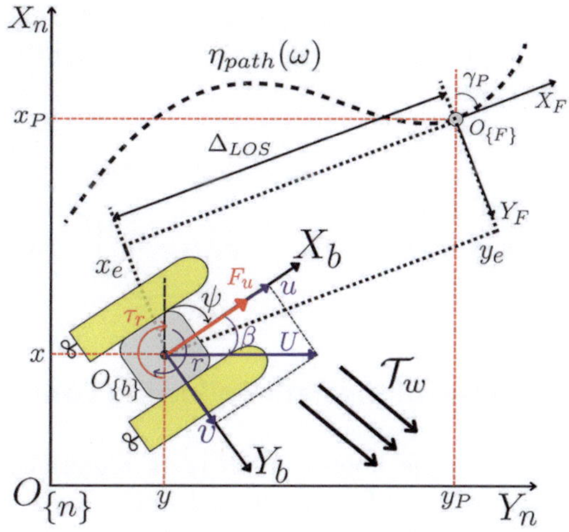

Fig. 17 Path-following control geometry and relevant coordinate frames

sen 2011), as defined in (2), causing the control actions of force and torque to affect the velocities in the three modelled degrees of freedom.

$$\mathcal{M} = \begin{bmatrix} m_{11} & 0 & 0 \\ 0 & m_{22} & m_{23} \\ 0 & m_{23} & m_{33} \end{bmatrix}. \tag{2}$$

The rotational matrix $\mathcal{R}(\psi)$ between the reference frames $\{b\}$ and $\{n\}$ is given by (3):

$$\mathcal{R}(\psi) = \begin{bmatrix} \cos(\psi) & -\sin(\psi) & 0 \\ \sin(\psi) & \cos(\psi) & 0 \\ 0 & 0 & 1 \end{bmatrix} = \begin{bmatrix} \mathcal{R}_2(\psi) & 0 \\ 0 & 1 \end{bmatrix}, \tag{3}$$

where $\mathcal{R}_2(\psi) \in \mathbb{R}^{2\times 2}$ is the upper left submatrix of $\mathcal{R}(\psi)$.

From (1), the lumped generalised disturbance vector $\sigma \equiv [\sigma_u, \sigma_v, \sigma_r]^T$ is defined in (4), grouping hydrodynamic effects, unmodelled internal dynamics, uncertainties, and external disturbances into a single term:

$$\sigma \equiv \mathcal{M}^{-1}\left[-C(\nu)\nu - \mathcal{D}(\nu)\nu - g(\eta) + \mathcal{T}_w\right]. \tag{4}$$

According to (4), the dynamics of the ASV can be rewritten as in (5):

$$\begin{cases} \dot{\eta} = \mathcal{R}(\psi)\nu, \\ \dot{\nu} = \mathcal{M}^{-1}\mathcal{T} + \sigma. \end{cases} \tag{5}$$

The ASV is assumed to be equipped with at least a GPS and a gyrocompass, which provide potentially noisy measurements $\mathcal{Y}_{meas}$ of the ASV position and orientation:

$$\mathcal{Y}_{meas} = \eta + \mathcal{N}, \tag{6}$$

where $\mathcal{N} = \left[n_x, n_y, n_\psi\right]^T = \left[\mathcal{N}_p^T, n_\psi\right]^T$ refers to the noise affecting the measured variables.

Overview of the Path-Following Control Strategy

The complete control strategy is represented in Fig. 18, where the estimated positions $\hat{x}$ and $\hat{y}$, orientation $\hat{\psi}$, and sway $\hat{v}$, outputs of the *State Recovery* block, are used as inputs for the surge-guided LOS kinematic controller implemented in the *Kinematic Control* block. Additionally, the PID speed control integrated into ArduPilot is used as the low-level or dynamic controller. Since the surge-guided LOS law only generates the references for the surge u_{ref} and orientation ψ_{ref}, and the ArduPilot low-level controller only receives references for the angular velocity, a first-order derivative filter is implemented to obtain the reference of the angular velocity r_{ref}. Furthermore, the ArduPilot low-level controller outputs are the PWM signals corresponding to the left and right propellers, $\mathbb{P}_L$ and $\mathbb{P}_R$, respectively.

Notice in Fig. 18 that the minimum desired surge $u_{d,min}$ is also an input for the surge-guided LOS controller, while the desired path $\eta_{path} = [x_P, y_P]^T$, parametrised through the so-called path variable ω, must be known in advance.

Non-linear State Observer

The objective of the *State Recovery* block in Fig. 18 is the estimation of the ASV state variables. In particular, the ESO implemented in the Yellowfish ASV is the one proposed in (Bejarano et al. 2022), which uses the previously presented model. One of the main features of this observer is that it only uses position (GPS) and orientation (compass) measurements to estimate the velocities and lumped generalised disturbances. As previously stated, the latter include hydrodynamic effects, unmodelled internal dynamics, uncertainties, and external disturbances, enabling a robust approach in an experimental validation test. However, this ESO considers only noise in the positional measurements.

This observer features a cascaded structure divided into two parts. The first part estimates the non-linear positional dynamics, as shown in (7), while the second part estimates the non-linear rotational dynamics, as described in (8). The state vector for the positional part is defined as $\mathcal{X}_p \equiv [\eta_p, \nu_p, \sigma_p]^T$, where $\eta_p \equiv [x, y]^T$, $\nu_p \equiv [u, v]^T$, and $\sigma_p \equiv [\sigma_u, \sigma_v]^T$. On the other hand, the state vector for the rotational part is defined as $\mathcal{X}_\psi \equiv [\psi, r, \sigma_r]^T$. It is important to note

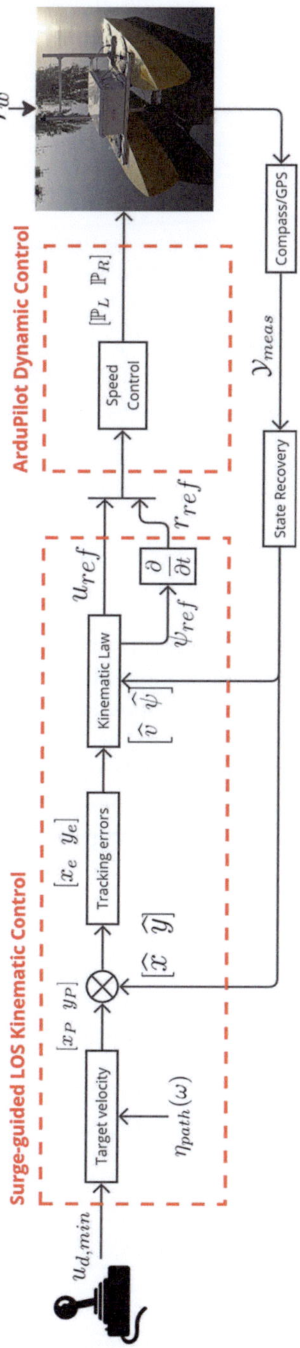

Fig. 18 Path-following complete control strategy

that σ_p and σ_r represent the generalised disturbances affecting linear and rotational accelerations, respectively, while $\mathcal{Y}_{meas,p}$ and $\mathcal{Y}_{meas,\psi}$ denote sensor position and orientation measurements, with only the former being considered affected by noise:

$$\dot{\hat{\mathcal{X}}}_p = A_p(\psi)\hat{\mathcal{X}}_p + B_p\mathcal{T} + L_p(\psi)(\mathcal{Y}_{meas,p} - C_p\hat{\mathcal{X}}_p). \tag{7}$$

$$\dot{\hat{\mathcal{X}}}_\psi = A_\psi\hat{\mathcal{X}}_\psi + B_\psi\mathcal{T} + L_\psi(\mathcal{Y}_{meas,\psi} - C_\psi\hat{\mathcal{X}}_\psi). \tag{8}$$

Matrices A_ψ, $A_p(\psi)$, C_ψ, and C_p can be obtained through direct manipulation of (5)–(6) (Bejarano et al. 2022). The positional and rotational measurement $\mathcal{Y}_{meas,p} \equiv \eta_p + \mathcal{N}_p$ and $\mathcal{Y}_{meas,\psi} \equiv \psi + n_\psi$ can be obtained by linear manipulation of $\mathcal{Y}_{meas}$ in (6). Meanwhile, B_ψ and B_p are matrices dependent on the terms of the inertia matrix $\mathcal{M}$, as shown in (9), where $\mathcal{M}_p^{-1}$ corresponds to the first two rows of the inverse of the inertia matrix, and $\mathcal{M}_\psi^{-1}$ corresponds to the last row:

$$B_p = \begin{bmatrix} 0_{2\times3} \\ \mathcal{M}_p^{-1} \\ 0_{2\times3} \end{bmatrix}, \quad B_\psi = \begin{bmatrix} 0_{1\times3} \\ \mathcal{M}_\psi^{-1} \\ 0_{1\times3} \end{bmatrix}. \tag{9}$$

However, since the values of the inertia matrix of the Yellowfish ASV are not available, an identification of the input gain $IG \equiv [IG_u, IG_v, IG_r]^T$ for each degree of freedom was performed. The input gain can be defined as the impact of the PWM signals on velocity change and was used in the observer to replace the unknown terms of B_p and B_ψ into global expressions of the whole terms $B_p\mathcal{T}$ and $B_\psi\mathcal{T}$ as functions of $\bar{\delta}$ and $\Delta\delta$, corresponding to the mean δ and difference $\Delta\delta$ of the normalised PWM signals from both propellers δ_L and δ_R, respectively, as shown in (10):

$$\begin{aligned} \bar{\delta} &= \frac{\delta_L + \delta_R}{2}, \\ \Delta\delta &= \delta_L - \delta_R, \\ B_p\mathcal{T} &= \begin{bmatrix} 0_{2\times1} \\ IG_u(\bar{\delta}, \Delta\delta) \\ IG_v(\bar{\delta}, \Delta\delta) \\ 0_{2\times1} \end{bmatrix}, \\ B_\psi\mathcal{T} &= \begin{bmatrix} 0 \\ IG_r(\bar{\delta}, \Delta\delta) \\ 0 \end{bmatrix}. \end{aligned} \tag{10}$$

The identification of the input gain was computed using the normalised PWM signals from the left and right thrusters as inputs, assuming that they are identical, and applying a quadratic static function as shown in (11), where $q \in \{u, v, r\}$, and a_q, b_q, c_q, and d_q are identified scalar parameters. The keen reader is referred to (Morel et al. 2025) for additional details about the identification of the actuation dynamics:

$$
\begin{aligned}
IG_q(k) = a_q \left(\bar{\delta}(k)^2 + \frac{\Delta\delta(k)^2}{4} \right) + b_q \bar{\delta}(k) \Delta\delta(k) \\
+ c_q \bar{\delta}(k) + d_q \frac{\Delta\delta(k)}{2}.
\end{aligned} \tag{11}
$$

Finally, the calculation of the constant rotational observer gain L_ψ is performed in (12), where $\mathcal{P}_\psi$ and $\mathcal{W}_\psi$ are part of the solution to an optimisation problem based on linear matrix inequalities (LMI), with one tuning parameter. Similarly, the calculation of the time-varying positional observer gain $L_p(\psi)$ is defined by (13), where the constant matrices $\mathcal{P}_p$ and $\mathcal{W}_p$ are part of the solution to another LMI-based optimisation problem, with the designer having three tuning parameters, and $T_p(\psi)$ is computed as in (14), considering $\mathcal{I}_2$ as the two-dimensional identity matrix. The keen reader is referred to the work (Bejarano et al. 2022) for additional details concerning the observer design and tuning:

$$
L_\psi = -\mathcal{P}_\psi^{-1} \mathcal{W}_\psi. \tag{12}
$$

$$
L_p(\psi) = -T_p(\psi)^{-1} \mathcal{P}_p^{-1} \mathcal{W}_p \mathcal{R}_2^T(\psi). \tag{13}
$$

$$
T_p(\psi) = \begin{bmatrix} \mathcal{R}_2^T(\psi) & 0 & 0 \\ 0 & \mathcal{I}_2 & 0 \\ 0 & 0 & \mathcal{I}_2 \end{bmatrix}. \tag{14}
$$

Surge-Guided LOS Kinematic Controller

The desired curvilinear path η_{path} (see Fig. 17) is considered to be parametrised by a time-dependent path variable ω, whose dynamics are also intended to be defined by the kinematic controller. The Frenet-Serret frame {F} is related to the virtual target $[x_P, y_P]^T$, which is a moving point on the path. Then, the axes of {F} are normal and tangent to the desired path at the virtual target, while {F} is rotated with angle γ_P relative to {n}, defined by (15):

$$
\gamma_P = \text{atan2}(\dot{y}_P, \dot{x}_P) = \text{atan2}(\partial y_P / \partial \omega, \partial x_P / \partial \omega), \tag{15}
$$

where $\text{atan2}(y, x)$ is the four-quadrant version of the trigonometric function $\arctan(y/x) \in [-\pi/2, \pi/2]$. The virtual target velocity u_{tar} is computed as in (16):

$$u_{\text{tar}} = \sqrt{\dot{x}_P^2 + \dot{y}_P^2} = \dot{\omega}\sqrt{(\partial x_P/\partial\omega)^2 + (\partial y_P/\partial\omega)^2}. \qquad (16)$$

The cross-track error y_e and the along-track error x_e are defined with respect to {F} as in (17):

$$\begin{bmatrix} x_e \\ y_e \end{bmatrix} = \begin{bmatrix} \cos(\gamma_P) & -\sin(\gamma_P) \\ \sin(\gamma_P) & \cos(\gamma_P) \end{bmatrix}^T \begin{bmatrix} x - x_P \\ y - y_P \end{bmatrix}. \qquad (17)$$

The surge-guided LOS (SGLOS) kinematic controller proposed by Wang et al. (2019a) is implemented, which aims to drive the along- and cross-track errors to zero and is divided into two parts. The surge reference is computed as in (18):

$$u_{ref} = k_u\sqrt{y_e^2 + \Delta_{LOS}^2}, \qquad (18)$$

where $k_u = u_{d,min}/\Delta_{LOS}$, with $u_{d,min}$ being the desired speed upon reaching the trajectory. The orientation law is governed by (19):

$$\psi_{ref} = \gamma_P - \beta_{ref} - \arctan\left(\frac{y_e}{\Delta_{LOS}}\right), \qquad (19)$$

where $\Delta_{LOS} \in \mathbb{R} > 0$ is the so-called look-ahead distance and $\beta_{ref} = \text{atan}(v/u_{ref})$. The first term γ_P follows the desired path curvature, the second one β_{ref} compensates the potential sideslip caused by disturbances and currents, while the last one is responsible for driving the cross-track error y_e to zero. Finally, the velocity of the virtual target u_{tar} is computed as in (20):

$$u_{\text{tar}} = k_{\text{tar}} x_e + U_{ref}\cos(\psi - \gamma_P + \beta_{ref}), \qquad (20)$$

where $k_{\text{tar}} \in \mathbb{R} > 0$ is a tuning parameter and $U_{ref} = \sqrt{u_{ref}^2 + v^2}$ is the total reference linear velocity. The velocity of the virtual target u_{tar} allows computing the dynamics of the path variable ω from (16).

PID Speed Control

One of the main advantages of using ArduPilot is the inclusion of a combination of two proportional-integral-derivative (PID) controllers: one for forward speed and another for turning speed. These controllers can be tuned experimentally and do not require an accurate mathematical model of the vehicle for their use. These features

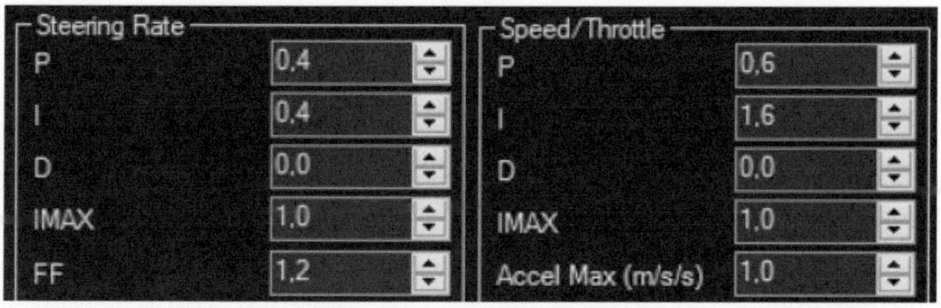

Fig. 19 Tuned parameters of the combined PID speed controller

give beginners an initial approach to speed control and understanding its main challenges. In addition, these controllers can be integrated with other controllers, reducing development times.

This controller was used in the control strategy described in section "Overview of the Path-Following Control Strategy" and tuned following the developers' guidelines. The calibration process was carried out through a trial-and-error approach, based on multiple repetitions and manual verifications of the system behaviour in order to assess the performance of each adjustment. First, the turning speed controller was adjusted, which has four parameters: the feedforward gain FF and the three PID gains P, I, and D. Then, the forward speed controller was tuned. This controller only includes the three PID gains P, I, and D; however, an additional parameter, the maximum acceleration $\mathcal{A}_{max}$, was also required for its configuration.

It is important to emphasise that proper tuning of this controller is crucial to determining how the vehicle responds to speed commands and its ability to adapt to changes in currents or operating conditions. Furthermore, it directly impacts performance during autonomous missions or other preconfigured flight modes in ArduPilot. Within the scope of this chapter, only the parameters shown in Fig. 19 were configured and tuned. The platform's official documentation is recommended for those interested in delving deeper into ArduPilot's autonomous control for more complex missions.

Implementation in ROS 2

Once the control strategy and its mathematical foundations have been explained, this subsection presents the motion control implementation in ROS 2 Humble using a node-based architecture. It is important to note that this subsection's primary objective is to describe each node's functionality and its interaction with topics and services without delving into specific code details. The aim is to provide the reader with sufficient tools to translate a theoretical control strategy into a modular, node-based architecture that facilitates development and scalability. Furthermore, this approach avoids biasing the implementation towards a specific programming language, allowing the reader the freedom to apply these concepts using the language they are most comfortable with.

This subsection addresses the project structure and explains the created packages and their purpose. Subsequently, each node is individually analysed, detailing the most important considerations during its programming and operation. Finally, a ROS 2 diagram is presented, grouping all nodes and illustrating their interaction within the proposed architecture.

The project consists of seven packages, each designed with a specific purpose to assign clear functions, improve code readability, and maintain an organised structure. Below is a brief description of the purpose and content of each package:

- **ASV_interfaces:** This package groups the custom interfaces created for the architecture, including the interface to publishing estimated states.
- **ASV_yellowfish:** This package contains the vehicle descriptions in Unified Robot Description Format (URDF) files and primarily aims to enable data visualisation in the RVIZ2 tool.
- **ASV_library:** This package stores classes and functions defined in C++ that are used by various packages, including common elements such as parametrised curve programming.
- **ASV_communication:** This package includes nodes related to the communication module, the data logging node, and those intended for reading new sensors or peripherals integrated into the vehicle.
- **ASV_observer:** This package contains nodes related to navigation, including the state observer previously described.
- **ASV_Control:** This package groups nodes related to motion control, including the implemented path-following algorithm.
- **ASV_bringup:** This package centralises the configuration required for node operation, ensuring consistency and interaction between packages. It contains launch files that specify each node's parameters and startup sequences, managing their dependencies and specific requirements.

This modular structure simplifies the project's development, debugging, and expansion, ensuring efficient integration of its components. Below, the nodes where the control and estimation algorithms were implemented are described, along with additional nodes related to coordination and communications.

ESO Node

The `obs_bejarano` node is part of the `asv_observer` package and is responsible for estimating velocities and disturbances that feedback into the control loop. As explained in section "Non-linear State Observer", this algorithm uses position, orientation, and PWM signals. Figure 20 presents a diagram of the node, along with its subscriptions and publications. The diagram shows that the node subscribes to three topics: local position, RC output channels with the PWMs of each motor, and the state of ArduPilot. The first two topics are used in each iteration to calculate the states, while the third acts as an indicator to start or stop the execution. This means

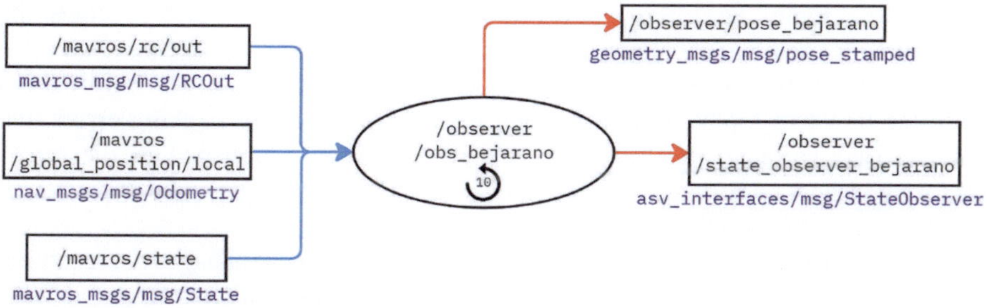

Fig. 20 Diagram of the ESO node

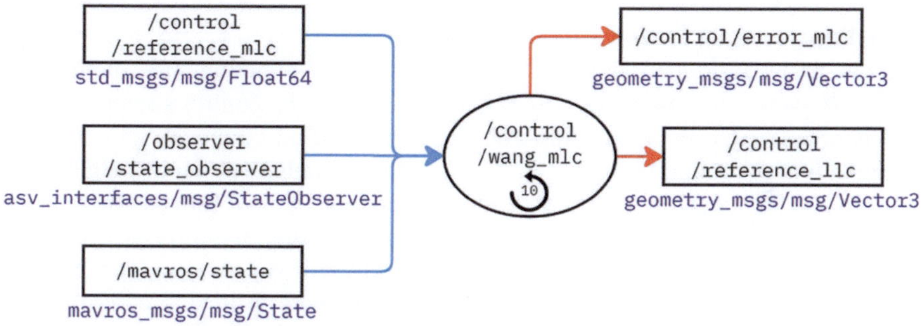

Fig. 21 Diagram of the kinematic control node

that the node only performs an estimation when the vehicle is armed, resetting all variables when disarmed.

Two execution threads were implemented in this node to optimise performance. The first thread handles the reading and reception of data from the subscribed topics, while the second focuses on state estimation. This second thread uses a timer configured to a 10 Hz execution frequency. Upon activation, it retrieves the most recent sensor data, estimates the states, and publishes them to a main topic named `/observer/state_observer_bejarano`. This topic contains nine data points organised into three vectors, each with three elements: the first vector corresponds to positions and orientation, the second includes linear and angular velocities, and the third represents disturbances in the three modelled degrees of freedom.

Additionally, a second topic, `/observer/pose_bejarano`, was added specifically for visualising the estimated position and orientation in the RVIZ2 software, facilitating the interpretation and analysis of the results.

Kinematic Control Node

The `wang_mlc` node, part of the `asv_control` package, generates velocity references in the body reference frame within a North-East-Down (NED) coordinate system. This node implements the algorithm described in section "Surge-Guided LOS Kinematic Controller" and has three inputs corresponding to the subscribed topics, as illustrated in Fig. 21.

The first topic, `reference_mlc`, is a floating point type that specifies the minimum desired speed along the path. The second topic corresponds to the vehicle's estimated states, and the third to the ArduPilot state. The information from the first two topics is used in each iteration to compute tracking errors, the parametrised variable, and generate velocity references published in the topics `error_mlc` and `reference_llc`, respectively. The third topic indicates whether to start or stop the node's execution.

This node runs with two execution threads. The first thread manages the reception of data from the subscribed topics, while the second uses a timer configured to 10 Hz to perform calculations in each iteration. This structure ensures computational efficiency and maintains a precise execution frequency.

Coordination and Communication Nodes

The two nodes presented above implement the main estimation and control algorithms. However, to ensure the suitable functioning of the control strategy in ROS 2, it is necessary to develop additional nodes focused on communication and test coordination. Although these nodes do not directly implement control algorithms, they are crucial in enabling them to operate effectively. In the following, these nodes are described along with their main functions and features.

The `apm_llc` node, part of the `asv_communication` package, is responsible for reading the velocity references generated by the path-following controller and transforming them into the ENU coordinate system. This ensures that references are published to the appropriate MAVROS topic. The main goal of this node is to ensure consistency between the information sent to the speed control and the proposed reference framework. Figure 22 illustrates the connection scheme of this node, along with the topics to which it is connected.

The `mux_obs` node, part of the `asv_observer` package, was designed to provide scalability to the proposed strategy. Although this chapter addresses a single state observer, this node-based architecture aims to enable the implementation of multiple estimators, giving users the option to select which one to use for each test. Additionally, this multiplexer allows for the simultaneous execution of several estimators, enabling direct comparison within the same test.

Figure 23 shows the connection diagram of this node, including its subscriptions and publications. Among the subscriptions is the topic containing the estimated

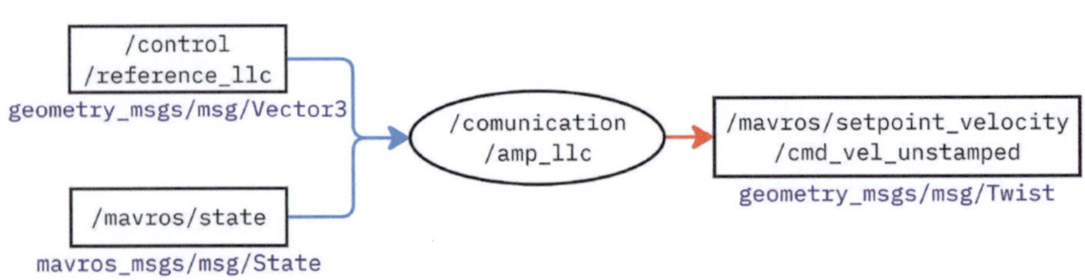

Fig. 22 Diagram of the speed command node for ArduPilot

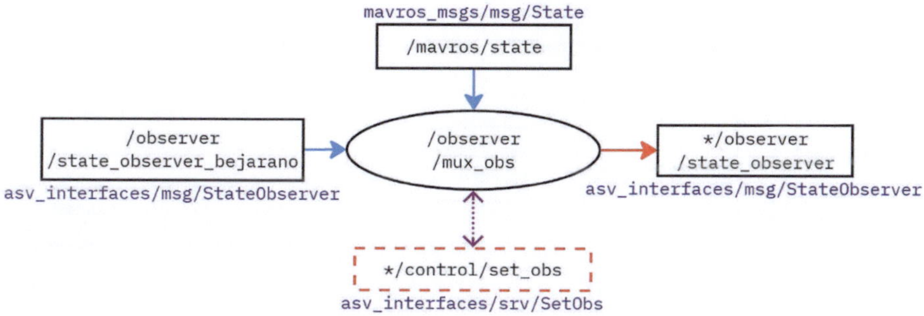

Fig. 23 Diagram of the estimated state multiplexer node

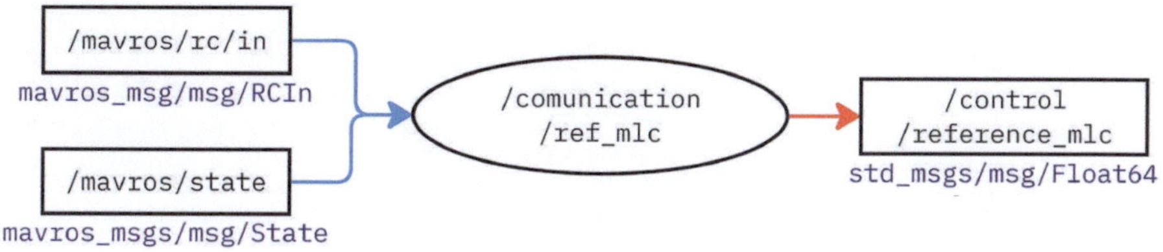

Fig. 24 Diagram of the reference generator node for path-following control

states, published by the estimation node and the ArduPilot state. The node, in turn, publishes the data to a single topic, to which the nodes implementing the control algorithms subscribe. Finally, the `/set_obs` service allows configuring which data received from the subscribers will be published on the output topic, providing flexibility in the selection of estimated data.

It is important to highlight that this structure facilitates the addition of new nodes intended for data estimation. However, another alternative is using the lifecycle concept included in ROS 2, which allows managing which node is active at a given time. Although this option offers more dynamic control, it requires a higher level of programming management. Readers interested in this concept are encouraged to explore it further if they find it useful for their needs.

The `ref_mlc` node, part of the `asv_communication` package, is responsible for generating the minimum velocity reference for the path-following control. As shown in Fig. 24, this node subscribes to the `rc/in` topic, retrieving the current state of the remote control joysticks. Using this information, it calculates the desired minimum velocity within the range of 0.3–1.0 m s^{-1}. Subsequently, it publishes this velocity reference at the same sampling rate as the data received from the RC controller. The main objective of this node is to allow the user to control this velocity via remote control during tests.

The `rc_handler` node, part of the `asv_communication` package, plays a fundamental role in the architecture and the operation of the control strategy. Its purpose is to monitor the state of the remote control channels by subscribing to the `rc/in` topic, to identify when the user intends to initiate a test. Once identified, the

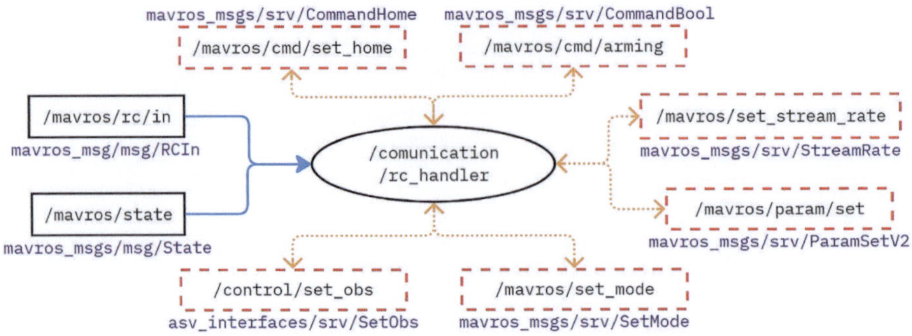

Fig. 25 Diagram of the coordination node for configuration verification

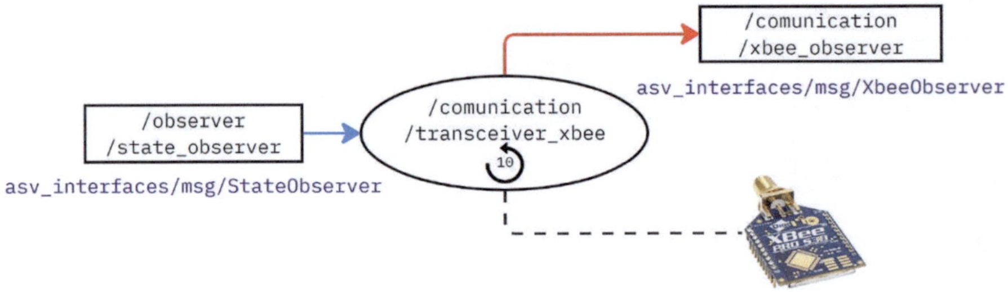

Fig. 26 Diagram of the XBee module transceiver node

node prepares and configures the necessary communications and options between the nodes by requesting various services. Figure 25 presents the connection diagram of this node, highlighting the use of services.

The configurations performed by this node include verifying the topic publication rate using the `set_stream_rate` service, checking the origin coordinates using the `set_home` service, switching to `GUIDED` mode through the `set_mode` service, and arming or disarming the vehicle via the `arming` service. These services enable the node to adjust and verify all the necessary conditions for the proper operation of other nodes. Additionally, this node has access to the `param/set` service, which allows it to modify ArduPilot parameters if needed, and the `set_obs` service, used to configure the estimated state multiplexer.

The `transceiver_xbee` node, part of the `asv_communication` package, does not play a direct role in the current control strategy but has been implemented to verify the functionality of the XBee modules. This node subscribes to the estimated state topic and transmits this information to all its neighbours whenever new data are received. It also functions as a receiver, obtaining information from other vehicles and publishing it to the `xbee_observer` topic at 10 Hz. This node is expected to be more critical in future coordinated control strategies. Figure 26 illustrates the connection diagram for this node.

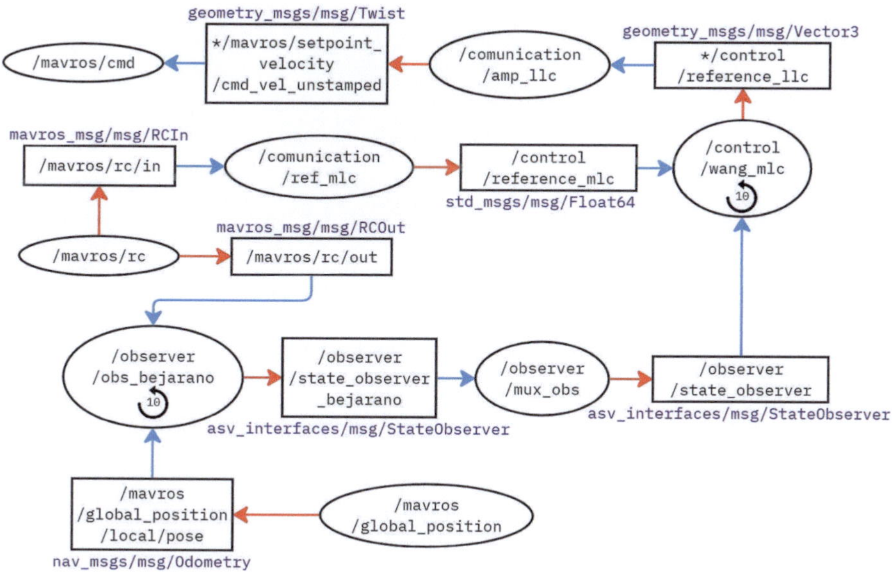

Fig. 27 General diagram with all topics and nodes involved in the control flow

To conclude this subsection, Fig. 27 presents a summarised diagram that includes all the nodes and topics involved in the control strategy. This diagram groups the previously described nodes and the MAVROS nodes responsible for publishing sensor and actuator information. Its purpose is to illustrate the integration of the different nodes and show how information flows through each, highlighting the interaction between components to achieve coordinated and efficient operation. This diagram is part of the repository [1]

Experiments and Results

This section presents the experimental results obtained with a Yellowfish ASV at the *Lago Mayor*, located in *Parque del Alamillo*, Seville, Spain. The tests were carried out in the morning under the influence of disturbances caused by wind and currents. GPS data were transformed into a local reference frame, using the point with coordinates latitude 37° 25′ 8.97″ N and longitude 5° 59′ 56.37″ W as the origin or home. These tests validated the implementation and confirmed the correct operation of the control strategy by completing a circuit composed of several path segments. Below are some of the most relevant results from these tests.

[1] https://github.com/manuelgantiva/ASV_Loyola_US_low_level/releases/tag/Version

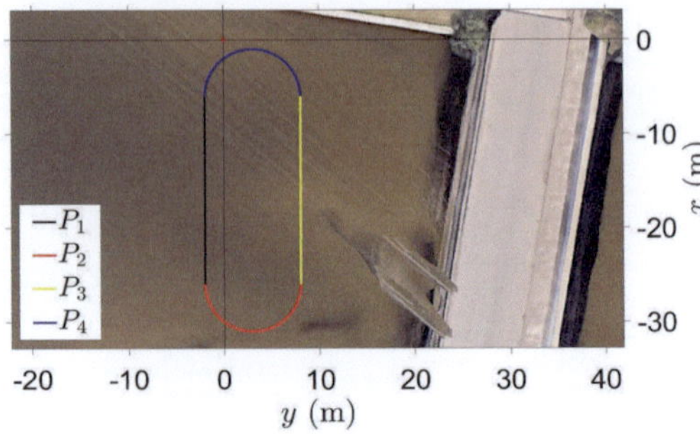

Fig. 28 Closed-loop reference circuit consisting of four paths

Table 5 Parametrised equations for each segment of the test circuit

Path	$\eta_{path}(\omega)$	ω_{min}	ω_{max}
P_1	$y_P(\omega) = -2.0$	0.0	4.0
	$x_P(\omega) = -6.0 - 5.0\omega$		
P_2	$y_P(\omega) = 3.0 + 5.0\cos(\omega - 0.85841)$	4.0	7.14158
	$x_P(\omega) = -26.0 + 5.0\sin(\omega - 0.85841)$		
P_3	$y_P(\omega) = 8.0$	7.14158	11.14159
	$x_P(\omega) = -26.0 + 5.0(\omega - 7.14158)$		
P_4	$y_P(\omega) = 3.0 + 5.0\cos(\omega - 11.14159)$	11.14159	14.28318
	$x_P(\omega) = -6.0 + 5.0\sin(\omega - 11.14159)$		

The circuit used, shown in Fig. 28, consists of four parametrised paths designed to ensure spatial continuity and validate the integrity of the control strategy. Table 5 details the parametrised equations corresponding to each of these paths, along with the intervals of the parametrised variable ω for each segment, where ω_{min} represents the lower limit and ω_{max} is the upper limit.

In summary, the circuit comprises two straight segments, each 20-metre long, and two semicircles with a 5-metre radius, resulting in a total circuit length of 71.42 m. To ensure circuit continuity, the parametrised variable ω was programmatically rescaled to the same interval upon completing the route, allowing multiple laps within a single test and thereby facilitating the development and analysis of the trials.

The exercise consisted of three complete laps around the circuit. During the first two laps, the minimum desired surge $u_{d,min} = 0.5$ m s^{-1} was configured, while for the final lap, the value was increased to $u_{d,min} = 1$ m s^{-1}. This setup allowed observation of how the vehicle, starting from a position off the path, approached it and how a change in reference speed could impact its performance in the same circuit.

The initial results detailed below correspond to the first lap of the circuit. Figure 29 presents the performance of the control strategy, showing the lake map alongside the desired circuit and the actual path followed by the vehicle. To facilitate understanding of this figure, an image of the ASV is included, depicting its position and orientation every 20 seconds during the same lap.

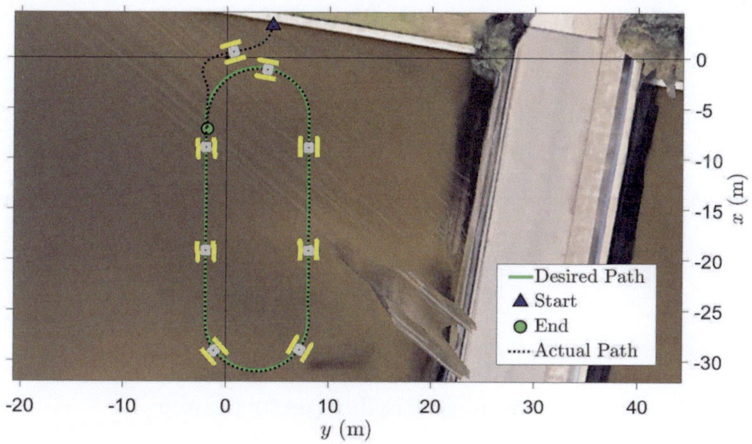

Fig. 29 Performance of the motion control strategy for path following at $u_{d,min} = 0.5$ m s^{-1}

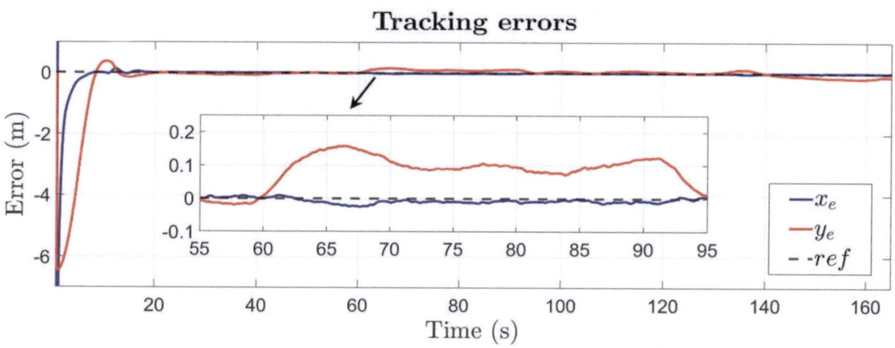

Fig. 30 Tracking errors at $u_{d,min} = 0.5$ m s^{-1}

On the other hand, Fig. 30 shows the evolution of the tracking errors during this first lap. A significant initial peak is observed, caused by the vehicle's starting position off the desired path. However, these errors rapidly decrease and tend to zero within approximately 15 seconds.

This and the following figures include a zoomed-in view of the time interval between 55 and 95 seconds, corresponding to the section where the vehicle navigates the first curve. In this zoomed-in view, a peak in the cross-track error of approximately 0.18 m is identified, caused by the change in orientation during the curve. Subsequently, the controller is observed to correct this error, reducing it again to a value close to zero.

Similarly, Figs. 31, 32, and 33 illustrate the performance of the ArduPilot speed controller in surge, yaw, and heading, respectively. During the initial moments of the exercise, the path-following control saturates the surge reference to reach the path as quickly as possible. However, once the path is reached, the references converge

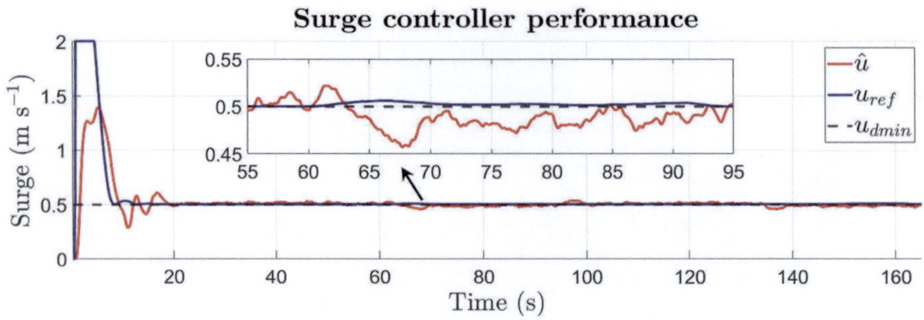

Fig. 31 Surge control performance along the circuit at $u_{d,min} = 0.5$ m s^{-1}

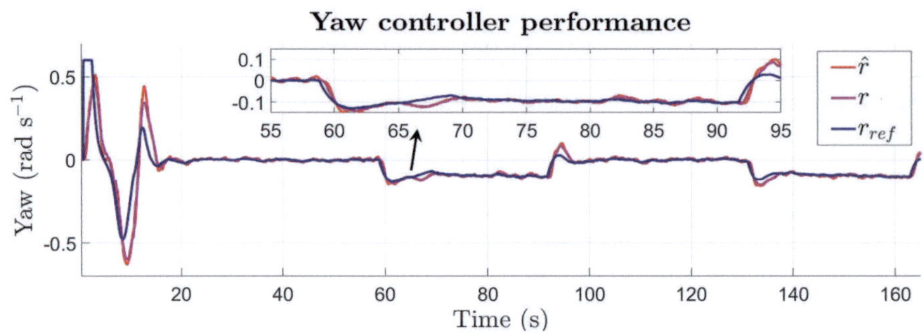

Fig. 32 Yaw control performance along the circuit at $u_{d,min} = 0.5$ m s^{-1}

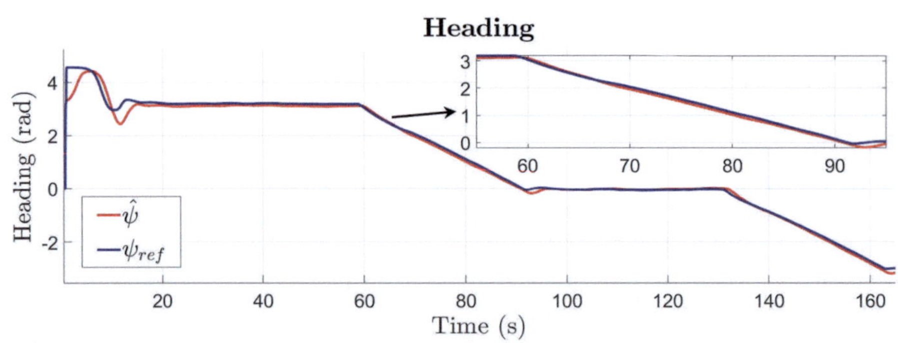

Fig. 33 Evolution of the heading along the circuit at $u_{d,min} = 0.5$ m s^{-1}

towards a value close to the desired speed along the trajectory. A similar behaviour is observed with the yaw reference, which initially saturates to achieve the desired orientation, stabilising at a value close to zero when following the straight-line segment.

Analysing the behaviour of these variables during the curved segment within the specified time window, it is observed that the surge reference remains stable.

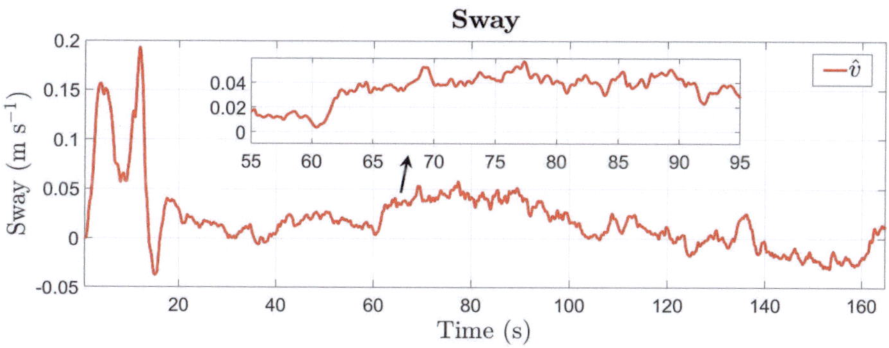

Fig. 34 Evolution of the sway along the circuit at $u_{d,min} = 0.5$ m s^{-1}

At the same time, yaw increases to adjust the trajectory, generating a response in orientation. Although this response shows a slight delay with respect to the reference, it is sufficient to keep the cross-track error close to zero. Additionally, it is noted that the variation in yaw slightly affects the surge, an effect caused by the coupling between control actions.

Another variable that, while not controlled, is important to analyse is the sway. The latter is primarily caused by external disturbances or changes in yaw when a positive surge is present. Figure 34 shows the evolution of this variable during the first lap of the circuit. It can be observed that the initial saturation of the references generates a significant increase in sway, which has a considerable impact in the early moments. However, once the vehicle reaches the straight-line segment, sway stabilises at a value close to zero, showing significant increases only during the curved segments. These increases mainly contribute to the tracking errors, as they tend to deviate the vehicle from the curve.

One of the main features of the implemented extended state observer is its ability to estimate generalised disturbances in the three modelled degrees of freedom. Although this information has not been included in the control loop, the exercise allowed for evaluating its performance and validating its potential for future integrations. Figure 35 shows the evolution of these estimates throughout the first lap of the circuit.

To conclude this section, the strategy's performance is presented on the same circuit but with two main differences: the vehicle starts from a point within the circuit, and the desired minimum speed is increased. Figure 36 shows the lake map along with the desired circuit and the actual path followed by the vehicle during this lap. In addition, a representation of the ASV is included, illustrating its position and orientation every 20 seconds along the route.

In general, the strategy responded well to this change, completing the circuit without deviating from the desired path and achieving similar results. However, the most significant impact of this change is observed in the curved segments, where in the previous exercise, a peak cross-track error of 0.18 m was recorded, while in this

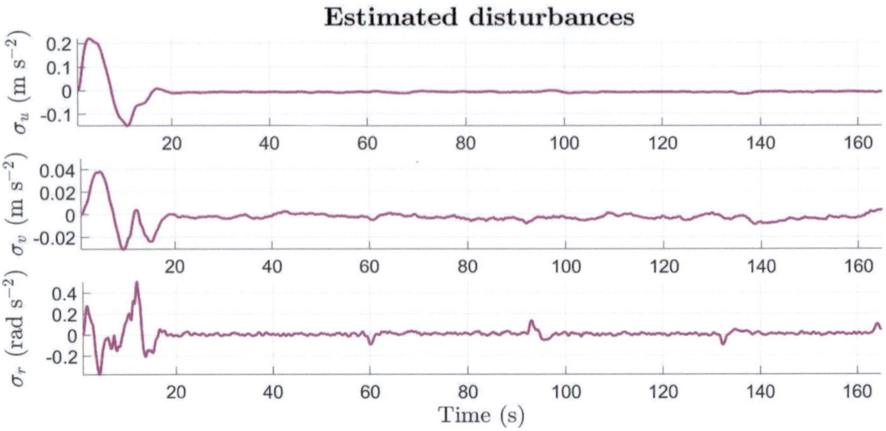

Fig. 35 Evolution of the estimated lumped generalised disturbances along the circuit at $u_{d,min} = 0.5$ m s^{-1}

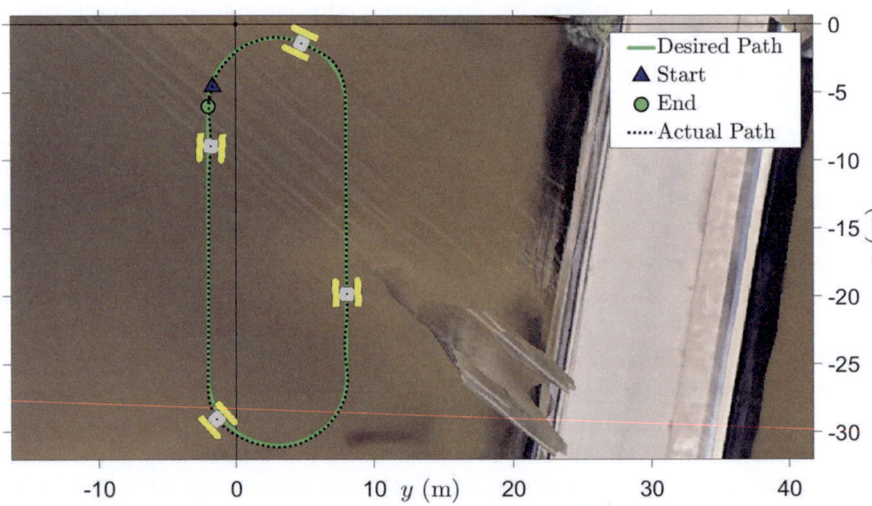

Fig. 36 Performance of the motion control strategy for path following at $u_{d,min} = 1$ m s^{-1}

exercise, the peak increased to 0.32 m, as shown in Fig. 37. This increase is primarily attributed to the sway effect, which is amplified due to the vehicle dynamics and the higher desired forward speed.

Conclusions and Advice for Future Developments

This chapter has addressed the motion control of an autonomous surface vehicle by implementing a path-following control strategy. Starting with the basic concepts

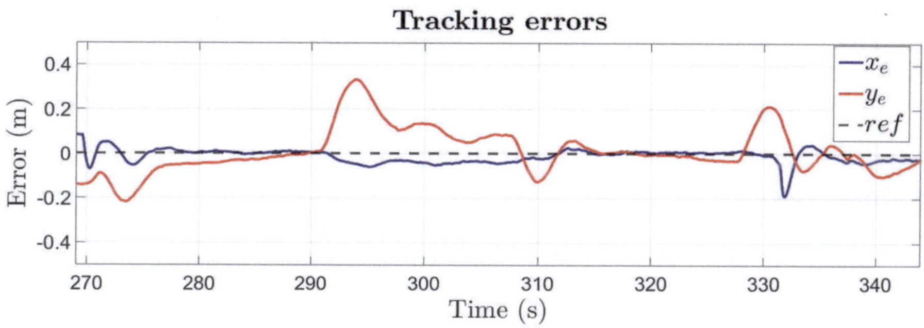

Fig. 37 Tracking errors at $u_{d,min} = 1$ m s^{-1}

of control and the unique characteristics of this type of vehicle, the reader has been provided with an introduction to understanding the essential elements for development. Subsequently, the Yellowfish ASV, a surface catamaran developed by the research group to which the authors belong, has been described, enabling the programming and experimental testing of various algorithms. Equipped with multiple sensors, this vehicle facilitates the experimental validation of control strategies. Additionally, the description of two cleaning tools that can be used in tandem by a pair of these vehicles has been included, broadening their range of practical applications.

An introduction to the open-source software used in this implementation has also been presented, highlighting its main features, capabilities, and the tools employed. Following the introduction of these tools, the mathematical foundations of the strategy have been outlined, including the path-following algorithm and the extended state observer, along with their features and assumptions. Finally, its implementation in ROS 2 has been explained, adopting a modular node-based structure.

The results of this chapter include the evaluation of the control strategy's performance in experimental tests conducted in an uncontrolled environment under unknown external disturbances, thereby validating its robustness and practical applicability. Additionally, it demonstrates the capability to implement theoretical proposals developed and validated in simulation on an actual vehicle, enabling their subsequent experimental deployment. This chapter guides the reader through each stage of implementing a control algorithm, highlighting the main challenges in reducing the transitional gap between simulation and experimentation in real-world environments.

Acknowledgments Special thanks are extended to Javier de Loma-Ossorio, Girish Naresh, Luciano Villarreal, Simone Viviani, Patrizio Ricciarelli, Sajjad Fakharshahreza, Ana Vera Guerrero, Laura de las Mercedes Pérez Soto, and Alexandre Isenbrandt for their valuable contributions to the development of this platform and their assistance in all the tests performed with it. Their expertise, dedication, and collaborative efforts have been fundamental to the progress of the work presented in this chapter on the control of the Yellowfish ASV.

Competing Interests This work was supported by the Agencia Estatal de Investigación (AEI) through Project AQUATRONIC under Grant PID2021-126921OA-C22, through Project ECO-PORT under Grant TED2021-131326A-C22.

The authors have no conflicts of interest to declare that are relevant to the content of this chapter.

References

Anderson A, Martin JG, Mougin J, Bouraqadi N, Duviella E, Etienne L, Fabresse L, Langueh K, Lozenguez G, Alary C, Billon G, Superville PJ, Maestre JM (2022) Water quality map extraction from field measurements targeting robotic simulations. IFAC-PapersOnLine 55(5):1–6. https://doi.org/10.1016/j.ifacol.2022.07.630

ArduPilot (2025) Ardupilot—open source autopilot. https://ardupilot.org/ardupilot/index.html. Accessed 1 Feb 2025

Arrichiello F, Heidarsson HK, Chiaverini S, Sukhatme GS (2011) Cooperative caging and transport using autonomous aquatic surface vehicles. Intell Service Robot 5(1):73–87. https://doi.org/10.1007/s11370-011-0104-9

Bai X, Li B, Xu X, Xiao Y (2022) A review of current research and advances in unmanned surface vehicles. J Marine Sci Appl 21(2):47–58. https://doi.org/10.1007/s11804-022-00276-9

Bejarano G, Sufiyan NY, Orihuela L (2022) Velocity and disturbance robust nonlinear estimator for autonomous surface vehicles equipped with position sensors. IEEE Trans Control Syst Technol 30(5):2235–2242. https://doi.org/10.1109/TCST.2021.3136334

Blue Robotics (2025) Blueboat. https://bluerobotics.com/store/boat/blueboat/blueboat/. Accessed 1 Feb 2025

Castano-Londono L, Marrugo Llorente SP, Paipa-Sanabria E, Orozco-Lopez MB, Fuentes Montaña DI, Gonzalez Montoya D (2024) Evolution of algorithms and applications for unmanned surface vehicles in the context of small craft: a systematic review. Appl Sci 14(21):9693. https://doi.org/10.3390/app14219693

CHCNAV (2025) Apache 3—chcnav geospatial. https://geospatial.chcnav.com/es/products/chcnav-Apache-3. Accessed 1 Feb 2025

Chiang HTL, Tapia L (2018) COLREG-RRT: an RRT-based COLREGS-compliant motion planner for surface vehicle navigation. IEEE Robot Autom Lett 3(3):2024–2031. https://doi.org/10.1109/lra.2018.2801881

Del-Rio-Rivera F, Ramirez-Rivera VM, Donaire A, Ferguson J (2020) Robust trajectory tracking control for fully actuated marine surface vehicle. IEEE Access 8:223897–223904. https://doi.org/10.1109/access.2020.3042091

El-Sheimy N, Youssef A (2020) Inertial sensors technologies for navigation applications: state of the art and future trends. Satellite Navigation 1(2). https://doi.org/10.1186/s43020-019-0001-5

Emlid (2025) Navio2—configuring raspberry pi. https://docs.emlid.com/navio2/configuring-raspberry-pi. Accessed 1 Feb 2025

Eugene CZ, Lim JJ, Nirmal U, Lau STW (2019) Battery powered RC boats: a review of its developments for various applications. Curr J Appl Sci Technol 33:1–29. https://doi.org/10.9734/cjast/2019/v33i530098

EvoLogics (2025) Sonobot 5 system. https://www.evologics.com/sonobot-5-system. Accessed 1 Feb 2025

Fossen TI (2011) Handbook of marine craft hydrodynamics and motion control. Wiley, London. https://doi.org/10.1002/9781119994138

Gantiva Osorio M, Bejarano G, Millán Gata P (2024a) Validación de ley de guiado para vehículos autónomos de superficie. Revista Iberoamericana de Automática e Informática Industrial. https://doi.org/10.4995/riai.2024.21949

Gantiva Osorio M, Ierardi C, Jurado Flores I, Pereira Martín M, Millán Gata P (2024b) Coordinated control of multiple autonomous surface vehicles: challenges and advances—a systematic review. Ocean Eng 312:119160. https://doi.org/10.1016/j.oceaneng.2024.119160

Ghazali MHM, Satar MHA, Rahiman W (2024) Unmanned surface vehicles: from a hull design perspective. Ocean Eng 312:118977. https://doi.org/10.1016/j.oceaneng.2024.118977
Gu N, Wang D, Peng Z, Wang J, Han QL (2023) Advances in line-of-sight guidance for path following of autonomous marine vehicles: an overview. IEEE Trans Syst Man Cybern Syst 53(1):12–28. https://doi.org/10.1109/tsmc.2022.3162862
Han J, Xiong J, He Y, Gu F, Li D (2017) Nonlinear modeling for a water-jet propulsion USV: an experimental study. IEEE Trans Ind Electron 64(4):3348–3358. https://doi.org/10.1109/tie.2016.2598322
Kathen MJT, Flores IJ, Reina DG (2021) An informative path planner for a swarm of ASVs based on an enhanced PSO with Gaussian surrogate model components intended for water monitoring applications. Electronics 10(13):1605. https://doi.org/10.3390/electronics10131605
Leong PY, Ahmad NS (2024) LIDAR-based obstacle avoidance with autonomous vehicles: a comprehensive review. IEEE Access 12:164248–164261. https://doi.org/10.1109/access.2024.3493238
Li C, Jiang J, Duan F, Liu W, Wang X, Bu L, Sun Z, Yang G (2019) Modeling and experimental testing of an unmanned surface vehicle with rudderless double thrusters. Sensors 19(9):2051. https://doi.org/10.3390/s19092051
Liu Z, Zhang Y, Yu X, Yuan C (2016) Unmanned surface vehicles: an overview of developments and challenges. Ann. Rev. Control 41:71–93. https://doi.org/10.1016/j.arcontrol.2016.04.018
Luo H, Jia S, Chen S, Ma L (2024) Embodied intelligence in unmanned surface vehicles: current applications and future perspectives. In: International conference on algorithms, high performance computing, artificial intelligence (AHPCAI), vol 13403. https://doi.org/10.1117/12.3051585
Macenski S, Foote T, Gerkey B, Lalancette C, Woodall W (2022) Robot operating system 2: design, architecture, and uses in the wild, vol 7. https://doi.org/10.1126/scirobotics.abm6074
Majid MHA, Arshad MR (2016) Design of an autonomous surface vehicle (ASV) for swarming application. In: 2016 IEEE/OES autonomous underwater vehicles (AUV), pp 230–235. https://doi.org/10.1109/auv.2016.7778676
MAVLink (2025) Mavlink communication protocol. https://mavlink.io/en/. Accessed 1 Feb 2025
Morel T, Orihuela L, Combastel C, Bejarano G (2025) Practical identification approach for the actuation dynamics of autonomous surface vehicles with minimal instrumentation. Ocean Eng 319:120098. https://doi.org/10.1016/j.oceaneng.2024.120098
Mouat A (2016) Using docker: developing and deploying software with containers, 1st ed. O'Reilly, Beijing
NX Innovation (2025) Nx100 pro. https://www.nxinnovation.com/nx100pro. Accessed 1 Feb 2025
OceanAlpha (2025) Sl20 unmanned surface vehicle. https://oceanalpha.com/product-item/sl20/. Accessed 1 Feb 2025
Paliotta C, Lefeber E, Pettersen KY, Pinto J, Costa M, de Sousa JTFB (2019) Trajectory tracking and path following for underactuated marine vehicles. IEEE Trans Control Syst Technol 27(4):1423–1437. https://doi.org/10.1109/tcst.2018.2834518
Pattiaratchi C, van der Mheen M, Schlundt C, Narayanaswamy BE, Sura A, Hajbane S, Wijeratne S (2022) Plastics in the Indian Ocean–sources, transport, distribution, and impacts. Ocean Sci 18(1):1–28. https://doi.org/10.5194/os-18-1-2022
Peng Z, Liu L, Wang J (2021a) Output-feedback flocking control of multiple autonomous surface vehicles based on data-driven adaptive extended state observers. IEEE Trans Cybern 51(9):4611–4622. https://doi.org/10.1109/tcyb.2020.3009992
Peng Z, Wang J, Wang D, Han QL (2021b) An overview of recent advances in coordinated control of multiple autonomous surface vehicles. IEEE Trans Ind Inform 17(2):732–745. https://doi.org/10.1109/tii.2020.3004343
PX4 (2025) Px4 autopilot software overview. https://px4.io/software/software-overview/. Accessed 1 Feb 2025
Qiao Y, Yin J, Wang W, Duarte F, Yang J, Ratti C (2023) Survey of deep learning for autonomous surface vehicles in marine environments. IEEE Trans Intell Transp Syst 24(4):3678–3701. https://doi.org/10.1109/tits.2023.3235911

Renard E (2024) ROS 2 for beginners, 1st edn. Packt Publishing, Birmingham

Rico FM (2022) A concise introduction to robot programming with ROS2. Chapman and Hall/CRC. https://doi.org/10.1201/9781003289623

Ringback R, Wei J, Erstorp ES, Kuttenkeuler J, Johansen TA, Johansson KH (2021) Multi-agent formation tracking for autonomous surface vehicles. IEEE Trans Control Syst Technol 29(6):2287–2298. https://doi.org/10.1109/tcst.2020.3035476

Sarda EI, Qu H, Bertaska IR, von Ellenrieder KD (2016) Station-keeping control of an unmanned surface vehicle exposed to current and wind disturbances. Ocean Eng 127:305–324. https://doi.org/10.1016/j.oceaneng.2016.09.037

Seafloor Systems (2025a) Hydrone. https://www.seafloorsystems.com/hydrone. Accessed 1 Feb 2025

Seafloor Systems (2025b) Tridrone. https://www.seafloorsystems.com/tridrone. Accessed 1 Feb 2025

Souissi S, Boukattaya M, Damak T, Nejim S (2020) Adaptive control for fully-actuated autonomous surface vehicle with uncertain model and unknown ocean currents. Ocean Eng 217:108147. https://doi.org/10.1016/j.oceaneng.2020.108147

Tabish N, Chaur-Luh T (2024) Maritime autonomous surface ships: a review of cybersecurity challenges, countermeasures, and future perspectives. IEEE Access 12:17114–17136. https://doi.org/10.1109/access.2024.3357082

Tanakitkorn K (2019) A review of unmanned surface vehicle development. Maritime Technol Res. https://doi.org/10.33175/mtr.2019.140730

Teledyne Marine (2025) Z-boat 1250. http://teledynemarine.com/brands/oceanscience/z-boat1250. Accessed 1 Feb 2025

Thombre S, Zhao Z, Ramm-Schmidt H, Vallet Garcia JM, Malkamaki T, Nikolskiy S, Hammarberg T, Nuortie H, Bhuiyan MZH, Sarkka S, Lehtola VV (2022) Sensors and AI techniques for situational awareness in autonomous ships: a review. IEEE Trans Intell Transp Syst 23(1):64–83. https://doi.org/10.1109/tits.2020.3023957

Universidad Loyola Andalucía (2025) Optimization and control of distributed systems. http://www.uloyola.es/oferta-cientifica/grupos-de-investigacion/optimization-and-control-of-distributed-systems. Accessed 5 Feb 2025

Vagale A, Oucheikh R, Bye RT, Osen OL, Fossen TI (2021) Path planning and collision avoidance for autonomous surface vehicles I: a review. J Marine Sci Technol 26(4):1292–1306. https://doi.org/10.1007/s00773-020-00787-6

Vitor AMJ, Granada R, Maidana RG, Jurak DA, Heck G, Negreiros APF, dos Santos DH, Gonçalves LMG, Amory AM (2019) A survey on unmanned surface vehicles for disaster robotics: main challenges and directions. Sensors 19(3):702. https://doi.org/10.3390/s19030702

Wang N, Sun Z, Yin J, Zou Z, Su SF (2019a) Fuzzy unknown observer-based robust adaptive path following control of underactuated surface vehicles subject to multiple unknowns. Ocean Eng 176:57–64. https://doi.org/10.1016/j.oceaneng.2019.02.017

Wang W, Gheneti B, Mateos LA, Duarte F, Ratti C, Rus D (2019b) Roboat: an autonomous surface vehicle for urban waterways. In: 2019 IEEE/RSJ international conference on intelligent robots and systems (IROS), pp 6340–6347. https://doi.org/10.1109/iros40897.2019.8968131

Wu N, Fan Y, Wang R, Wang M, Wu C, Wang Y (2025) Development and field testing of an unmanned surface vehicle system for guidance control and target tracking. Ocean Eng 316:119924. https://doi.org/10.1016/j.oceaneng.2024.119924

Zhang S, Yang S, Xiang X (2019) Formation control of autonomous surface vehicle and experimental validation. IFAC-PapersOnLine 52(24):278–282. https://doi.org/10.1016/j.ifacol.2019.12.421

Zhang X, Wang C, Jiang L, An L, Yang R (2021) Collision-avoidance navigation systems for maritime autonomous surface ships: a state of the art survey. Ocean Eng 235:109380. https://doi.org/10.1016/j.oceaneng.2021.109380

Zhou W, Wang Y, Ahn CK, Cheng J, Chen C (2020) Adaptive fuzzy backstepping-based formation control of unmanned surface vehicles with unknown model nonlinearity and actuator saturation. IEEE Trans Veh Technol 69(12):14749–14764. https://doi.org/10.1109/tvt.2020.3039220

Model-Based Online Planning Strategy for Environmental Disaster Scenarios with Autonomous Vehicles

Samuel Yanes Luis, Sergio Toral Marín, and Daniel Gutiérrez Reina

Contents

Abstract

This chapter explores the use of Gaussian processes (GPs) for modeling physicochemical variables in aquatic environments and the Covariance Matrix Adaptation Evolution Strategy (CMA-ES) for planning sampling trajectories. Gaussian processes allow obtaining accurate estimates and quantifying uncertainty, crucial

S. Yanes Luis (✉) · S. Toral Marín · D. Gutiérrez Reina
Universidad de Sevilla, Sevilla, Spain
e-mail: syanes@us.es; storal@us.es; dgutierrezreina@us.es

D. Gutiérrez et al. (eds.), *Smart Water Quality Monitoring*,
https://doi.org/10.1007/978-3-032-03049-8_12

for real-time decision-making. The definition of uncertainty in Gaussian processes is not data-dependent, which facilitates the implementation of a Receding Horizon algorithm to plan trajectories based on predicted uncertainty. The CMA-ES algorithm optimizes the sampling trajectory, maximizing the information collected. Compared to other algorithms, the CMA-ES proves to be more effective in reducing uncertainty and adapting to the geometry of the search space.

Keywords

Evolutionary strategies · Adaptive informative planning · Smart monitoring · Gaussian processes · Model-based optimization

Introduction

The conservation of hydrological resources, such as rivers, reservoirs, or lakes, is fundamental for the economic and social development of the communities that live from their waters. A fundamental aspect of conservation is the monitoring of the state of water quality. In any body of water, there are physicochemical indicators that can be measured with sensors and that indicate whether the biological status is adequate. Such variables should be measured as frequently as possible to have an up-to-date status of water quality (Baron et al. 2002). However, water sampling at multiple sites is a physically demanding and economically costly task (see Fig. 1 as an illustrative example of such task), as it requires continuous travel to specific areas, often remote or difficult to access. A specific problem of sampling is also the lack of flexibility in the manual planning of sampling points. Teams must collect samples without being able to dynamically adapt to local water conditions, which means that any anomalies, unexpected contamination, or significant variations in water quality go unnoticed until much later. Water samples are collected and analyzed later in the laboratory, which does not allow the sampling to be readapted in real time according to intermediate indicators that allow the importance of a sampling point to be predicted.

Fig. 1 Example of a manual operation of water sampling in a water reservoir. This type of operation is physically demanding and economically costly. The lack of flexibility in the manual planning of sampling points makes it difficult to adapt to local water conditions. Image source: https://dilus.com

In the face of this challenge, it is possible to employ autonomous surface vehicles (ASVs) as an innovative and efficient solution to address the limitations of manual sampling (Sánchez-García et al. 2018). These uncrewed vehicles exist, thanks to the extension of the autonomy of lithium batteries and the advancement in the energy efficiency of embedded systems. For the task of environmental monitoring, it is possible to equip them with physicochemical water quality sensors, so they can measure in real time key parameters such as pH, turbidity, temperature, or the presence of specific pollutants. However, to take full advantage of their potential, it is essential to have a system that accurately models pollution variables, using historical data and current measurements to predict patterns and areas of high sampling interest (Baron et al. 2002). In addition, it is imperative to implement an intelligent navigation system that allows ASVs to dynamically adapt such model of the environment, autonomously guiding them to regions where models indicate the highest probability of anomalies or hazards. With this integration, ASVs not only automate the data collection process but also optimize resources, improve the capacity to respond to environmental emergencies, and promote more efficient and sustainable monitoring.

A clear example of a disaster situation, where these systems are highly useful, is in the problem of cyanobacterial blooms. Cyanobacteria are a type of photosynthetic microorganism closely related to eutrophication, a process caused by excess nutrients such as nitrates and phosphates in water bodies (Álvarez Rogel and al 2020). When eutrophication conditions are favorable, these bacteria experience massive blooms that form dense layers on the water surface, blocking sunlight, lowering dissolved oxygen levels, and releasing toxins that endanger aquatic life and human health (Codd et al. 2020). A dramatic example of this phenomenon can be seen in Mar Menor, in Murcia (Spain) (see Fig. 2), where the influx of nutrients (mainly phosphates and nitrates) from intensive agriculture and other human activities has triggered recurrent episodes of environmental deterioration (Álvarez Rogel and al 2020). These blooms have led to periods of massive fish kills and an ecological crisis that affects not only the biodiversity of the area but also local economic and tourism activities.

In this context, autonomous surface vehicles (ASVs) are presented as extremely useful tools for early monitoring and problem management, as demonstrated throughout this book. Equipped with advanced sensors, these vehicles can detect changes in physicochemical parameters such as nutrient levels, dissolved oxygen, and photosynthetic pigments associated with cyanobacteria, providing an early warning system before blooms reach critical levels. It is not difficult to see that, given a precise model of the physicochemical variables, it can be extrapolated as an indicator of the value of cyanobacteria (Kim et al. 2020). This methodology relies on the ability to measure chemical bioindicators such as conductivity or pH to establish where blooms will appear. Thanks to their ability to move autonomously, ASVs can cover large areas, as in the case of the Mar Menor. This would make it possible to implement preventive measures more effectively, such as reducing nutrient inputs, and significantly improve the capacity to respond to such environmental disasters.

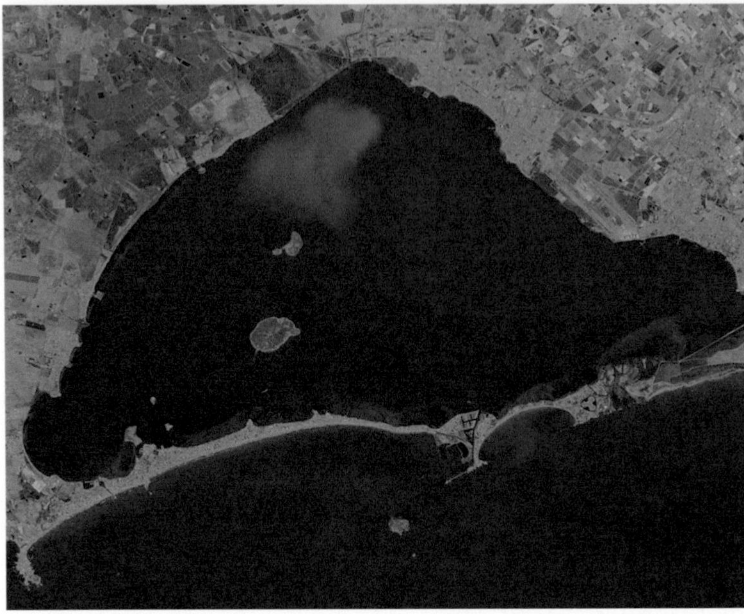

Fig. 2 Sentinel-2 satellite image from Mar Menor (Murcia, España). Cyanobacteria blooms can be seen at first sight in the water surface

To model the water quality status of the resource itself, sampling should be carried out in areas of interest, where bioindicators are anomalous and favor the occurrence of blooms. The vehicle must be able to store the water samples in individual containers for further treatment in the laboratory. However, due to the limited cargo capacity and space in autonomous vehicles, it is necessary that sampling be selective and efficient, maximizing the information collected with a limited number of samples and within a limited time and travel distance.

To ensure that the information collected is sufficient, the intelligent guidance system must be able to plan a sufficiently informative route. Given the impossibility of visiting all points of the hydrological resource in a reasonable time, it is necessary to estimate the distribution of physicochemical variables in the water with a few samples. This model will serve to guide decision-making and optimize the collection of more data, prioritizing sampling in areas of greater uncertainty or of greater hydrological interest. However, modeling physicochemical variables in aquatic environments is a complex challenge due to several reasons. First, these variables often present arbitrary distributions, subject to intrinsic sensor noise and with nonlinear spatial relationships, not known a priori. Secondly, it is not enough to predict the distribution of the variables; it is essential to quantify the uncertainty associated with the predictions to ensure reliable decisions. This implies developing models that, in addition to estimating magnitudes, provide confidence intervals that reflect the inherent variability of the system. Finally, the model must be able to be updated in real time as new sensor data is received, so that decision-making can be optimized as the monitoring mission progresses.

In this chapter, we propose the use of Gaussian processes (GPs) as a tool for modeling physicochemical variables in aquatic environments, based on their proven success in previous works such as Peralta et al. (2021) or Kathen et al. (2023). GPs are particularly well suited for this type of problem for several reasons:

1. Ability to model arbitrary nonlinear relationships by imposing an uncertainty structure called a kernel.
2. GPs not only provide a point prediction of the values of the variables but also generate an explicit estimate of the uncertainty associated with each prediction. This feature allows the identification of areas where data are sparser or where uncertainty is greater.
3. GPs are adaptive, meaning that they can be easily updated in real time as new sensor observations are incorporated. This allows the model to improve its accuracy and reflect dynamic changes in the system for online adaptation of ASV behavior.

On the other hand, to solve the sequential planning problem in sampling, we propose an approach based on planning through optimization by Receding Horizon Mattingley et al. (2011), over the Gaussian model. This proposed strategy is based on sequential decision-making, using the model as a reference for future planning. GPs allow modeling the uncertainty associated with the samples without taking into account the real values of the variables to be estimated. In this way, an optimization algorithm can be used to solve the problem of which points to sample at each step to minimize model uncertainty without having to wait for the actual measurements to be made. Once the trajectory is optimized for a control horizon, the first decision is made and the model is updated with the new information. This process is repeated at each step, allowing dynamic adaptation to changes in the model and real-time decision-making. This planning strategy is called receding horizon online planning (RHO) or model predictive optimization (MPO) and is widely used in control and sequential planning problems. Even when the model (GP) does not perfectly fit the data, making only the first decision of the control horizon ensures that the next step can be replanned with the new information allowing dynamic adaptation to changes in the model.

To solve the scheduling problem, there are numerous algorithms capable of solving optimization tasks: Monte Carlo Tree Search (MCTS)-based scheduling (Świechowski et al. 2023), Evolutionary Strategies-based scheduling (Arzamendia et al. 2019), and greedy scheduling (Krause and Golovin 2014), among others. These algorithms, in one way or another, are able to efficiently explore the decision space, considering both the expected return and the associated risk.

A robust and convergence-guaranteed approach is found in Evolutionary Strategies (ES) (Hansen and Ostermeier 2001). An Evolutionary strategy is an optimization algorithm based on the principles of biological evolution, which uses mutation and recombination operators to explore the solution space. ES are especially suitable for nonlinear and nonconvex optimization problems, where the topology of the search space is unknown or complex. For this chapter, we propose the use of the CMA-ES (Covariance Matrix Adaptation Evolution Strategy) strategy, a variant of Evolutionary Strategies that has been shown to be effective in nonlinear and nonconvex optimization problems (Popovic et al. 2020). The CMA-ES algorithm will be in charge of finding a set of optimal sampling points, given the state of a GP (properly adjusted to the sampled data).

In summary, the contributions of this chapter are the following:

- The use of ASVs equipped with physicochemical sensors for water quality monitoring in aquatic environments is proposed, with the aim of improving the efficiency and sustainability of sampling campaigns.
- The modeling of pollution in aquatic environments is approached using GPs to obtain explicit estimates of the associated uncertainty.
- The use of the CMA-ES strategy for online sampling planning, allowing dynamic adaptation to model changes and real-time decision-making.

This chapter is organized as follows: in section "Statement of the Problem", the problem of water quality monitoring in aquatic environments is presented and the need for an automated approach is justified. In section "Methodology", the proposed methodology is described, including the use of GPs for modeling physicochemical variables and online planning algorithms for decision-making based on CMA-ES. In section "Online Planning Using Receding Horizon", the experiments performed and the results obtained are presented, demonstrating the effectiveness of the proposal. Finally, in section "Experiments and Results", the conclusions are presented and possible lines of future work are discussed.

Statement of the Problem

The path planning problem for information gathering can be formulated as a sequential problem. It starts by defining a graph $G(E, V)$ within the navigable space of the ASV, where $v_i \in V$ represents a vertex of the graph and a location to take a water quality sample. Each edge $e_i \in V \times V$ represents a connection between two vertices and the distance between them. The objective is to find a sequence of vertices $\psi = v_1, v_2, \ldots, v_n$ that maximizes the information $\mathcal{I}(\psi)$ collected, taking into account the uncertainty associated with the model predictions of physicochemical variables. That is, the search for an optimal path ψ^* such that:

$$\psi^* = \arg\max_{\psi} \left[\frac{\mathcal{I}(\psi)}{C(\psi)} \right] \tag{1}$$

where $C(\psi)$ is the cost associated with that path. The agent p has a maximum distance d' that it can travel before returning to the base to recharge batteries or discharge the collected samples. Thus, the main constraint will be:

$$\sum_{i=1}^{n-1} \|v_i - v_{i+1}\|_2 \leq d' \tag{2}$$

Definition of the Information

The amount of information provided by a ψ path can be defined in multiple ways. A common approach is to consider the model entropy as an indicator of the uncertainty associated with the predictions. If we consider that a physicochemical variable y follows a Gaussian distribution of mean 0, the entropy of the variable can be calculated as:

$$H(y) = \frac{1}{2}\log(2\pi e\sigma^2) \tag{3}$$

where σ^2 is the variance of the distribution. The joint entropy of a set of variables $Y = \{y_1, y_2, \ldots, y_n\}$, one for each vertex v_i, can be calculated as the sum of the individual entropy:

$$H(Y) = \sum_{i=1}^{n} H(y_i) \tag{4}$$

A more accurate model, i.e., less random, will lead to lower entropy and, therefore, more information collected. The measure of information collected by a path ψ can be defined as the difference between the initial entropy of the model and the entropy of the model after collecting the samples at vertices v_i:

$$\mathcal{I}(\psi) = H(Y_0) - H(Y_p) \tag{5}$$

where Y_0 is the set of physicochemical variables before collecting samples (the prior), and Y_p is the set of physicochemical variables after collecting samples at vertices v_i (the posterior). See Fig. 3 for a graphical representation of the entropy decrease procedure in the model.

Definition of the Cost

The cost associated with a ψ path can be defined in multiple ways, depending on the constraints and objectives of the problem. In general, one can consider a cost function that penalizes the distance traveled, the time spent, or the amount of energy

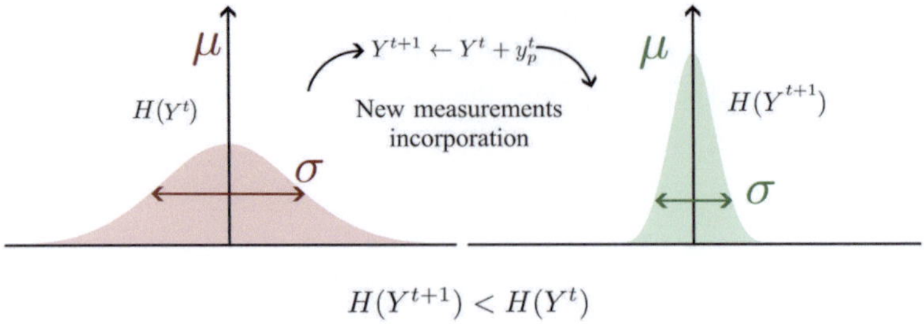

Fig. 3 Entropy decrease procedure in the model. An initial model with high entropy (left) is adjusted as samples are collected at the vertices of the path, decreasing the uncertainty and increasing the information collected (right). When a new observation is incorporated, the entropy conditional on that new observation decreases with respect to the initial entropy

consumed by the ASV. In this work, we consider that each movement between two vertices v_i and v_j has an associated cost equal to the Euclidean distance between them:

$$C(\psi) = \sum_{i=1}^{n-1} \|v_i - v_{i+1}\|_2 \tag{6}$$

If it is imposed that all movements between vertices have the same duration, the cost can be simplified and taken as constant.

Methodology

In this section, the model used for monitoring physicochemical variables in aquatic environments, based on Gaussian processes (GPs), will be described. The concepts of kernel and hyperparameters will be discussed, as well as the dynamic update of the model with new data. Subsequently, the online planning approach based on the CMA-ES algorithm, which allows dynamic adaptation to model changes and real-time decision-making, will be presented.

Gaussian Processes

A Gaussian process is a probabilistic model that describes a distribution over stochastic functions. It is characterized by the fact that any finite subset of points within the domain of the function follows a joint Gaussian distribution, i.e., a multivariate normal distribution (see Fig. 4). This means that each point x of the function to be estimated, $f(x)$, is defined as a Gaussian distribution:

$$f(x) \sim \mathcal{N}(\mu_x, \sigma_x), \quad x \in X \tag{7}$$

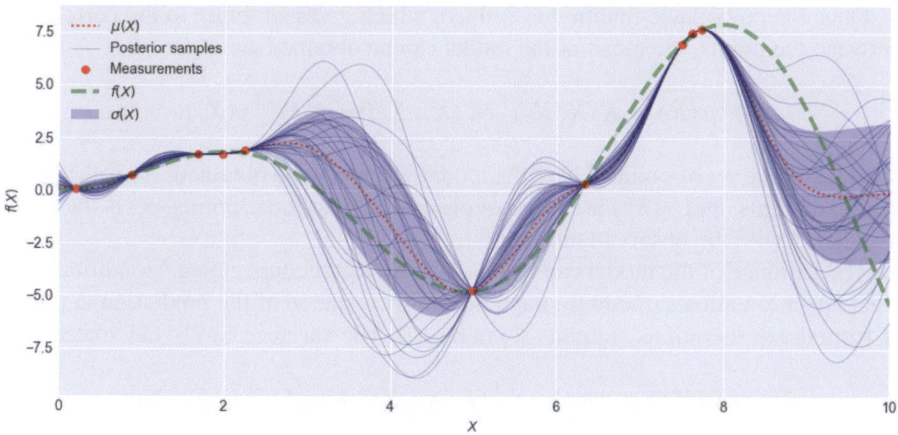

Fig. 4 Graphical representation of a Gaussian process for a one-dimensional function. The red dots represent the samples. The red line represents the mean $\mu(x)$ of the GP. The blue shades represent the uncertainty associated with the model predictions $\pm\sigma(x)$

Formally, a Gaussian process is denoted as follows:

$$f(x) \sim \mathcal{GP}(\mu(x), K(x, x')) \tag{8}$$

where $\mu(x) := [\mu_{x_1}, \ldots, \mu_{x_n}]$ represents the mean at each point of all random functions and $K(x, x')$ is a covariance matrix defining the correlation between the points. $X := [x_1, \ldots, x_n]$ is the set of all possible points at which $f(x)$ is defined.

GPs serve to estimate the function $f(x)$ from a set of observations $y = [y_1, \ldots, y_n]$, where each observation y_i is associated with a point x_i of the domain X through the observation of new data. Each sample y_i is considered a random variable following a Gaussian distribution:

$$y(x) = f(x) + \mathcal{N}(0, \sigma_n^2) \tag{9}$$

where σ_n models the noise associated with the measurement. This noise is assumed to be Gaussian, internal to the sensor, and can be calculated a priori or estimated from the data.

The covariance matrix K can be defined as the pairwise evaluation of a covariance function $k(x, x')$ at all points in the domain X:

$$K = [k(x_i, x_j)]_{i,j=1}^{n} \tag{10}$$

The covariance matrix K is symmetric and positive semidefinite. It represents the correlation between each point in the domain.

Once the covariance function is defined, which gives structure to the correlation between the points, the mean of the model can be obtained as:

$$\mu(X) = K(X, X_s)^T (K(X_s, X_s) + \sigma_n^2 I)^{-1} y(X_s) \tag{11}$$

where X is the set of points where the model value is to be obtained, X_s is the set of sampled points, and $y(X_s)$ is the set of observations at those points. σ_n^2 is the noise associated with the observations.

The variance of the model can be obtained by a procedure called *conditioning*. This operation allows obtaining the a posteriori variance of the prediction at points X from the observations at points X_s (a priori). The variance can be calculated as:

$$\sigma^2(X) = K(X, X) - K(X, X_s)(K(X_s, X_s) + \sigma_n^2 I)^{-1} K(X_s, X) \tag{12}$$

Model variance is a measure of the uncertainty associated with model predictions. The larger the variance, the greater the uncertainty associated with the prediction. The variance calculation does not involve the model mean, which means that the uncertainty can be calculated independently of the point prediction. This property is useful for sample planning, since it allows evaluating the uncertainty associated with the predictions without necessarily visiting any point.

Kernel and Hyperparameter Optimization

The kernel is a function that defines the correlation structure between two points. Imposing a kernel is a key step in sequential monitoring, because it determines how the a posteriori variance of the model behaves as a function of the distance between the points. A kernel function can be selected based on prior knowledge of the properties of the function to be estimated. In the task of monitoring physicochemical variables, a common choice is the radial basis function (RBF) kernel (Peralta et al. 2021). This kernel imposes no prior constraint on the function and allows for arbitrary nonlinear modeling. It is defined as:

$$k(x, x') = \exp\left(-\frac{||x - x'||^2}{2\theta^2}\right) \tag{13}$$

where θ is a hyperparameter that controls the correlation length between points. A small value of θ implies local spatial correlation, while a large value implies broader spatial correlation. This is consistent with the assumption that physicochemical variables follow a smooth spatial correlation: the farther two points are, the less influence they have on each other's estimation.

The value of the hyperparameter θ is not arbitrarily chosen. In GPs, the lengthscale θ directly controls the smoothness of the modeled functions. Determining an appropriate lengthscale is central to capturing either complex patterns or smoother variations in the data. To obtain the optimal value of θ, the marginal likelihood is used. This evaluates the likelihood of the observed data under the GP model, integrating over all possible latent functions $f(x)$:

$$p(y \mid X, \theta) = \int p(y \mid f)\, p(f \mid X, \theta)\, df \tag{14}$$

Maximizing the marginal likelihood provides the value of θ that best fits the data. For a GP with an RBF kernel, the marginal likelihood has a closed-form expression:

$$p(y \mid X, \theta) = \frac{1}{(2\pi)^{n/2} \det(K + \sigma_n^2 I)^{1/2}} \exp\left(-\frac{1}{2} y^\top (K + \sigma_n^2 I)^{-1} y\right) \tag{15}$$

To simplify the optimization, the logarithm of the marginal likelihood is used:

$$\log p(y \mid X, \theta) = -\frac{1}{2} y^\top K^{-1} y - \frac{1}{2} \log |K| - \frac{n}{2} \log(2\pi) \tag{16}$$

This log-likelihood function balances data fit and model complexity. It is maximized numerically using optimization algorithms such as gradient descent or quasi-Newton methods, enabling estimation of an optimal θ that avoids both overfitting and underfitting (Fig. 5).

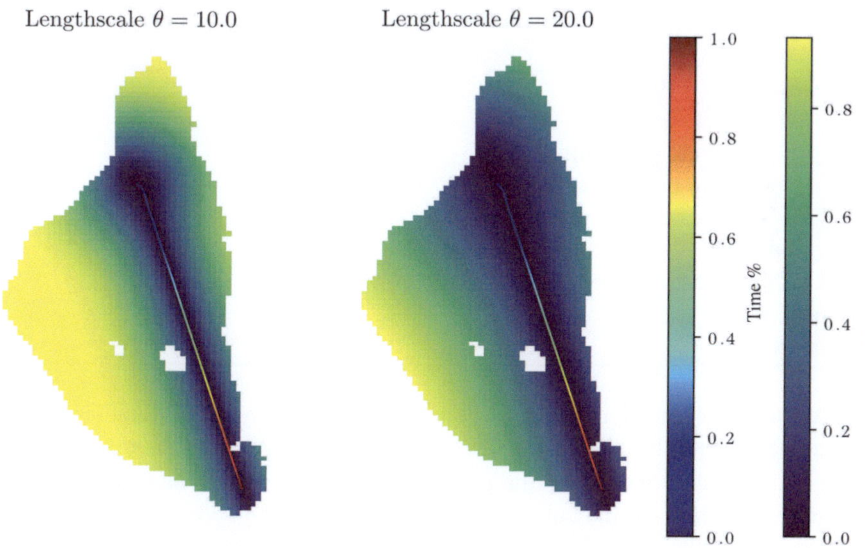

Fig. 5 Representation of two uncertainty maps from GPs with different values of θ for an RBF kernel, using the same data. Left: small lengthscale, implying local spatial correlation. Right: large lengthscale, implying extended spatial correlation

Lengthscale Optimization

In this chapter, we use the stochastic gradient descent (SGD) algorithm to optimize the lengthscale of the RBF kernel. SGD is an iterative optimization method that updates parameters in the direction of the negative gradient to find the minimum of an objective function.

The loss function to optimize the lengthscale is the negative log marginal likelihood:

$$L(\theta) = -\log p(y \mid X_s, \theta) \tag{17}$$

SGD allows approximating the optimal value of the lengthscale without requiring a closed-form derivative of the loss function.[1] Although SGD is stochastic, convergence to the global minimum can be guaranteed by selecting an appropriate learning rate and number of iterations.

Algorithm 1: Stochastic gradient descent for lengthscale optimization

Require: $X, y, \theta_i, \theta_{\min}, \theta_{\max}, \alpha, \epsilon$

```
if θ not initialized then
    Initialize θ ~ U(θ_min, θ_max)
else
    Initialize θ ← θ_i
end if
Initialize Δ ← ∞
while Δ > ε do
    for i = 1, ..., n do
        Compute gradient: ∇_θ ← −(∂/∂θ) log p(y | X_s, θ)
        Update parameter: θ_{k+1} ← θ_k − α∇_θ
        Clip parameter: θ_{k+1} ← max(θ_min, min(θ_max, θ_{k+1}))
    end for
    Compute convergence: Δ ← |θ_{k+1} − θ_k|
end while
return θ
```

In Algorithm 1, the lengthscale is updated sequentially, using gradient steps after each new sample is added to the set X_s. Each new update starts from the most plausible value of θ given the previous sample set $X_s^{(t-1)}$ and moves along the negative gradient for the current X_s^t. The optimization process stops when the change in θ is smaller than a threshold ϵ.

Clipping θ to a bounded range is necessary. To avoid extreme values, the lengthscale is constrained within $[\theta_{\min}, \theta_{\max}]$. A lengthscale greater than the optimum

[1] The gradient of the loss function has a closed analytical expression in GPs. However, the SGD algorithm is used to avoid calculating the inverse of the covariance matrix K.

leads to overconfident predictions—samples appear more correlated than they are—which can cause overfitting. Limiting the upper bound of θ helps to promote exploration of the domain.

A practical guideline is to set the upper bound as a fraction of the maximum expected correlation:

$$\theta_{\text{max}} = \beta_\theta \cdot \max \mathbb{E}[\theta] \tag{18}$$

where β_θ is a value between 0 and 1. A common choice is $\beta_\theta = 0.1$ (Peralta et al. 2023). In any case, θ will be updated dynamically as new samples are incorporated. If the value reaches θ_{max}, it can serve as an indicator to expand the feasible range of values for θ.

Lengthscale Behavior During Sampling

To illustrate the behavior of the lengthscale during the sampling process, a simple example is proposed. We assume a straight-line trajectory from north to south over a map of some water quality variable. This variable behaves smoothly, following a function with local maxima and minima. Samples are taken approximately every 1.5 km. At each sampling point, the lengthscale of the RBF kernel is updated using the SGD algorithm, and a new model is generated.

Figure 6 shows the behavior of the Gaussian process model as samples are taken along a trajectory. The normalized standard deviation decreases near the sample locations. The fit of the kernel lengthscale evolves according to Fig. 7. It can be observed that the lengthscale stabilizes and reaches a peak near the local

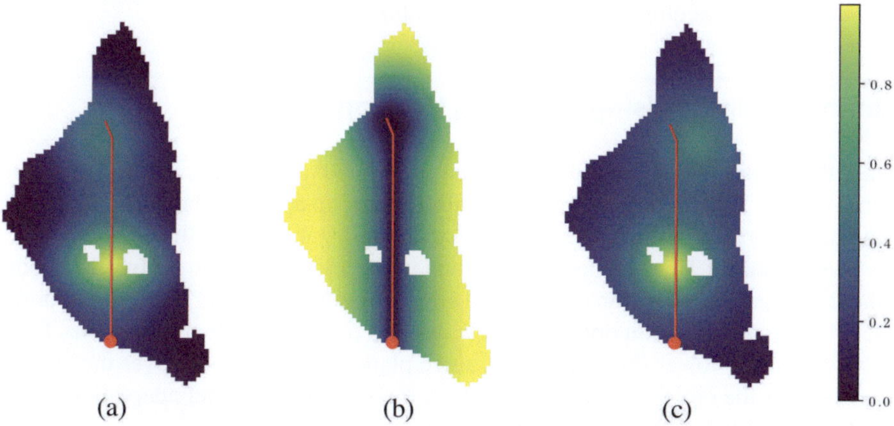

Fig. 6 GP resulting from performing the trajectory marked in red. It can be seen in (**a**) the mean $\mu(x)$, in (**b**) the uncertainty $\sigma(x)$, and in (**c**) the ground truth function. All three plots are presented normalized

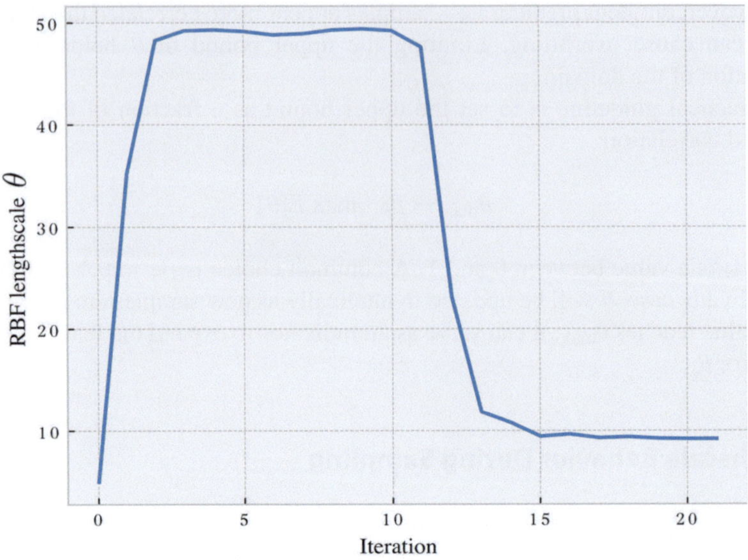

Fig. 7 Evolution of the lengthscale θ during the sampling process in Fig. 6. The lengthscale stabilizes and peaks near the local maximum, after which the GP adapts and θ decreases.

maximum point, around location 50.0. This happens because during the first half of the trajectory, the data show high smoothness and correlation, which is not enough to capture the function's variability. After sampling the local maximum (Fig. 6c), the model incorporates the new information and the lengthscale decreases, enabling the GP to adapt to the variability present in the second half of the trajectory. This behavior highlights the importance of constraining the range of θ values to prevent the model from becoming overly optimistic about spatial uncertainty.

Online Planning Using Receding Horizon

Once the Gaussian process (GP) model has been fitted to the data, sampling planning can proceed. This section proposes an online planning approach based on Receding Horizon control combined with the CMA-ES optimization algorithm.

The Receding Horizon technique divides the planning into control horizons, where a sequence of actions is planned for a horizon of size h. This breaks down the problem of finding an optimal path ψ into a series of simpler subproblems. Each subproblem has lower dimensionality, since planning is done only over the current horizon, not the entire budget. This approach also enables model adaptation, since the GP is updated at each step with new data.

At each time step, the optimal path ψ_h^* is computed over a control horizon h using a sequential optimization method (CMA-ES, described in the following section). Only the first movement of this optimal subpath is executed, and the GP is updated

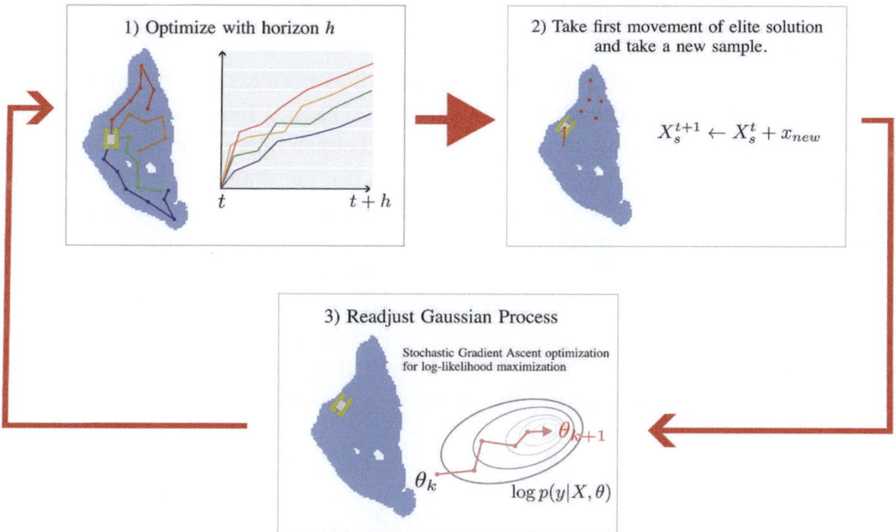

Fig. 8 Representation of two uncertainty maps from two GPs with different θ values for an RBF kernel and the same data. Left: small lengthscale implying local spatial correlation. Right: large lengthscale implying high spatial correlation

with new information. This process is repeated at every step, allowing dynamic adaptation to changes in the environment and real-time decision-making. Even if the GP is not perfectly fitted, executing only the first decision in the control horizon ensures re-planning is possible as new information becomes available (see Fig. 8). This makes Receding Horizon planning a common choice for sequential decision problems in uncertain environments (Popovic et al. 2020).

Algorithm 2: Receding Horizon algorithm

Initialize:
$\boldsymbol{p}_0$ — Initial position
$d \leftarrow 0$ — Distance
h — Control horizon
while $d < d_{\max}$ **do**
  $\boldsymbol{p}_{t+1} \leftarrow$ CMA-ES$(\boldsymbol{X}_t, \boldsymbol{y}_t, h)$
  $d \leftarrow d + \|\boldsymbol{p}_{t+1} - \boldsymbol{p}_t\|_2$
  $\boldsymbol{p}_{t+1} \leftarrow \boldsymbol{p}_t$ — Update position
end while

Reward and Uncertainty

Once the planning algorithm has been defined, it is necessary to define a reward function to guide decision-making. In general, the reward function can be defined in multiple ways, depending on the objectives and constraints of the problem. In the context of monitoring physicochemical variables, it is common to consider a reward function that maximizes the information gathered and minimizes the uncertainty associated with the model predictions.

As described in section "Introduction", the amount of information collected by a path ψ can be defined as the difference between the initial model entropy and the model entropy after collecting samples at vertices v_i. However, the estimation of the a posteriori entropy entails a high computational cost, since it requires the computation of the determinant of the covariance matrix K:

$$\mathcal{I}(\psi) = H(X^t) - H(X^{t+1}) \tag{19}$$

$$H(X) = \frac{1}{2}\log(|\Sigma|) + \frac{\dim(X)}{2}(1 + \log(2\pi)) \tag{20}$$

To alleviate the computational cost associated with this calculation, an approximation of the a posteriori entropy can be taken by the model variance. Since the covariance matrix K is a positive semidefinite Gramm's matrix, the a posteriori entropy can be calculated as:

$$H(X) \sim \frac{1}{2}\sum_j \lambda_j \tag{21}$$

where λ_j are the eigenvalues of the covariance matrix K. This implies that the a posteriori entropy decrease is realized through the decrease of the eigenvalues of the covariance matrix. One criterion for the decrease in a posteriori entropy is the decrease in model variance. Since the eigenvalues decrease with the values of the diagonal of the covariance matrix, one can consider the decrease in variance as a reward criterion for decision-making. This criterion is known as the A-optimality criterion. Thus:

$$\mathcal{I}(\psi) = H(X_\psi^t) - H(X_\psi^{t+1}) \propto \sigma(X_\psi^t) - \sigma(X_\psi^{t+1}) \tag{22}$$

CMA-ES Algorithm

The CMA-ES (Covariance Matrix Adaptation Evolution Strategy) is an evolutionary optimization algorithm used to find minima or maxima in nonlinear, nonconvex, noisy (Hansen and Ostermeier 2001) functions. Its usefulness lies in its ability to adapt the covariance matrix of the solution space, which allows it to learn and exploit the structure of the objective function. We will assume that one seeks to minimize an objective function $f : \mathbb{R}^n \rightarrow \mathbb{R}$:[2]

$$x^* = \arg\min_x f(x) \tag{23}$$

The algorithm is based on an evolutionary strategy, updating a population of candidate solutions to approximate the optimum. First, a population $\mathbf{P}_k := \{\mathbf{x}_0, \ldots, \mathbf{x}_\lambda\}$ of candidate solutions is proposed, which is generated as a sample from a multivariate Gaussian distribution:

$$\mathbf{p}_k^{(g+1)} \sim \mathcal{N}(\mathbf{m}^{(g)}, \sigma^2 \mathbf{C}^{(g)}) \tag{24}$$

where g is the generation number, $\mathbf{m}$ is the mean of the distribution (also called the centroid of the solutions), σ is the step size, and $\mathbf{C}$ is the covariance matrix of the distribution.

The CMA-ES algorithm consists of three fundamental steps:

1. Selection/recombination of solutions from the population
2. Update of the step size σ
3. Update of the covariance matrix $\mathbf{C}$

Selection and Recombination

To select the set of solutions, once a population $\mathbf{P}^{(g)}$ has been obtained, the fitness of each of its $\mathbf{x}_k$ individuals must be evaluated. The best μ individuals will be selected, and the new population mean will be recalculated according to the following equation (Hansen and Ostermeier 2001):

$$\mathbf{m}^{(g+1)} = \sum_{i=1}^{\mu} w_i \mathbf{x}_i^{(g+1)} \tag{25}$$

This update does not treat all individuals equally. The weights w_i are assigned according to the fitness of the objective function $f(x)$. The weights w will be calculated such that they are monotonically decreasing with fitness according to the following formula:

$$w_i = \log(\mu + 0.5) - \log(i), \quad i \in [1, \mu] \tag{26}$$

Updating the σ_p Step Size

To adapt the step size, an estimator called "Evolution Path" (p_σ) is used. This indicator is a measure of the population shifts within the search space. The estimator

[2] This objective can be converted to a maximization problem if $f_{\max}(x) = -f_{\min}(x)$

is calculated as follows (starting from an initial value $p_\sigma^{(0)} = 0$):

$$\mathbf{p}_\sigma^{(g+1)} \leftarrow (1 - c_\sigma)\mathbf{p}_\sigma^{(g)} + \sqrt{c_\sigma(2 - c_\sigma)\mu_{\text{eff}}}\mathbf{C}^{(g)^{-1/2}} \left(\mathbf{m}^{(g+1)}\right) - \frac{\mathbf{m}^{(g+1)} - \mathbf{m}^{(g)}}{\sigma^{(g)}} \tag{27}$$

This estimator incrementally updates the evolution vector p_σ, accumulating information about the new direction and magnitude of the solutions. The last term $\frac{\mathbf{m}^{(g+1)} - \mathbf{m}^{(g)}}{\sigma^{(g)}}$ represents the weighted change between populations (the new evolution direction) without normalizing. The value c_σ is a parameter that determines the rate of adaptation. μ_{eff} represents the number of effective samples performed. Once p_σ is updated, a new step can be calculated as:

$$\sigma^{(g+1)} \leftarrow \sigma_g \exp\left(\frac{c_\sigma}{d_\sigma}\left(\frac{||\mathbf{p}_\sigma||}{\mathbb{E}[\mathcal{N}(0, 1)]} - 1\right)\right) \tag{28}$$

The cumulative path length $||\mathbf{p}_\sigma||$ is measured and compared to the expected length $\mathbb{E}[\mathcal{N}(0, 1)]$ of a Gaussian random vector in normalized space. If the path is longer than expected, it means that the search is moving fast, so σ is increased (exploration). If the path is shorter, the search is stabilizing, so σ is reduced (exploitation). The c_σ and d_σ parameters control the speed of adjustment.

Updating the Covariance Matrix C

The covariance matrix is adapted to capture the geometry of the search space according to the following equation:

$$\mathbf{C}^{(g+1)} = \underbrace{(1 - c_1 - c_\mu)\mathbf{C}^{(g)}}_{\text{Previous cov. matrix}} + \underbrace{c_1 \mathbf{p}_c^{(g+1)} \mathbf{p}_c^{(g+1)T}}_{\text{Evolutionary path direction}} \tag{29}$$

$$\underbrace{+ c_\mu \sum_{i=1}^{\mu} w_i \frac{(\mathbf{X}_{1:\lambda}^{(g+1)} - \mathbf{m}^g)(\mathbf{X}_{1:\lambda}^{(g+1)} - \mathbf{m}^g)^T}{\sigma^{(g)2}}}_{\text{Elite direction}} \tag{30}$$

It can be seen in Eq. that the covariance matrix is updated as a function of three terms:

1. The previous covariance matrix
2. The evolution vector $\mathbf{p}_c$
3. In the direction of the best individuals

The c_1 term controls the speed of adaptation of the covariance matrix. The term c_μ controls the importance of elite solutions in updating the covariance matrix. The term w_i are the weights assigned to each elite solution as a function of its fitness (see).

By iterating these three steps, the CMA-ES algorithm allows adapting the population of solutions to the objective function by adjusting the mean, the step size, and the covariance matrix. A fundamental advantage of this algorithm is its ability to adapt to the geometry of the search space with a reduced number of hyperparameters. In addition, the algorithm is able to define the strategy without regard to the dimension or magnitude of the objective function or the individuals. In Algorithm 3, the pseudocode of the CMA-ES algorithm is presented.

Algorithm 3: CMA-ES algorithm

Data: λ, μ, σ_o^2, c_{cov}
Result: $\mathbf{x}_{\text{best}}$
initialize:
$\lambda \leftarrow \mu \leftarrow$ population size
σ_o^2 Initial variance
c_{cov} Covariance learning rate
for $g \in \{1, \ldots, N_g\}$ **do**
  Step 1: Update the Covariance Matrix
  $\mathbf{C}^{(g+1)} \leftarrow (1 - c_{\text{cov}})\mathbf{C}^g + \frac{c_{\text{cov}}}{\mu_{\text{cov}}} \mathbf{P}_c^{(g+1)} \mathbf{P}_c^{(g+1)T}$
  $+c_{\text{cov}} \left(1 - \frac{1}{\mu_{\text{cov}}}\right) \sum_{i=1}^{\mu} w_i \frac{(\mathbf{X}_{1:\lambda}^{(g+1)} - \mathbf{m}^g)}{\sigma^g} \left(\frac{(\mathbf{X}_{1:\lambda}^{(g+1)} - \mathbf{m}^g)}{\sigma^g}\right)^T$
  Step 2: Update the step size σ^g
  $\sigma^{(g+1)} \leftarrow \sigma_g \exp\left(\frac{c_\sigma}{d_\sigma}\left(\frac{||\mathbf{P}_\sigma||}{\mathbb{E}[\mathcal{N}(0,1)]} - 1\right)\right)$
  Step 3: Generate sample population for next generation:
  $\mathbf{x}_k^{(g+1)} \sim \mathcal{N}(\mathbf{m}^{(g)}, (\sigma^{(g)})^2 \mathbf{C}^{(g)}) \quad \text{for } k = 1, \ldots, \lambda$
  Step 4: Update the mean for generation $g + 1$:
  $\mathbf{m}^{(g+1)} \leftarrow \sum_{i=1}^{\mu} w_i \mathbf{x}_{i:\lambda}^{(g+1)}$
  Step 5: Update best solution:
  $\mathbf{x}_{\text{best}} \leftarrow \arg\max_x f(\mathbf{x}_k^{(g+1)})$
return $\mathbf{x}_{\text{best}}$

Definition of Individuals

An individual will be defined in the context of the optimization of the CMA-ES algorithm as a vector of displacements about the current position of the vehicle p_t:

$$\mathbf{x}_k := \{\Delta x_0, \Delta y_0, \ldots, \Delta x_h, \Delta y_h\} \tag{31}$$

To generalize over different scenarios, the magnitude of the displacements is measured in units of metric given the graph $G(V, E)$. Thus, according to Eq. , it is

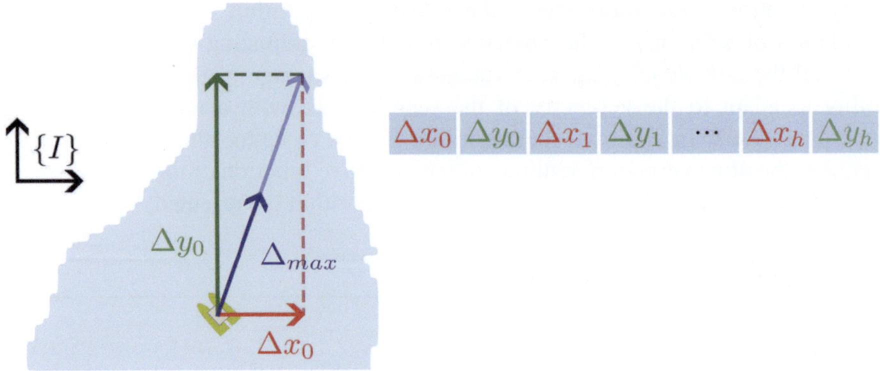

Fig. 9 Definition of the individuals in the CMA-ES algorithm. Each individual consists of a vector of displacements in the axis directions according to the fixed inertial system I. Each motion is fixed by the displacement angle and the magnitude of the displacement saturated to a maximum of $\Delta_{\max}$

composed of h pairs of Cartesian motions. Each motion is defined by a displacement in the directions of the axes according to the fixed inertial system I.

To avoid solutions that move too far away from the current position (Peralta et al. 2021), an upper bound on the magnitude of the displacements, $\Delta_{\max}$, can be imposed. An example of the definition of individuals in the CMA-ES algorithm is shown in Fig. 9. In addition, any movement beyond the boundary of the graph $G(V, E)$ will produce a movement to the nearest boundary in that direction.

Experiments and Results

This section will present the results obtained in the application of the algorithms described in the previous section. To validate the performance of the proposed system, a monitoring scenario based on the Mar Menor with an ASV equipped with a dissolved oxygen sensor detecting the abnormality of concentrations for the detection of algal spikes as proposed in (Karaköse and Bal 2025). The results obtained with the CMA-ES algorithm will be compared with those obtained with a planning algorithm based on the Expected Improvement acquisition function (Peralta et al. 2021) and an intensive search greedy algorithm. The effectiveness of both algorithms will be evaluated in terms of the prediction error on the variable and the predicted uncertainty associated with the model. All experiments have been performed in a Python 3.12 simulated environment. For the implementation of the optimization algorithms, the Python library *Deap* (Fortin et al. 2012) has been used. For the implementation of the Gaussian process, the *GPytorch* (Gardner et al. 2018) library has been used.

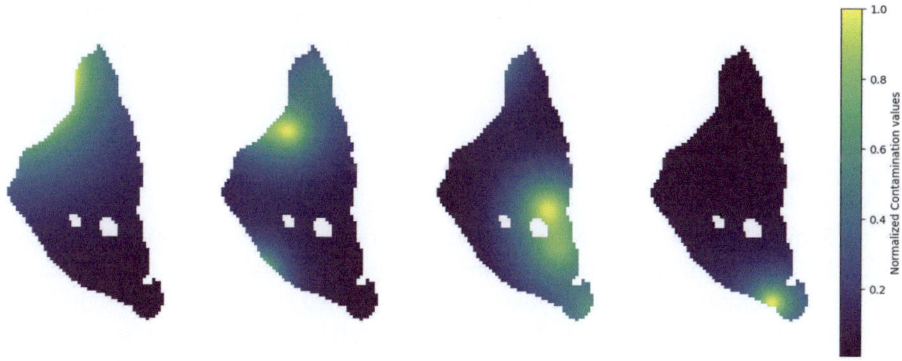

Fig. 10 Different examples of ground truths generated by the Shekel function with a variable number of local maxima and minima and random position

A multimodal generator function of the Shekel[3] type has been chosen to represent the contamination maps. This function allows the generation of contamination maps with an arbitrary number of randomly distributed maxima and minima. The generating function will have the following expression:

$$f_{\text{Shekel}}(\mathbf{x}) = \sum_{i=1}^{M} \left(\sum_{j=1}^{N} (\mathbf{x}_j - \mathbf{A}_{ij}^2 + C_i)^{-1} \right) \tag{32}$$

where M and N are the dimensions of the map in cells, $\mathbf{A}$ is a matrix indicating the position of local maxima and minima, and $\mathbf{C}$ is a vector of constants. The dimension of $\mathbf{A}$ is chosen randomly between 1 and 6 (at most 6 maxima or minima), and the dimension of $\mathbf{C}$ is chosen randomly between 0 and 1. Since f_{Shekel} only generates maxima, arbitrarily, it is selected if a maximum reverses its position to become a minimum. Different examples of maps generated with the Shekel function are shown in Fig. 10. This function is based on the hypothesis that the dissolved oxygen concentration in water behaves similar to a multimodal generating function with a smooth correlation at least locally (Kathen et al. 2023).

With respect to the experiment, a population size of $20 \times H$ has been chosen, where H is the prediction horizon and the size in moves of each individual. The minimum number of generations has been set at 25. However, the CMA-ES algorithm will continue to optimize as long as it detects an improvement in the fitness of the individual with respect to the average of ten previous generations. If the algorithm does not improve after ten consecutive generations, a local maximum is considered to have been reached. Additionally, optimization is enforced to stop if the maximum time available for the computation is exceeded, which is considered

[3] https://www.sfu.ca/~ssurjano/shekel.html

to be 2 minutes to ensure the efficiency of the algorithm and real-time decision-making. However, the CMA-ES algorithm is able to converge in less than 2 minutes in all cases.

Hyperparameterization

A fundamental prior step to obtain the best results with a Receding Horizon algorithm is the parameterization of the algorithm. In this case, a tabular search for the best combination of the hyperparameters related to the optimization horizon H and control horizon C over ten different functions has been performed. The optimization horizon represents how many steps will be planned in each iteration, i.e., the depth of the search within the solution space. In the context of the CMA-ES algorithm, it represents the size of the individual, since each individual consists of a vector of displacements in the axis directions according to the fixed inertial system. A prediction control horizon implies short-term planning, while a large prediction horizon implies long-term planning. The control horizon represents how many steps of the optimized trajectory will be taken before re-optimizing with the new model. A large control horizon implies slower adaptation to changes in the model, while a small control horizon implies faster adaptation.

Figure 11 shows the average uncertainty $\sigma(x)$ of the reward as a function of the control and optimization horizon. It can be seen that an excessively short optimization horizon can produce suboptimal solutions due to lack of long-term

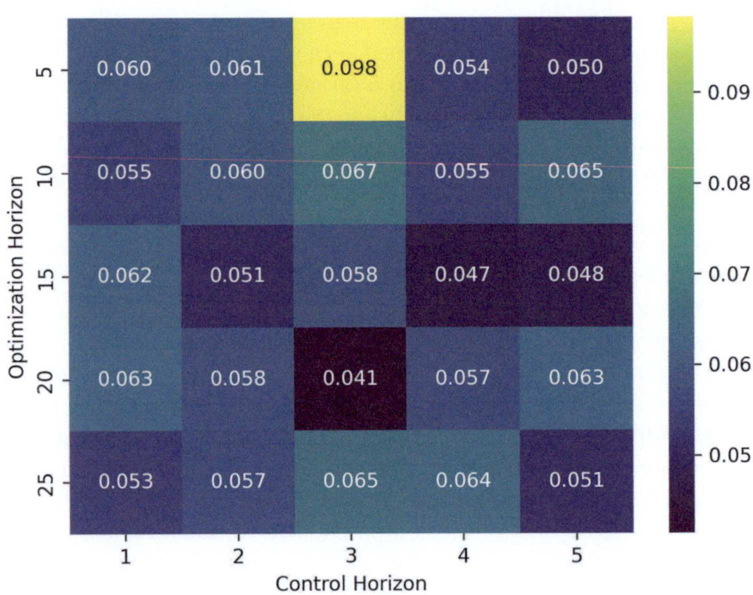

Fig. 11 Average uncertainty among ten simulations for each value of H and C

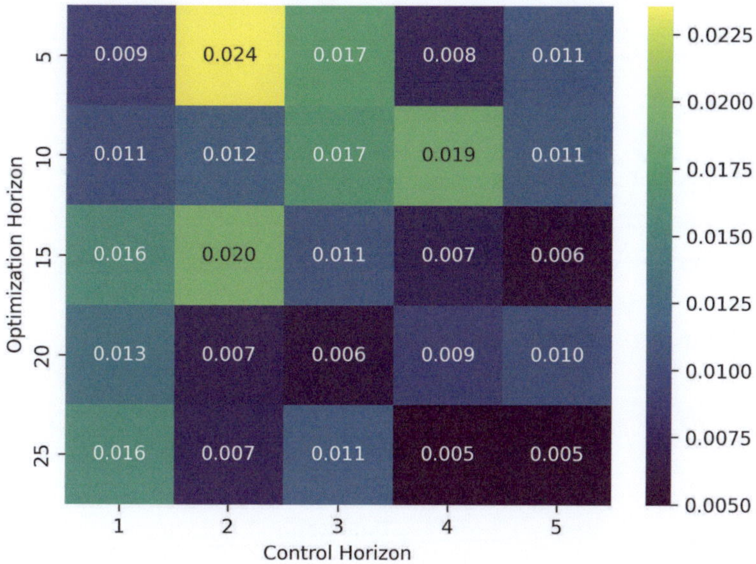

Fig. 12 Mean square error among ten simulations for each value of H and C

planning. On the other hand, above a certain level of optimization depth, no significant improvement is observed. An excessively long optimization horizon may only contribute to the computational burden. The control horizon allows to alleviate this burden, since only every C steps will be optimized with the CMA-ES algorithm. In Fig. 11, it is observed that a moderate control horizon allows for better planning. It can be explained by the fact that a larger control horizon allows slower adaptation to changes in the model. The model, in the initial instants, is not too adequate, so more short-term re-planning is needed. Once a mature model has been reached from the point of view of finding the optimal GP hyperparameters, a slower adaptation can be allowed.

In Fig. 12, we observe the effect of the H and C parameters this time on the mean square error of the reward. A similar effect as in Fig. 11 can be observed. The model benefits from a longer optimization and control horizon. To avoid overfitting the model, the choice is made to take the value that most minimizes the mean uncertainty, since the MSE depends directly on ground truth. In this case, an optimization horizon of 20 and a control horizon of 3 have been selected following the results shown in Fig. 11.

Uncertainty Analysis

Figure 13 shows the progression of the mean uncertainty $\sigma(x)$ as the trajectory progresses for a set of ten simulations. It can be seen that the uncertainty decreases

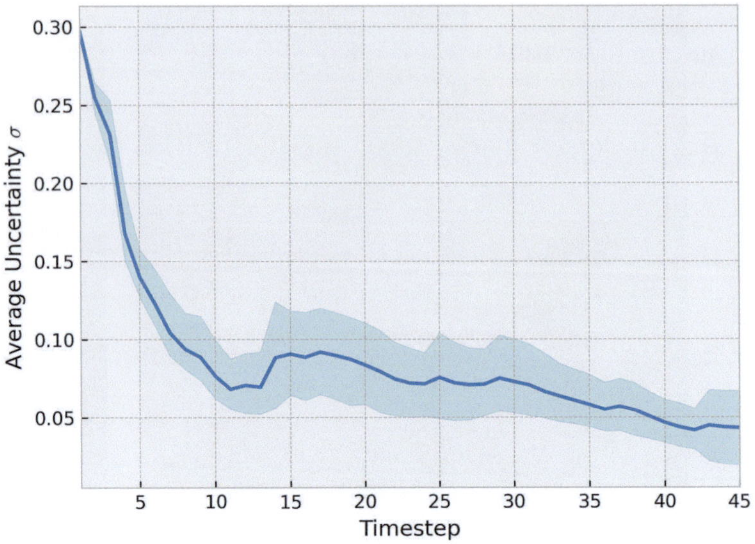

Fig. 13 Average uncertainty $\sigma(x)$ over ten simulations along the timesteps of a path. The uncertainty decreases as the path progresses. The shadowed area represents the standard deviation of $\sigma(x)$

as information is collected. On average, the mean uncertainty is reduced by up to 5%. The largest decrease occurs at the beginning, where taking the first samples suddenly brings the uncertainty down. An example of the final path followed by the vehicle can be seen in Fig. 14. In the initial instants, the solution focuses on minimizing the uncertainty at the center of the water mass. Once sufficient information has been collected, the vehicle moves toward the areas of higher uncertainty by a circular path around the edges. This behavior is expected and appropriate, since sampling at the edges is less informative than in the center of the water body. It can be observed that the resulting path at the edge maintains an almost constant distance to the edge, indicating that the model has learned that information at the edges is less informative. In addition, because the smooth correlation hypothesis of the data holds, the optimal strategy is to collect zones in the inner zone of the lake (*inner-flanging*), rather than at the edges (where only one-sided uncertainty is decreased).

In Fig. 15, you can see the final uncertainty map $\sigma(x)$. It can be seen that the uncertainty is higher at the edges of the map. The Receding Horizon strategy is able to readapt the path as the GP is adjusted. It can be seen how because the optimal lengthscale θ is moderately large (approximately 10 for the Fig. 14 experiment), optimal coverage does not require entering angled zones, such as the Playa Honda zone (lower-right lobe in Fig. 14). Entering these angled zones does not reduce the entropy sufficiently to assume an optimal decision. This demonstrates that the

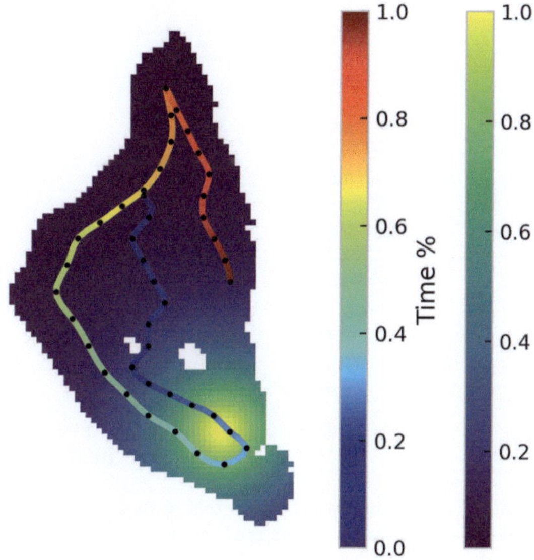

Fig. 14 Resulting path of the vehicle after 42 samples/steps. In the background, it is represented the contamination ground truth

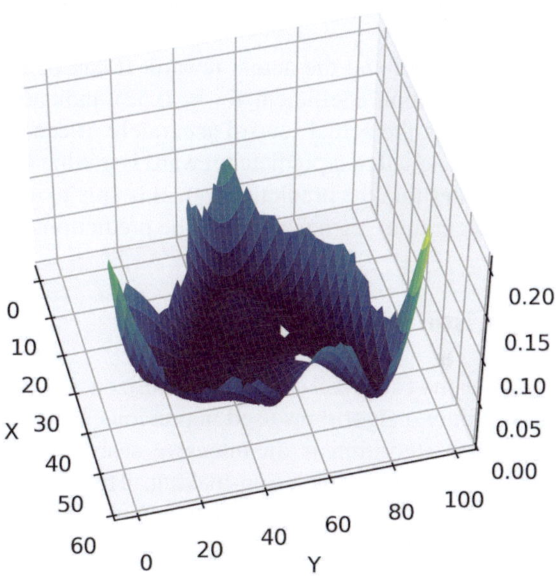

Fig. 15 Final uncertainty $\sigma(x)$ after 42 samples/steps following the path generated in Fig. 14. The uncertainty is higher in the borders of the map, where the model has less information due to the *inner-flanging* trajectory

CMA-ES algorithm is able to adapt to the geometry of the search space and the spatial correlation of the data.

To validate that the uncertainty prediction in the optimization horizon is accurate, the reward obtained with the model for each prediction step (assuming the length-scale θ above as valid) has been compared with the actual reward for the resulting trajectory above (obtaining the decrease in uncertainty with the a posteriori data). In Fig. 16, one can observe the correlation between the reward obtained with the

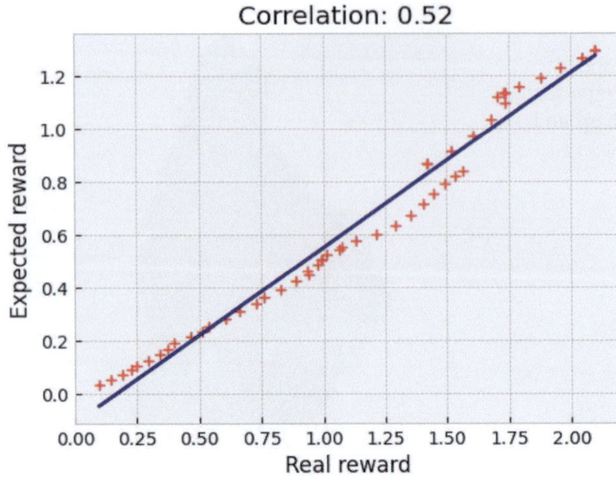

Fig. 16 Predicted reward versus real reward correlation. The predicted reward assumes the prior lengthscale θ as valid for the prediction horizon. The real reward is calculated as the decrement in uncertainty given the posterior data generated from the next timestep

prediction model and the actual reward. It can be seen that the correlation is high (with a correlation coefficient $R^2 \approx 0.52$), indicating that the prediction model is able to predict the actual reward accurately. It can be observed that the divergence between the actual and predicted reward is gradually readjusted as data is collected. This indicates that the prediction model is able to adapt to the information collected and the uncertainty associated with the predictions.

Comparison with Other Algorithms

To validate the CMA-ES + Receding Horizon algorithm, we propose to compare the results with two algorithms used in previous work for similar objectives.

The first algorithm is an intensive search greedy algorithm, which is based on immediate reward decision-making. This algorithm will evaluate all possible trajectories within computational time limits. The greedy algorithm is based on, once the reduction of uncertainty associated with all possible trajectories within that horizon has been evaluated, selecting the one that reduces uncertainty the most. This algorithm theoretically provides a solution that tends to the lower bound of the optimal trajectory, following the principle of entropy submodularity—the entropy reduction is greater in the first steps (Krause and Golovin 2014).

The second algorithm is a planning algorithm based on the Expected Improvement (EI) (Peralta et al. 2021) acquisition function. This algorithm is based on taking a sample following the expected gain maximization principle. The function $EI(x)$ is a measure of the difference between the expected value of the function at a point and the expected value of the function at the point of highest value. The EI

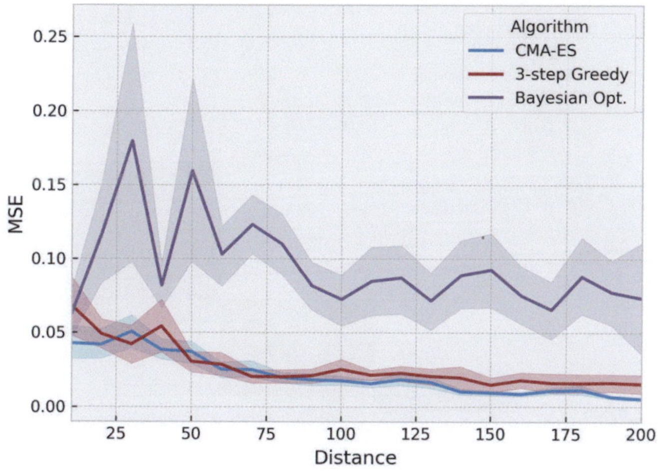

Fig. 17 Mean squared error produced by the three algorithms. The CMA-ES algorithm produces the best results, followed by the greedy algorithm and the Expected Improvement algorithm

algorithm can be computed analytically for a GP as:

$$EI(x) = (\mu(x) - f(x^*) - \chi)\Phi(Z) + \sigma(x)\phi(Z) \tag{33}$$

where $\mu(x)$ is the mean of the GP at point x, $f(x^*)$ is the expected value of the function at the point of greatest value, $\sigma(x)$ is the standard deviation of the function at point x, $\Phi(Z)$ is the cumulative distribution function of the standard normal distribution, and $\phi(Z)$ is the density function of the standard normal distribution. This algorithm will simply evaluate the function $EI(x)$ and sample at the point of highest value. To avoid a large offset and to account for uncertainty, we limit the travel distance to a maximum of Δ_{max} as proposed in (Peralta et al. 2021).

In Fig. 17, it can be seen how the mean square error varies with distance. It can be seen that the $EI(x)$ function-based approach does not produce as good results as the CMA-ES algorithm. This is because the $EI(x)$ algorithm tends to prioritize movements toward the highest expected values, without taking into account the uncertainty associated with the prediction. The greedy algorithm produces good results, but the solutions lack long-term planning, causing a suboptimal model. The final error improvement over the algorithm using $EI(x)$ is 0.069 (a 12.5-fold improvement) and over the greedy algorithm is 0.009 (a 2.5-fold improvement). In Fig. 18, you can see the final distribution of the error committed with each model fed with the data of each proposed algorithm for ten different scenarios.

If we analyze the trajectories generated in a particular case (see Figs. 19 and 20), it can be seen that the algorithm based on $EI(x)$ is quite effective in finding the local maxima, but it is not able to explore the areas of higher uncertainty. To maximize the exploratory behavior, a value of $\chi = 1$ has been imposed on the $EI(x)$ function. However, the agent still prioritizes the highest expected values.

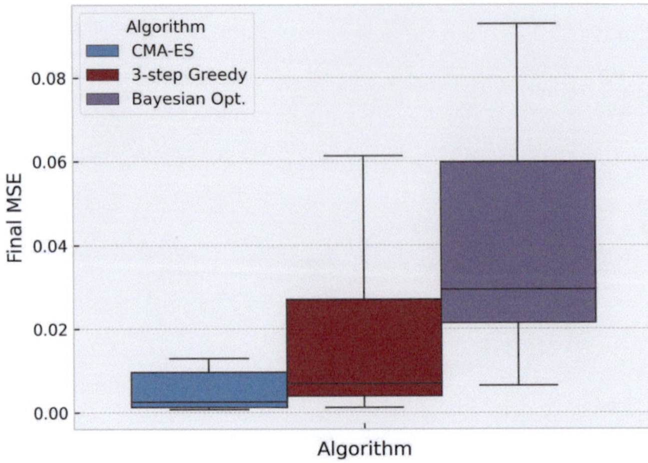

Fig. 18 Boxplot with the final mean squared error produced by the three algorithms.

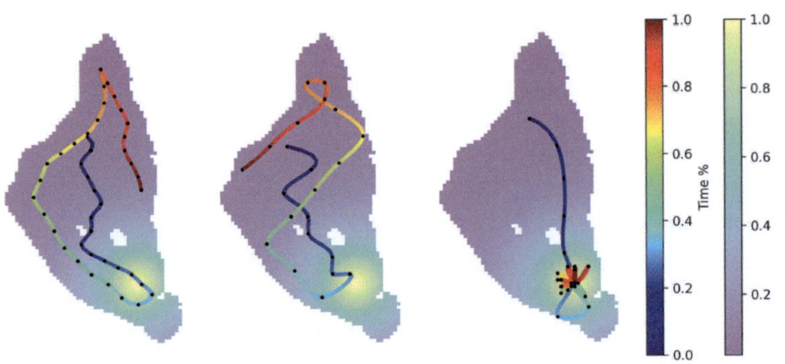

Fig. 19 Example of the resulting path using different algorithms. In the background, it is represented the contamination ground truth

The greedy algorithm, on the other hand, is able to explore areas of higher uncertainty, but is not able to plan for the long term. This is observed in the effect of internal trajectory crossings. An optimal trajectory should not cross areas of low uncertainty, as they do not provide additional information. This occurs because the prediction horizon is so short that it does not allow the algorithm to consider escape trajectories from non-convex areas of the map. Intensive search, although it allows finding moderately good solutions, cannot scale to large prediction horizons due to the computational load.

The CMA-ES algorithm, on the other hand, while involving considerable computational complexity, is able to adapt to the geometry of the search space and the spatial correlation of the data. This can be seen in the resulting trajectory, which maintains a constant distance from the edges of the map.

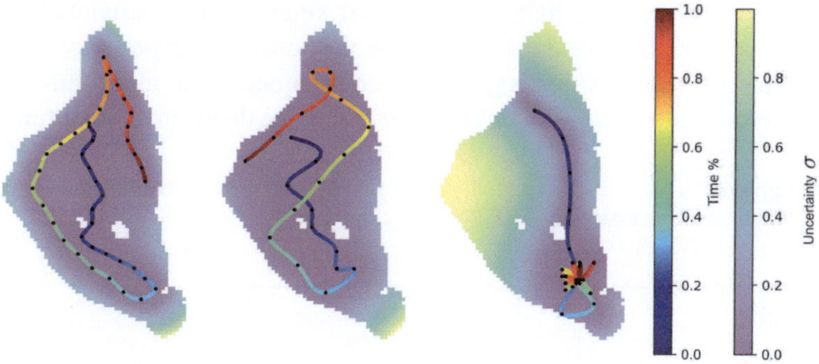

Fig. 20 Example of the resulting path using different algorithms. In the background, it is represented the predicted uncertainty

Conclusions and Future Work

This chapter has explained how GPs can be used to model physicochemical variables in aquatic environments and how the CMA-ES algorithm can be used for planning sampling paths. GPs allow obtaining accurate estimates and quantifying the uncertainty associated with predictions, which is essential for real-time decision-making. An advantage of GPs is that the definition of the uncertainty does not depend on the data, thanks to the definition of the kernel. This makes it possible to implement a Receding Horizon algorithm that plans sampling trajectories based on the predicted uncertainty of the model. It has been proposed to use the CMA-ES algorithm to optimize the sampling trajectory as a function of the predicted model uncertainty. The CMA-ES algorithm, combined with the Receding Horizon technique, allows planning robust trajectories that maximize the information collected by minimizing the predicted model uncertainty. The algorithm has been compared with an intensive search greedy algorithm and an algorithm based on the Expected Improvement acquisition function. The results obtained show that the CMA-ES algorithm is able to find optimal solutions in terms of uncertainty and reward reduction. The greedy algorithm produces moderately good solutions but is not capable of long-term planning. The algorithm based on the Expected Improvement acquisition function, while able to find local maxima, is not able to explore the areas of higher uncertainty. The CMA-ES algorithm, on the other hand, is able to adapt to the geometry of the search space and the spatial correlation of the data.

Future lines of work include the implementation of the algorithm in a real ASV equipped with a dissolved oxygen sensor. This proposal should consider in future studies multiple sensors and magnitudes at the same time. Multi-output Gaussian processes can be considered to manage multiple profiles of uncertainty, with a different kernel definition for each one. In addition, the implementation of the algorithm in a real ASV will allow validating the algorithm in real-time decision-making. In a real implementation, additional sources of uncertainty should be considered, such as the presence of noise in the sensors, the presence of external agents, or the presence of obstacles in the environment. The algorithm should be able to adapt to these sources of uncertainty and make decisions in real time.

Acknowledgments This work was partially supported by I+D+i grant REINFORCED funded by Research Plan of the University of Seville.

References

Álvarez Rogel J, al GGBe (2020) The case of Mar Menor eutrophication: state of the art and description of tested Nature-Based Solutions. Ecol Eng 158:106086

Arzamendia M, Gregor D, Gutierrez-Reina D, Toral S (2019) An evolutionary approach to constrained path planning of an autonomous surface vehicle for maximizing the covered area of Ypacarai Lake. Soft Comput 23(5):1723–1734

Baron JS, Poff NL, Angermeier PL, Dahm CN, Gleick PH, Hairston NG, Jackson RB, Johnston CA, Richter BD, Steinman AD (2002) Meeting ecological and societal needs for freshwater. Ecol Appl 12(5):1247–1260

Codd GA, Testai E, Funari E, Svirčev Z (2020) Cyanobacteria, cyanotoxins, and human health, chapter cyanobacteria, cyanotoxins, and human health, pp 37–68

Fortin FA, Rainville FMD, Gardner MA, Parizeau M, Gagne C (2012) Deap: evolutionary algorithms made easy. J Mach Learn Res 13:2171–2175

Gardner JR, Pleiss G, Bindel D, Weinberger KQ, Wilson AG (2018) Gpytorch: blackbox matrix-matrix gaussian process inference with GPU acceleration. In: Advances in neural information processing systems

Hansen N, Ostermeier A (2001) Completely derandomized self-adaptation in evolution strategies. Evol Comput 9(2):159–195

Karaköse P, Bal C (2025) Efficient chlorophyll prediction and sampling in the sea: a real-time approach with UCB-based path planning. IEEE Access 13:8127–8139

Kathen MJT, Peralta F, Johnson P, Jurado Flores I, Gutiérrez Reina D (2023) Performance evaluation of AquaFeL-PSO informative path planner under different contamination profiles. In: Rivera G, Cruz-Reyes L, Dorronsoro B, Rosete A (eds) Data analytics and computational intelligence: novel models, algorithms and applications. Springer, Cham, pp 405–431

Kim S, Kim S, Mehrotra R, Sharma A (2020) Predicting cyanobacteria occurrence using climatological and environmental controls. Water Res 175:115639

Krause A, Golovin D (2014) Submodular function maximization. In: Tractability

Mattingley J, Wang Y, Boyd S (2011) Receding horizon control. IEEE Control Syst Mag 31(3):52–65

Peralta F, Reina DG, Marín SLT, Gregor DO, Arzamendia M (2021) A Bayesian optimization approach for water resources monitoring through an autonomous surface vehicle: the Ypacarai Lake case study. IEEE Access 9(1):9163–9179

Peralta F, Reina DG, Toral S (2023) Water quality online modeling using multi-objective and multi-agent Bayesian Optimization with region partitioning. Mechatronics 91:102953

Popovic M, Vidal-Calleja T, Hitz Gea (2020) An informative path planning framework for UAV-based terrain monitoring. Auton Robot 44:889–911

Sánchez-García J, García-Campos J, Arzamendia M, Reina D, Toral S, Gregor D (2018) A survey on unmanned aerial and aquatic vehicle multi-hop networks: wireless communications, evaluation tools and applications. Comput Commun 119:43–65

Świechowski M, Godlewski K, Sawicki B, Mańdziuk J (2023) Monte Carlo Tree Search: a review of recent modifications and applications. Artif Intell Rev 56(3):2497–2562

Unmanned Underactuated Surface Vehicle Formation Control Using Deep Reinforcement Learning

Luciano Villarreal, Federico Peralta, Pablo Millán, and Guillermo Bejarano

Contents

Abstract

Designing controllers can be highly dependent on the knowledge of the underlying model, especially in aquatic environments where the hydrodynamics parameters of the vehicles are almost always unknown. This work leverages this difficulty by using a machine learning-based guidance subsystem that produces valid control outputs, not only for one unmanned surface vessel but also for a group of them so that they can efficiently move while maintaining a formation. The formation control network was trained using proximal policy optimization

L. Villarreal · F. Peralta (✉) · P. Millán · G. Bejarano
Universidad Loyola Andalucía, Seville, Spain
e-mail: lvillarrealalava@al.uloyola.es; fdperalta@uloyola.es; pmillan@uloyola.es; gbejarano@uloyola.es

D. Gutiérrez et al. (eds.), *Smart Water Quality Monitoring*,
https://doi.org/10.1007/978-3-032-03049-8_13

(PPO) and through a simulator considering the well-known Cybership-II ship. Results show that the actor-critic deep reinforcement learning network produces an efficient policy to control a fleet of unmanned surface vehicles (USVs) and maintains formation whilst moving towards a goal.

Keywords

Formation control · Deep reinforcement learning · Autonomous surface vehicles · Policy optimization · Fleet coordination

Introduction

Unmanned surface vehicles (USVs) have been deployed for numerous applications where human operators cannot participate in situ, either for security or availability reasons. These applications may include search and rescue scenarios, where vehicle deployment needs to be fast (Wang et al. 2023), or monitoring missions where the repetitive tasks of the vehicles can be automated, such as water quality monitoring (Vasudevan and Baskaran 2021), patrolling (Oleynikova et al. 2010), and/or deep-sea animal observation (Verfuss et al. 2019). Current trends focus on the autonomous operation of vehicles through a modular architecture, integrating hardware, software, and control systems. These modules are implemented to fully control a USV without human interaction, considering (i) navigation, (ii) guidance, and (iii) control, which focus on determining states, goals, and actuation, respectively.

The control system is in charge of performing manoeuvres through the actuation system, where the objective is to use the actuators (generally electric motors) to perform changes in the state of the USV (position, orientation, velocity, etc.) so that a predefined goal can be achieved. For example, a predefined goal could be reaching a specific waypoint or following a trajectory. It is important to define that this predefined goal must be achievable from the control system's point of view, i.e. the system must be capable of transforming a given objective into effective signals to feed the motors. Another more realistic example can be described in systems such as Dash et al. (2021); while the goal of a water quality monitoring mission might be to measure the physicochemical parameters of a water body, the control system's goal is to use the motors to reach a particular measurement location. In this sense, the control system is generally unaware of the overall purpose of the USV's operation, which is a good characteristic for modularized systems, where each subsystem can efficiently solve a well-defined problem that considers the known characteristics and behaviours of the specific part.

Several studies suggest that segmenting tasks and objectives in modular architectures is fundamental for efficient system design. The main advantage of this paradigm is its potential to connect with artificial intelligence (AI) techniques, where modern methods can enhance autonomous control or solve unaddressed problems with specific objectives. However, a key challenge in control systems is that achieving predefined goals depends heavily on how accurately system dynamics are modelled or observed. USV dynamical models, particularly the well-known model described by Fossen in (Fossen 2011), rely on a high number of hydrodynamic parameters that must be obtained through parameter estimation techniques or approximations.

Navigation subsystems consist of several key components that efficiently process real-world data to determine position (related to an Earth-fixed reference frame), velocity, and/or external disturbances (such as wind and wave). Usually, these data are sufficient to fully determine the state of a USV. Still, additional information can be obtained to increase the vehicle's knowledge of itself and the environment, such as pressure, current directions, and water depth.

Finally, the guidance subsystem can be considered the bridge between the mission objective, the navigation information, and the control subsystem. Guidance subsystems generally translate information and provide waypoints to be reached by the control subsystem. Generally speaking, USVs, such as the Yellowfish (Morel et al. 2022), shown in Fig. 1, implement the mentioned subsystems to perform autonomous motion and control.

In this work, the proposed machine learning-based guidance subsystem will utilize information from the navigation subsystem to produce valid control outputs for a group of USVs to move efficiently while maintaining a formation. Machine learning (ML) techniques have been applied to fully control the system without explicitly determining unknown hydrodynamic parameters of the USVs. Moreover, some works have controlled USVs for specific high-level goals using

Fig. 1 Dual-hulled unmanned surface vessel

deep learning techniques, such as a patrolling mission using double deep Q-learning (DDQL) (Barrionuevo et al. 2024) and multi-formation control using deep deterministic policy gradient (DDPG) (Zhao et al. 2021).

Formation control refers to the process where several, usually similar, agents use their respective actuators to follow a path while maintaining a set distance between them and avoiding obstacles. In this work, the goal is to follow a predetermined path with a set of USVs in a closed regular polygon structure, i.e. each USV is located at the vertices of an equilateral triangle when using three vehicles. We propose that the formation control actions are decided by a neural network agent that is trained to minimize formation error and distance error. Some advantages of leveraging the decision to a trained neural network include considering a black-box model that learns how to interconnect these two subsystems without explicitly knowing the actual system behaviour or its model. This approach prevents many assumptions or approximations usually done in this topic, such as hydrodynamic parameters.

The proposed approach utilizes a proximal policy optimization (PPO) architecture that can be easily integrated and used to perform formation control considering multiple USVs without knowledge of their respective hydrodynamic parameters. The rest of the chapter includes the definition of the underlying USV model and formation control in section "Problem Statement". The proposed approach defines how we utilize a deep reinforcement learning (DRL) approach in section "Proposed Approach" and state and action definitions. Afterwards, the simulation results are shown in section "Results", and finally, section "Conclusion and Future Work" presents the conclusions of the work.

Problem Statement

Modelling of Unmanned Surface Vehicle

USVs are marine vehicles capable of moving on the surface of any water body. In (Fossen 2011), the author presents a complete 6-DOF model. However, for practical applications, a common simplification is the three-degree-of-freedom (3-DOF) representation, which neglects heave, roll, and pitch, making it suitable for conventional marine crafts operating in calm waters at low speeds. The dynamic model considers an Earth-fixed North-East-Down (NED) frame and a local body frame with the centre as the origin. Figure 2 shows an illustration of a dual-hulled vessel, where global {g} and local {b} frames are displayed. The position and orientation of the vehicle are grouped in vector $\eta = [x, y, \psi]^T$ related to the global frame, while the local velocities of the USV (surge, sway, and yaw rate) are grouped in vector $\nu = [u, v, yaw_{rate}]^T$.

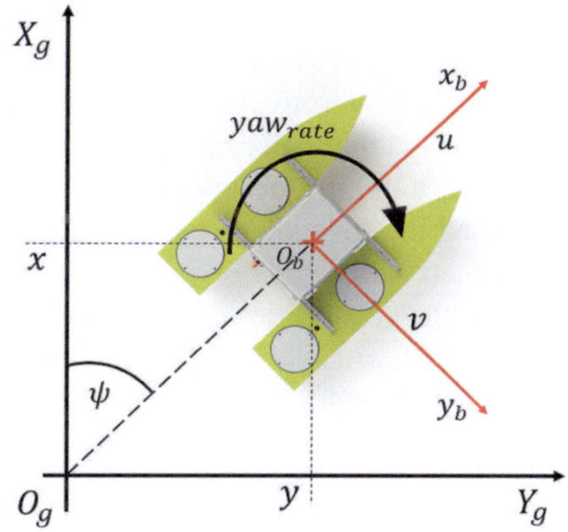

Fig. 2 Illustration of the USV showing the global and local frames

The hydrodynamic model can be described considering the presented position and velocity vectors, Coriolis, added mass, damping effects, restoring forces, and environmental disturbances, as indicated in (1):

$$\begin{aligned} \dot{\eta} &= Rot(\psi)\nu, \\ M\dot{\nu} &= \tau - C(\nu)\nu - D(\nu)\nu - g(\eta) - g_0 + \tau_w, \end{aligned} \tag{1}$$

The rotation $Rot(\psi)$, inertia M, which includes the added mass matrix, Coriolis $C(\nu)$, and damping $D(\nu)$ matrices are defined based on some variables and hydrodynamic parameters, as shown in (2). Moreover, τ_w refers to the environmental forces and torque due to wind, waves, and currents, while $g(\eta)$ describes the gravitational and buoyancy forces, the hydrostatic forces are represented by g_0 but, to the extent of this work, are assumed negligible:

$$\begin{aligned} Rot(\psi) &= \begin{bmatrix} \cos(\psi) & -\sin(\psi) & 0 \\ \sin(\psi) & \cos(\psi) & 0 \\ 0 & 0 & 1 \end{bmatrix}, \\ M &= \begin{bmatrix} m_{11} & 0 & 0 \\ 0 & m_{22} & m_{23} \\ 0 & m_{32} & m_{33} \end{bmatrix}, \\ C(\nu) &= \begin{bmatrix} 0 & 0 & c_{13} \\ 0 & 0 & c_{23} \\ -c_{13} & -c_{23} & 0 \end{bmatrix}, \\ D(\nu) &= \begin{bmatrix} d_{11} & 0 & 0 \\ 0 & d_{22} & d_{23} \\ 0 & d_{32} & d_{33} \end{bmatrix}. \end{aligned} \tag{2}$$

Note that the inertia matrix is assumed to be symmetric, whereas the Coriolis matrix is skew-symmetric. These two matrices, as well as the damping matrix, comprise all hydrodynamic parameters of each USV. Therefore, estimating these parameters is mandatory to use this 3-dof model.

Furthermore, by observing (1), it can be seen that the control input is vector τ, expressed in the local frame. This control problem is underactuated, since the USV has two actuators (one on each hull of the boat) that can exert forces on each hull, yielding the capabilities of two possible actuations: forward force $F = F_{\text{left}} + F_{\text{right}}$ and torque $T = (F_{\text{left}} - F_{\text{right}}) \cdot d/2$, where d is the distance between both propellers. So, the input vector of the USV will always have the form indicated in (3):

$$\tau = \begin{bmatrix} F_{\text{left}} + F_{\text{right}} \\ 0 \\ (F_{\text{left}} - F_{\text{right}}) \cdot d/2 \end{bmatrix}. \tag{3}$$

In this work, the 3-dof vessel model is used for simulation purposes, since a simulated environment is almost crucial for a DRL framework. The main reason is that, as explained in the following sections, numerous state data need to be used to train neural networks properly, and this data gathering can be more efficiently obtained through a well-designed simulation environment.

Formation Control

In a formation control scenario, as seen in Fig. 3, there are n vehicles that need to follow a certain pattern while following a certain spatial path $p(\omega) = (x_\omega, y_\omega)$, parametrized by a so-called path variable ω. In the case of this work, the vehicles are set to follow a real or virtual USV leader, such that its position $p_v = (x_v, y_v)$

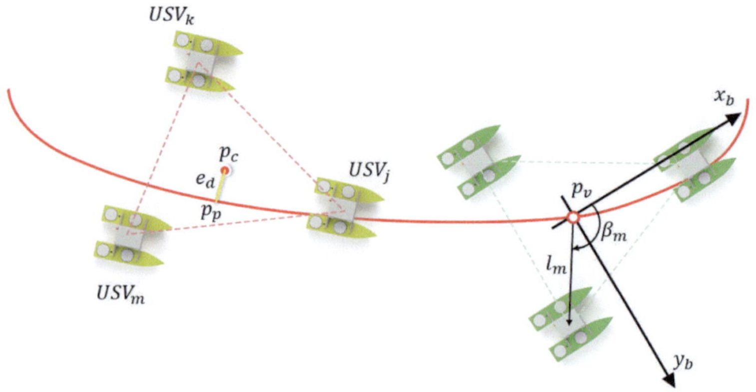

Fig. 3 Formation control. Yellow vehicles correspond to their current positions, and green vehicles correspond to their desired positions

is always part of the predetermined path $p(\omega)$, i.e. the position of the virtual leader $p_{v|\omega=\omega_0} = (x_{v|\omega=\omega_0}, y_{v|\omega=\omega_0})$ is given by a parametrized path $p(\omega)$ at ω_0. Then, the i-th vehicle has the desired expected location $p_{i\,d} = (x_{i\,d}, y_{i\,d})$ expressed in (4), related to the virtual location at a particular value ω_0:

$$
\begin{aligned}
x_{i\,d} &= x_{v|\omega=\omega_0} + l_{i\,d} \cos(\alpha_{\omega=\omega_0} + \beta_{i\,d}), \\
y_{i\,d} &= y_{v|\omega=\omega_0} + l_{i\,d} \sin(\alpha_{\omega=\omega_0} + \beta_{i\,d}),
\end{aligned} \tag{4}
$$

where $\alpha_{\omega=\omega_0}$ is the angle of the predefined path with respect to the global frame at ω_0, while $l_{i\,d}$ and $\beta_{i\,d}$ are control parameters that vary from USV to USV to define different desired locations.

The objective of formation control is accomplished if the vehicles can safely manoeuvre to reach their desired locations. These can be verified considering the mean distance between the n vehicles and their desired locations, namely, the formation error, defined in (5):

$$
e_f = \frac{\sum_{i=1}^{n} \text{dist}(p_i, p_{i,d})}{n}, \tag{5}
$$

where $\text{dist}(a, b)$ refers to the distance function between two two-dimensional locations a and b.

Another useful value is known as the cross-track error. This value evaluates the distance between the current centre of the formation and the preset trajectory. It is typically used to determine if the fleet is travelling such that its centroid aligns with the trajectory regardless of the formation. The representation of this error is $e_c = \text{dist}(\bar{p}_i, p(\omega))$, where $\bar{p}_i$ represents the centroid of the USV positions.

Additionally, the orientation error provides insight into how well the USVs are aligned with their desired orientations. The average orientation error for the fleet is computed as in (6):

$$
e_\psi = \frac{1}{n} \sum_{i=1}^{n} \left| \psi_i - \psi_{i,d} \right|, \tag{6}
$$

where $\psi_{i,d}$ is the desired orientation of the i-th USV. This metric ensures that the USVs not only reach their target positions but are also adequately oriented to maintain the desired formation or trajectory. Furthermore, another metric of interest is the final orientation error, which evaluates the overall alignment of the USVs with a common target orientation, as defined in (7):

$$
e_{\psi f} = \frac{1}{n} \sum_{i=1}^{n} \left| \psi - \psi_{\text{target}} \right|. \tag{7}
$$

In this case, ψ_{target} represents the desired global orientation for the fleet. This metric is handy when USVs must maintain a collective heading during formation control.

Proposed Approach

In underactuated aquatic systems, one of the greatest challenges lies in the uncertainty associated with the hydrodynamic parameters that describe them. Elements such as the Coriolis and damping matrices are often unknown or, at best, difficult to estimate accurately, significantly limiting predictive and control capabilities. Additionally, identifying and modelling external disturbances require extended state observers that, while effective, impose a considerable computational burden. These challenges become even more pronounced in scenarios requiring the coordination of multiple USVs, as maintaining an organized formation depends on fast and reliable inter-vehicle communication—a requirement that cannot always be guaranteed in dynamic maritime environments characterized by highly variable and often adverse conditions (Ten Kathen et al. 2024).

To address these issues, machine learning emerges as a promising tool by offering systems that not only can learn and adapt from data but also eliminate the need for explicit programming for each specific task. As a fundamental branch of artificial intelligence, this discipline enables designing algorithms to identify complex patterns and extrapolate solutions from large datasets. This is particularly relevant in contexts such as autonomous vehicle control, where adaptability is essential for handling unpredictable and variable conditions. Among ML methodologies, reinforcement learning stands out, focusing on optimizing decision-making through direct interaction with the environment and a reward system that guides the agent's behaviour toward optimal outcomes.

Reinforcement Learning

Reinforcement learning (RL) is particularly well-suited for formation control in USVs due to its inherent ability to operate in contexts where correct responses are not predefined. Unlike supervised approaches, which rely on pre-labelled datasets and are therefore limited in unforeseen scenarios, or unsupervised methods, which, while helpful in exploring data patterns, lack clear action guidelines, RL allows systems to learn from direct experience (Andrychowicz 2021). In this framework, agents progressively adapt their behaviour, optimizing their decisions by accumulating rewards associated with each action taken in the environment.

In this work, RL has been selected as the methodological foundation for addressing the complexity of USV coordination. This technique not only enables vehicles to adjust dynamically to changing maritime conditions but also provides the system with fundamental robustness to handle uncertainties and operational contingencies. Rather than treating each vehicle independently, a centralized approach is adopted through a "meta-agent", which assumes the responsibility of calculating and distributing the forces and torques required to maintain the formation. This design allows each USV to respond differently according to environmental demands and specific mission requirements, fostering a more precise and efficient response.

Deep Reinforcement Learning

DRL combines the principles of RL with the computational power of deep neural networks (Andrychowicz 2021). This approach leverages the networks' ability to efficiently represent complex state spaces, which is crucial for high-dimensional problems such as fleet control of USVs. In the context of this work, the environment is modelled using partially observable Markov decision processes (POMDPs), incorporating key elements such as state, observations, and actions. This theoretical framework addresses the uncertainty inherent in limited observations and dynamic environmental changes, providing a solid foundation for the agents' adaptive learning.

The simulation environment designed for this work is an integral part of the system, as it provides the information necessary for the meta-agent to compute the control actions required for each vehicle in formation. This environment not only replicates the actual dynamics of the maritime medium but also incorporates continuous feedback, enabling models to adjust their strategies in real time. The constant interaction between the agent and the environment, structured through an iterative cycle of learning and optimization, endows the system with the ability to adapt and overcome the dynamic and challenging conditions of the aquatic environment.

Within this framework, the fundamental RL concepts—such as the agent, the environment, actions, state, and reward—play a central role. The agent, represented in this case by the centralized meta-agent, acts as the system responsible for making decisions based on environmental perception and received feedback. Each executed action generates a reward, measuring the degree of success or failure of that action relative to the system's objectives. This reward drives the learning process, guiding the agent towards more effective and better-adapted strategies. Meanwhile, the system state, defined by the current environmental conditions, provides the basis for the agent to evaluate its options and make informed decisions. The continuous interaction among these elements ensures that the system evolves and optimizes its performance in response to operational demands.

Figure 4 provides the workflow for collecting information in a RL scenario. RL applied to USV coordination begins when the meta-agent receives information about the initial state of the environment, represented as s_0 in Fig. 5. This state may include data such as the initial positions of the USVs, the presence of obstacles, and the conditions of the maritime environment. Based on s_0, the meta-agent selects an initial action a_0, such as redirecting the USVs towards a specific direction to avoid an obstacle. In response, the environment transitions to a new state s_1, reflecting the consequences of the chosen action—for instance, the new positions of

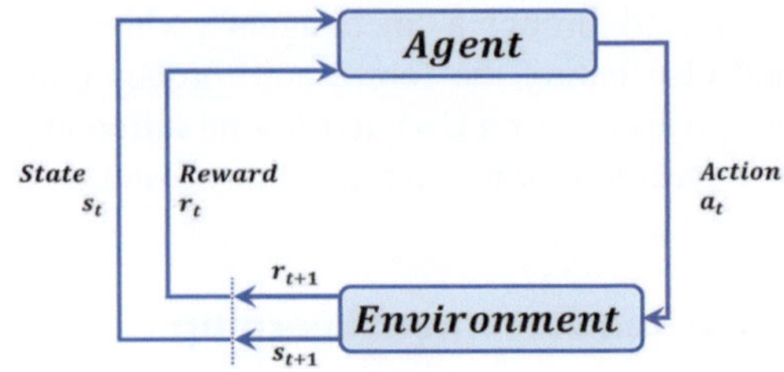

Fig. 4 Agent and environment interaction process

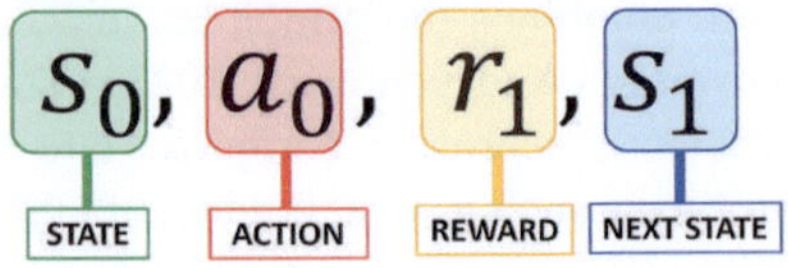

Fig. 5 Components of the Markov decision process in reinforcement learning

the vehicles after the movement. Along with this change, the meta-agent receives a reward r_1, which could be positive if the manoeuvrer was successful, such as avoiding collisions and advancing towards the objective without incidents.

As shown in Fig. 4, this continuous loop of state, action, reward, and subsequent state allows the meta-agent to adjust and refine its strategies progressively, improving the effectiveness of USV coordination in maritime missions. This approach not only ensures safer and more efficient performance but also equips the system with the adaptability required to tackle complex and dynamic operational challenges. However, relying solely on immediate rewards and focusing exclusively on the next step limits the system's ability to plan and execute long-term strategic manoeuvres. To overcome this limitation, the objective shifts towards optimizing the cumulative sum of future rewards over an extended horizon, computed as R_t in (8):

$$R_t = \sum_{k=0}^{\infty} r_{t+k+1}. \tag{8}$$

This formulation ensures that decisions are not only aligned with immediate benefits but also maximize long-term performance, enabling the planning and coordination necessary for complex manoeuvres in USV formations. Rewards are directly related to the observations of the environment, which are derived from the states.

Formation Control Using DRL Techniques

In this context, a state represents a complete and detailed description of the environment at a given moment, providing the agent with all the necessary information to make optimal decisions. For example, in USV control, the state would include data such as the position of each vehicle, speeds, and the conditions of the surrounding water. In contrast, observations represent only a subset of state information. These may be limited by the sensory capabilities of the system or by noise in the measurements, which reduces the precision of the perception of the environment. This distinction is critical in systems like the proposed one, where learning must account for both the observations' partiality and the environment's inherent uncertainty.

Focusing on the control problem, each USV could be equipped to monitor only its immediate surroundings via its sensors, such as sonar, global positioning system (GPS), cameras, or inertial measurement unit (IMU). These observations might include the proximity of other USVs, obstacles in the water, and observable weather conditions. Taking into account the considerations above and now aiming to define the state from a USV's perspective, it is crucial to know the current position vector η and the velocity vector ν. Furthermore, it is equally essential to be aware of the desired position $p_{i\,d} = [x_{i\,d}, y_{i\,d}]$, as it marks the destination to which the USV must navigate.

However, to complete the definition of a state, it is necessary to include other relevant data, such as the desired position of the virtual leader p_v. It is important to note that the virtual leader may not always be on the predefined trajectory, requiring a continuous evaluation of the associated errors. These include the cross-track error e_c and the formation error e_f, which help to determine how accurately the virtual leader follows the planned route and maintains the desired formation. Taking the above into account, the state is defined as follows:

$$s = [x_1, y_1, \psi_1, u_1, v_1, yaw_{rate_1}, x_{1\,d}, y_{1\,d}, \ldots x_n, y_n, \psi_n,$$

$$u_n, v_n, yaw_{rate_n}, x_{n\,d}, y_{n\,d}, x_v, y_v, e_f, e_c].$$

Thus, to define the state of a fleet, at least 12 specific data points are required to enable effective management. Furthermore, each USV is known to be controlled by two distinct control actions F and T, which will be used to compute the control efforts as shown in (3), obtaining an action vector as follows: $a_t = [F_1, T_1, \ldots, F_n, T_n]$.

To accurately evaluate the success of the USV control formation, an effective reward function must be designed. The first step is to ensure that each USV is properly oriented towards its final desired position. Once the proper orientation is established, the next step is to regulate the USVs' velocities to optimize their movement towards the reference points within the formation. The reward function must encourage both efficient progress towards the objectives and the minimization of lateral deviations, which could compromise precision and travel time. For this purpose, a reward function based on the USV's velocities is introduced in (9):

$$r_v = u_i \cos(\beta_i) - v_i \sin(\beta_i). \tag{9}$$

Here, $\nu_i = [u_i, v_i, w_i]$ represents the USV's velocity, while β_i denotes the movement direction angle towards the reference point. The term $u_i \cos(\beta_i)$ promotes progress in the direction of the objective by maximizing the velocity component in that direction. Conversely, the second term, - $v_i \sin(\beta_i)$, penalizes lateral deviations, minimizing the velocity component orthogonal to the objective. This design ensures that by maximizing r_v, the USVs are incentivized to maintain a direct and efficient trajectory towards their goals, reducing any lateral deviations that might slow down the formation process.

However, while this function prioritizes efficiency in the movement direction, it may lead to a problem by excessively encouraging speed. In extreme cases, USVs could stray from their coordinated reference point, compromising formation accuracy and leading the system into an undesired local optimum. In this scenario, USVs might move quickly but fail to approach their specific objectives. The reward function must be complemented by a term that penalizes deviations from the reference point to prevent this issue. This penalty is introduced through a reward function based on the formation error, presented in (10):

$$r_d = -\frac{e_f}{e_{f\max}}. \tag{10}$$

In (10), e_f represents the formation error, defined as the mean distance between the n vehicles and their designated positions. This term penalizes USV deviations from the desired formation, ensuring that the fleet maintains its intended structure throughout the operation. The value $e_{f_{\max}}$ is the maximum allowable formation error, acting as a normalization factor to appropriately scale the penalty. This ensures that the reward function remains well-calibrated and provides a robust metric for evaluating formation deviations. Thus, r_d remains negative, and its magnitude increases as the mean formation error grows, reinforcing the incentive to minimize deviations and maintain the formation structure. By combining both functions, the total reward is defined as a weighted linear combination, as shown in (11).

$$r = k_v r_v + k_d r_d, \tag{11}$$

where the terms k_v and k_d are weighting factors that adjust the relative importance of velocity and direction (r_v) versus the formation error (r_d). This approach balances the objectives of rapid progress towards the destination with maintaining proximity to the reference points, ensuring that USVs operate efficiently and precisely within the desired formation.

Finally, two different modifications to the reward function were introduced to enhance the precision of formation control. Firstly, the total reward was adjusted to include an additional condition:

If the formation error, e_f, is less than a threshold value r_β, an additional reward term is applied. Otherwise, the reward remains unchanged. This condition can be expressed as follows:

$$r_{\text{modified}_1} = \begin{cases} r + 1, & \text{if } e_{\text{f}} < r_\beta, \\ r, & \text{otherwise.} \end{cases} \tag{12}$$

By incorporating this condition, the reward function explicitly incentivizes achieving a low mean formation error. This modification strengthens precise formation control, particularly in challenging scenarios where alignment across all USVs is critical. Furthermore, the introduction of this condition not only improves formation accuracy but also facilitates the development of more robust policies during training.

Another modification to the reward function significantly refines the control objectives by incorporating formation and orientation errors into the total reward. This formulation evaluates not only how closely the USVs adhere to their desired positions but also how accurately they are oriented towards the target direction. The modified reward function is structured under the conditions indicated in (13):

$$r_{\text{modified}_2} = \begin{cases} r - 1, & \text{if } e_{\psi} > \frac{\pi}{3}, \\ r + 2, & \text{else if } e_f < 2, \\ r + 3, & \text{else if } e_f < 2 \text{ and } e_{\psi f} < 0.15, \end{cases} \tag{13}$$

where e_{ψ} represents the average orientation error, defined as the mean deviation of the USVs' orientations from their desired orientations. e_f denotes the formation error, which measures the average positional deviation of the USVs from their target positions within the formation. $e_{\psi f}$ evaluates the collective alignment of the USVs with a global desired orientation. This version of the reward function introduces penalties for significant orientation errors, rewards robust formation control, and simultaneously optimizes both formation and orientation errors. In cases where $e_f < 2$ and $e_{\psi f} < 0.15$, the episode terminates early, reflecting successful and satisfactory fleet control.

While this function does not achieve the same level of precision in formation as the previous version, it offers greater robustness by prioritizing the USVs' arrival at their target positions while maintaining a suitable orientation. This approach ensures that the USVs strike a balance between positioning and orientation, particularly in scenarios where the overall stability of the group takes precedence over strict formation accuracy.

Once the reward functions and the dynamic model of the USV are established, it is crucial to select an appropriate methodology to implement DRL. Solving a RL problem involves modelling an agent capable of maximizing its expected cumulative reward over time. This requires a clear understanding of the internal mechanisms governing the agent's behaviour and its interaction with the environment.

In this context, the policy π is the central component of the RL agent. This policy can be considered the agent's "brain", as it defines the action to be taken based on the system's current state. Learning the policy π relies on the continuous interaction between the agent and its environment. The agent performs actions based on the perceived state, receives rewards, and observes how the environment evolves due to its decisions. This feedback allows the agent to adjust its policy, learning from accumulated experiences to improve performance. Ultimately, the goal is to find the optimal policy π^*, which maximizes the expected return in all possible states.

The ability to iteratively refine π^* is what gives RL its adaptive power, allowing the agent to dynamically adjust to changing conditions and optimize its behaviour in complex scenarios.

The selection of the most suitable DRL algorithm for the formation control of USVs requires a thorough analysis of the system's characteristics and the particularities of each methodological approach. This process involves evaluating different categories of algorithms and justifying the choice of the one that best addresses the specific challenges of this application.

Initially, model-based algorithms were considered to construct an explicit representation of the environment's dynamics to predict the system's future transitions based on the actions taken. While this capability is useful in scenarios where accurately modelling the environment's behaviour is crucial, it is unsuitable for USVs. Here, the objective is not to understand how dynamic variables influence the internal system but to develop a mechanism that optimizes operational decisions in real time. Therefore, model-free methods were chosen, allowing the agent to interact directly with the environment without requiring an explicit representation of its internal dynamics.

Among model-free methods, value-based algorithms such as deep Q-learning (DQN) (Fan et al. 2020) present significant limitations. This approach focuses on learning a value function $Q(s, a)$, which estimates the expected reward when taking an action a from a state s. While appropriate for systems with discrete and bounded action spaces, such as video games or simple environments, its applicability to systems with continuous states and actions, like USVs, is limited. The need to discretize these spaces with sufficient granularity exponentially increases computational complexity and compromises the model's capacity to generalize, rendering this approach unfeasible in high-dimensional scenarios.

There are also two key subcategories within model-free methods: on-policy and off-policy methods (Alibabaei et al. 2022). Off-policy algorithms can leverage data generated by policies other than the one being optimized, which is beneficial in applications where historical data is available. However, in the case of USVs, relevant data can only be obtained through direct, real-time interactions with the environment, limiting the applicability of these approaches. On the other hand, on-policy methods optimize the policy based on current interactions, making them more suitable for this type of problem. Nevertheless, they often encounter difficulties regarding stability and efficiency, especially in complex environments with high-dimensional states.

Proximal Policy Optimization

Faced with these limitations, policy-based methods, particularly those employing actor-critic architectures, stand out as a robust and effective solution. This approach combines two essential components: an actor, which learns a probabilistic policy for selecting actions, and a critic, which evaluates these actions using a value function (Andrychowicz 2021). This structure enables continuous feedback between both components, reducing variance in the gradient estimation of the underlying unknown black-box model and facilitating more stable and efficient learning. Moreover, many modern algorithms integrate multiple approaches to harness the strengths of each. For instance, proximal policy optimization (PPO) (Schulman et al. 2017) combines policy-based techniques with advanced optimization strategies to balance stability and efficiency.

The PPO algorithm represents a significant evolution in the realm of DRL. It inherits advantage actor-critic (A2C) capabilities and can be applied in a distributed and asynchronous learning scenario (Schulman et al. 2017) while incorporating from trust region policy optimization (TRPO) an approach that constrains policy updates within a safe range, avoiding abrupt oscillations that could compromise model stability. One of the key innovations of PPO is the introduction of a clipped objective in its optimization function, which simplifies the tuning process and reduces computational requirements without sacrificing performance. This design enables it to effectively address the challenges associated with complex and dynamic environments, such as those faced by USVs.

The choice of PPO is not merely based on its modernity as an algorithm but on its ability to balance stability, efficiency, and ease of implementation. These characteristics are crucial in systems where states and actions are continuous and environmental conditions vary significantly. PPO operates using an actor-critic architecture, where the agent comprises two core components: the actor, which represents the agent's policy and determines actions based on observed environmental states, and the critic, which evaluates these actions by calculating a value function that estimates expected rewards. Mathematically, the actor models a parametrized policy $\pi_\theta(a|s)$, representing a probability distribution over actions a given a state s. The critic estimates the expected value of a state s using the value function $V_\phi(s)$, with ϕ denoting the parameters of the critic's network.

The learning process in PPO involves the agent interacting with the environment to collect data in the form of trajectories composed of states, actions, rewards, and transitions. These data are used to compute an advantage estimate, a crucial metric that measures the quality of a specific action relative to the expected outcome for a given state. The advantage estimate is defined as in (14):

$$\hat{A}(a, s) = Q(s, a) - V(s), \tag{14}$$

where $Q(s, a)$ represents the action-value function estimating the cumulative reward expected after taking action a in state s, and $V(s)$ is the state-value function estimating the expected reward starting from s. To balance bias and variance, $\hat{A}(s, a)$ is often approximated using generalized advantage estimation (GAE), which incorporates a decay parameter λ.

Policy updates in PPO are driven by the policy gradient, which determines how the actor's neural network parameters, θ, should be adjusted to maximize the expected reward. This gradient is expressed as in (15):

$$\nabla_\theta J(\pi_\theta) = \mathbb{E}_{\pi_\theta}[\nabla_\theta \log \pi_\theta(a|s) \cdot \hat{A}(s, a)], \tag{15}$$

where $\nabla_\theta \log \pi_\theta(a|s)$ captures the sensitivity of the policy to variations in its parameters. This mechanism guides updates by increasing the likelihood of advantageous actions while reducing the possibility of suboptimal ones. A key innovation of PPO lies in its method for limiting policy updates and preventing abrupt changes that could destabilize learning. PPO achieves this by introducing a probability ratio, defined in (16):

$$ratio_t(\theta) = \frac{\pi_\theta(a_t|s_t)}{\pi_{\theta_{\text{old}}}(a_t|s_t)}, \tag{16}$$

which measures the relative change in the probability of an action between the new policy, π_θ, and the previous policy, $\pi_{\theta_{\text{old}}}$. The objective function in PPO combines this ratio with the advantage estimate and incorporates a clipping mechanism to restrict the extent of change in $ratio_t(\theta)$ within a controlled range. The objective function is given in (17):

$$L^{CLIP}(\theta) = \mathbb{E}_t[\min(ratio_t(\theta)\hat{A}_t, \text{clip}(ratio_t(\theta), 1-\epsilon, 1+\epsilon)\hat{A}_t)] \tag{17}$$

where $\text{clip}(r_t(\theta), 1-\epsilon, 1+\epsilon)$ constrains $ratio_t(\theta)$ to the interval $[1-\epsilon, 1+\epsilon]$, with ϵ as a hyperparameter controlling the level of clipping, and $\mathbb{E}_t[\cdot]$ is the average value over t samples. By taking the minimum value, PPO ensures the policy remains within a stable update range, promoting conservative adjustments.

Simultaneously, the critic's network is optimized to minimize the error between the estimated value of a state and the observed cumulative reward. This optimization employs the loss function $L^V(\phi)$, defined in (18):

$$L^V(\phi) = \mathbb{E}_t[(V_\phi(s_t) - R_t)^2], \tag{18}$$

where ϕ represent the critic's neural network parameters, R_t denotes the cumulative reward from time t, and $V_\phi(s_t)$ is the critic's reward estimate for state s_t. This process enhances the critic's evaluation accuracy, which in turn supports more effective updates for the actor.

PPO integrates these principles with first-order optimization techniques, ensuring computational efficiency and ease of implementation. Its adaptability to both continuous and discrete action spaces makes it particularly effective in dynamic and high-dimensional environments. In the context of USV formation control, where states and actions are continuous and environmental conditions are highly variable, PPO balances exploring new strategies with consolidating effective policies. This robustness solidifies its position as a benchmark algorithm in DRL, particularly for complex and real-world applications.

The PPO algorithm employed for this task follows an iterative training process composed of two main phases: interaction with the environment and policy updating. During the interaction phase, the agent employs the current policy to make decisions based on observed states, generating transitions that include the initial state (s_t), the action taken (a_t), the reward received (r_t), the resulting state (s_{t+1}), and an indicator of whether the episode has concluded. These transitions are stored in a buffer for a fixed number of steps defined by the parameter N_{steps}. This phase focuses exclusively on collecting sufficient data for training without performing model updates.

Once the interaction phase is complete, the stored data are used during the policy updating phase. The buffer is divided into batches of a size determined by the parameter B, and each batch is processed over several epochs specified by the parameter N_{epochs}. This exhaustive utilization of the collected data ensures its maximum impact on model improvement. Each training cycle, defined by the parameter N_{envs}, involves the sequential execution of these two phases, allowing the agent to explore new strategies while progressively refining existing ones.

The PPO training process was configured using key hyperparameters to ensure stability and efficacy. These include the following:

- T_{max}: The maximum number of steps per episode, controlling interaction duration with the environment
- N_{envs}: The number of parallel environments used for efficient data collection
- T_{total}: The total number of training steps, defining the overall scope of the process
- α: The rate at which model parameters are adjusted during optimization
- γ: The discount factor determining the significance of future rewards
- λ_{GAE}: A parameter for temporal advantage estimation using the GAE method
- c_{ent}: The entropy coefficient encouraging exploration during training
- N_{steps}: The number of timesteps executed before the model updates
- B: The size of the data batches used in each update
- N_{epochs}: The number of epochs over which the data are processed before each update

A crucial aspect of the training process was the normalization of the state space, ensuring that all input variables were scaled uniformly to enhance stability and accelerate convergence. Variables such as USV positions, velocities, and orientations were normalized to standard ranges, typically $[0, 1]$ or $[-1, 1]$. This adjustment mitigated disparities between variable magnitudes and reduced instability during training. The normalization process is mathematically defined as in (19):

$$z_{\text{norm}} = \frac{z - z_{\text{min}}}{z_{\text{max}} - z_{\text{min}}}, \tag{19}$$

where z_{min} and z_{max} represent the minimum and maximum permissible values for each variable z.

By combining this rigorous training framework with the inherent robustness of the PPO algorithm, the developed system demonstrated significant efficacy in addressing the challenges of formation control for USVs.

Results

The following section presents the simulation, environment, and model setup. The simulated environment, the utilized neural network architecture, and the training were done in Python 3.11 running on a 3.3 GHz processor laptop computer with 16GB RAM and 1.78 GHz-6GB VRAM graphics processor. Regarding the USVs, the hydrodynamic parameters of the Cybership-II ship (Skjetne et al. 2004) were utilized. Recall that these values were only used to simulate the evolution of each USV agent so that new observations could be obtained. Hyperparameters for the RL framework were tuned to fit *continuous actions to continuous observations* starting from values commonly found in the literature.

Training Setup

The training of models for the formation control of unmanned surface vehicles (USVs) was conducted using the Gymnasium library (Towers et al. 2024), enabling the creation of bespoke virtual environments tailored to the specific characteristics of the problem. This approach was essential for simulating and evaluating control strategies in a controlled and reproducible setting. Algorithm 1 defines the customized environment's implementation, where three different functions (initialization, reset, and step) are described.

Multiple initial configurations and reward strategies were implemented in the developed environment to address diverse training objectives. These configurations include the following:

- **Formation 1:** USVs start in random positions that align with a predefined formation, with their orientations directed towards the target. This setup enhances the initial cohesion of the formation, ensuring that USVs occupy strategic positions to optimize navigation.
- **Formation 2:** USVs begin in random positions within a formation but with orientations assigned randomly. This configuration introduces additional complexity, requiring vehicles to adjust both their positions and orientations to achieve a stable formation aligned with the desired trajectory.

The environment also supports selecting different reward strategies tailored to specific training objectives. Adjusting parameters k_v and k_d allows the weighting of various reward components to be fine-tuned. For example, prioritizing k_v emphasizes the alignment of USV velocities with the desired direction, whereas increasing k_d accentuates penalties for formation errors, promoting greater precision. This flexibility ensures the training process remains adaptable to the system's priorities.

Algorithm 1: Gymnasium-based environment for multiples USVs

1 **Initialization:**
- Set n USVs.
- Define action space: $[-1, 1]^{2n}$.
- Define observation space: normalised [0, 1].
- Set trajectory parameters and rendering mode.

Reset Function:

- Initialize USV positions:
  - If formation mode: arrange in a circular or linear formation.
  - Else: set random positions.
- Compute initial normalised observations.
- Return the initial state.

Step Function:

- For each USV:
  - Apply action $a_t = [F, T]$.
  - Update state using dynamics.
- Calculate rewards r_d and r_v for all USVs:

$$r_d = -\frac{\text{error}}{\text{max error}}, \quad r_v = u\cos(\beta) - |v\sin(\beta)|.$$

- Check termination conditions:
  - If average formation error < threshold or max steps reached, terminate.
- Return next state, reward, and done flag.

The learning algorithm utilized in this work is shown in Algorithm 2. The complete PPO framework, including the computation of advantage estimates and optimizing policy networks with tailored functions, is also shown.

During the conducted simulations/training, multiple control models were trained for USV formation, utilizing different reward formulations. Among these models, the most robust and efficient implemented modified reward number 2 (R_{modified_2}). This model excelled in handling complex scenarios where USVs started from random positions and orientations while maintaining the desired formation. Notably, this capability was not achieved with modified reward 1, which, although it prioritized greater initial formation accuracy, exhibited limitations in adaptability to more challenging initial configurations.

Simulation, Results, and Discussion

The performance of the mentioned model was analysed through key metrics recorded during the training process, providing insight into its learning effective-

Algorithm 2: Training process of PPO for USV formation control

Initialization: Import required libraries and define hyperparameters;
Set up the custom environment with n USVs and initialize observation and action spaces;
Configure PPO model with actor and critic networks;
while *not converged* **do**
  for *each episode* **do**
    Reset the environment and obtain the initial state s_0;
    for *each time step t in the episode* **do**
      for *each USV* **do**
        Select an action a_t using the policy network $\pi_\theta(a_t|s_t)$;
        Execute the action a_t in the environment;
        Observe the next state s_{t+1}, reward R_t , and termination flag *done*;
        Store the transition (s_t, a_t, R_t, s_{t+1}) in the buffer;
      end
      if *termination flag is true for all USVs* **then**
        Break the loop;
      end
    end
    for *each update step* **do**
      Sample a batch of transitions from the buffer;
      Compute advantage estimates $\hat{A}_t$ using:

$$\hat{A}_t = R_t + \gamma V(s_{t+1}) - V(s_t)$$

      Optimize the policy network using:

$$L^{\text{policy}} = \mathbb{E}\left[\min\left(ratio_t(\theta) \cdot \hat{A}_t, \text{clip}(ratio_t(\theta), 1-\epsilon, 1+\epsilon) \cdot \hat{A}_t\right)\right]$$

      Optimize the value network using:

$$L^{\text{value}} = \mathbb{E}\left[\left(V_\phi(s_t) - R_t\right)^2\right]$$

      Update actor network θ and critic network ϕ;
    end
    Clear the buffer;
  end
end
Save the trained PPO model;
Optionally, render the trained model for visualization;

ness. The average reward is a critical measure reflecting the agent's progress in policy optimization. A comparison between the obtained rewards with and without modified reward functions (Eq. (13)) is shown in Fig. 6. It is observed that with the modified reward function, the agent proposes actions that quickly converge into an optimal policy, whereas with the initial reward function (Eq. (11)), the agent converges but with a slower rate. Moreover, when considering the modified function, during the training, and, as shown in Fig. 6a, it was observed that, following an initial

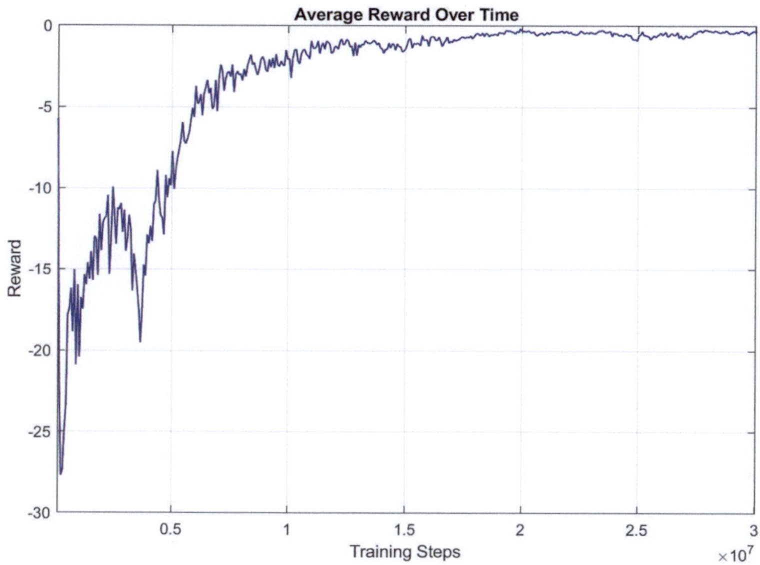

(a)

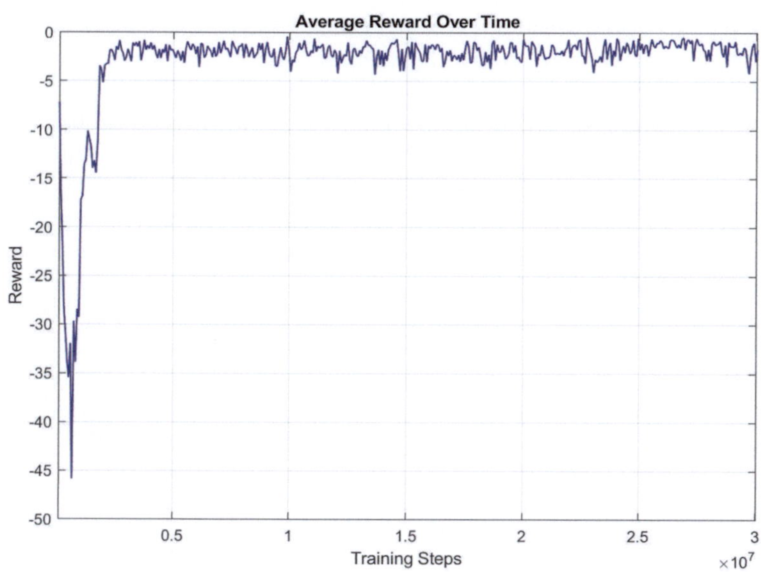

(b)

Fig. 6 Comparison of the obtained rewards with and without the modified reward function. (**a**) Reward obtained in the best model with the initial reward configuration. (**b**) Reward obtained in the best model with the modified reward configuration

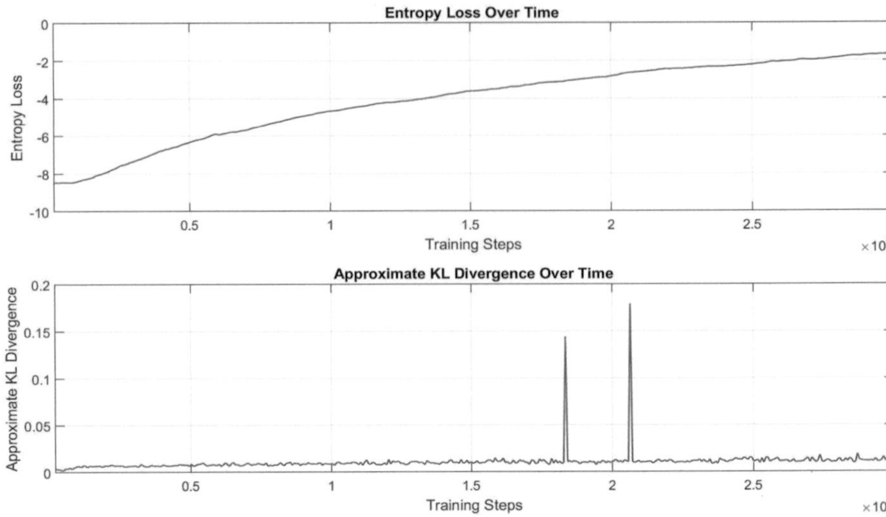

Fig. 7 Entropy loss and Kullback-Liebler divergence measure plot during the training of a model

phase characterized by significant fluctuations in reward values, the model achieved stabilization after approximately 1.0×10^7 training steps. This behaviour indicates that the model successfully reached a consistent and efficient policy, even under high variability scenarios.

Entropy loss evaluates the randomness inherent in the agent's decisions during training. It consists of the mean of the probabilities of choosing an action a given a policy π_θ, $H(a|\pi_\theta) = \log \text{prob}(a)$. The results shown in Fig. 7 depicts that this metric exhibited a progressive decline as training advanced, reflecting a controlled transition from exploration to the exploitation of a specific policy. This reduction is a positive indicator that the model increasingly identifies optimal actions, minimizing the need to explore less efficient alternatives. Finally, in the same figure, the approximate Kullback-Leibler divergence is shown. This value, $KL(\pi_\theta, \pi_{\theta_{old}}) = \log \text{prob}(a_{\pi_\theta} - a_{\pi_{\theta_{old}}})$, measures the deviation between the model's current policy and its previous one. This metric remained largely stable during training, with occasional peaks indicating more significant policy adjustments. This behaviour suggests that updates were performed in a controlled manner, preserving the model's stability while refining its performance.

The combined analysis of rewards and errors observed during the simulations provides a comprehensive assessment of the performance of the trained model and the effectiveness of the implemented reward function. On the one hand, Fig. 8 illustrates the progression of the total reward R alongside its individual components, r_v and r_d. The values of k_v and k_d were equally set to 1. At the beginning of the simulations, the partial rewards exhibit significant fluctuations as the agent explores optimal strategies to maximize its performance. However, around step 300, the curves begin stabilizing, indicating that the model has

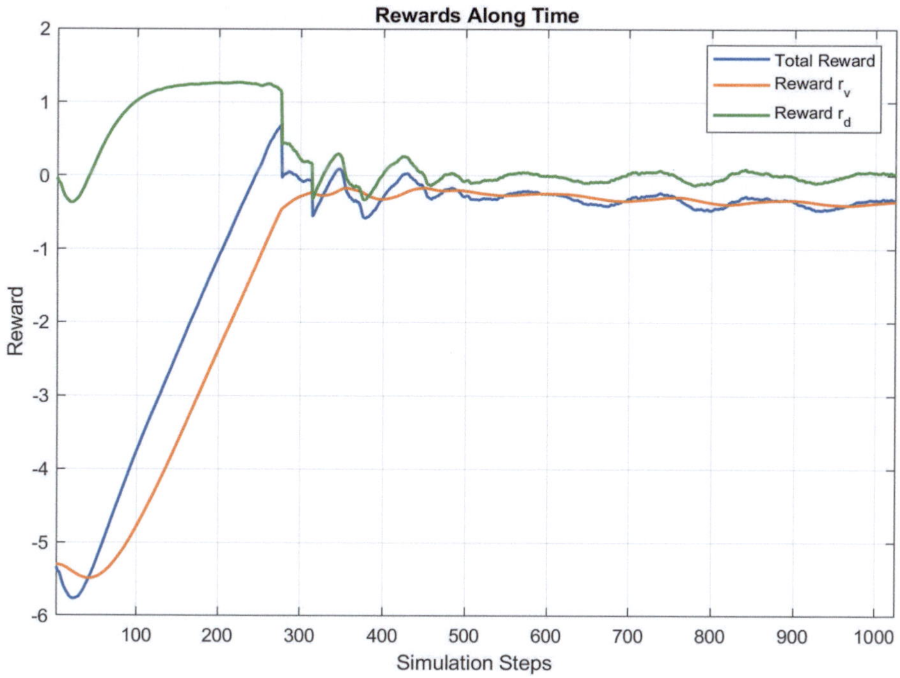

Fig. 8 Rewards obtained at each step during validation

successfully consolidated a policy that balances progress towards the objectives with the minimization of formation errors. This behaviour reflects effective learning, where the model prioritizes both velocity in the desired direction and precision in maintaining formation. The convergence of the total reward further underscores this balanced learning process.

On the other hand, Fig. 9 depicts the evolution of formation and lateral deviation errors. The formation error e_f quantifies the average distance between the current positions of the USVs and their target positions, while the cross-track error e_c measures the discrepancy between the centroid of the formation and the desired trajectory. At the start of the experiment, both errors display elevated values due to the random initial positions and orientations of the USVs. Then, after correctly setting the orientation of the vehicles, the formation and cross-track error are quickly decreased towards 0. From step 200 onwards, both errors stabilize at low values, indicating that the USVs have successfully aligned with the desired formation while accurately following the target trajectory. Additionally, the generated offset is due to the fact that the actor network is trying to minimize formation error while maximizing the surge velocities, as expected when considering Eq. (9).

This performance demonstrates the efficacy of the implemented reward function, which integrates terms designed to penalize deviations and reward gradual alignment towards the objectives. The rapid reduction in formation error, coupled with

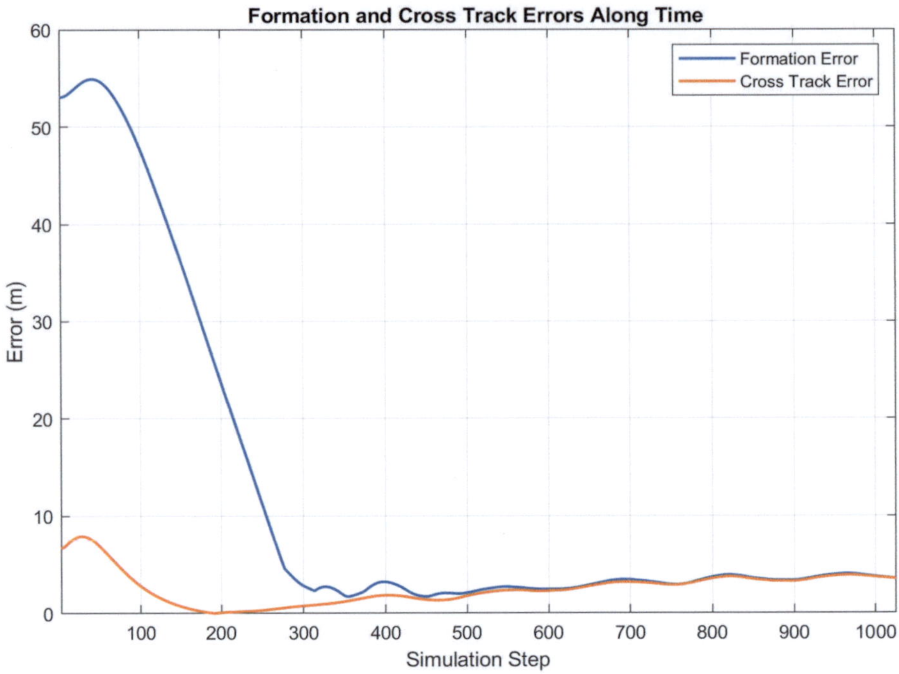

Fig. 9 Observed formation and cross-track errors during validation

the stabilization of the lateral deviation error, confirms the model's ability to handle challenging initial configurations and converge towards robust solutions. Moreover, the stabilization observed in the total reward and its components signifies that the model has achieved an optimal balance between exploration and exploitation, consolidating efficient policies for formation control in dynamic scenarios.

The actor-trained neural network provided efficient control actions for any number of vehicles. Figure 10 represents the trajectories followed by the USVs. Figure 10a shows the initial trajectories, where it can be observed that the policy quickly decided to perform manoeuvrers to orient the USVs towards their goal while maintaining a local formation similar to that of the desired positions. It is noteworthy to state that the initial orientation of all the USVs was set to 0, as well as their velocities.

It also can be observed in Fig. 10b that the USVs maintained formation through their trajectories moving towards their desired locations. The final locations of the USVs were very similar to their desired ones.

In order to strengthen the validity of the proposed approach and to provide an objective comparison of the obtained results, a second model was trained using the initial configuration corresponding to Formation 2—that is, with randomized initial positions and orientations for each USV, as described in Algorithm 2—but in this case employing exclusively the original reward function defined in Eq. (11). This

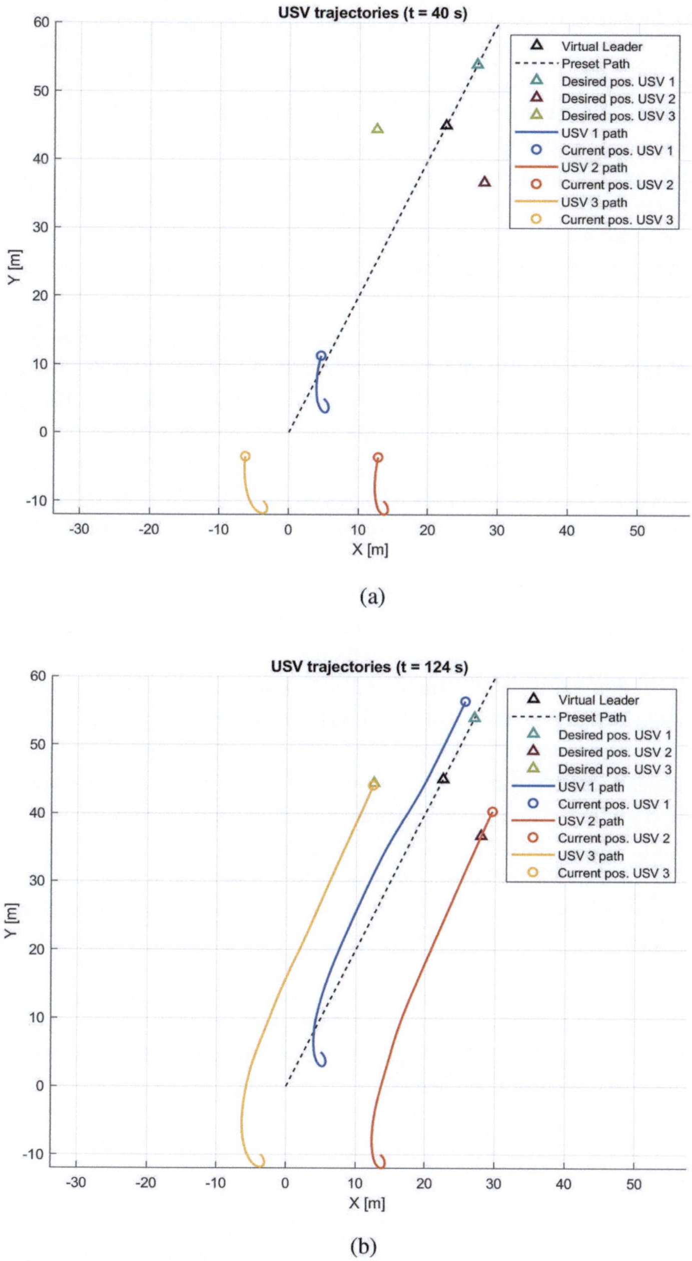

Fig. 10 Evolution of the formation control of USVs using a PPO framework. (**a**) Trajectories at $t = 40$ s. (**b**) Trajectories at $t = 124$ s

formulation lacks additional terms designed to penalize angular deviations or to reward progressive alignment with the desired formation. The results obtained by this baseline model are presented below to contrast, both quantitatively and visually, its performance against the previously analysed model that incorporates a modified reward structure. This comparison enables a clearer assessment of the robustness of the proposed approach under complex and highly variable initial conditions

Conclusion and Future Work

This work presented an actor-critic DRL implementation to control a fleet of USVs. While it is known that the hydrodynamics of surface vehicles are complex to approximate and control using robust controllers, the trained neural network has proven to work even with these underlying highly non-linear functions and provided efficient control actions to promote a formation control following a predefined path. The vehicle's actions were computed by a centralized network. However, since the network output is deterministic, the system can be easily adapted for a decentralized implementation.

The model was deeply studied and trained for three vehicles but can be easily extended to a larger fleet since the action principle is the same. In future works, the proposed system can be adapted to be executed on real scenarios, and, considering that hydrodynamic parameters are usually unknown, the simulation environment can be adapted to further increase the reliability of the simulator by incorporating input gain data as proposed in (Morel et al. 2025). Additional work must be done to include seamless integration of vehicles during execution that can overcome issues of random USVs being unavailable. Finally, to maintain formation, it was proposed in (Zhao et al. 2021) that the controller continuously updates the location of the virtual leader so that the USVs always maintain a certain distance to their desired locations; in this sense, the trajectory will be continuously updated during training as well.

Competing Interests This work was supported by the Agencia Estatal de Investigación (AEI) through Project AQUATRONIC under Grant PID2021-126921OA-C22 and through Project ECO-PORT under Grant TED2021-131326A-C22. The authors have no conflicts of interest to declare that are relevant to the content of this chapter.

References

Alibabaei K, Gaspar PD, Assunção E, Alirezazadeh S, Lima TM, Soares VNGJ, Caldeira JMLP (2022) Comparison of on-policy deep reinforcement learning A2C with off-policy DQN in irrigation optimization: a case study at a site in Portugal. Computer 11(7):104

Andrychowicz M, Raichuk A, Stanczyk P, Orsini M, Girgin S, Marinier R, Hussenot L, Geist M, Pietquin O, Michalski M et al (2021) What matters for on-policy deep actor-critic methods? A large-scale study. In: International conference on learning representations

Barrionuevo AM, Yanes Luis S, Reina, DG Marín SLT (2024) Informative deep reinforcement path planning for heterogeneous autonomous surface vehicles in large water resources. IEEE Access 12:71835–71852

Dash P, Moorhead RJ, Herman J, Beshah W, Sankar MS, Moorhead J, Chesser GD, Lowe W, Simmerman J, Turnage G et al (2021) Evaluation of water quality data collected using a novel autonomous surface vessel. In: OCEANS 2021: San Diego–Porto. IEEE, Piscataway

Fan J, Wang Z, Xie Y, Yang Z (2020) A theoretical analysis of deep Q-learning. In: Learning for dynamics and control. PMLR, pp 486–489

Fossen TI (2011) Handbook of marine craft hydrodynamics and motion control. John Willy & Sons, London

Morel TA, Manzano Crespo JM, Pellicer GB, Orihuela Espina DL et al (2022) Modelling and identification of an autonomous surface vehicle: technical report

Morel T, Orihuela L, Combastel C, Bejarano G (2025) Practical identification approach for the actuation dynamics of autonomous surface vehicles with minimal instrumentation. Ocean Eng 319:120098

Oleynikova E, Lee NB, Barry AJ, Holler J, Barrett D (2010) Perimeter patrol on autonomous surface vehicles using marine radar. In: OCEANS'10 IEEE Sydney, OCEANSSYD 2010

Schulman J, Wolski F, Dhariwal P, Radford A, Klimov O (2017) Proximal policy optimization algorithms. arXiv preprint arXiv:1707.06347

Skjetne R, Smogeli Ø, Fossen TI (2004) Modeling, identification, and adaptive maneuvering of cybership II: a complete design with experiments. IFAC Proc Vol 37(10):203–208

Ten Kathen MJ, Peralta F, Johnson P, Flores IJ, Reina DG (2024) AquaFeL-PSO: an informative path planning for water resources monitoring using autonomous surface vehicles based on multi-modal PSO and federated learning. Ocean Eng 311:118787

Towers M, Kwiatkowski A, Terry J, Balis JU, De Cola G, Deleu T, Goulão M, Kallinteris A, Krimmel M, Arjun KG et al (2024) Gymnasium: a standard interface for reinforcement learning environments. arXiv preprint arXiv:2407.17032

Vasudevan SK, Baskaran B (2021) An improved real-time water quality monitoring embedded system with IoT on unmanned surface vehicle. Ecol Inf 65:101421

Verfuss UK, Aniceto AS, Harris DV, Gillespie D, Fielding S, Jiménez G, Johnston P, Sinclair RR, Sivertsen A, Solbø SA, Storvold R, Biuw M, Wyatt R (2019) A review of unmanned vehicles for the detection and monitoring of marine fauna. Mar Pollut Bull 140:17–29

Wang Y, Liu W, Liu J, Sun C (2023) Cooperative USV–UAV marine search and rescue with visual navigation and reinforcement learning-based control. ISA Trans. 137:222–235

Zhao Y, Ma Y, Hu S (2021) USV formation and path-following control via deep reinforcement learning with random braking. IEEE Trans Neural Netw Learn Syst 32(12):5468–5478

Multitask Multiagent Deep Reinforcement Learning for a Fleet of Autonomous Surface Vehicles in Environmental Cleanup Missions

Dame Seck Diop, Samuel Yanes Luis, Manuel Perales Esteve, Daniel Gutiérrez Reina, and Sergio L. Toral Marín

Contents

Abstract

Plastic pollution in aquatic environments is a serious ecological concern that poses a significant threat to marine life and has a detrimental effect on ecosystems. Autonomous surface vehicles (ASVs) offer a scalable and safe solution for environmental cleanup missions, particularly in dynamic and partially observable aquatic environments. By decoupling the cleanup missions into an initial

D. S. Diop (✉) · S. Yanes Luis · M. P. Esteve · D. G. Reina · S. L. T. Marín
Department of Electronic Engineering, University of Sevilla, Sevilla, Spain
e-mail: dseck@us.es; syanes@us.es; mperales@us.es; dgutierrezreina@us.es; storal@us.es

D. Gutiérrez et al. (eds.), *Smart Water Quality Monitoring*,
https://doi.org/10.1007/978-3-032-03049-8_14

exploration phase followed by a cleaning phase, this chapter addresses the challenge of locating highly dynamic trash items. This chapter introduces a multitask multiagent deep reinforcement learning (MT-MADRL) framework to optimize ASV coordination for both exploration and waste collection, enhancing scalability by enabling knowledge transfer across multiple agents using a single policy trained using deep reinforcement learning (DRL). The proposed method is evaluated across multiple cleanup scenarios, demonstrating its ability to efficiently adapt to dynamic environments. Additionally, a Pareto front analysis is constructed to evaluate trade-offs between two competing objectives, such as exploration and cleaning.

Keywords

Multi-task multi-agent deep reinforcement learning · Multi-objective optimization · Autonomous surface vehicles · Environmental monitoring · Partially observable Markov games

Introduction

Plastic pollution poses several challenges to humans, animals, and the planet. It has significant negative impacts on marine ecosystems, the economy, and the climate. The United Nations estimates that over 800 marine species are negatively affected by marine trash, which is mostly made of plastic (United Nations 2016). Marine animals, including birds, turtles, mammals, and fish, mistake plastics for food and ingest them, causing intestinal blockages, internal damage, and even death. Additionally, these plastics gradually fragment into microplastics and can enter the marine food chain and eventually be consumed by humans through shellfish and fish.

Plastic pollution in aquatic environments has been attributed to a variety of sources, many of which are linked to human activities. A significant contributor is plastic waste generated on land, particularly in urban and coastal areas, which is then transported to the sea via rivers due to issues with urban waste management and illegal dumping (Schwarz et al. 2019). Regarding ports and fishing, coastal areas with high fishing activity tend to have high macroplastic pollution levels, though proximity to ports does not show a strong correlation with this pollution (Masiá et al. 2021). As noted in (Masiá et al. 2021), addressing the problem of marine plastic pollution requires a comprehensive approach, involving collaboration between fishers, maritime ports, and beach services to effectively manage waste and reduce environmental impact.

Autonomous surface vehicles (ASVs) are unmanned vessels designed to operate autonomously on water, equipped with sensors like GPS, sonar, cameras, and sensors to navigate efficiently without constant supervision (see Fig. 1). Additionally, they can map predefined routes and adapt to dynamic conditions, avoiding obstacles and optimizing movement. They are increasingly used for coastline surveillance,

Fig. 1 A fleet of two ASVs navigating at the Alamillo Lake (Seville)

ocean monitoring, multivehicle cooperation, and environmental tasks like marine waste collection (Qiao et al. 2023). ASVs are particularly useful for cleaning up plastic waste and other trash. Traditional waste collection relies on human labor and large vessels with high fuel consumption, driving up costs. ASVs, by contrast, reduce the need for manual labor, are energy-efficient, and have lower operational expenses. They can operate in areas that are impractical or dangerous for human workers, such as near shorelines, in estuaries, and in ports, without compromising safety. Through coordinated efforts and real-time data sharing, ASVs can distribute tasks across a fleet, ensuring efficient coverage of large areas and improving the overall effectiveness of waste collection operations.

Missions aimed at cleaning up plastic pollution in aquatic environments face challenges due to the dynamic nature of plastic trash and the uncertainty of its distribution. A primary challenge is the inability to know the precise locations of plastic waste prior to initiating a cleanup mission. Additionally, plastic in water is highly dynamic, influenced by a variety of factors such as water currents, tides, weather conditions, and human activities. This unpredictability poses significant challenges in the planning and execution of plastic collection operations. Consequently, a two-phase approach, comprising exploration and cleaning, is proposed, as in (Diop et al. 2024), to enhance the efficacy of cleanup operations. The exploration phase is essential for identifying the location of plastic trash. This is achieved by gathering information about the spatial distribution and density of the trash. ASVs equipped with sensors, such as cameras, and the utilization of advanced artificial intelligence (AI) algorithms can facilitate the detection and classification of trash. The exploration phase is instrumental in ensuring the efficient allocation of resources. By identifying areas with the highest concentrations of plastic, ASVs can focus their efforts where they are most needed. Once a sufficient understanding of the plastic's distribution has been obtained, the mission transitions into the collection phase. In this phase, ASVs actively remove the identified plastic trash from the water. Additionally, the collection phase is not static and as plastic trash is collected, water conditions and trash locations may shift. ASVs can adapt their strategies in real time, leveraging the information gathered during the mission to optimize their operations.

It is important to highlight that the tasks of exploration and cleaning are not executed simultaneously but sequentially. Therefore, the critical factor lies in the choice of duration of each phase. Excessive exploration time reduces the time available for cleaning, which can lower overall performance. Conversely, insufficient exploration may degrade the effectiveness of the cleaning phase, as the agents would lack sufficient information about trash locations. It is therefore essential to find the right balance in order to optimize both tasks. The specific objectives of the end user may vary, for instance, prioritizing the maximization of trash collection, ensuring all observed trash is cleaned, collecting a minimum of 80% of the trash, or visiting a minimum of 70% of the map, etc. Therefore, it is essential to ensure that all user-defined requirements are satisfied, enabling the end user to choose the most suitable balance between exploration and cleaning according to their specific objectives.

Problem Formulation

Following the work of Diop et al. (2024), the mission is decoupled into two distinct phases: an exploration phase and a cleaning phase:

- The exploration phase aims to maximize the coverage of the environment, ensuring that all navigable areas are visited to locate trash items.
- The cleaning phase focuses on maximizing the collection of trash once its locations are identified.

The main objective of this chapter is to develop optimal strategies for both tasks while maintaining efficient coordination among ASVs. However, exploration and cleaning represent conflicting objectives, e.g., spending more time to explore improves trash localization but reduces the time available for cleaning and vice versa. To address this trade-off, a multiobjective learning approach is adopted, allowing the training of policies that balance these competing objectives. This methodology enables the construction of a Pareto front, providing insights into the trade-offs between prioritizing exploration and prioritizing cleaning. This front consists of Pareto optimal (also called non-dominated) solutions, meaning each solution on the front achieves a trade-off where no other solution is better in all objectives simultaneously. For example, one policy might excel in exploration but perform slightly worse in cleaning, while another policy might balance the two more evenly. Mathematically, a solution x is Pareto optimal if it is impossible to find another solution x' that improves one objective without degrading another. This is expressed as:

$$f_i(x') \geq f_i(x)\,\forall i \quad \text{and} \quad f_j(x') > f_j(x) \text{ for at least one } j$$

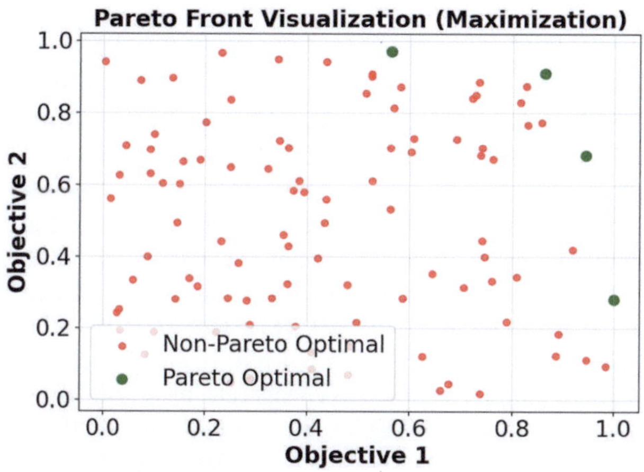

Fig. 2 An example of a Pareto front of various policies trying to maximize two objectives. The Pareto optimal points (green dots) clearly represent the best trade-offs when choosing a policy

Here, $f(x) = (f_1(x), f_2(x), \ldots, f_n(x))$ represents a vector of n objective function values (such as exploration and cleaning metrics) being n the number of objectives. This condition ensures that x cannot be improved in any one objective function f_i without sacrificing performance in another f_j. The Pareto front, representing the best trade-offs among multiple conflicting objectives, helps decision-makers to select the most balanced solutions (see Fig. 2).

Moreover, the problem of determining optimal routes for a large number of ASVs required for continuous monitoring of large bodies of water is a challenging one due to its high complexity. As the size of the problem increases, such as the number of ASVs or the size of the water body to be monitored, the computational effort required to solve it increases exponentially. This complexity makes traditional optimization methods infeasible for large-scale instances. Specifically, this challenge is classified as NP-hard (Luis et al. 2020a; Yanes Luis et al. 2021), which indicates that there is no known algorithm capable of solving all instances of this problem efficiently (i.e., in polynomial time). As suggested by Yanes Luis et al. (2021), a deep reinforcement learning (DRL) approach is recommended for addressing the high dimensionality of this problem, where ASVs must navigate large, partially observable aquatic environments while optimizing multiple objectives. Furthermore, the problem is inherently multitask, as the experiences gained while learning one objective (exploration) can directly contribute to improving performance in the other objective (cleaning). The ability to leverage knowledge transfer between these tasks enhances overall learning efficiency and accelerates policy optimization.

Framework

The aquatic environments chosen to make the missions are the Alamillo Lake (Fig. 3) and the Malaga Port (Fig. 4. The Alamillo Lake is a relatively small artificial lake located in the Parque Alamillo in Seville (southern Spain) with a large number of flora and fauna. It presents a more controlled environment for testing ASV operations. In contrast. the port of Malaga is a Spanish seaport located in the bay of Malaga (southern Spain). It is a commercial, passenger, cruise, marina, and fishing port with a water surface area of 6884 Ha. These bodies of water have been discretized as matrices $M[i, j] \in \{0, 1\}$ of size $H \times W$, where $M[i, j] = 0$ indicates navigable positions and $M[i, j] = 1$ represents obstacles or non-navigable areas (see Fig. 5). In simpler terms, ASVs operate in a grid-like eight-connected environment, meaning that the ASVs can move in any of the eight directions (up, down, left, right, and diagonally). Each movement happens by stepping to an adjacent cell in the grid. If a vehicle tries to move into a non-navigable area, it will be prevented from doing so by a masking system, ensuring it only moves to areas where movement is possible. As proposed in (Luis et al. 2020b), when multiple ASVs aim for the same spot, a consensus algorithm ensures they adjust their paths to prevent crashes. Battery life limits how far the ASVs can travel, setting a maximum distance calculated based on a fully charged battery for each mission. Moreover, ASVs have a limited capacity for the collection of trash. By default, all ASVs share an equal collection capacity, calculated by dividing the total number of trash items by the number of ASVs. Once an ASV reaches its maximum capacity, it is programmed to retire from the mission. ASVs use cameras to "see" within a defined radius Γ around them. This radius determines how far around the ASV the camera can detect objects.

At the start of each mission, K pieces of trash are randomly placed in the environment. These trash items are distributed across one to three spots, chosen

Fig. 3 Satellite view of the port of Malaga, a seaport in the city of Malaga and therefore prone to pollution from plastic waste

Fig. 4 Satellite view of the Alamillo Lake. It is the largest lake in the Alamillo Park (Seville) and is frequently visited by locals and tourists

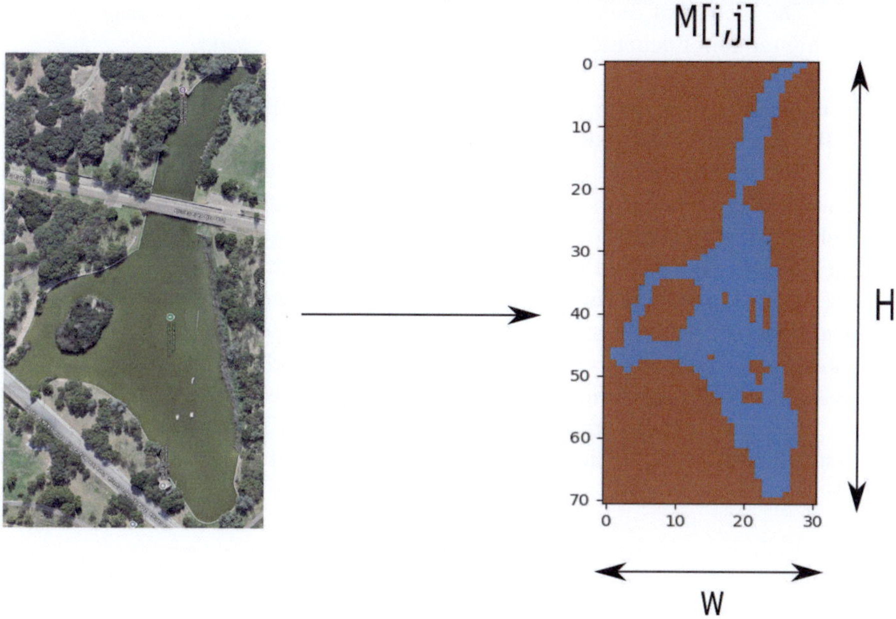

Fig. 5 Discretization of the Alamillo Lake to its matrix $M[i, j] \in \{0, 1\}$. Brown pixels indicate non-navigable zones, assigned the value of 0, while blue pixels denote navigable zones, assigned the value of 1. It is noteworthy that the internal obstacles within the lake have been taken into account

randomly. The placement of the trash at each spot follows a multivariate normal distribution, which means that the trash clusters together in certain areas, mimicking how waste tends to accumulate in the real world. This clustering makes the simulation more realistic, as trash is not evenly spread but tends to gather in particular locations. To simulate how trash moves in the water due to wind and random fluctuations, the position of each piece of trash is updated at each step using the following formula:

$$b_k^{t+1} = b_k^t + \Delta t\,(w_{wind} \cdot \mu_{wind} + w_{rand} \cdot \mu_{rand})$$

This means the position of each piece of trash at time $t+1$ is based on its position at time t, plus a change in position influenced by the following:

- Wind effect: The wind velocity is denoted by a vector μ_{wind}, which is randomly chosen at the start of the mission. It can fluctuate between -1 and 1. The weight of the wind's influence is controlled by a factor $w_{wind} = 0.1$, thereby making wind a moderate influence.
- Random fluctuations: Each piece of trash also has a random movement component, with a random variation μ_{rand} sampled independently for each item at each step. The weight of these random movements is modulated by $w_{rand} = 0.01$, making random fluctuations a smaller influence than the wind.

To track where the trash is located, each element in the matrix $Y[i, j]$ of size $H \times W$ represents how many pieces of trash are located within the region that corresponds to pixel (i, j) on the grid, helping to visualize where the trash is accumulating over time. In order to facilitate visualization, Fig. 6 illustrates an example of a matrix $Y[i, j]$ of trash located in the Alamillo Lake map. Additionally, for the Malaga Port map, Fig. 7 shows an example of the evolution of a matrix $Y[i, j]$ of trash located in the map. Communication between agents is assumed to be ideal, ignoring delays, range limitations, or processing overhead. This simplifies the modeling of collaborative behavior. In this framework, the agents collectively maintain and share a trash model, which serves as a model of the trash distribution in the environment. The trash model, denoted as $\hat{Y}[i, j]$, represents the fleet's understanding of trash locations. This matrix is initialized to zero at the start of the

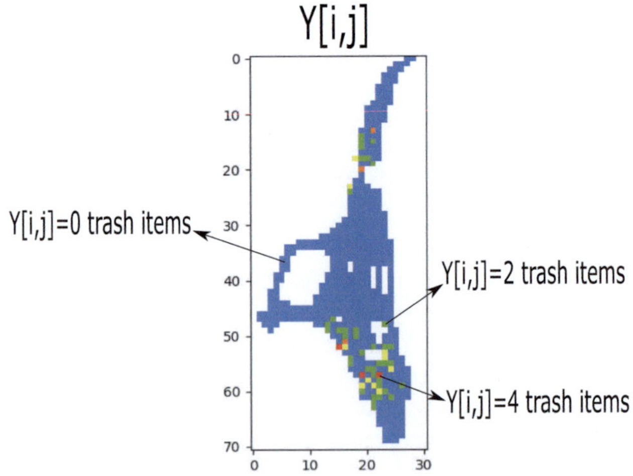

Fig. 6 An example of a matrix $Y[i, j]$ of trash located in the Alamillo Lake map. At each step, the matrix is updated in accordance with the movement of the trash, thereby facilitating a more accurate representation of the discretized distribution of the trash

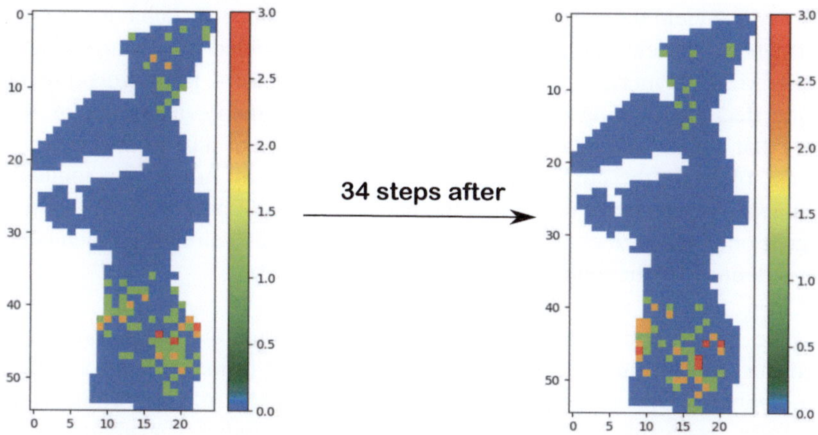

Fig. 7 An example of a matrix $Y[i, j]$ of trash located in the Malaga Port map. Due to the influence of winds and random fluctuations, trash items move over time. In this case, after 34 steps, it is observable that the trash has predominantly moved toward the southwest region of the map

mission and dynamically updated with the values of $Y[i, j]$ for positions within their detection range (Γ). This formulation acknowledges the partial observability of the environment. If a trash item moves within an area no agent is currently monitoring, the agents cannot account for this change until they revisit the area. So the agents can only partially update $\hat{Y}[i, j]$ and the trash model remains unaltered for regions outside their sensory coverage. Despite this limitation, the agents' ability to share their observations ensures that the fleet as a whole has the most comprehensive and up-to-date model of the environment possible.

Methodology

In this section, the methods employed in the development of a strategy for exploration and cleaning are outlined, with a clear rationale provided for their selection. Furthermore, we will discuss the modifications and enhancements made to these methods in order to align them with the specific requirements of our framework.

Reinforcement Learning

Reinforcement learning (RL) (Sutton et al. 1998) is a subfield of machine learning that focuses on training agents to make sequential decisions by interacting with an environment. The agent learns by exploring the environment, trying different actions, and learning through a process of trial and error. Over time, it identifies which actions yield the most favorable outcomes and adapts its behavior to improve performance. The environment in RL is typically modeled as a Markov decision process (MDP) (Puterman 1990), which is a mathematical framework utilized to describe decision-making problems. The Markov property, a feature of MDPs, implies that the environment is memoryless, i.e., the future state of the environment depends solely on the current state and the action taken, not on the sequence of past states. This property simplifies the decision-making process by ensuring that the agent only needs to consider the present state to make an optimal decision. An MDP is defined by the tuple (S, A, P, R, γ):

- S: The set of all possible states, representing the environment's configurations and the relevant information required for the agent to make decisions. In fully observable MDPs, the agent has complete knowledge of the current state; in partially observable ones, only partial or noisy information is available.
- A: The set of all possible actions is defined as the set of decisions or control inputs that an agent can make in order to influence the state of the environment. Actions can be classified as either discrete (e.g., move left/right, stop) or continuous (e.g., adjust speed or heading angle).
- $P(s'|s, a)$: The transition function, describing the probability of moving to state s' from state s after taking action a, capturing the environment's dynamics.
- $R(s, a)$: The reward function is defined as the numerical reward received by the agent for taking an action a in state s. The purpose of the reward function is to provide feedback for actions in order to encourage desirable behavior and discourage undesirable ones.
- γ: The discount factor, determining the importance of future rewards. Values close to 1 emphasize long-term rewards, while lower values focus on immediate rewards.

The policy is defined as the strategy that determines how the agent selects actions based on the current state. At each time step t, the agent observes a state $s_t \in S$ and selects an action $a_t \in A$ based on the policy $\pi(a_t|s_t)$. The environment produces a reward of r_t based on the reward function $r_t = R(s_t, a_t)$ and progresses to the next state s_{t+1} through the transition probability $P(s_{t+1}|s_t, a_t)$. Figure 8 provides a visual representation of this process. The agent's long-term policy return ($J(\pi)$) is defined as the expected discounted accumulated rewards:

$$J(\pi) = \mathbb{E}_{a_t \sim \pi}\left[\sum_{t=0}^{\infty} \gamma^t r_t\right] \tag{1}$$

Therefore, the agent must learn an optimal behavior policy π that optimizes the expected performance $J(\pi)$. The policy is optimized during training to maximize the agent's cumulative reward. Figure 8 shows this process in a more visual way. Value functions (state-value $V^{\pi}(s)$ and action-value $Q^{\pi}(s, a)$) help guide the agent's decision-making process by providing feedback on the long-term benefit of

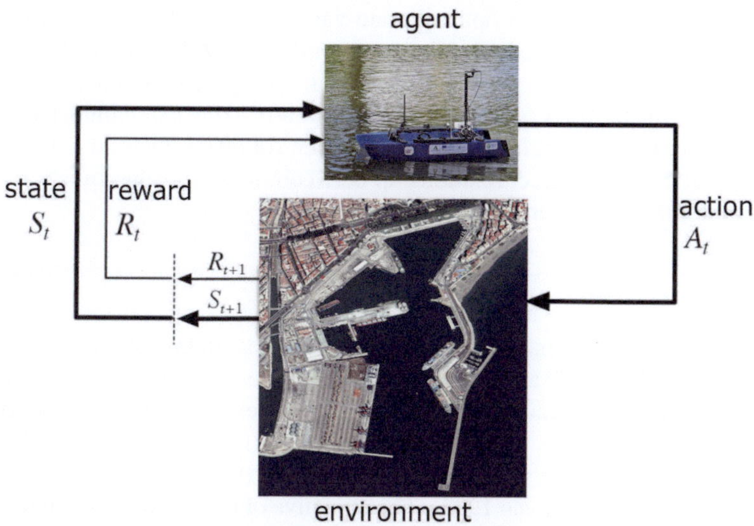

Fig. 8 Illustrations of the key components of the RL framework, including the agent, environment, state representation, action space, and reward function. The agent (in this case, an ASV) interacts with the environment (the port of Malaga) by taking actions (e.g., moving up/down, right/left, etc.) based on its policy (the algorithm employed by the ASV), receiving rewards (or penalties), and updating its state (i.e., the information that the ASV is able to observe)

its actions or states. The state-value function $V^{\pi}(s)$ gives the expected cumulative reward the agent can achieve starting from state s and following a given policy π. It quantifies how good is it to be in state s under policy π.

The action-value function $Q^{\pi}(s, a)$ (also called Q-function) gives the expected cumulative reward the agent can achieve starting from state s, taking action a, and then following policy π. It quantifies how good is it to take action a in state s under policy π. The Bellman equations provide a recursive relationship for calculating the value functions. The Bellman equation for the state-value function is given by:

$$V^{\pi}(s) = \mathbb{E}_{\pi}\left[R_t + \gamma V^{\pi}(S_{t+1}) \mid S_t = s\right] \tag{2}$$

In other words, the value of state s is the expected reward for taking an action and transitioning to the next state, plus the discounted value of that next state. The Bellman equation for the action-value function is:

$$Q^{\pi}(s, a) = \mathbb{E}_{\pi}\left[R_t + \gamma \sum_{s'} P(s'|s, a) V^{\pi}(s')\right] \tag{3}$$

which means that the action-value of state s is the expected reward for taking an action and transitioning to the next state, plus the discounted action-value of that next state. The corresponding value functions for the optimal policy are called optimal value functions. The optimal state-value function $V^*(s) = \max_\pi V^\pi(s)$ gives the maximum expected cumulative reward achievable from state s, starting with the optimal policy π^*. On the other hand, the optimal action-value function $Q^*(s,a) = \max_\pi Q^\pi(s,a)$ gives the maximum expected cumulative reward achievable from state s, taking action a, and then following the optimal policy π^*. Furthermore, there is a direct relationship between the state-value function and the action-value function under the optimal policy:

$$V^*(s) = \max_a Q^*(s,a) \tag{4}$$

This means that the optimal value of a state is the maximum action-value over all possible actions. The RL algorithms that focus on estimating the value functions are denoted value-based methods. The agent uses these value estimates to derive a policy by selecting actions that maximize the value (or Q-value). A popular value-based method for learning the action-value function (Q-function) is the Q-learning algorithm (Watkins and Dayan 1992). The objective of this algorithm is to iteratively update the Q-function as the agent interacts with the environment. The temporal difference (TD) error is used to update the Q-value as it serves as a correction term. It quantifies the difference between the current Q-value estimate $Q(s_t, a_t)$ and the target Q-value derived from the following:

1. The immediate reward r_{t+1} from taking the action a_t.
2. The best estimated future reward $\gamma \max_{a'} Q(s_{t+1}, a')$ based on the agent's current knowledge of the environment. Formally, it is the maximum Q-value over all possible actions a' in the next state s_{t+1}:

$$TD_{error} = r_{t+1} + \gamma \max_{a'} Q(s_{t+1}, a') - Q(s_t, a_t) \tag{5}$$

The update rule is derived from the Bellman equation for action-value function. The adjustment to the Q-value is proportional to the TD error and scaled by the learning rate (α), which controls the weight of recent experiences relative to previous ones (typically $0 < \alpha \leq 1$). The update rule is expressed as:

$$Q(s_t, a_t) \leftarrow Q(s_t, a_t) + \alpha * TD_{error} \tag{6}$$

A key advantage of using Q-learning algorithm is that it is an off-policy algorithm, i.e., it updates the Q-values using a greedy approach, using the action with the maximum Q-value $\max_{a'} Q(s_{t+1}, a')$, regardless of the policy used to select the action a_t that generated the experience. In other words, the policy the agent follows during the learning stage, which may include random actions, does not prevent Q-learning from converging to the optimal policy. In order to facilitate exploration and ensure that the agent does not become trapped in local optima, the epsilon-greedy algorithm is frequently employed in conjunction with Q-learning. In the epsilon-greedy approach, the agent selects a random action with probability

ϵ, thereby encouraging exploration of the environment. Therefore, with probability $1 - \epsilon$, the agent selects the action with the highest Q-value to exploit its current knowledge. As the agent learns and gains more knowledge about the environment, ϵ is often decayed over time. This allows the agent to explore more in the early training stages and gradually shift toward exploiting the learned knowledge as it converges toward an optimal policy. This exploration ensures that the agent explores the environment sufficiently to discover potentially better actions while still exploiting its current knowledge to maximize reward.

Deep Reinforcement Learning

As the number of states and actions increases, the number of state-action pairs grows exponentially. Such rapid growth makes it impractical to store and update the Q-values for every possible state-action pair, particularly in high-dimensional environments. This phenomenon is referred to as the curse of dimensionality. Rather than employing a Q-table, the use of function approximation techniques, such as deep neural networks, has emerged as a more efficient alternative for approximating the Q-function. A notable example is deep Q-learning (Mnih et al. 2015), which employs the deep Q-network (DQN) to estimate the Q-values for a given state-action pair. With the DQN, the current state of the environment serves as the input, while the output is the Q-values for all possible actions the agent can take in that state. This approach enables the agent to generalize across states and reduces the necessity to store every possible pair explicitly while estimating the corresponding Q-values using parameterized weights θ. For a given state s, action a, received reward r, and next state s', the Q-value update in DQL is expressed as:

$$Q(s, a; \theta) = Q(s, a; \theta) + \alpha * (r + \gamma * \max_{a'} Q(s\prime, a'; \theta) - Q(s, a; \theta)) \tag{7}$$

However, when an agent interacts with the environment, consecutive experiences are often highly correlated, which can negatively affect the training process of deep neural networks. Experience replay addresses this by storing past experiences $(s, a, r, s\prime)$ in a memory buffer and sampling them randomly. By presenting uncorrelated samples during training, the agent learns more robust and stable policies. Also, experience replay allows the agent to reuse past experiences multiple times, leading to more efficient use of available data and prevents catastrophic forgetting. In a variation called prioritized experience replay, experiences are not sampled uniformly. Instead, experiences that have a higher temporal difference (TD) error are given higher priority. This method helps the agent focus on experiences that have a higher learning potential, improving the learning efficiency. The neural network used to approximate the Q-values, $Q(s, a; \theta)$, is also called the online network. However, instead of using the current Q-values for future states (which are still in the process of being learned), a stable target is used from the target network $\max_{a'} Q_{target}(s\prime, a'; \theta^{-})$:

$$Q(s, a; \theta) = Q(s, a; \theta) + \alpha * (r + \gamma * \max_{a'} Q_{target}(s\prime, a'; \theta^{-}) - Q(s, a; \theta)) \quad (8)$$

The target network is a delayed copy of the online network, periodically updated by copying the weights of the online network after a fixed number of steps or episodes. This separation between the target and online networks is crucial for mitigating instability and reducing variance in training. That is because using the online network itself to also compute the target Q-values for training can lead to instability and high variance due to the network continuously updating. With a separated target network, the Q-values are updated using a stable reference, which reduces these fluctuations. Also, by updating the target network less frequently, the Q-value estimates become more consistent, which leads to more stable convergence of the learning. However, when calculating the target Q-value as $\max_{a'} Q_{\text{target}}(s', a'; \theta^{-})$, a bias arises when the target network is used both to select the best next action and to evaluate it, overestimating the action values. This overestimation bias was addressed by the introduction of double DQN (Van Hasselt et al. 2016). By updating the target network less frequently, the Q-value estimates become more consistent, resulting in a more stable convergence of the learning process. However, a bias can arise when the target network overestimates action values because the same network is used for both selecting the best action and evaluating it. This issue was addressed with the introduction of double DQN (Van Hasselt et al. 2016). In double DQN, the process of action selection and evaluation is decoupled. While the target network is still used to compute the target Q-value, the action with the highest Q-value is selected using the online network. This modification reduces overestimation bias, as the action selection leverages the most recent updates in the online network, while the evaluation relies on the more stable target network. Specifically, instead of calculating the target Q-value as $\max_{a'} Q_{\text{target}}(s', a'; \theta^{-})$, double DQN computes it as:

$$Q_{\text{target}}(s', \arg\max_{a'} Q_{\text{online}}(s', a'; \theta); \theta^{-})$$

Furthermore, Wang et al. (2016) proposed the *dueling architecture*, an enhancement to the traditional DQN that improves learning efficiency, especially in environments where the importance of actions varies across states. This architecture separates the estimation of the state-value function $V^{\pi}(s)$ and the advantage function $A^{\pi}(s, a)$. The latter represents the relative importance or "advantage" of taking a specific action a in state s, compared to other actions. In this dueling architecture, two streams representing $V^{\pi}(s)$ and $A^{\pi}(s, a)$ (for now called V and A for simplicity) share a common feature learning module and are then combined through a unique

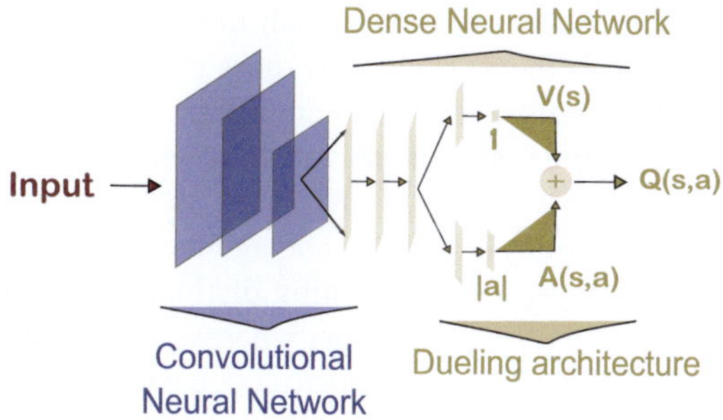

Fig. 9 Illustration of the DQN with a dueling architecture. The input images are processed by convolutional neural networks (CNNs), which are well-suited for 2D image data, to extract features from the bidimensional input, followed by a dense network with a dueling architecture at the final layers

aggregation layer to obtain an estimate of the state-action value function Q using the following general formula:

$$Q(s, a; \theta) = V(s; \theta') + \left(A(s, a; \theta'') - \frac{1}{|A|} \sum_{a'} A(s, a'; \theta'') \right) \tag{9}$$

where θ' and θ'' are the separate parameter sets used to estimate V and A, respectively. In Eq. 9, the advantage function is normalized by subtracting its mean. The normalization ensures that V is the baseline for all actions, and A only describes the deviation of each action's value relative to that baseline. This clear separation improves learning stability, as the network can better distinguish between state and action-specific contributions to the Q-value (they are identifiable). Figure 9 shows how the final layers of the DQN are separated into separate dense parts, where the final layer of each head corresponds to V and A. These are then aggregated, as shown in Eq. 9.

Multiagent Deep Reinforcement Learning

Multiagent systems (MASs) (Weiss 1999) consist of multiple autonomous agents working within a shared environment to achieve individual or collective goals. By leveraging principles of parallelism, decentralization, and cooperation, MAS offers advantages like robustness, flexibility, and efficiency over single-agent systems. In our case, a group of ASVs collaborates to explore and clean an environment, which is a cooperative setting. This allows faster task completion and scalability to larger or more complex areas by simply adding more ASVs. Agents can dynamically adapt their roles and compensate for one another, ensuring effective coverage and cleaning. In this framework, the environment is partially observable, and agents must explore to construct a model of the unknown trash locations at the start of the mission. For example, an ASV may detect trash in a specific location, influencing the navigation decisions of others, leading to more efficient coordination. MAS is particularly effective in partially observable environments, as agents combine their limited views to create a more comprehensive environmental model.

Multiagent reinforcement learning (MARL) (Tan 1993) extends the principles of RL to scenarios involving multiple agents. The agents may operate in cooperative, competitive, or mixed settings, each presenting distinct challenges. In cooperative settings, agents share objectives and rewards, whereas in competitive settings, they may have conflicting goals, leading to adversarial behavior. Mixed settings combine elements of both, where agents must decide when to cooperate or compete. Partially observable Markov games (POMGs) model a specific class of sequential decision problems in which multiple agents interact with a partially observable environment (Liu et al. 2022). Formally, a POMG is defined by the tuple $(\mathcal{N}, \mathcal{S}, \mathcal{A}, \mathcal{O}, \mathcal{T}, \mathcal{R}, \gamma)$, where:

- $\mathcal{N} = \{1, 2, \ldots, N\}$ is the set of $N > 1$ agents.
- $\mathcal{S}$ is the state space, representing the set of global but unobserved states.
- $\mathcal{A}$ is the action space, denoting the collection of individual action spaces $\mathcal{A}_n$ for each agent $n \in \mathcal{N}$.
- $\mathcal{O}$ is the observation space, denoting the collection of individual observation spaces $\mathcal{O}_n$.
- $\mathcal{T}$ represents the transition probability function, modeling the dynamics of the environment.
- $\mathcal{R}$ is the collection of reward functions, where each agent n has an associated reward function $\mathcal{R}_n$.
- $\gamma \in [0, 1)$ is the discount factor, representing the agent's preference for immediate rewards over future rewards.

At each time step t, each agent $n \in \mathcal{N}$ observes $o_n^t \in \mathcal{O}_n$ and selects an action $a_n^t \in \mathcal{A}_n$ based on its policy $\pi_n : \mathcal{O}_n \rightarrow \mathcal{T}(\mathcal{A}_n)$. The environment then transitions to the next state $s_{t+1} \sim \mathcal{T}(\cdot|s^t, \boldsymbol{a}^t)$ under the joint action $\boldsymbol{a}^t = (a_1^t, \ldots, a_N^t)$. Each agent subsequently receives immediate feedback $\mathcal{R}_n^t$ corresponding to the state transition. The agents must learn an optimal joint behavior policy as the collection of all individual policies, i.e., $\pi = \{\pi_1, \ldots, \pi_N\}$. This interdependence means that the effectiveness of an agent's policy depends on the actions of others, leading to a nonstationary learning environment where policies evolve over time.

However, the extension of DRL to the multiagent case (MADRL) enhances the performance of the single-agent case but brings new inherent challenges such as computational complexity and credit assignment (Wong et al. 2023). The proposed approach employs parameter sharing (PS) (Gupta et al. 2017), which shares a single policy among all agents. It is possible because the agents are homogeneous, i.e., they share the same capabilities, observation spaces, and action spaces. Unlike independent learning, where each agent optimizes its own policy separately, PS allows agents to learn from a broader set of experiences, improving sample efficiency and convergence speed. PS enables simultaneous training with the experiences of all agents, significantly reducing computational complexity and scaling efficiently to any number of agents.

Credit assignment in cooperative multiagent systems is a challenge where agents must determine how their actions contribute to the overall team performance when rewards are shared. Improper credit assignment can lead to inefficient learning, causing agents to misjudge their impact on the team's success. For instance, in a trash collection task, if rewards are only given for collective cleanliness, an individual ASV might not differentiate between its effective actions and those of its teammates. This could lead to suboptimal behaviors, such as following other ASVs rather than exploring efficiently. This work addressed it by employing a decoupled reward system, as suggested by Yanes Luis et al. (2023). Each agent receives a reward based solely on its individual contributions rather than a shared team reward. This approach ensures that each agent's actions are clearly linked to their performance metrics, improving credit assignment.

Multitask Deep Reinforcement Learning

The execution of dual tasks within a single environment, such as exploration and cleaning, involves the acquisition of knowledge that is advantageous across tasks. For instance, while exploring, the agent learns efficient ways to traverse the environment, which can be directly applied to reaching and cleaning specific areas. Furthermore, the ability to navigate toward goal locations while minimizing energy usage or time, along with an understanding of navigational obstacles, is a transferable skill that can be applied to both tasks. This interdependence is a key motivation for adopting a multitask learning (MTL) (Caruana 1997) approach. MTL consist of training a model to perform multiple related tasks by exploiting the underlying relationships and commonalities among tasks to improve performance on all tasks. By training on multiple tasks, the model learns shared features that are beneficial across tasks. However, tasks need to have some degree of relatedness for MTL to be effective (Liu et al. 2016). Unrelated tasks can introduce noise into the shared representation, reducing performance.

When using MTL in reinforcement learning, there is a collection of Z tasks $\mathcal{Z} = \{z_i\}_{i=0}^{Z}$. Each task z_i has a unique environment, modeled by POMGs in this setting. In our case, while all tasks share the same state-action space, they differ in their transition probabilities and reward functions. Therefore, at each step, the environment produces a vectorized reward function for feedback where each element corresponds to the reward for a specific task. The objective is to maximize the sum of the expected cumulative discounted rewards for all tasks. Sharing experience across tasks reduces the number of interactions with the environment needed to learn effectively. The natural extension of the DQN algorithm to a multitask setting, as proposed in (Liu et al. 2016), is to make a separate output layer (head) for each task. In this Multitask DQN (MDQN), each head estimates a separate Q-function for each task, Q_e for exploration and Q_c for cleaning, and therefore trained independently using the reward feedback specific to its corresponding task. As a result, the shared

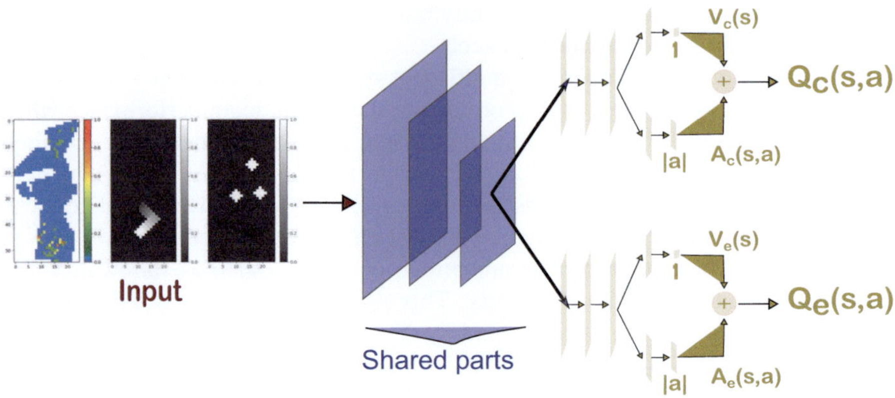

Fig. 10 Architecture of the neural network trained. The proposed PSMA-MDQN architecture uses shared CNN layers to capture common features of the tasks from the input state. Each task is assigned an individual head with a three-layer dense neural network and a dueling architecture. The final Q-values of each task, $Q_c(s, a)$ for cleaning and $Q_e(s, a)$ for exploration, are obtained by combining corresponding V and A streams. All activation layers use the ReLU function

part of the policy network is trained simultaneously with all the rewards and, if the tasks are sufficiently related, it will successfully learn useful common features across the tasks. Sharing experience and network components across tasks reduces the computational and memory requirements compared to training separate models for each task. In this chapter, we use the Parameter-Sharing Multiagent MDQN (PSMA-MDQN) presented in (Diop et al. 2024), which extends the MDQN to the multiagent paradigm (see Fig. 10). This approach leverages the homogeneity of the agents, allowing the training of a single policy $\pi^1 = \pi^2 = \ldots = \pi^N$ that serves multiple agents without increasing the network parameters as compared to a single-agent setup. Since the goal is to learn more than one policy, our proposal utilizes only $Z \times |A|$ neurons in the final layer, regardless of the number of agents N or tasks Z. To effectively distinguish between the exploration and cleaning phases of the mission, a variable ν is introduced to define the duration of these two phases, as in (Diop et al. 2024). The variable ν represents the probability of selecting an action from the exploratory policy (Q_e) versus the cleaning policy (Q_c) at any given moment. Actions are chosen probabilistically, with a proportion ν favoring exploration and $1 - \nu$ favoring cleaning When $\nu = 1$, the mission is fully in the exploration phase, meaning all actions are chosen according to the exploratory policy Q_e. Conversely, when $\nu = 0$, the mission is fully in the cleaning phase, with all actions chosen from the cleaning policy Q_c. In situations where $0 < \nu < 1$, the mission is in a transition phase, where actions are chosen probabilistically. As ν decreases, cleaning actions are more likely to be taken, but exploratory actions are still occasionally taken (see Fig. 11). This gradual shift allows the system to adapt from exploration to cleaning as more information about the environment becomes available. The combined policy $\pi_\nu(s)$ integrates both the exploratory and cleaning

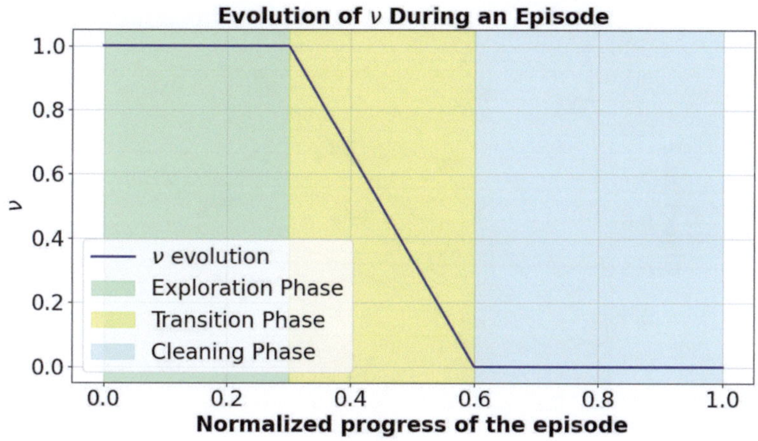

Fig. 11 Example of the evolution of ν in an episode. In this case, the exploration phase lasts until the 30% of the episode, followed by a transition phase until the 60% of the episode, and finally the cleaning phase continues until the end of the episode

policies, determining the action a to take in a given state s depending on the value of ν. The policy is defined as follows:

$$\pi_\nu(s) = \begin{cases} \arg\max_a Q_e(s,a), & \text{with probability } \nu, \\ \arg\max_a Q_c(s,a), & \text{with probability } 1-\nu. \end{cases} \tag{10}$$

Multiobjective Approach

Our multitask policy can be trained using a multiobjective approach, since multitask learning is inherently a multiobjective problem when different tasks conflict, necessitating a trade-off (Sener and Koltun 2019). Multiobjective optimization provides a framework for addressing this challenge by optimizing a set of potentially conflicting objectives simultaneously. In our case, exploration and cleaning share similarities as the agents coordinate their efforts across both phases. However, exploration focuses on covering as much of the environment as possible and minimizing time spent in already visited areas, while cleaning requires repeated visits to specific areas with high contamination. To speed up the learning process and increase the efficiency of sampling, we relax the requirement that agents must use the same policy for both exploration and cleaning, acknowledging the conflicting nature of these goals. In a multiobjective problem, there is no single optimal solution that simultaneously maximizes all objectives. Instead, the focus is on identifying a set of solutions known as the Pareto front. These solutions are Pareto optimal, meaning that, for any other solution, there must be at least one

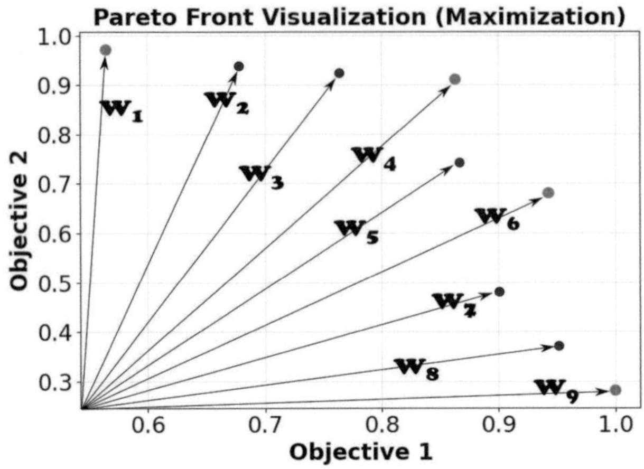

Fig. 12 This is an example illustrating the solutions encountered during training using a decomposition strategy with weights (w_1,w_2, ..., w_9). This approach breaks down the problem into multiple subproblems, each optimized with a different weight combination. Once solutions are obtained, the Pareto front is constructed by selecting the non-dominated solutions, representing the best trade-offs among the objectives

objective where the new solution is worse or equal, and at least one objective where it is strictly better. In order to find the Pareto front in a DRL setting, multiobjective techniques such as decomposition are employed (Zhang et al. 2020). Decomposition methods address the multiobjective problem by decomposing it into a set of scalar optimization subproblems, which are then solved collaboratively. Each subproblem corresponds to a specific trade-off between objectives, defined by weight vectors, and solving these scalar optimization problems typically leads to Pareto-optimal solutions. Upon the resolution of all subproblems, the desired Pareto front can be obtained (see Fig. 12). In our case, scalarization is not applied to the reward function but rather to the selection of the variable ν, which determines the duration of each phase. It is important to note that exploration and cleaning are executed sequentially. So there is no need to prioritize one task over the other by assigning weights within the reward function. An excessive amount of exploration will reduce cleaning time, while insufficient exploration will limit knowledge of where the trash is, hindering performance. Therefore phase duration is a critical factor and it is essential to balance both phases to optimize overall performance. By presenting a Pareto front with different phase durations and their associated best trade-offs, a complete solution can be provided. This approach ensures that all user-defined requirements can be met, allowing the end user to select the most appropriate balance between exploration and cleaning based on their objectives. It has been observed that two neighboring subproblems, which have adjacent weight vectors, frequently yield similar optimal solutions (Zhang and Li 2007). This finding suggests that the solutions to one subproblem can be utilized to assist in the resolution of neighboring subproblems. Specifically, the neighborhood-based parameter-transfer strategy is employed, as in (Zhang et al. 2020), wherein the optimal network parameters obtained for one subproblem are employed as the initialization point for the subsequent subproblem. This sequential transfer of parameters significantly reduces the computational burden. Without parameter transfer, the independent training of a large number of subproblems would demand an immense amount of time and computational resources. Instead, the parameter-transfer mechanism ensures efficient convergence across subproblems while still achieving a comprehensive and high-quality Pareto front.

Throughout training, at regular intervals, the current policy is evaluated over a set of episodes to track its performance. Three distinct policies are saved as candidates for being Pareto optimal: The policy achieving the best performance in one objective, the one achieving the best performance in the other objective and the final policy, since this policy has been exposed to the most training iterations, it may generalize better across unseen scenarios. Once all subproblems have been trained, the saved policies from each subproblem are subjected to an evaluation phase. Each candidate policy is tested over multiple episodes using an independent evaluation set to ensure a fair and comprehensive assessment. The mean values of both objective metrics across these evaluation episodes serve as performance indicators (objective function values). They serve as the criteria for determining the Pareto optimality of the selected policies and therefore to construct the Pareto front.

State

In this framework, the environment is partially observable, which means agents cannot directly perceive all the information about trash locations within the environment. At the beginning of the mission, the locations of trash are entirely unknown, and agents must rely on their exploration capabilities to construct a model of the environment. This model, referred to as the trash model $\hat{Y}[i, j]$, represents the estimated positions of trash based on the observations made by the agents. As agents explore the environment and detect trash within their sensor ranges, they update this model and share their observations with the entire fleet. However, due to the limited detection ranges of the agents, they can only reliably know about trash in their immediate vicinity. If trash moves in an uncovered area, agents cannot account for this change until they revisit it, creating a partially observable and evolving state of the environment. The communication between agents regarding the trash model and their positions is assumed to be ideal, meaning there are no delays, losses, or errors in sharing information. This shared knowledge is integrated into the state representation, allowing agents to make informed decisions based on the best available information. To ensure scalability and efficient use of computational resources, we propose an egocentric formulation of the state representation. This approach ensures that the state is relative to each agent's perspective, enabling all agents to share a single neural network for decision-making. This design ensures scalability and adaptability to scenarios with varying numbers of agents.

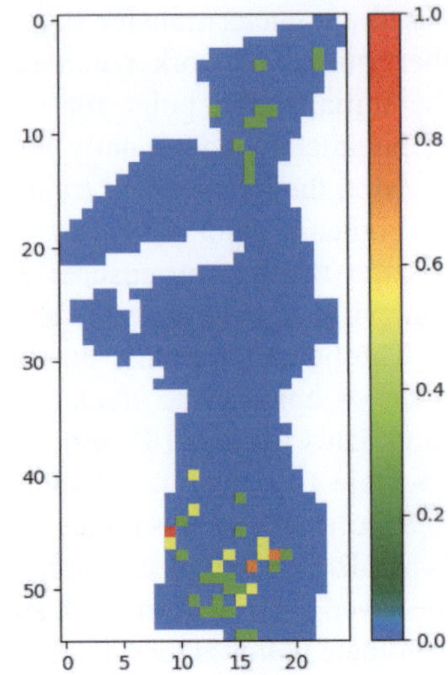

Fig. 13 Example of a trash model $\hat{Y}[i, j]$ min-max normalized, where values range from 1 (representing the known cell with the maximum amount of trash) to 0 (indicating a cell with no trash). This matrix is shared across all agents' states and serves as a cooperation mechanism to enhance the cleaning process

The state representation is constructed as a set of three-channel images, each channel providing critical information necessary for effective navigation and decision-making. These images are min-max normalized, meaning that all pixel values fall within the range [0, 1]. The channels are as follows:

1. Trash Model Channel: This channel contains the trash model (min-max normalized), represented as a matrix $\hat{Y}[i, j]$, which is shared among all agents (see Fig. 13). This matrix provides the estimated locations of trash based on the collective observations of the fleet. The sharing of this map enables agents to operate within a cooperative multiagent paradigm, leveraging the strengths of collaboration. The channel enables agents to focus their efforts on areas with higher trash density while avoiding redundant efforts in already explored or cleaned areas. The trash model forms the foundation for effective decision-making, helping the fleet prioritize regions requiring attention.
2. Agent Position Trail Channel: This channel encodes the current and recent positional history of the observing agent (see Fig. 14). It is structured such that cells included in the agent's detection mask during the current time step (t) are assigned a value of 1, while the cells corresponding to the detection masks of previous time steps decrease in value incrementally by 0.1 per step, down to a minimum of 0.1 at $t - 9$. This decreasing value array reflects the agent's movement over the past ten steps. The values of all other cells are zero. By explicitly encoding the agent's position and movement history within the state representation, this channel provides essential context for decision-making. It enables the neural network to understand the agent's current location at any given moment relative to the environment. Moreover, the decreasing value gradient helps the network distinguish between the agent's current position and areas it has

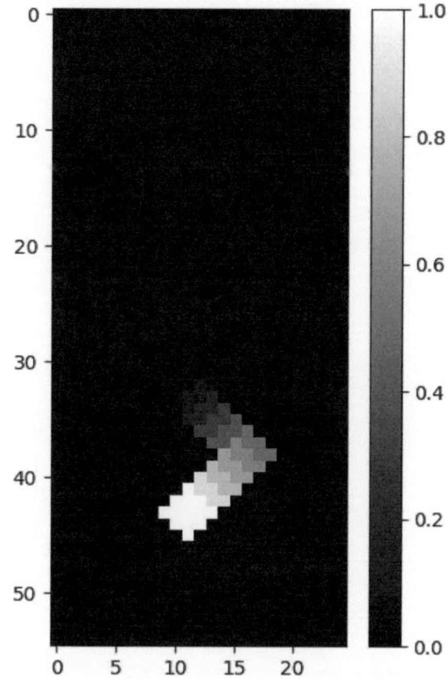

Fig. 14 This figure shows a visualization of the Agent Position Trail Channel. It appears as a fading trail, where positions recently visited by the agent are highlighted, helping to reduce redundancy in exploration

recently traversed. This minimizes redundancy by discouraging revisiting already explored or cleaned regions.

3. Other Agents' Position Channel: This channel captures the positions of all other agents in the fleet, as seen from the perspective of the observing agent (see Fig. 15). In this representation, cells that fall within the detection range of other active agents are assigned a value of 1, while all other cells are set to 0. Including this information ensures that the observing agent is aware of the fleet's overall distribution and can coordinate its actions effectively. For instance, an agent can avoid overlapping with others' areas of exploration or cleaning, maximizing the fleet's collective efficiency and avoiding redundant work. This channel is essential for fostering cooperation and ensuring that agents operate in a decentralized yet harmonized manner.

Rewards

Designing an effective reward function is crucial for guiding agents toward achieving their objectives in a multitask, multiagent environment. In this approach, two distinct rewards are crafted, each corresponding to a different phase of the task. These rewards are designed with penalties to discourage inefficient behavior, ensuring that agents avoid suboptimal solutions and are motivated to perform tasks

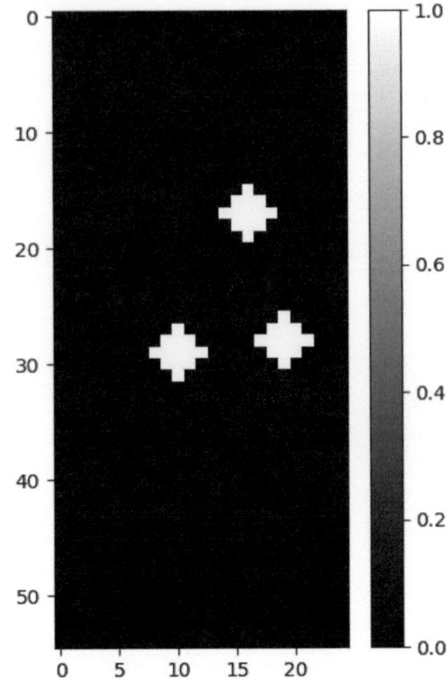

Fig. 15 This figure shows the positions of other agents along with their detection mask, presented in an egocentric view relative to the observing agent

efficiently. Formally, at each step t, the agents receive a vector of two rewards $R_t = [R_{exp}, R_{cleaning}]$, corresponding to the exploration and cleaning tasks, regardless of the policy followed to take the action that generated the reward. Then, each policy is trained using its corresponding reward. Therefore, this approach allows a single experience to contribute to training two different policies, leveraging the off-policy nature of Q-learning.

Exploration Reward

The exploration reward aims to encourage agents to discover previously unvisited areas in an environment while discouraging redundancy and inefficient behavior. The design of the exploration reward balances positive reinforcement for discovery and penalties for redundant or nonproductive actions. The agent is penalized if it repeatedly explores areas it has already visited or if it spends too much time without making discoveries. Therefore, the reward is composed of three main components: a positive reward for discovery, a penalty for inactivity, and a penalty for redundancy:

$$R_{exp} = r_{discovery} - p_{inactivity} - p_{redundancy} \tag{11}$$

The term $r_{discovery}$ rewards the agent for exploring previously unvisited areas. It is defined as the ratio of newly discovered cells to redundant fleet visits in the current step:

$$r_{discovery} = \frac{\omega}{\rho} \tag{12}$$

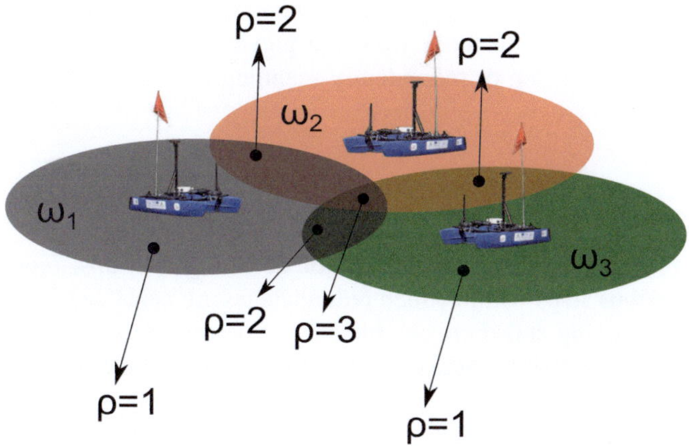

Fig. 16 Visualization of detection masks of three agents overlapping and the values of ρ

where ω is the number of newly discovered cells in the current step and ρ accounts for the overlap between the detection mask of the agent receiving the reward and the detection masks of other agents (see Fig. 16). Therefore, if multiple agents discover the same cell simultaneously, the reward for that cell is shared equally among the discovering agents. This penalizes overlapping exploration, encouraging agents to focus on different areas. The inactivity penalty $p_{inactivity}$ discourages agents from spending time without discovering anything new. The penalty increases with the number of consecutive non-discovery steps and resets when a discovery is made:

$$p_{inactivity} = \frac{1}{\kappa} \cdot \xi \tag{13}$$

where ξ is the term for the ineffective steps, i.e., the number of consecutive steps in which the agent does not discover any new area. Each time the agent fails to discover a new cell, ξ is incremented by 1. When a new cell is discovered, ξ is reset to 0, and no penalty is applied. Mathematically, we define ξ as:

$$\xi = \begin{cases} \xi + 1, & \text{if no new cells are discovered} \\ 0, & \text{if a new cell is discovered} \end{cases}$$

Furthermore, κ is a normalization factor that ensures penalties are proportional to the agent's movement capacity. The value of κ is half the number of steps required to complete an episode when the agent uses all of its battery capacity. Note that the value of $\frac{1}{\kappa}$ serves as a scaling factor to adjust the magnitude of the penalty. This scaling factor have been empirically tuned to provide the desired behavior. The tuning process involves adjusting this factor so that penalties for inefficiency are neither too harsh nor too lenient.

The redundancy penalty $p_{redundancy}$ discourages agents from overlapping their current detection mask with those from the previous steps. This penalty ensures that agents avoid revisiting or redundantly scanning the same areas within a short timeframe:

$$p_{redundancy} = \frac{1}{\psi \cdot \tau} \cdot \beta(\tau) \tag{14}$$

where τ is the time window considered for detecting overlaps, representing the number of previous steps within which redundancies are penalized. $\beta(\tau)$ is the total number of overlapping cells between the current detection mask and the detection masks (previously discovered areas) from the past τ steps. We can write the formula for $\beta(\tau)$ as follows:

$$\beta(\tau) = \sum_{t=t-\tau}^{\tau} \sum_{j=1}^{N} \sum_{\text{cell}_k \in \text{Mask}_i} \mathbf{1}(\text{cell}_k \in \text{Mask}_j)$$

First, within a time window defined by τ, the detection mask of the agent currently receiving the penalty i is compared against the detection masks of all the agents j, including the one currently receiving the penalty. For each step in the time window, the formula examines every cell cell_k in the current detection mask of the agent Mask_i. It checks whether each cell also appears in the detection mask of another agent Mask_j. This comparison is done using an indicator function $\mathbf{1}(\text{cell}_k \in \text{Mask}_j)$, which returns 1 if the cell is found in another agent's detection mask and 0 if it is not. The sum of these indicator function values across all cells, agents, and time steps gives the total count of overlapping cells. The result, $\beta(\tau)$, quantifies how many cells the agent's detection mask has redundantly covered, considering both its own previous scans and the scans of other agents. By penalizing overlaps between agents, they implicitly learn to divide the environment and focus on unique regions, maximizing the overall exploration. Also, without this penalty, agents might repeatedly scan the same areas, leading to wasted computational resources and reduced coverage.

Moreover, ψ is the total number of cells in the agent's current detection mask and is used to normalize the penalty relative to the agent's sensing capacity.

Cleaning Reward

The primary objective of the agents is to clean all the trash within the environment. To achieve this, the reward system incorporates both sparse and dense feedback mechanisms to incentivize effective and efficient trash collection while addressing the challenges of learning in sparse reward environments.

The trash collection reward is defined as the number of trash items r_{trash} collected by the agent in a single step. This reward directly promotes the primary objective and ensures that the agent receives positive reinforcement whenever it successfully collects trash. However, relying solely on trash collection rewards can result in sparse feedback, particularly in environments where trash is scarce or widely distributed. To address this, a dense reward component is introduced, which is based on the proximity of the agent to trash items. This component is calculated by considering every cell cell_k within the agent's detection mask and summing the inverses of the Dijkstra distances to all trash items in the environment. Mathematically, the dense reward component is expressed as follows:

$$R_{dense} = \sum_{\text{cell}_k \in \text{Mask}} \sum_{b_k \in \text{B}} \frac{1}{\text{DijkstraDistance}(\text{cell}_k, b_k)} \tag{15}$$

Here, $\text{DijkstraDistance}(\text{cell}_k, b_k)$ denotes the shortest path distance between cell cell_k and trash item b_k, computed using Dijkstra's algorithm. This reward mechanism encourages the agent to move closer to trash items, as cells nearer to trash yield higher reward values due to the inverse relationship with distance. By summing over all cells in the detection mask and all trash items, the dense reward captures the agent's overall proximity to trash, providing a strong signal for navigation toward areas of interest. In order to ensure that the agent does not receive improper rewards based solely on proximity to trash-rich areas, but instead is penalized for its failure to approach such areas, the dense reward is integrated into a penalty component:

$$p_{\text{distance}} = 1 - \frac{1}{\psi \cdot \rho} \cdot R_{dense} \tag{16}$$

In this formula, ψ is s the total number of cells in the agent's current detection mask and ρ counts the overlap between the detection mask of the agents. Note that p_{distance} is greater when R_{dense} is lower, thereby discouraging agents from moving into areas with minimal or no trash, guiding them toward regions with higher trash density.

Additionally, a trash model update reward is introduced to promote a more dynamic and adaptive cleaning strategy. Whenever an agent collects trash or detects its movement, it receives a reward proportional to the change in the trash model. The reward is calculated by comparing the trash model at the current time step t, $\hat{Y}[i, j]_{\text{current}}$, with the model at the previous time step $t - 1$, $\hat{Y}[i, j]_{\text{previous}}$, but only within the region covered by the agent's detection mask (Mask_i):

$$r_{\text{model}} = \hat{Y}[i, j]_{\text{current}}(\text{Mask}_i) - \hat{Y}[i, j]_{\text{previous}}(\text{Mask}_i) \tag{17}$$

This reward mechanism is designed to motivate agents not only to collect trash but also to detect changes in trash locations, thereby keeping the trash model updated and reinforcing behaviors that actively contribute to the cleaning process. The total cleaning reward is thus a combination of these components:

$$R_{\text{cleaning}} = r_{\text{trash}} + r_{\text{model}} - p_{\text{distance}} \tag{18}$$

Results

As mentioned above, the selected environments for the missions are Alamillo Lake and Malaga Port, chosen based on their shapes and navigability rather than their real-world dimensions. A satellite image (e.g., from Google Maps) is processed to create a discretized representation suitable for input into the neural network. This discretization ensures that the model focuses on the navigability of the water bodies and their overall layout, abstracting away real-world scale differences, which are ignored. The main differences between the scenarios are the following:

- The Alamillo Lake has 563 navigable cells, with more confined space with tighter navigation areas. There are also numerous internal obstacles (islands, vegetation, or artificial structures) that create bottlenecks and require precise navigation strategies.
- The Port of Malaga has 705 navigable cells, 25% more than the Alamillo Lake. It is also a more open area with fewer navigational constraints, offering greater connectivity between navigable zones, which simplifies coverage but requires more travel distance per mission.

In the mission framework, different metrics are used to evaluate performance during the exploration and cleaning phases. These metrics are essential for assessing how effectively the policies achieve their goals:

- Percentage of the Map Visited (PMV): This metric measures how much of the map has been visited at least once during the exploration phase. A higher percentage indicates a more exhaustive coverage of the environment, ensuring sufficient information is gathered for the subsequent cleaning phase.
- Percentage of Trash Cleaned up (PTC): This metric quantifies the proportion of trash that has been effectively eliminated from the environment, relative to the total amount of trash initially present. A higher percentage indicates a more efficient and thorough cleaning operation.

Training and Evaluation Settings

An MDQN is used to train the agents for both tasks, with each task assigned to a separate network head. The decomposition strategy for multiobjective optimization is employed by assigning different values to the variable ν. The chosen values lie within the range [1.0, 0.0] with a step size of 0.1, forming the sequence:

$$\nu = 1.0 - 0.1k, \quad k \in \{0, 1, 2, \ldots, 10\}$$

where k represents the step index, ensuring a uniform decrement from 1.0 to 0.0. Ranging from $\nu = 1$, only exploration phase and $\nu = 0$, only cleaning phase. For

Table 1 Environment parameters and training hyperparameters

Parameters	Values
Number of ASVs	4
Detection area radius (Γ)	2 cells
Distance budget	200 units
Movement length	1 cell
Maximum trash capacity	25 items
Buffer replay size	1×10^6
Batch size	128
Learning rate	1×10^{-4}
Target network hard update	2000
ϵ interval	$[1, 0.05]$
ϵ threshold	50%
Discount factor γ	0.99

simplicity, no transition phase is modeled and the ASVs switch directly from one phase to the other. In the first subproblem ($\nu = 1$), agents are trained for 10,000 episodes, as the initial network weights are completely random. For subsequent subproblems, training is reduced to 5000 episodes due to the neighborhood-based parameter-transfer strategy. This strategy initializes the network parameters for each new subproblem using the optimal parameters obtained from the previous subproblem. This sequential transfer reduces computational costs by nearly half.

The environment parameters and training hyperparameters are summarized in Table 1. The setup involves four ASVs, each with a detection radius of 2 (covering two cells horizontally and vertically and one cell diagonally). In each episode, each ASV can travel a maximum of 200 distance units. Moving one cell horizontally or vertically counts as one unit of distance, while moving diagonally counts as $\sqrt{2} \approx 1.414$ units of distance. Each ASV moves one cell per step. ASVs are deployed at four fixed positions on the map at the beginning of each episode. Each ASV has the same maximum trash collection capacity, thereby, in scenarios with 100 trash items, any ASV that collects 25 or more items leaves the mission area. The prioritized experience replay buffer can store up to 1×10^6 experiences. Batches of 128 samples are drawn from this buffer to train the neural network with a learning rate of 1×10^{-4}. The network is updated every seven steps, and a hard update of the target network occurs every 2000 episodes. During training, the epsilon-greedy exploration strategy begins with $\epsilon = 1$, gradually decreasing to $\epsilon = 0.05$ over the first half of the episodes. For the remaining episodes, ϵ remains fixed at 0.05.

In the training process, it is imperative to periodically evaluate the policies to monitor their progress and identify the best performing policies before potential degradation occurs. To this end, every 200 episodes, the current policy is evaluated over a set of 100 episodes. For each subproblem, three distinct policies are saved as candidates for being Pareto optimal: The policy achieving the best performance in the PMV metric ($BestPMV$), the one achieving the best performance in the PTC metric ($BestPTC$), and the final policy (FP). Once all subproblems have been trained, the selected policies from each subproblem undergo an evaluation

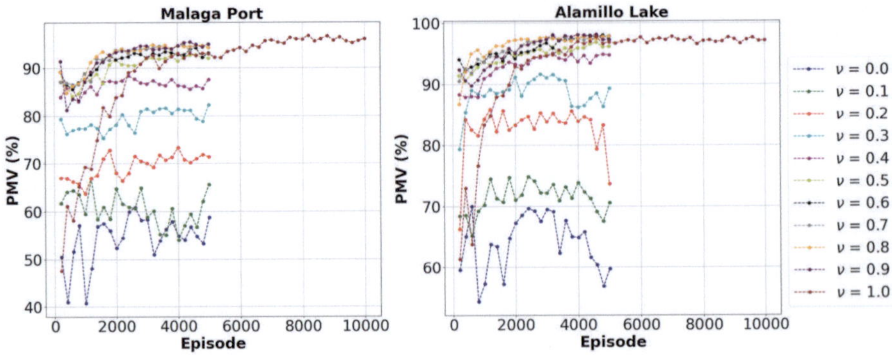

Fig. 17 Illustration of the evolution of PMV in the training of the policies. Each 200 episode, the mean values of PMV are computed over 100 evaluation episodes. Note that the first policy $\nu = 1$ is trained for 10,000 episodes, while the rest are trained for 5000 episodes

phase. Each candidate policy is evaluated for 200 episodes on a dedicated evaluation set, ensuring a fair and comprehensive assessment of its performance. The mean values of both PMV and PTC metrics over these 200 episodes are recorded as the performance indicators, forming the basis for determining Pareto optimality.

The proposed algorithm has been implemented in Python[1] 3, using the PyTorch[2] library for the construction of the neural network. The Gym[3] library served as base for the simulation environment, and numerical and matrix operations were conducted with the NumPy[4] and SciPy[5] libraries. The Pareto optimality of the policies is done using Pymoo.[6] The algorithm's code and results are accessible in a GitHub repository.[7] All simulations were conducted utilizing an Intel Xeon Gold 5220R CPU operating at 2.20 GHz with 187 GB of RAM. Additionally, a NVIDIA GeForce RTX 3090 GPU with 24 GB of VRAM was employed to speed up the neural network training.

Training Results

To monitor the evolution of policy performance during training, policies are evaluated every 200 episodes on the PMV and PTC metrics. Specifically, at each 200-episode interval, the mean values of PTC and PMV are computed over 100 evaluation episodes, serving as performance indicators. Figure 17 presents the

[1] https://www.python.org/

[2] https://pytorch.org/

[3] https://gymnasium.farama.org/

[4] https://numpy.org/

[5] https://scipy.org/

[6] https://pymoo.org/

[7] https://github.com/dsdiop/MultiAgentPlasticCleaningProblem

evolution of PMV for every policy and map throughout training. As expected, PMV decreases as ν decreases, highlighting the intended trade-off between exploration and cleaning. Additionally, the PMV curves become increasingly irregular over time, forming wavelike patterns. This irregularity suggests growing variability in performance, which may be attributed to environmental dynamics and policy adaptations. It is interesting to note that when only cleaning is done ($\nu = 0$), a certain percentage of the map must still be visited in order to locate trash. This underlines the necessity of at least some implicit exploration and reinforces the importance of balancing exploration and exploitation in the cleanup strategy. Furthermore, the impact of knowledge transfer between successive training policies is significant. The first policy trained with $\nu = 1$ begins with a relatively low PMV (47% for Malaga Port and 61% for Alamillo Lake) due to the fact that the initial network has only undergone 200 episodes of training. However, when transitioning to the next policy ($\nu = 0.9$), the training does not start from scratch. Instead, the policy is initialized with the previously trained network, allowing it to leverage prior learning. As a result, the initial PMV values for $\nu = 0.9$ jump significantly to approximately 91.5% for both maps. This knowledge transfer mechanism accelerates the overall training process by building upon previously acquired knowledge rather than relearning from the beginning.

Figure 18 illustrates the evolution of PTC across policy training for all ν's and maps. Unlike PMV, the PTC curves are not smoother as ν decreases, but instead it stays in a fluctuating wavelike patterns in every policy. These fluctuations reflect the stochastic nature of the initial trash distribution in each environment. A noteworthy observation is that, even when there is only exploration ($\nu = 1$), a considerable amount of trash is still collected. This occurs because ASVs inevitably encounter trash during their exploration, which then automatically collect upon entering a cell containing trash. This phenomenon underlines the fact that exploration, while primarily intended for mapping, is unintentionally contributing to the cleaning

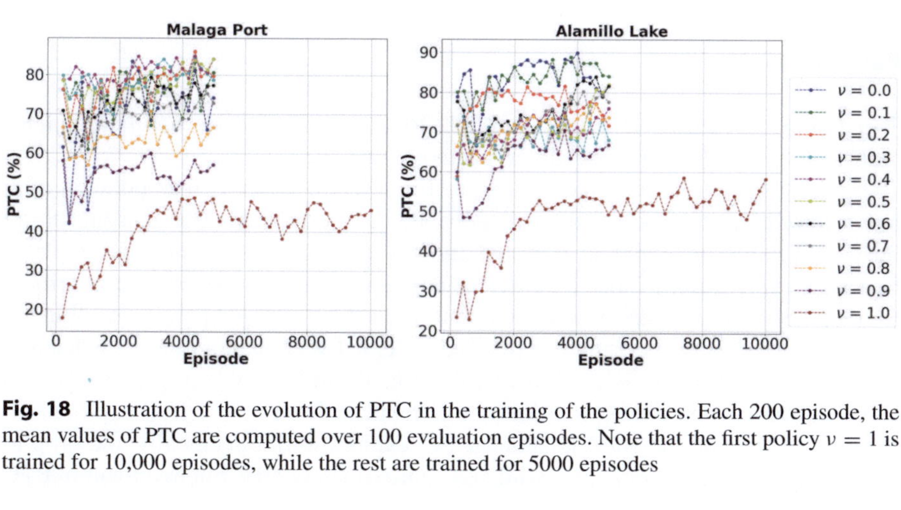

Fig. 18 Illustration of the evolution of PTC in the training of the policies. Each 200 episode, the mean values of PTC are computed over 100 evaluation episodes. Note that the first policy $\nu = 1$ is trained for 10,000 episodes, while the rest are trained for 5000 episodes

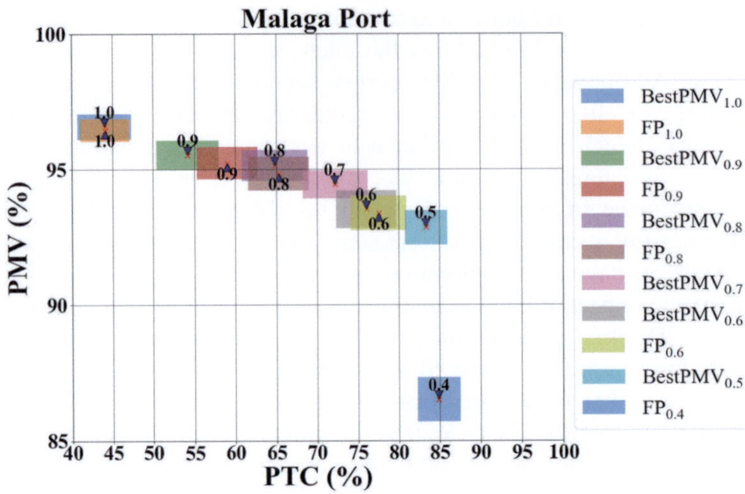

Fig. 19 Pareto front of the policies alongside their 99% confidence intervals in both metrics for the Malaga Port map

process. Furthermore, in Malaga Port, the best performing PTC policy is not the one with the lowest ν, suggesting that an initial exploration phase is crucial for successful cleaning. This highlights the need for structured exploration in larger, more complex environments where trash is widely distributed and difficult to locate. In contrast, in Alamillo Lake, PTC steadily increases as ν decreases, suggesting a less dependent relationship between exploration and cleaning. This observation indicates that in a smaller and less complex environment, the necessity for an extensive exploration phase is reduced, enabling cleaning-focused policies to be effective with minimal prior exploration.

Pareto Fronts

The figures presented (Figs. 19 and 20) illustrate the Pareto optimal policies alongside their 99% confidence intervals in both metrics.

Looking specifically at the results in Fig. 19, we observe that policies with low exploration rates (ν) fail to achieve Pareto optimality. The reason behind this is the large navigational space of the Malaga Port map, which necessitates thorough exploration to locate sufficient amounts of trash. Exploration is critical for discovering trash locations, and without it, the ASVs cannot perform well in the cleaning phase. When we analyze the trade-offs between exploration and cleaning in more detail, it becomes evident that there is a minimal drop in performance in terms of PMV when the exploration rate is reduced from $\nu = 1$ (full exploration) to $\nu = 0.5$ (half of the episode dedicated to exploration). Specifically, the drop in PMV is approximately only 4%, while an approximately 40% improvement in PTC

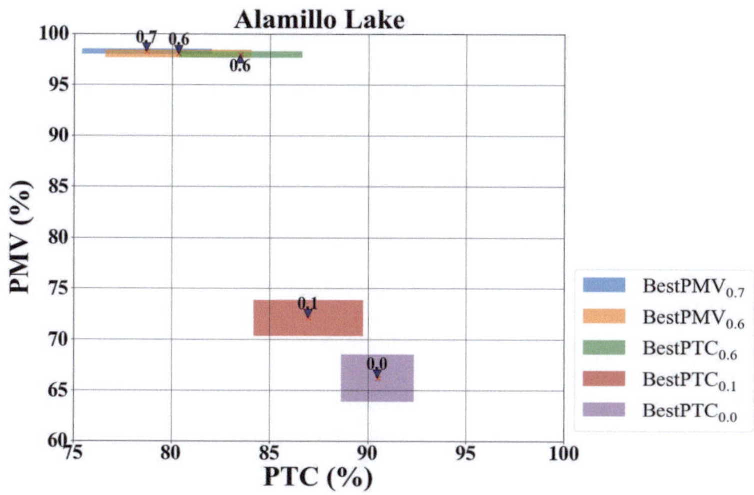

Fig. 20 Pareto front of the policies alongside their 99% confidence intervals in both metrics for the Alamillo Lake map

is observed when the exploration rate is reduced. This implies that a large portion of the map is covered rapidly during the first half of the episode, which means that after this initial phase, there are diminishing returns from further exploration. As a result, focusing more on exploration after covering a large portion of the map yields only limited gains in terms of discovering new trash. Furthermore, more time used to cleaning directly increase the amount of trash collected (PTC). This suggests that in larger environments like Malaga Port, a policy that seeks a balance between exploration and cleaning can achieve better overall performance, provided that the ASVs focus on cleaning once the initial exploration phase has adequately covered the environment. Additionally, it is worth noting that all the Pareto optimal policies are $BestPMV$ or FP. This reinforces the idea that in extensive environments such as the Malaga Port, effective exploration is key to identifying trash locations early on. Without sufficient exploration, the cleaning phase is compromised, resulting in suboptimal performance across both PMV and PTC metrics.

Contrary to Malaga Port, the situation in Alamillo Lake, depicted in Fig. 20, reveals a very different dynamic. Given the relatively small navigational space of the Alamillo Lake environment, a dedicated exploration phase is not as critical. The ASVs in this environment are able to adopt a more greedy strategy, collecting trash as they encounter it without the need for a comprehensive exploration phase. The policy can follow a straightforward approach where, once a trash item is discovered, the ASVs immediately move to collect it. The small map size and the relatively less complex navigation space allow the ASVs to clean effectively by simply focusing on trash spots as they become visible. In fact, the ASVs were able to clean 90% of the trash by only visiting 65% of the map, showcasing the efficiency of this greedy strategy in the context of smaller environments. One additional element to consider is the presence of internal obstacles within the Alamillo Lake map. Despite the simplicity of the environment, these obstacles introduce complexity to the navigation process. However, the policy appears to manage these obstacles well. Once the ASVs detect the obstacles, they can choose the right action, making use of the easily identifiable optimal actions to bypass or navigate around them. This demonstrates the adaptability of the policy, which efficiently handles obstacles within the context of the map's limitations.

Conclusions

In this chapter, we proposed an environmental cleanup strategy using a fleet of autonomous surface vehicles (ASVs) operating in a multitask scenario. The problem was approached by decoupling the mission into two distinct phases: An exploration phase to locate trash items, followed by a cleaning phase to collect. To enable ASVs to cooperate effectively, we trained a deep Q-network (DQN) with two heads, where each head estimates the Q-value for a specific task. A key feature of our approach is parameter sharing, where all ASVs share the same policy due to an egocentric state formulation. This scalability mechanism reduces the number of trainable parameters while maintaining high performance, as proved in (Diop et al. 2024). As exploration and cleaning represent conflicting objectives, a multiobjective learning approach was adopted, with 11 different policies trained for varying phase durations. This approach enabled the construction of a Pareto front, offering insights into the trade-offs between prioritizing exploration (effective location of waste) and prioritizing cleaning (maximizing waste collection). The knowledge transfer mechanism played a fundamental role in accelerating training, as each policy was initialized with the final parameters of the previously trained policy. This progressive learning approach significantly reduced the required training time.

In the case of Malaga Port, the larger navigational area of the environment required an intensive initial exploration phase to ensure efficient subsequent cleaning. The policies trained favoring exploration showed superior performance in locating trash, which in turn improved the effectiveness of the cleaning phase. This finding highlights the importance of dedicated exploration in larger maps where a lack of prior knowledge about trash distribution can significantly compromise cleaning efficiency. In contrast, the smaller environment of Alamillo Lake enabled cleaning policies to achieve satisfactory performance even with minimal exploration. The results imply that, in more confined spaces, ASVs can depend on exploratory behavior during cleaning, thereby diminishing the necessity for an explicit exploratory phase. This outcome indicates that the effectiveness of a cleaning strategy is highly dependent on the size and complexity of the environment and that the proposed framework can adapt to different environmental conditions by appropriately tuning the duration of exploration and cleaning phases. Furthermore, the policies demonstrated the ability to navigate efficiently in environments with inner obstacles while maintaining safe operation through built-in collision avoidance mechanisms that prevent ASVs from entering non-navigable zones.

While the present study has demonstrated the effectiveness of multitask multiagent deep reinforcement learning for environmental cleanup missions, there remain several areas for further exploration and improvement. One promising direction for future research is the development of more computationally efficient methodologies to construct the Pareto front. Our approach is computationally expensive, requiring multiple independent training sessions. Future work could explore fast Pareto front approximation techniques, such as multiobjective evolutionary algorithms (MOEAs). These are optimization techniques designed to solve multiobjective problems by evolving a diverse set of solutions over successive generations Another interesting area for future work is the inclusion of time minimization as an additional objective in the Pareto front. By introducing mission duration as a third objective, future studies could construct a 3D Pareto front, visualizing the trade-offs between exploration efficiency, cleaning efficiency, and time minimization. Additionally, future studies could investigate adaptive policies that dynamically adjust the exploration-cleaning balance in real time based on environmental conditions. For example, an ASV fleet could start with a strong exploration focus and gradually transition to cleaning based on the density of detected trash, rather than following a predefined phase schedule. This could be achieved using hierarchical reinforcement learning (HRL), where a high-level controller decides the exploration-cleaning trade-off, while low-level policies execute individual exploration and cleaning tasks.

Acknowledgments This work was funded by the following projects: TED2021-131326B-C21 of the Spanish Ministry of Science MCIN/AEI/10.13039/501100011033, and the European Union with "NextGenerationEU/PRTR" funds; in part by "Junta de Andalucía: Consejería de Universidad, Investigación e Innovación" through the project "Monitorization of Environmental Dangers with Unmanned Surface Agents: (MEDUSA)" under Grant PCM_00019; in part by the European Union with "Next Generation EU/PRTR" funds, the Spanish Ministry of Science PID2021-126921OB-C21, under Grant TED2021-131326A-C22.

References

Caruana R (1997) Multitask learning. Mach Learn 28:41–75

Diop DS, Luis SY, Esteve MP, Marín SLT, Reina DG (2024) Decoupling patrolling tasks for water quality monitoring: a multi-agent deep reinforcement learning approach. IEEE Access 12:75559–75576.

Gupta JK, Egorov M, Kochenderfer M (2017) Cooperative multi-agent control using deep reinforcement learning. In: Autonomous agents and multiagent systems: AAMAS 2017 workshops, best papers, São Paulo, Brazil, May 8–12, 2017, revised selected papers 16. Springer, Berlin, pp 66–83

Liu LT, Dogan U, Hofmann K (2016) Decoding multitask DQN in the world of minecraft

Liu Q, Szepesvári C, Jin C (2022) Sample-efficient reinforcement learning of partially observable Markov games. In: Advances in neural information processing systems, vol. 35, pp 18296–18308

Luis SY, Reina DG, Marín SLT (2020a) A deep reinforcement learning approach for the patrolling problem of water resources through autonomous surface vehicles: the Ypacarai Lake case. IEEE Access 8:204076–204093

Luis SY, Shutin D, Gómez JM, Reina DG, Marín ST (2020b) Deep reinforcement multi-agent learning framework for information gathering with local Gaussian processes for water monitoring

Masiá P, Ardura A, Gaitán M, Gerber S, Rayon-Viña F, Garcia-Vazquez E (2021) Maritime ports and beach management as sources of coastal macro-, meso-, and microplastic pollution. Environ Sci Pollut Res 28:30722–30731

Mnih V, Kavukcuoglu K, Silver D, Rusu AA, Veness J, Bellemare MG, Graves A, Riedmiller M, Fidjeland AK, Ostrovski G et al (2015) Human-level control through deep reinforcement learning. Nature 518(7540):529–533

Puterman, ML (1990) Markov decision processes. In: Handbooks in operations research and management science, vol. 2, pp 331–434

Qiao Y, Yin J, Wang W, Duarte F, Yang J, Ratti C (2023) Survey of deep learning for autonomous surface vehicles in marine environments. IEEE Trans Intell Transp Syst 24(4):3678–3701

Schwarz A, Ligthart T, Boukris E, van Harmelen T (2019) Sources, transport, and accumulation of different types of plastic litter in aquatic environments: a review study. Marine Pollut Bull 143:92–100. https://doi.org/10.1016/j.marpolbul.2019.04.029

Sener O, Koltun V (2019) Multi-task learning as multi-objective optimization. https://arxiv.org/abs/1810.04650

Sutton RS, Barto AG et al (1998) Introduction to reinforcement learning

Tan M (1993) Multi-agent reinforcement learning: independent vs. cooperative agents. In: Proceedings of the tenth international conference on machine learning, pp 330–337

United Nations (2016) Marine debris: understanding, preventing and mitigating the significant adverse impacts on marine and coastal biodiversity. https://api.semanticscholar.org/CorpusID:135303347

Van Hasselt H, Guez A, Silver D (2016) Deep reinforcement learning with double q-learning. In: Proceedings of the AAAI conference on artificial intelligence, vol. 30

Wang Z, Schaul T, Hessel M, Hasselt H, Lanctot M, Freitas N (2016) Dueling network architectures for deep reinforcement learning. In: International conference on machine learning, pp. 1995–2003. PMLR

Watkins CJ, Dayan P (1992) Q-learning. Mach Learn 8:279–292

Weiss G (1999) Multiagent systems: a modern approach to distributed artificial intelligence. MIT Press, Cambridge

Wong A, Bäck T, Kononova AV, Plaat A (2023) Deep multiagent reinforcement learning: challenges and directions. Artif Intell Rev 56(6):5023–5056

Yanes Luis S, Gutiérrez D, Toral S (2021) A multiagent deep reinforcement learning approach for path planning in autonomous surface vehicles: the Ypacarai-Lake patrolling case. IEEE Access. https://doi.org/10.1109/ACCESS.2021.3053348

Yanes Luis S, Perales Esteve M, Gutiérrez Reina D, Toral Marín S (2023) Deep reinforcement learning applied to multi-agent informative path planning in environmental missions. Stud Comput Intell 1090:31–61. https://doi.org/10.1007/978-3-031-26564-8_2

Zhang Q, Li H (2007) Moea/d: a multiobjective evolutionary algorithm based on decomposition. IEEE Trans Evol Comput 11(6):712–731 (2007). https://doi.org/10.1109/TEVC.2007.892759

Li K, Zhang T, Wang R (2020) Deep reinforcement learning for multiobjective optimization. IEEE Trans Cybern 51(6):3103–3114

Deep Reinforcement Learning and Informative Path Planning: Diving into the Cooperation of Heterogeneous Aquatic Surface Vehicles

Alejandro Mendoza Barrionuevo, Samuel Yanes Luis, Daniel Gutiérrez Reina, and Sergio L. Toral Marín

Contents

Abstract

This chapter focuses on an analysis of a study in which a fleet of autonomous surface vehicles collaborate to find and clean plastic debris from aquatic environments. The fleet is divided into two teams with distinct roles. On one side is the scout team, which is in charge of updating and finding the trash items

This work was funded by "Junta de Andalucía: Consejería de Universidad, Investigación e Innovación" through the project "Monitorization of Environmental Dangers with Unmanned Surface Agents: (MEDUSA)" under Grant PCM_00019.

A. M. Barrionuevo (✉) · S. Yanes Luis · D. G. Reina · S. L. T. Marín
Department of Electronic Engineering, University of Sevilla, Sevilla, Spain
e-mail: amendoza1@us.es; syanes@us.es; dgutierrezreina@us.es; storal@us.es

D. Gutiérrez et al. (eds.), *Smart Water Quality Monitoring*,
https://doi.org/10.1007/978-3-032-03049-8_15

in the environment, thanks to its long-range cameras and high speed. On the other hand, there is the cleaner team, with more basic and short-range cameras to limit economic costs. These are also slower, so they cover half the space in the same time, since they carry the waste collection system. This chapter introduces reinforcement learning and how it can be applied to solve this problem, using a previous work as an example. In addition, an ablation and parameter tuning study of two key components of reinforcement learning is conducted: the reward function and the observation function.

Introduction

Water is one of the Earth's most vital resources, sustaining life, ecosystems, and human activities. Oceans, rivers, and lakes not only are home to a great diversity of aquatic life but also play a fundamental role in regulating the climate of the planet, provide food, and serve as a means of transportation. However, the health of these bodies of water is under serious threat due to the increasing presence of pollution, mostly in the form of floating waste. Plastic debris, discarded fishing nets, and other types of trash are accumulating at an alarming rate, endangering marine species, ecosystems, and human livelihoods.

The impact of this pollution scales to many levels. Marine organisms, including fish, turtles, and seabirds, often mistake plastic waste for food, leading to ingestion and, in many cases, death. This pollution also affects biodiversity and enters human food chains, posing significant health risks. A recent report from the World Economic Forum, Ellen MacArthur Foundation and McKinsey & Company (2016) warns that, without urgent intervention, the weight of plastic in the oceans could surpass that of fish by 2050. The scale of the problem is immense, with millions of tons of waste entering water bodies each year, and addressing it is both a global and an urgent challenge.

Traditional methods of removing wastes from the water involve manual collection or large-scale cleanup operations with boats and nets. However, these methods are often costly, laborious, and ineffective, especially in large and remote areas. In addition, they can pose risks to worker safety and even damage aquatic habitats if not carefully managed. To address this urgent problem, the use of autonomous surface vehicles (ASVs) presents a promising solution. These autonomous robotic systems are capable of navigating bodies of water, detecting and collecting floating debris with minimal human intervention. Thanks to advanced cameras, intelligent path-planning algorithms, and cooperative strategies, ASVs can effectively cover areas with high concentrations of debris, making cleanup operations more efficient and sustainable. The deployment of electric ASVs offers a scalable and adaptable approach to address water pollution, ensuring cleaner aquatic environments.

There are several types of unmanned robots designed specifically for collecting wastes in bodies of water, each adapted to different needs and environments. For example, to capture small-sized debris, some ASVs are equipped with nets to

Fig. 1 Prototypes of ASVs equipped with stereo cameras and differential GPS. Depth camera triangulation, GPS and attitude allow them to locate the global position of the wastes

collect floating objects, such as Kamarudin et al. (2021), which incorporates a net to collect trash. It is designed primarily for cleaning small lakes and narrow rivers, but encountered difficulties due to interference with plants. There are other more sophisticated robots, such as the one proposed in (Akib et al. 2019), where a mechanical hand pushes the floating waste toward a conveyor belt and dumps it in a bin attached at the back of the vehicle. This robot is designed to operate in larger bodies of water such as canals, lakes, rivers, and even oceans. The prototype can collect up to 10 kg of garbage and clean an area of 3000 cm^2, with a battery autonomy of 4 hours. Depending on the characteristics of the vehicles, they can operate in different areas. Compact ASVs are ideal for lakes or narrow channels, while larger and more robust models can handle rough conditions in open seas.

Advances in sensors and perception technologies have enabled many ASVs to include advanced detection systems, such as cameras, LIDAR, or radar, to identify and locate waste with greater precision. Some models are even designed to work in coordinated fleets, maximizing efficiency and covering wider areas. This variety reflects the growing innovation in technological solutions to address the global problem of contamination in water bodies.

For example, an ASV designed specifically for smart water quality monitoring applications is presented in (Díaz et al. 2024). This vehicle is equipped with a variety of high-quality sensors that enable it to measure important water quality parameters such as pH, conductivity, temperature, and turbidity. Additionally, it is equipped with a sonar for seabed bathymetry tasks. A remarkable aspect of this prototype is its ability to detect and locate macroplastics in real aquatic environments. This is achieved through a stereo camera system and the implementation of advanced deep-viewing models, such as recent versions of YOLO.[1] These vehicle types can be seen in Fig. 1, illustrating their practical design and deployment capabilities.

[1] https://github.com/ultralytics/ultralytics

The use of detection networks such as YOLO has become a common practice in applications that require real-time object identification and localization, including those related to ASVs, as seen in works like (Díaz et al. 2024; Shang et al. 2024). This type of convolutional neural network is especially valued for its ability to process images quickly and with high accuracy, which is crucial in aquatic environments where conditions change dynamically. In the context of ASVs for waste collection, YOLO makes it possible to identify floating trash, macroplastics, or other objects of interest while the vehicle is in motion. Furthermore, by integrating modern versions of YOLO with stereo cameras, it is possible not only to detect objects but also to estimate their relative distance and GPS position, which significantly improves the ability of the ASVs to plan their trajectories and optimize collection.

As seen in Fig. 2, the method used in (Díaz et al. 2024) employs a stereo camera setup combined with YOLO to perform both object detection and spatial localization. The stereo camera captures two slightly offset images of the same scene, which allows the system to compute depth information by disparity mapping. YOLO processes these images to identify and classify objects of interest, such as floating plastics, while the depth information from the stereo setup provides the position relative to the ASV of the detected objects. This combined approach allows

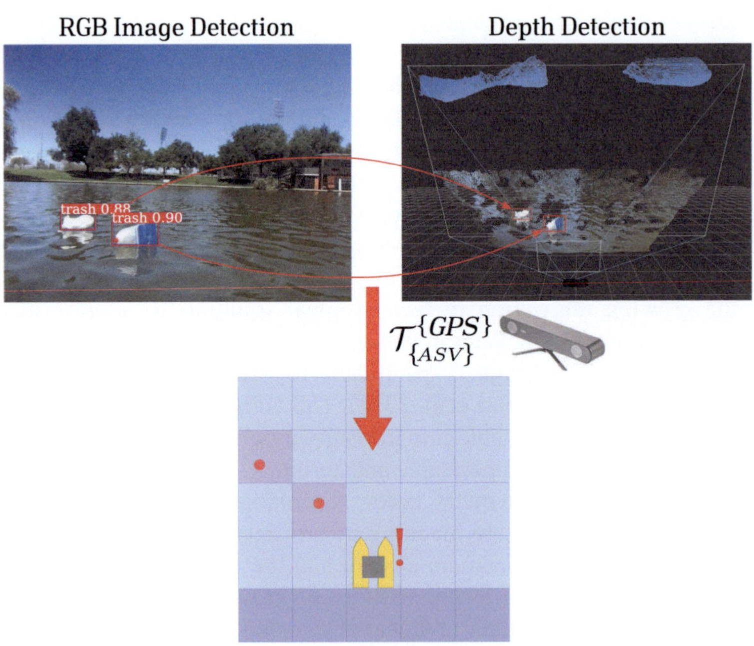

Fig. 2 Visual representation of trash item detections and their GPS positioning using detection neural networks, such as YOLO, and stereo camera. The detection bounding box is marked in red in the RGB image, as well as its predictive confidence. Using the GPS positioning and attitude of the ASV, it is possible to locate the macro-plastics in a global frame

the ASV to generate an accurate map of the location of trash in its environment. When combined with GPS data, the system can geolocate each detected object, creating a detailed spatial representation of the distribution of waste in the monitored area.

Problem Description

This chapter will start from the previous research conducted in (Barrionuevo et al. 2024b), where a fleet of heterogeneous ASVs must cooperate to locate and collect plastic waste in aquatic environments. The system proposed in that study divides the fleet into two specialized teams: scouts, in charge of exploring and mapping areas of contamination, and cleaners, responsible for collecting the detected waste. This division of roles takes advantage of the unique capabilities of each type of vehicle, maximizing operational efficiency. Coordination between the two teams is achieved through a framework based on deep reinforcement learning (DRL), which allows the agents to learn strategies to maximize cleanup efficiency. In particular, the scout team is faster and uses advanced cameras to provide an updated model of the waste distribution, while the cleaner team has basic cameras and is slower, as it carries the trash collection systems. Both teams must adapt their trajectories based on the information obtained in real time to maximize the total amount of trash collected, following the models obtained.

This approach emphasizes the ability to optimize cooperation between teams by designing a specific reward function that balances local and global objectives and a representation of the states that provides the agents with all the information they need. This chapter will focus on explaining and analyzing all the parties involved in the task, showing special attention to how cooperation between agents is established and executed. It will explore what are the mechanisms that allow agents from different teams to work in a coordinated way to achieve common goals and what role do state and reward functions play in the interaction between teams, evaluating how changes in them can influence the overall fleet performance.

Methodology

This section describes the approach used to address the problem presented in the previous work Barrionuevo et al. (2024b). Section "Heterogeneous Fleets of Autonomous Surface Vehicles" introduces the concept and advantages of employing heterogeneous fleets of ASVs for plastic waste collection in aquatic environments. Section "Informative Path Planning" explains the informative path planning (IPP) strategy, emphasizing the importance of information gain when guiding ASVs, which is essential to maximize efficient cooperation. Finally, section "Deep Reinforcement Learning" explores the application of DRL techniques, showing an overview of the different algorithms within this framework. In particular, the Q-learning algorithm is explained in detail, analyzing its mechanics to achieve efficient decision-making.

Heterogeneous Fleets of Autonomous Surface Vehicles

ASVs are a class of unmanned surface vehicles designed to operate in bodies of water, performing tasks autonomously or semi-autonomously. Figure 1 shows two types of ASV prototypes, with different designs and capacities, and equipped with various types of sensors and measuring devices. A heterogeneous ASV fleet refers to a group of these vehicles with different physical characteristics, operational capabilities, or functional purposes. Unlike homogeneous fleets, where all vehicles are identical, heterogeneous fleets consist of ASVs optimized for different but complementary tasks. This diversity allows for a more versatile, adaptable, and efficient approach to complex missions in aquatic environments. Due to the complexity of performing the intended tasks automatically, ASVs often rely on intelligent algorithms to navigate and perform their work efficiently. These algorithms allow them to process sensor data, make decisions, and adapt to changing environmental conditions.

In a heterogeneous fleet, differences may be evident in aspects such as size, speed, maneuverability, and energy autonomy. For example, smaller and faster ASVs may be ideal for initial exploration tasks, such as locating points of interest or identifying general patterns over a wide area. In contrast, larger and slower vehicles may be designed for heavier or longer tasks, such as collecting samples, transporting materials, or taking measurements with sensors. This physical and functional diversity allows the fleet to cover a wider range of needs and adapt to changing conditions more easily.

Another area of heterogeneity may lie in the technological capabilities of the vehicles. ASVs in a fleet may be equipped with different sensors and tools, depending on the mission objectives. For example, one vehicle might carry an advanced sonar system for mapping the seafloor, while another might be equipped with high-resolution cameras for visual inspection of floating objects or underwater structures. As data collected by different vehicles can be combined, the fleet can generate a more complete and diverse view of the environment, overcoming the limitations of individual vehicles. This data integration also enables more effective collaboration between vehicles, which can use this shared information to optimize their actions.

Moreover, a heterogeneous fleet not only benefits from specialization but also from functional redundancy. If a specialized vehicle is unavailable due to failure, other vehicles with similar, though not identical, capabilities can take over its role, ensuring that the mission can continue with minimal impact. In this sense, heterogeneity not only optimizes efficiency but also increases the robustness of the fleet and reduces the risk associated with relying entirely on a single vehicle type.

Informative Path Planning

Informative path planning (IPP) is the process of designing routes or trajectories for autonomous agents, such as robots, drones, or ASVs, with the specific goal of maximizing the information collected along their paths. According to published works Binney and Sukhatme (2012), Hitz et al. (2017), IPP is characterized by a lack of a priori knowledge about the environment, which may be total or partial. Unlike usual path planning methods, which focus on just reaching destinations or optimizing parameters such as time or energy, IPP prioritizes the collection of valuable data. This is especially useful in applications such as environmental monitoring, scientific exploration, and surveillance, where understanding the environment is a primary goal.

The primary goal of IPP is to determine where and how vehicles should move to obtain the most relevant information about an area, taking into account constraints such as energy, time, and sensor limitations. For example, an ASV equipped with measurement sensors for water quality parameters could use IPP to explore regions where contamination levels are expected to be high or to vary, ensuring that its trajectory maximizes the scientific information obtained about the water body (Peralta et al. 2021; Yanes Luis et al. 2024). This can be seen in Fig. 3, where a vehicle takes actions iteratively based on information acquired during the trip.

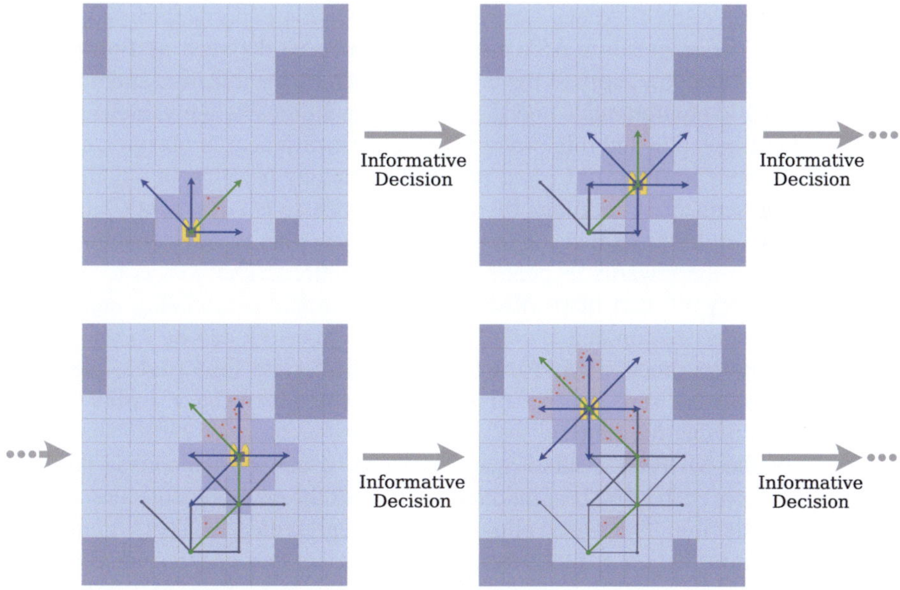

Fig. 3 Visual representation of the informative path planning process. At each step, the ASV iteratively evaluates the available options (blue arrows) based on the information currently known and selects the path that maximizes information gain. The chosen path is highlighted in green, while the alternative paths that were not selected are shown in gray

The problem becomes considerably more complicated when multiple vehicles are involved. In the case of a fleet of ASVs, each vehicle must be assigned complementary routes to avoid redundancy and ensure efficient use of resources. This chapter addresses heterogeneous vehicle fleets, which have varying capabilities, such as different speeds, sensor types or ranges, or carrying capacities. This makes path planning even more difficult, as all of these aspects must be considered when assigning tasks, and the individual strengths of each ASV must be leveraged to maximize information gathering. Moreover, the dynamic and uncertain nature of environments, such as changing currents, unpredictable weather conditions, or the movement of objects of interest, adds another layer of complexity to the planning process.

Deep Reinforcement Learning

Given the complexity of the task to be solved, planning these paths is not a trivial task, and it is necessary to find optimization algorithms with the potential to handle such large amounts of information. Although several algorithms have been tested for IPP, such as rapidly exploring random tree star (RRT*) (Schmid et al. 2020), Bayesian optimization (Peralta et al. 2023), or particle swarm optimization (PSO) (Kathen et al. 2023), among others, the use of deep reinforcement learning (DRL) has proven to be particularly effective for solving the challenges inherent in IPP, as seen in works such as Yanes Luis et al. (2024). This is particularly noticeable in the case of heterogeneous ASV fleets, as seen in works such as Barrionuevo et al. (2024a). The complexity of the problem, driven by the dynamic and uncertain nature of aquatic environments and the diverse capabilities of the vehicles, makes traditional optimization and rule-based approaches insufficient in many cases.

Reinforcement learning (RL) is a branch of machine learning in which agents learn to make sequential decisions by interacting with an environment (Sutton and Barto 2018). Similar to humans, agents learn on their own through trial and error, based solely on the rewards or penalties of their actions. The goal is to discover an optimal policy π^* that maps observations to actions, maximizing cumulative rewards over time (Sutton and Barto 2018). The framework can be formulated as a Markov decision process (MDP), where at each discrete time step t, the agent observes the current state of the environment, which may be a partial or full representation of the true state s.

To adapt the MDP to a multi-agent scenario, it can be defined as partially observable where each agent n individually perceives the environment using its own observation function. The observation function, represented as $O(s)$, transforms the raw state information s into a format that an agent can understand. Based on this observation, an agent selects an action a from a set of possible actions A. The environment then transitions to a new state s', and the agent receives a scalar feedback value, the reward r, which reflects how positive the action chosen in that context has been for the final objective.

The inherent limitation of traditional reinforcement learning, which relies on simple tables or models to represent value functions and policies, becomes evident when facing problems with high-dimensional state spaces and actions. This restriction has motivated the introduction of deep reinforcement learning (DRL) (Mnih et al. 2015). Using deep neural networks (DNNs), DRL enables a much more flexible and powerful representation of value and policy functions, which facilitates the resolution of complex and large-scale problems.

Within the field of DRL, there is a wide variety of algorithms designed to address different types of environments, problem structures, and computational constraints. Most DRL algorithms are based on a model-free approach (Huang 2020; Otto 2021), in which agents directly learn to optimize a policy or value function from interactions with the environment, without the need to explicitly know or model the dynamics of the environment, such as transitions or rewards. This approach is widely used due to its simplicity and versatility, as it allows addressing complex and unknown problems without requiring an explicit model of the environment, especially effective in certain domains. Through repeated interactions, model-free methods use environment exploration strategies, such as epsilon-greedy or Gaussian noise (Houthooft et al. 2017). This provides a balance between discovering new high-value actions and exploiting what is known to yield high rewards. The reward guides the learning process to adjust the policy, allowing the agent to associate certain actions with favorable objectives. The policy of the agent gradually approaches the optimal policy π^*, which specifies the best action to take in any given state to maximize the expected cumulative reward. A schematic representation of the DRL process is shown in Fig. 4, which highlights the interaction loop between the agents and the environment.

In general, model-free DRL algorithms can be divided into three main categories (Ding et al. 2020): value-based, policy-based and actor-critic methods. Value-based

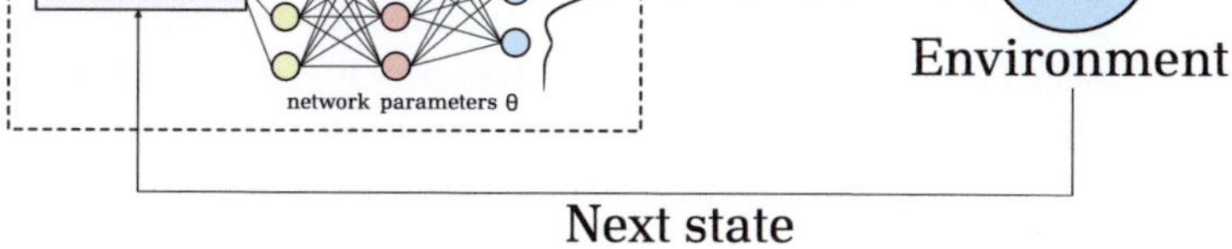

Fig. 4 Illustration of the basic principle of deep reinforcement learning process. Agents interact with the environment through state observation by taking actions based on their current policy, receiving feedback in the form of rewards, and transitioning to new states. The policy, represented as a DNN, is updated iteratively allowing the agents to learn and optimize their behavior over time

methods focus on estimating the value of states or state-action pairs. These methods aim to learn a value function, such as the action-value function $Q(s, a)$, which represents the expected cumulative reward from a given state-action pair. The agent obtains its policy indirectly by selecting actions that maximize these values. Examples include deep Q-networks (DQN) (Mnih et al. 2015) and variants such as double DQN (Van Hasselt et al. 2016) or dueling DQN (Wang et al. 2016). Policy-based methods take a different approach by directly optimizing the policy $\pi(a|s)$, which maps states to probabilities of selecting actions. Algorithms, such as proximal policy optimization (PPO) (Schulman et al. 2017), optimize the policy using techniques such as gradient ascent on the expected reward. Finally, actor-critic methods introduced in (Sutton et al. 1999), such as advantage actor-critic (A2C) further popularized by Mnih (2016), combine two networks, the "actor," which represents the policy, and the "critic," which estimates a value function to provide feedback to the actor. Each algorithm is suited to specific types of problems. For example, value-based methods excel in discrete action spaces and stability, while policy-based methods are better suited for high-dimensional and stochastic continuous action space. Actor-critic methods strike a balance and are often used in complex environments that combine aspects of both.

The diversity of DRL algorithms reflects the versatility of the framework, which enables solutions to a wide range of real-world challenges, from robotics to resource optimization in uncertain environments. In this case, due to the discrete nature of the scenario and the finite set of possible actions for the ASVs, value-based methods are well suited to optimize decision-making. More specifically, the cooperation of agents will be studied in this chapter with the dueling double deep Q-learning (DDDQL) algorithm (Wang et al. 2016), based on the DQN approach.

Dueling Double Deep Q-Learning

As mentioned, DQN (Mnih et al. 2015) is a reinforcement learning algorithm based on a value function called $Q(s, a)$, which measures the expected cumulative long-term reward of taking an action a in a state s and following the policy subsequently. At the beginning of training, the agent does not know which is the best action in each state. In DQN, the Q-function is represented by a DNN, initialized with random parameters. The goal of training is to approximate the optimal value function $Q^*(s, a)$ as best as possible through the DNN so that the agent can make near-optimal decisions in most states. The Bellman equation that defines the optimal value function can be expressed as a recursive relationship: the value of a decision at a given state depends on the immediate reward and the value of the optimal decisions in the subsequent states. Formally, the equation is written as:

$$Q^*(s, a) = r + \gamma \max_{a'} Q^*(s', a') \tag{1}$$

where r is the immediate reward received after taking action a in state s; $\gamma \in [0, 1)$ is the discount factor, which balances the importance future rewards; and s' and a' are the next state and next selected action, respectively.

To improve its estimation and gradually approach optimal behavior, a DQN agent must actively explore its environment. This involves performing random actions and observing the resulting rewards and states. From these experiences, it must learn how its actions affect the environment and associated rewards and what are the best decisions in the given circumstances. One of the main problems of DQN during training lies in the overestimation of the Q-value, which generates suboptimal policies and unstable updates.

For this reason, double DQN (Van Hasselt et al. 2016) is introduced, which seeks to mitigate the overestimation bias. The concept lies in the separation of the choice of the best action from the evaluation of that action. For this purpose, two estimators of the Q-function are introduced, represented by two neural networks. The first is the Q-network, responsible for selecting the best action based on current knowledge. The second is the target network, which serves as a stabilized reference by providing a delayed version of the Q-network. This provides stable estimates for the evaluation of the selected action. The parameters of the target network are updated less frequently, effectively decoupling the action selection from its evaluation and reducing the risk of feedback loops that can destabilize training. The technique of updating them is usually divided into two main methods. The first is periodic updating, where the target network parameters are copied directly from the main Q-network at fixed intervals, either every certain number of steps or after a certain time. The second method is Polyak averaging (Polyak 1987), which provides a smoother update. In this method, the target network parameters are updated as a weighted combination between the current parameters of the main Q-network and the previous parameters of the target network. This weighting, controlled by a smoothing factor, determines how fast the parameters of the target network are modified.

Since training requires a multitude of experiences to effectively approximate the optimal Q-function, it is essential to ensure that the data used for learning are both diverse and relevant. Standard experience replay buffers store transitions (s, a, r, s') uniformly and sample them randomly during training, which can lead to inefficiencies. Some transitions have little influence on the learning process, while others can provide much more information to update the model. The expected learning progress of experiences is measured by the magnitude of their temporal-difference (TD) error, which measures how informative a transition is for the agent. To address this, prioritized experience replay (PER) (Schaul et al. 2016) was introduced as an enhancement. In PER, as can be seen in Fig. 5, transitions are sampled based on their TD error magnitude, prioritizing experiences that the agent is struggling to predict accurately. By focusing more frequently on such transitions, the network can learn faster and more effectively from critical experiences, improving convergence. These transitions are collected as the agent interacts with the environment using an epsilon-greedy policy. By storing and reusing past experiences with the experience replay method, it prevents the agent from "forgetting" previous knowledge and allows for more stable and convergent learning. This policy is designed to balance exploration, trying random actions to discover new knowledge about the environment, and exploitation, choosing actions

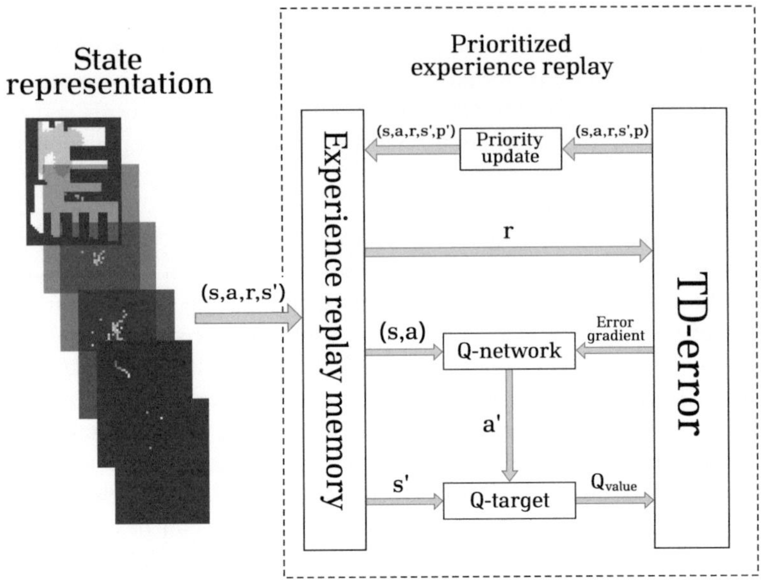

Fig. 5 Diagram of the process of storing and updating priorities with prioritized experience replay (Schaul et al. 2016) and double deep Q-learning. The experience memory stores each of the transitions, which consist of the observation of the state (s), the reward obtained in the transition (r), the action taken (a), and the next state (s'). Through the Q-network and Q-target, the TD error magnitude is calculated, which is used to estimate the priorities of each experience. Each time an experience is selected as a sample of the training batch, its priority is updated, giving more weight to the most informative transitions.

that maximize the predicted Q-value based on the current model. During the early stages of training, the agent heavily relies on random actions to ensure sufficient exploration. As training progresses and the model improves, the agent gradually shifts toward exploiting the learned policy by selecting actions with higher expected rewards.

During training, the parameters of the Q-network are iteratively updated using the experiences to minimize the discrepancy between the network prediction and the actual reward obtained, following:

$$
\begin{aligned}
Q(s, a; \theta) &\leftarrow Q(s, a; \theta) \\
&+ \alpha \left(r + \gamma \cdot Q^{target}(s', \underset{a'}{\text{argmax}}\, Q(s'; \theta); \theta^{-}) - Q(s, a; \theta) \right)
\end{aligned}
\tag{2}
$$

where θ are the trainable parameters of the Q-network that are updated by this update equation, θ^{-} are the frozen parameters of the target network that are updated less frequently, α is the learning rate, r is the immediate reward for taking action a in state s, and γ is the discount factor that weights future rewards. Thus, $Q(s, a, \theta)$ is the Q-network expected cumulative reward for taking action a in state s, while the term of Q^{target} refers to the expected cumulative reward computed by the target network for taking in the next state s' the best action a' according to the Q-network.

Therefore, to update the Q-network parameters, samples are drawn from the PER buffer to form training batches. By error backpropagation, updates are performed using stochastic gradient descent to minimize the loss function L, effectively reducing the Bellman error (Bellman 1957):

$$L(\theta) = \left[r + \gamma \cdot Q^{target}(s', \underset{a'}{\operatorname{argmax}}(Q(s'; \theta)); \theta^{-}) - Q(s, a; \theta) \right]^2 \qquad (3)$$

The loss function is constructed from the quadratic difference between the current estimate of Q-network and the target network. Reducing the loss, the function estimation gradually approaches the true expected reward.

Once double DQN is introduced, which mitigates the overestimation problem by separating the selection and evaluation of actions, it is possible to further optimize the DQN architecture. Despite the improvements brought by double DQN, an inherent challenge remains: not all actions are equally important in each state. In other words, not all states have the same importance in terms of the decision to be made by the agent. In some states, the differences between the available actions have a significant impact on future rewards, while in other states, most actions have similar effects, and what really matters is the value of the state itself.

To address this issue, dueling DQN arises (Wang et al. 2016), a variant that introduces a new way of estimating Q-function. The main idea behind dueling DQN, as seen in Fig. 6, is to decompose $Q(s, a)$ into two components: the value of the state $V(s)$ and the advantage of taking each action in that state $A(s, a)$. This separation allows the neural network to differentiate between the intrinsic importance of being in a given state and the advantages of choosing among the actions available in that state. Thereby, the network can learn to take actions that lead it to beneficial states, where the potential reward will be, regardless of the action to be taken.

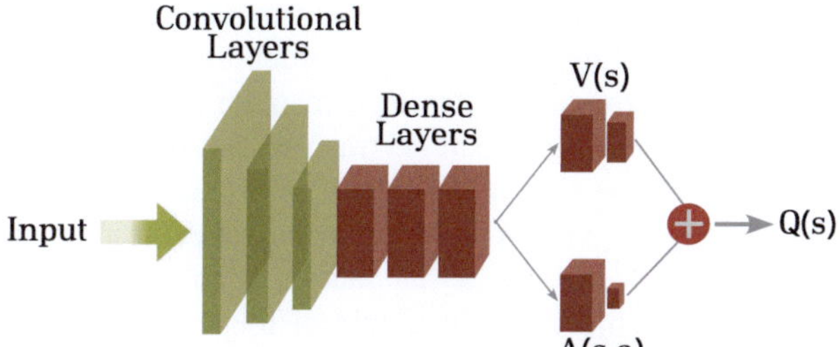

Fig. 6 Conceptual representation used to estimate Q-function in dueling DQN architecture. The architecture begins with an initial convolutional stage acting as a feature extractor, succeeded by a fully connected layer stage. The final stage corresponds to the dueling part, which is divided into two branches: one computes the advantage function $A(s, a)$, while the other evaluates the value function $V(s)$

Since there are many combinations of $V(s)$ and $A(s, a)$ that could produce the same Q-value, which can make learning unstable or ambiguous, a normalization process is used on the advantage component. Instead of using $A(s, a)$ directly, it is adjusted by subtracting its average over all possible actions $\hat{a}$ on a state s. Thus, the formula for $Q(s, a)$ in dueling learning is described as:

$$Q(s, a) = V(s) + \left(A(s, a) - \frac{1}{|\mathrm{A}|} \sum_{\hat{a} \in \mathrm{A}} A(s, \hat{a}) \right) \tag{4}$$

In dueling DQN, the neural network is divided into two branches after the initial shared layers, which correspond to each of the two components. This approach makes the architecture more stable and efficient during training, allowing a better representation of the relationships between states and actions.

Proposed Approach

DRL has been found to offer a powerful alternative to allow agents to learn behaviors by seeking optimal actions through interaction with their environment, making it a compelling solution for several reasons. This proposed approach builds on the work presented in (Barrionuevo et al. 2024b), where the focus will be on making modifications to the reward function and the observation model to investigate the key elements that drive effective cooperation between teams. By modifying these components, the aim is to understand their impact on the overall performance and efficiency of the heterogeneous ASV fleet.

The approach consists of a fleet of ASVs divided into two teams with two distinct roles, where the objective is to clean up plastic debris in aquatic environments. The first team, the scouting team, has high-quality detection cameras at a wide range and can reach high speeds of movement. This team will be in charge of obtaining and updating as best as possible a contamination model that shows the positions of the trash items on the map. The second team, the cleaning team, will be in charge of collecting the waste. In order to reduce costs, it does not incorporate high-quality cameras, so its range of trash detection is much smaller. In addition, by incorporating the mechanical trash collection system, it cannot reach the same speeds, so it will be able to cover half the space at the same time as the scouting team. The garbage elements in the environment exhibit dynamic movement patterns, which are influenced by external factors such as wind, water currents, or interactions with other objects. An example of these dynamics is illustrated in Fig. 7, where the trajectories and interactions of the garbage elements are visualized over time. These dynamics are modeled to reflect realistic scenarios, ensuring that the agent learns to handle complex and unpredictable environments. To coordinate the cooperation between these vehicles and optimize plastic waste collection, this research employs DRL. Specifically, a deep Q-learning approach is implemented using two DNNs, each shared among the agents within the same team. The proposed

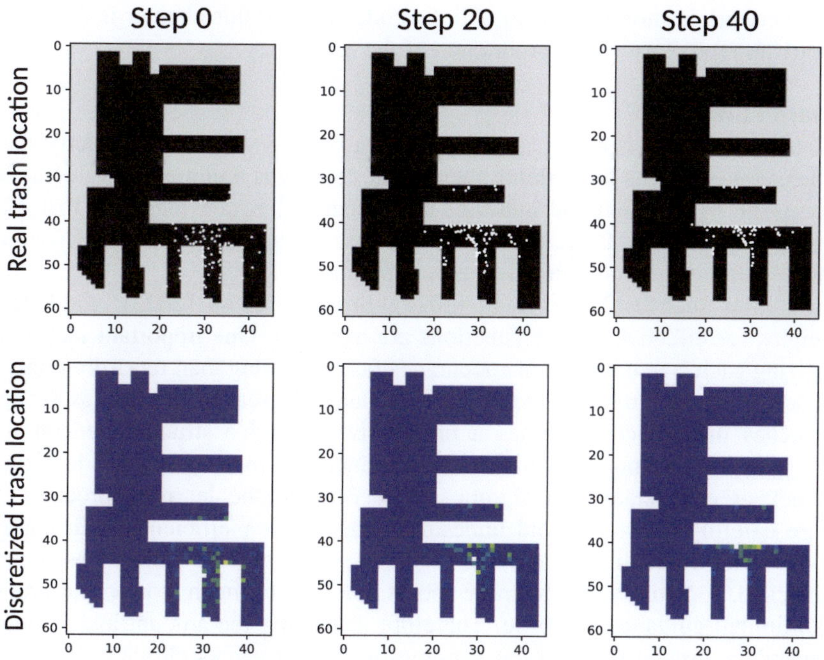

Fig. 7 Example of the movement dynamics of garbage elements in the environment over time. The top row shows the real-world positions of the garbage elements, represented as continuous spatial maps. The bottom row displays the corresponding discretized maps, which must be discovered by the agents to update the model and serve as input to the neural network. This discretization process enables efficient representation and processing of the environment's state, facilitating the agent's decision-making process

DNN architecture includes a first stage composed of three convolutional layers (64, 32, and 16 filters, respectively), inspired by the convolutional encoder originally proposed in (Mnih et al. 2015). This stage serves as a feature extractor, transforming the raw state observations into a representation of spatial features. The output of the three convolutional layers is fed into three fully connected layers, each with 256 units. These dense layers combine the features extracted by the convolutional layers, learning complex patterns and relationships between them. Finally, the architecture incorporates a dueling Q-network structure, which separates the learning process into two streams: the advantage and the value stream. The advantage stream consists of two fully connected layers with 64 and 8 units, respectively. The final layer outputs a vector of size 8, corresponding to the advantage values for each of the eight possible actions. The value stream comprises two fully connected layers with 64 and 1 unit, respectively. The final layer outputs a single scalar value representing the estimated value of the current state. The outputs of these two streams are then combined to produce the final Q-values using Eq. 4. A visual representation of this neural network is shown in Fig. 6. All the code used for this chapter is available in the GitHub repository[2] for reproduction of the results.

Reward Function

The reward function is a critical component in DRL systems, as it directly shapes the behavior of agents by defining their goals. It assigns a numerical value to each action taken by an agent and indicates how beneficial or detrimental is that action in the specific state. In DRL, this function is essential, since it is the only source of feedback for the agent to guide its learning process.

In studies such as Knox et al. (2023), Sowerby et al. (2022), good design principles for effective reward functions are explored. One important idea is that penalizing each step taken often encourages faster learning than rewarding only the final achievement of a goal. Additionally, gradually increasing rewards as the agent approaches the objective creates a progressive incentive structure, encouraging steady progress. Although dense rewards, which are those that provide frequent feedback at each state, can significantly accelerate the learning process, they require careful design to avoid undesired behavior or inefficiencies. In contrast, sparse rewards, which only provide feedback when specific milestones or goals are reached, sometimes can be more robust, but lead to much slower learning due to the limited guidance available. Therefore, balancing the amount and frequency of rewards is crucial to bring the learning process of agents closer to the desired outcomes. In the context of heterogeneous fleets of ASVs, the reward function must balance individual and team goals, ensuring that scouts and cleaners work in synergy to achieve the shared mission of locating and collecting plastic waste.

The reward function proposed in (Barrionuevo et al. 2024b) for each team is defined following:

$$
\begin{aligned}
R_C(s, a_n) &= c_\alpha \cdot r_{\alpha,n} + c_\delta \cdot r_{\delta,n} \\
R_S(s, a_n) &= c_\alpha \cdot r_{\alpha,n} + c_\beta \cdot r_{\beta,n} + c_\gamma \cdot r_{\gamma,n}
\end{aligned}
\tag{5}
$$

where $R_C(s, a_n)$ and $R_S(s, a_n)$ represent the reward functions for the cleaner and scout team, respectively. The parameters c_α, c_β, c_γ, and c_δ will be analyzed to determine their impact on effective learning during training. This work will focus on systematically adjusting these parameters to evaluate their influence on achieving the overall mission objectives.

Both teams share a common penalty term, $r_{\alpha,n}$, which is defined as the negative distance to the nearest known trash item for agent n, provided that the agent has not collected trash during that step. This penalty is expressed as:

$$
r_{\alpha,n} = -\min(\|p_n, B\|) \quad \text{if} \quad C(p_n, Y) > 0. \tag{6}
$$

[2] https://github.com/AMendB/HeterogeneousTrashCleanup/tree/chapterbook

where p_n is the position of the agent being evaluated, B is the set of positions of all known trash items, and $C(p_n, Y)$ is the function that indicates how much trash has the agent cleaned according to the distribution map Y. The trash distribution map is defined as a matrix Y of size $H \times W$ where each value $Y[i, j]$ indicates the number of trash items within the area corresponding to each position at each time step. This reward mechanism is designed to encourage agents to move toward trash items, ensuring that they prioritize reaching areas with waste. Additionally, it helps prevent agents from receiving minimal or negligible rewards when they are far from any trash, thereby reinforcing behaviors that contribute to waste collection or updating. For the cleaner team, an additional reward $r_{\delta,n}$ is provided based on the number of trash items collected by the agent at its current position, expressed as:

$$r_{\delta,n} = C(p_n, Y). \tag{7}$$

The scout team, on the other hand, receives two additional rewards aimed at balancing exploration and exploitation. To promote exploration during the early stages, a reward $r_{\beta,n}$ is given for covering previously unexplored areas:

$$r_{\beta,n} = \sum U^t - U^{t-1}. \tag{8}$$

where U is a matrix of size $H \times W$ updated at each time step t. Each value $U[i, j] = 1$ indicates that that positions have been covered at least once during the episode, and zero otherwise. This reward naturally decreases as the map becomes progressively explored, subsequently prioritizing the identification of high-contamination zones.

The second reward for the scout team, $r_{\gamma,n}$, is designed to incentivize improving the contamination model. It measures the updates made to the model within the area of vision of each vehicle and is calculated as the discrepancy between two consecutive model states:

$$r_{\gamma,n} = \sum_{ij} \frac{\hat{Y}_{ij}^t - \hat{Y}_{ij}^{t-1}}{\eta(v, \mathcal{P})}, \tag{9}$$

where $\hat{Y}$ is the model, represented as a matrix of size $H \times W$ that estimates the Y real distribution map, and $\eta(v, \mathcal{P})$ represents the number of vehicles p_n simultaneously covering a node v_{ij}.

The scenario used will be the map of a sports dock with intricate corridors outlined in (Barrionuevo et al. 2024b), which is represented by a matrix of $H = 62 \times W = 46$, as can be seen in Fig. 8. In that work, the ponderation parameters of the reward function were determined by trial and error. They are shown in Table 1. Next, a study will be carried out in which each of these parameters will be modified independently to see how they affect the final objective. When a parameter is modified, the other parameters remain constant as in the original work.

Fig. 8 Representation of the discretized scenario map used in this approach. Initial deployment positions are marked in blue

Table 1 List of original ponderation parameters of the reward function

Ponderation parameter	Value
c_α	1
c_δ	50
c_β	2
c_γ	1

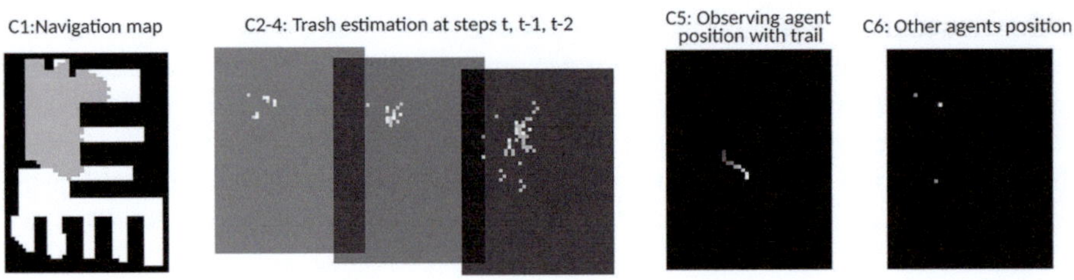

Fig. 9 Representation of all the six original channels that make up the state representation

Observation Function

The observation function is another essential cornerstone in DRL, as it defines the information available to the agents to make decisions. In the context of heterogeneous fleets of ASVs, the observation function must provide agents with sufficient data to coordinate effectively while minimizing irrelevant or redundant information that could hinder learning effectiveness.

In this approach, the state is represented as an image-like structure, which has demonstrated strong performance when combined with DNNs (Yanes Luis et al. 2024; Barrionuevo et al. 2024a). As shown in Fig. 9, this representation consists of six min-max normalized matrices of size $H = 62 \times W = 46$.

The first matrix represents the navigation map, where non-navigable positions are marked with a value of 0, while navigable positions are set to 1. Those navigable positions that have been visited at least once take a value of 0.5. This state is intended to provide agents with relevant spatial information about the accessibility of the environment, allowing them to differentiate between explored and unexplored areas. In this way, agents can make more informed decisions about their movement, avoiding redundancies in exploration and optimizing the coverage of the working area.

The next three matrices represent the estimated trash distribution over time, capturing the temporal evolution of the trash model. Specifically, these matrices store the predicted trash density at three consecutive time steps: the current time step t, the previous time step $t-1$, and two steps prior at $t-2$. By incorporating this temporal information, the intention is to enable agents to predict trends in the accumulation and movement of trash, allowing them to make more informed decisions about their cleanup strategy.

The fifth matrix is designed to track the recent path of the observing agent, providing a temporal reference of its movements. The last ten positions visited by the agent form a fading trail with values ranging from 0 to 1. The most recently visited position is assigned a value of 1, while previous positions gradually decrease in intensity, down to a small value at the tenth step. This fading effect enables the agent to retain short-term memory of its past movements, helping it avoid redundant exploration.

The last matrix encodes the positions of other agents within the environment, providing essential spatial awareness for coordination. Each position occupied by a scout agent is assigned a value of 0.5, while positions occupied by cleaner agents are set to 1. By incorporating this information into the state representation, agents can make more informed decisions based on the positioning of the other agents, also taking into account the roles to which they belong, facilitating better cooperation.

Experiments

This section will conduct a series of experiments to analyze the key elements in the proposed method. The reward function, which guides agent behavior, will be analyzed to identify how different weighting components influence the balance between individual team task optimization and overall mission success. Similarly, the observation model, which represents the information available to the agents at each decision step, will be evaluated to assess its role in enabling smooth coordination between teams. Through a series of experiments, this analysis aims to clarify how these factors contribute to maximizing fleet cooperation and achieving mission objectives.

As in the original work, the training has a duration of 60,000 episodes, where in each episode, each agent will be able to perform a maximum of 150 movements. In each episode, agents are deployed in four fixed positions on the map and can move in any of the eight cardinal directions in each step. The length of movement of scout agents is twice that of the cleaners, i.e., they go at twice the speed. In each of the episodes, the number of trash items, their positions, and the currents that move them are randomly generated using normal distributions. The prioritized

Table 2 List of hyperparameters for training DQL algorithm

Learning hyperparameter	Value
Training episodes	60,000
Training episodes	1×10^5
Learning rate	1×10^{-4}
Prioritized buffer replay size	1×10^6
Randomness of the sample in buffer	0.2
Bias compensating constant in buffer	0.6
Batch size	128
Target network hard update	6000
ϵ interval	$[1, 0.05]$
ϵ threshold	50%
Discount factor γ	0.99

experience replay is capable of storing up to 1×10^6 experiences, from which batches of 128 samples are drawn to train the neural network with a learning rate of 1×10^{-4}, 0.5 value for epsilon-greedy training, and 6000 steps to perform the hard update of the target network. The training parameters are summarized in Table 2. The selected policy is the one with the highest average performance in the metrics that are presented below. During the evaluation process, the average results were recorded over 100 distinct episodes for each configuration of the algorithm, which have never been seen during the training process.

Evaluation Metrics

To evaluate the performance of the proposed system, a set of evaluation metrics is employed to analyze both the efficiency of the cleaning process and the accuracy of the contamination model. The metrics will be the same as those used in the original work Barrionuevo et al. (2024b), which correspond to the following:

- **Percentage of trash cleaned (PTC):** This metric represents the proportion of trash successfully collected by the cleaner team relative to the total waste initially present in the environment. A higher percentage indicates better cleaning performance and a more effective strategy in coordinating between scout and cleaner teams.
- **Mean squared error (MSE) of the contamination model:** The MSE is used to evaluate the accuracy of the contamination model maintained by the scout team. It measures the difference between the predicted contamination map and the ground truth at each step. A lower MSE reflects a more accurate model, which is crucial for guiding the cleaning efforts effectively.

Together, these metrics provide a detailed understanding of the strengths and weaknesses of the system, allowing a thorough evaluation of modifications to the reward and observation functions. In addition, they help identify desired set conditions, such as whether to prioritize cleaning speed over final cleanliness or model accuracy.

Reward Function Experiments

To investigate the impact of reward function weights on performance, several modifications will be tested, focusing on how each component of the reward function influences the behavior of the scout and cleaner teams and whether how the changes affect the evaluation metrics. The reference reward function, based on the work of Barrionuevo et al. (2024b), includes components such as reward for trash items cleaned, penalty for distance to nearest trash, reward for discovering zones, or reward for model changes. The weights of each of these components determine how each of the tasks intended to be rewarded is prioritized. To analyze their influence, several variants of the original function will be studied. The performance metrics for comparison of these modifications will be the PTC and the MSE, as previously described.

? How Does c_α Impact on Overall Performance?

The parameter c_α is common to both teams and affects on the negative distance penalty to the nearest known trash item. What would happen if we increase or decrease that parameter?

In the original work, this parameter was set to $c_\alpha = 1$. Two new configurations have been tested, which correspond to ten times less and ten times more, such that $c_\gamma = 0.1$ and $c_\gamma = 10$. The metrics obtained with these configurations can be seen in Fig. 10. It can be seen how both new configurations significantly worsen the performance of the agents, being especially noticeable the low percentage of trash cleaned with $C_\alpha = 0.1$. It is therefore seen that the negative distance penalty to the nearest trash is an essential element to achieve good results, since at values close to zero, the performance is much lower. Despite this, an excessively high value such as $C_\alpha = 10$ also worsens the performance metrics, suggesting that too strong penalties can lead to suboptimal behaviors. In particular, with $C_\alpha = 10$, agents appear to excessively prioritize reducing the distance to known trash, rather than exploring new areas that might contain more trash. This may result in overfitting to initial areas of higher contamination, leaving other regions unexplored. In the original $C_\alpha = 1$ configuration, an equilibrium is achieved that allows agents to make decisions that bring them closer to known trash positions while retaining some flexibility to explore new areas. This is evidenced in the results of the cleanup percentage metric, which is significantly higher compared to the modified configurations, and in the MSE, as well as lower variability between episodes, as evidenced by the narrower confidence intervals.

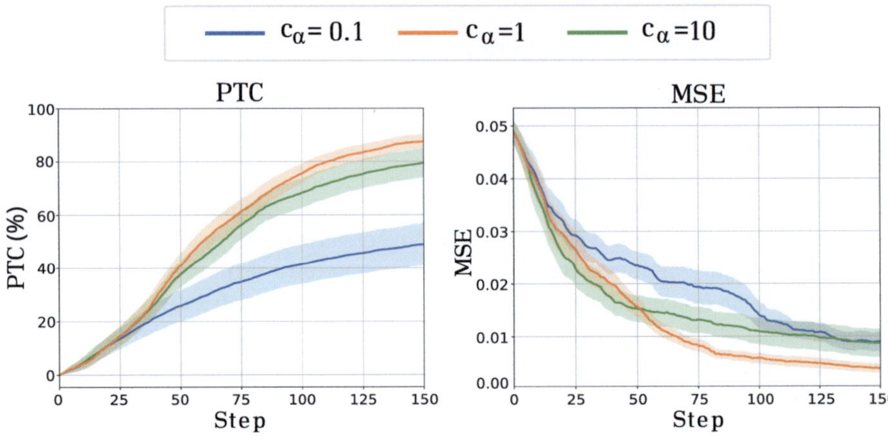

Fig. 10 Comparison of average performance over 100 evaluation episodes of PTC and MSE metrics for different configurations of parameter c_α. Confidence intervals are represented by the shaded regions

? How Does c_δ Impact on Overall Performance?

The c_δ parameter affects only the cleaner team. It is an enhancer of the reward per item cleaned, so its adjustment is essential. What would happen if this parameter is increased or decreased?

In the original work, this parameter was set to $c_\delta = 50$. Several new configurations have been tested, which correspond to ten, five, and two and a half times less and two times more. The metrics obtained with these configurations can be seen in Fig. 11. In them, it can be seen how the original performance of the algorithm in the PTC metric has been surpassed with the configuration $c_\delta = 10$, both in speed and in final cleaning percentage at $c_\delta = 50$. The performance in the MSE metric remains at very close values, although it is not the most important metric in the analysis of this parameter, since it is the parameter that rewards the trash collection and not the decrease of MSE in the model. It can be observed how the performance and tested values of the c_δ parameter do not follow a clear trend, with $c_\delta = 10$ being the only value that outperforms the original in PTC, while with $c_\delta = 20$ and $c_\delta = 5$, the metric worsens slightly. The $c_\delta = 100$ configuration stands out the most for its poor cleanup performance, which may be due to the fact that agents are exclusively prioritizing immediate trash collection, overlooking other essential aspects for efficient cooperation, such as strategic positioning or adequate coverage of the contaminated area. In contrast, the results obtained with $c_\delta = 5$ and $c_\delta = 20$, although slightly lower than the original value, indicate that small variations in the parameter do not drastically alter the performance. This could be because the overall design of the algorithm still allows for some adaptability in the

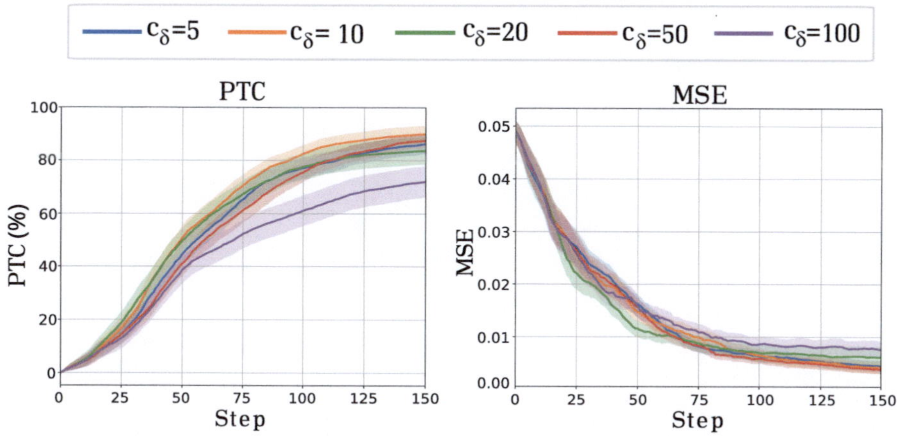

Fig. 11 Comparison of average performance over 100 evaluation episodes of PTC and MSE metrics for different configurations of parameter c_δ. Confidence intervals are represented by the shaded regions

face of moderate changes in reward. This underscores the importance of carefully tuning this parameter to maximize system performance.

? How Does c_β Impact on Overall Performance?

The parameter c_β is unique to the scout team and is a weighting of the initial exploration of each agent. That is, it weights the number of still unexplored nodes that are covered for the first time with the action taken. How do changes in this parameter affect the final result?

In the original work, this parameter was set to $c_\beta = 2$. Several new configurations have been tested, which correspond to ten times less and ten and twenty times more. The metrics obtained with these configurations can be seen in Fig. 12. In the graphs, it can be seen how the final cleaning percentage is very close with almost all configurations, except with $c_\beta = 0.2$, where the PTC metric drops noticeably with a larger confidence interval. This suggests that the cleaners are not very significantly affected by this parameter, which influences scanning in the scouts. It is only noticeably worse when the parameter is too low and there is no incentive to cover nodes not yet explored. In the MSE metric, on the other hand, greater differences can be observed. The worst result is obtained with the value of $c_\beta = 40$, which seems to indicate that the error in the model rises because agents focus excessively on exploring and covering unseen nodes instead of updating the model with the displacements of already known trash items.

For the values $c_\beta = 0.2$ and $c_\beta = 20$, the model error at the end of the episode takes very close values in both cases, but significant differences in the convergence

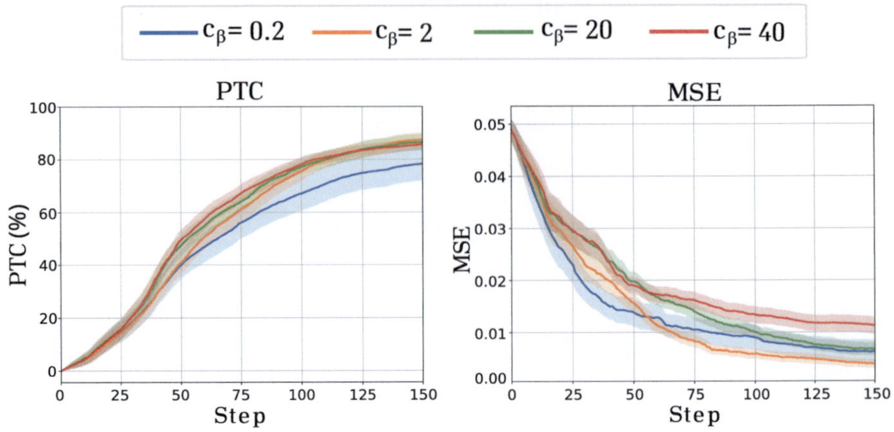

Fig. 12 Comparison of average performance over 100 evaluation episodes of PTC and MSE metrics for different configurations of parameter c_β. Confidence intervals are represented by the shaded regions

speed can be appreciated. With the value $c_\beta = 0.2$, the MSE takes values in step 50 that with $c_\beta = 20$ are not reached until steps close to 90. This indicates that with very low values of the parameter, since exploration is not enhanced, the agents focus on thoroughly characterizing the first contamination source they find, which causes the overall MSE of the model to drop rapidly. On the other hand, with higher values, the trash items take longer to be discovered and updated all over the map, but there is a greater exploration that can be useful in certain cases where the trash is very scattered. It is clear that the best option is an intermediate value between the two, since the original performance stands out with the value $c_\beta = 2$, which manages to obtain by far a lower MSE at the end of the episodes with a smaller confidence interval. Furthermore, the convergence of the model balances speed and a low final error, being also the parameter configuration that slightly gives the best result in the PTC metric. Finally, it is demonstrated that this reward is fundamental to obtain the final objective and an efficient cooperation between both teams, since with values close to zero, which would be the absence of reward, the scouts do not cover new nodes and the PTC of the cleaners decreases notably.

? How Does c_γ Impact on Overall Performance?

The parameter c_γ is unique to the scout team and is a weighting of the changes that each agent produces in the model. That is, it is a multiplier of the number of garbage elements it has updated in the model. How do changes in this parameter affect the final result?

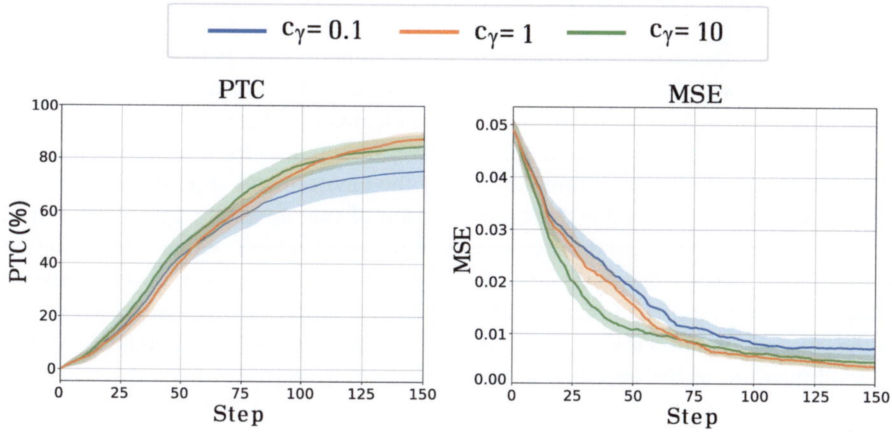

Fig. 13 Comparison of average performance over 100 evaluation episodes of PTC and MSE metrics for different configurations of parameter c_γ. Confidence intervals are represented by the shaded regions

In the original work, this parameter was set to $c_\gamma = 1$. Two new configurations have been tested, which correspond to ten times less and ten times more, such that $c_\gamma = 0.1$ and $c_\gamma = 10$. The metrics obtained with these configurations can be seen in Fig. 13. It can be seen that the worst results are clearly for $C_\gamma = 0.1$, obtaining much lower mean cleanup percentages than the other two configurations at the end of the episodes, as well as a higher MSE metric. In addition, its confidence interval is wider for both metrics, indicating greater variability between episodes. With a reduced weight for changes in the model, agents are not incentivized enough to meaningfully update the contamination model. This can lead to suboptimal behaviors, such as prioritizing actions with higher immediate rewards but less relevant to the overall goal (e.g., moving to already explored areas or performing non-useful actions). This explains the low PTC rate and higher MSE. In contrast, the other two configurations obtain more similar results at the end of the episode, with the configuration with $c_\gamma = 0.1$ standing out in its faster cleanup and especially in reducing the model error very early. Despite this, at the end of the episodes, the original configuration obtains slightly better results. This intermediate value seems to offer a better balance between incentives to update the pollution model moderately early and prioritizing actions that benefit the ultimate goal. A high value gives excessive weight to updates in the model, which may induce agents to prioritize actions that maximize this reward in the short term. This may be useful initially to reduce model error quickly, but may penalize longer-term behaviors more focused on identifying and guiding teams when there are few trash items left. As a result, the model improves faster at the beginning, but the overall results are not consistently better at the end of the episode.

Observation Function Experiments

To investigate the impact of the observation function on performance, several modifications will be tested, focusing on what information each part of the observation function provides to the scout and cleaner teams, and whether it aids and encourages cooperation. The reference observation function from work Barrionuevo et al. (2024b) includes data from the map of the environment and visited areas, the model in three different time steps, the position of the observing agent with a trail of its previous positions, and the positions of the other agents with a differentiation according to the team they belong to. To address how objectives and cooperation are achieved through information sharing via the observation function, several variants of the original function are studied. The performance metrics for the comparison of all modifications of the observation function can be seen in Fig. 14.

No Map Channel

This modification removes the environment map and covered areas from the observation function. Consequently, agents lose direct information about the areas they have traveled through, which may reduce their ability to effectively plan future paths.

However, as can be seen in Fig. 14, the performance metrics remain very close to those of the original algorithm, even significantly exceeding the convergence times in the PTC metric, although with a wider confidence interval. At the end of the episodes, the average percentage is slightly below, losing that advantage due to the initial speed with which it started. In MSE, something similar happens; it starts being the configuration with the shortest convergence time, but gradually worsens to end up being a configuration that does not stand out for its final average performance. This increase in convergence speed but final performance loss seems to indicate that, although the absence of the initial map channel encourages a more dynamic and aggressive exploration in the early stages, the lack of information on previously covered areas hinders the optimization of trajectories toward the end of the episodes. Without the spatial reference provided by the map, agents find it more difficult to identify areas of contamination not yet covered, leading to a deterioration in the final average results, both in PTC and MSE. This seems to be driving exploration behavior that is less dependent on visit history, causing agents to focus more on their immediate environment and less on long-term planning. This may be advantageous at the beginning, when the environment is not yet well covered and there is a high probability of finding new contaminated areas. However, as the episode progresses and areas of contamination become sparser, the lack of accumulated information limits the ability of agents to direct their movements to locations of greatest need. This may explain why, although the configuration excels in the early stages, its final average performance is negatively affected.

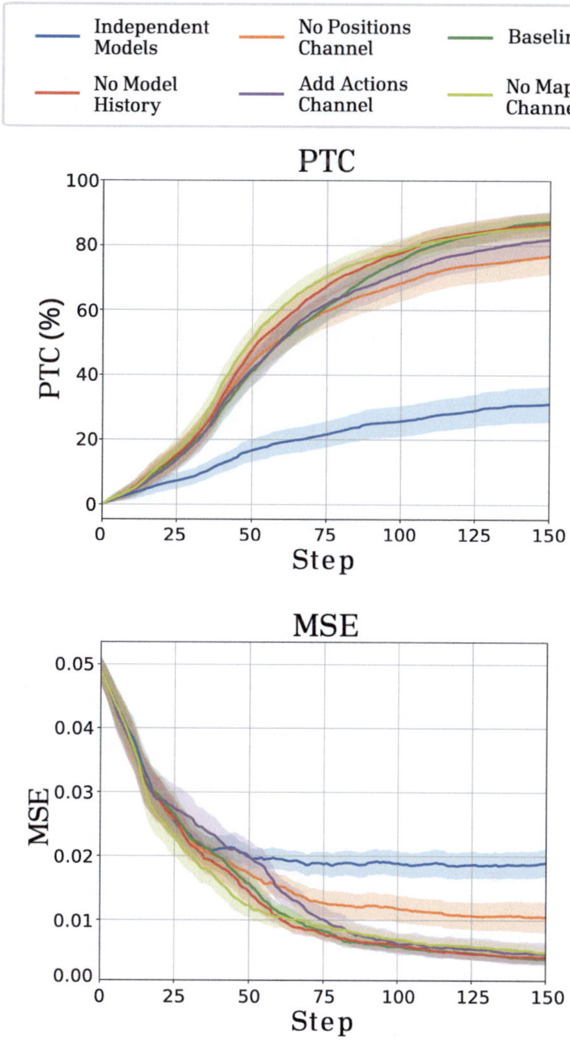

Fig. 14 Comparison of average performance over 100 evaluation episodes of PTC and MSE metrics for different configurations of observation function. Confidence intervals are represented by the shaded regions

Independent Models

In this configuration, the model sharing mechanism is disabled and each team maintains its own separate contamination model. No acquired information is shared between teams, only the positions. This modification examines how important the trash map provided by the scouts is to the cleaners, and whether sharing a unified model is essential for effective cooperation between teams.

The results depicted in the graphs in Fig. 14 clearly show that this observation function produces the worst performance on all metrics by a significant margin. This indicates that the lack of a shared contamination model severely hinders inter-team cooperation, as the cleaners cannot know the trash positions on their own and do not take advantage of the information collected by the scouts. This demonstrates emphatically that the observation function is the nexus of cooperation between the two teams and that this cooperation exists in the original work and is the key to the success of the project. In particular, the PTC metric shows a pronounced decline, highlighting that cleaners have difficulty locating and prioritizing areas of high contamination without information from scouts. Similarly, the MSE metric remains remarkably high, reflecting a significant inability of the scouts to update and refine the overall contamination model. This can be attributed to the fact that, as the cleaners collect very few trash items, the amount of waste present in the environment remains high for much of the episodes. Such can be seen by the fact that the MSE metric is very similar and drops in all configurations until about step 40, when the difference in trash collection is still not as drastic. This scenario makes the task of the scouts more difficult, as they must handle a model that becomes more complex to update and keep accurate due to the abundance of data and changes in the environment.

The lack of synchronization between the detection of new contaminated areas and the ability of the cleaners to reduce the level of contamination generates a negative cumulative effect on the quality of the overall model. These results highlight the critical role of sharing models to encourage cooperation among heterogeneous teams. Without shared information, scouts and cleaners act in isolation, resulting in ineffective and poorly coordinated strategies.

No Positions Channel

In this scenario, agents do not have access to the positions of their teammates or the opposing team. This test evaluates how critical it is for agents to have spatial awareness of the other agents in order to coordinate their actions effectively and avoid overlaps in their tasks. Figure 14 shows how the metrics significantly worsen to the base case, both in PTC and MSE. This deterioration in performance is accompanied by a wider confidence interval in both metrics, indicating higher variability in the results obtained between the different episodes evaluated.

The absence of information on the positions of other agents significantly compromises the ability of teams to coordinate effectively. Scouts, who lack knowledge about areas already covered by their teammates, may become redundant in exploring previously covered areas, reducing the overall efficiency of the team and making it difficult to explore new areas. This lack of spatial representation of their partners makes strategic task sharing difficult, resulting in less consistent updates and a less accurate and reliable contamination model.

On the other hand, cleaners are also negatively affected by model deterioration, as they rely heavily on the updated information provided by scouts, as seen in . Furthermore, without the reference of scout positions, cleaners lose a key indicator to identify potentially contaminated areas, which limits their ability to prioritize tasks and optimize their performance.

In general, this disconnection between agents and the overlapping of tasks deteriorates the strategic planning of the cooperative system, especially in complex episodes. This experiment highlights the importance of sharing as much information as possible between fleets, as long as it contributes to coordination between them. In this case, position information is an essential component to ensure effective cooperation and reduce redundancies in the tasks assigned to the teams.

No Model History

This modification excludes the history of the three-channel contamination model, providing only the current state of the contamination model. The objective is to evaluate whether having temporal information about the model significantly influences decision-making and cooperative strategies. Figure 14 shows the result obtained during the evaluation process of this observation function, where a great performance in both metrics can be observed. Very close results are obtained to that of the initial configuration, even increasing the convergence times in the PTC metric, where the percentage rises faster, although at the end of the episodes, this improvement is diminished and ends up being surpassed by the original work. Additionally, in the PTC metric, the confidence interval is also larger, especially in the final section of steps, which indicates that there are episodes in which it can behave better than the initial work, but there is inconsistency in this improvement and the average ends up worsening. Meanwhile, the MSE also improves initially in convergence time to the original work, although to a lesser extent than in the cleaning metric. When they are reaching the end of the episodes, the differences are minimal, even in the confidence interval.

From these results, it can be deduced that the omission of the model history has a differentiated impact depending on the stage of the episode and the team evaluated. During the initial phases, scouts manage to build an effective contamination model without having the movement history, allowing a competitive performance and even making cleaners improve notably in convergence times in the PTC metric. However, this initial advantage is not maintained toward the end of the episodes, in which the lack of historical information seems to penalize, especially the cleaners. The analysis suggests that cleaners rely more on model history to identify and prioritize cleanup areas when trash items are less abundant and more dispersed. This dependence is reflected in the more pronounced drop in the PTC metric toward the end, contrasting with the more stable performance of the explorers in the MSE metric. In addition, the observed inconsistency in the confidence intervals of the PTC metric could indicate that, in some episodes, the no-history strategy achieves superior results, but its overall unreliability affects the average.

It is difficult to draw a final conclusion with these results as to which option is better, and depending on the objectives, one may be more effective than the other. While omitting model history may speed up initial progress in cleaning and model building, its absence is detrimental in the long run, especially in scenarios where element identification is more complicated. If the goal is to obtain the highest final cleanup rate regardless of time, this analysis reinforces the importance of having historical information to ensure balanced and consistent performance on both metrics. On the other hand, if a slight decrease in the final cleanup percentage is accepted in exchange for faster cleanup for the remainder of the episode, it appears that the option without model history is a better choice. This may be because the network is not able to clearly deduce what this history is providing, and it is consuming network resources without being absolutely essential information.

Add Actions Channel

This modification incorporates an additional channel containing the latest actions taken by all agents. The objective is to provide a temporal context to the decision-making process, allowing agents to anticipate the behavior of their teammates and adapt their actions accordingly. By integrating this historical data into the observation function, the aim is to improve the efficiency of cooperation and trajectory optimization across the fleet.

The results obtained with this modification observed in Fig. 14 indicate a deterioration of metrics and confidence intervals, mainly in PTC. This deterioration could be due to information overload in the decision-making process, especially if agents fail to effectively interpret historical action data. Instead of facilitating cooperation and trajectory optimization, this additional information seems to introduce noise or unnecessary complexity in the evaluation of states.

The PTC metric shows a decrease in both the percentage of final cleanup achieved and inter-episode consistency, reflected in the wider confidence intervals. This suggests that agents are not taking advantage of the additional channel to coordinate effectively and may be making suboptimal decisions in some situations. Furthermore, the MSE metric also shows a lower performance, especially in convergence times, which could indicate that the integration of historical data of actions does not contribute significantly to the construction of a more accurate model. On the contrary, it could be hindering the identification of clear patterns in contamination, by biasing the focus toward past behaviors of the agents instead of the current conditions of the environment. However, despite the decrease in PTC and the consequent difficulty to maintain an updated model, the negative impact on the MSE is significantly lower at the end of the episode.

Overall, these results suggest that, although the idea of providing a temporal context may seem promising, its implementation in this case has not had the expected effect. This could be related to the way agents process information or to the lack of an explicit mechanism linking past actions to current goals. Therefore, it would be necessary to explore other adjustments in the design of this channel or in the decision algorithms to evaluate whether it is possible to extract benefits from this additional information. To conclude, a summary of the results of the evaluation metrics for each of the observation function tests is shown in Table 3, as well as a summary of the analysis of the impact of the changes in Table 4.

Table 3 Metric comparison between different configurations of observation functions for DDDQL training. The evaluation set consists of 100 episodes over which the mean is calculated. The highlighted values correspond to the best performance value

	PTC		MSE	
Observation function	Mean	CI 95%	Mean	CI 95%
Baseline	**87.41**	2.74	**0.0038**	0.0009
Independent models	31.11	5.45	0.0190	0.0021
No positions channels	76.84	5.71	0.0105	0.0024
No model history	86.79	3.67	0.0042	0.0010
Add actions channel	81.95	5.66	0.0048	0.0018
No map channel	85.95	3.71	0.0049	0.0010

Conclusions

This chapter has explored methodologies, experiments, and results around the optimization of heterogeneous ASV fleets using DRL techniques. Through the analysis of the impact of different reward functions, observation function configurations, and parameter settings, some key factors influencing the performance and cooperation of ASV teams have been discovered. The results contribute to the understanding of effective strategies for collaborative autonomous systems.

The analysis of the reward function parameters has provided significant information on the incentives needed to meet the desired objectives. Proper parameter tuning is essential to encourage agents to prioritize the actions with the greatest impact, either by effectively cleaning up contaminated areas or by accurately updating the contamination model.

One of the key findings is the critical role that inter-team cooperation and information sharing play in completing tasks efficiently and effectively. The experiments highlight the importance of maintaining a unified contamination model and sharing relevant positional and temporal data between agents. Configurations that restrict or eliminate these interactions show a significant decrease in performance, underscoring the need for coordinated strategies in multiagent systems.

The correct design of the observation function has proven to be essential and has a profound effect on agent behavior. Despite this, the power of adaptability of the DRL has been seen, for example, with the removal of specific channels such as the map or positional data from other vehicles. The agents are able to adapt to overcome the lack of information, obtaining results not too far from the original work despite having less data available.

Finally, the experiments showed that it is difficult to trade off short-term gains against long-term performance. Configurations that emphasized fast convergence tended to have difficulty maintaining their advantage in later phases of the task, highlighting the need for algorithms that balance initial exploration with sustained exploitation of collected information. A future work may focus in a dynamic reward function, which adjusts the rewards given as a function of the number of steps taken. This could ensure that the initial advantage gained by fast convergence configurations does not decay toward the end of the episode. This work focuses on validating the effectiveness of the model in a controlled and simulated environment

Table 4 Summary table of the impact analyzed for each of the changes made to the observation function

Observation function	Objective	Positive impact	Negative impact
Baseline	Provide a balanced set of environmental, spatial, and temporal data to facilitate coordination between scouts and cleaners	Enables robust decision-making, thanks to rich information, achieving the best overall results in both performance metrics	A larger number of channels may lead to increased computational cost and processing requirements
Independent models	Disable model sharing so that each team gets its own contamination model, sharing only positional information	Not observed	Results in poor coordination and significantly degraded performance in all metrics, due to the loss of a shared contamination model
No positions channel	Removes the channel of the observation function that indicates the positions of the other agents	Not observed	Severely limits spatial awareness, leading to inefficient exploration and task allocation, resulting in overall reduced performance
No model history	Eliminate the historical evolution of the contamination model, providing only the current state	Accelerates initial convergence by reducing the noise from historical data, allowing faster early performance improvements	Over time, the lack of temporal context leads to less accurate long-term model updates and suboptimal coordination, as agents lose valuable trend information
Add actions channel	Incorporate an additional channel containing the latest actions taken by all agents to provide additional context	Not observed	The agents are not able to interpret this information, or it does not contribute anything to the decision-making process, introducing noise. The results worsen and increase the complexity of the observation function
No map channel	Remove the environmental map and information on covered areas from the observation function	Promotes a more dynamic and aggressive exploration strategy in the early stage, speeding up initial convergence in metrics	The lack of spatial reference provided by the map makes it more difficult for agents to identify areas of contamination not yet covered, resulting in a deterioration of the final average results

to isolate and analyze the impact of each modification on the observation and reward function. However, another aspect to be addressed in future work is adaptation to real-world conditions, where factors such as uncertainty in the environment, sensor failures, or variations in weather conditions can affect performance.

References

Akib A, Tasnim F, Biswas D, Hashem MB, Rahman K, Bhattacharjee A, Fattah SA (2019) Unmanned floating waste collecting robot. In: TENCON 2019—2019 IEEE region 10 conference (TENCON), pp 2645–2650

Barrionuevo AM, Yanes Luis S, Reina DG, Marín SLT (2024a) Informative deep reinforcement path planning for heterogeneous autonomous surface vehicles in large water resources. IEEE Access 12:71835–71852

Barrionuevo AM, Yanes Luis S, Reina DG, Marín SLT (2024b) Optimizing plastic waste collection in water bodies using heterogeneous autonomous surface vehicles with deep reinforcement learning

Bellman R (1957) Dynamic programming, 1st edn. Princeton University Press, Princeton University

Binney J, Sukhatme GS (2012) Branch and bound for informative path planning. In: 2012 IEEE international conference on robotics and automation. IEEE, Piscataway, pp 2147–2154

Díaz LM, Yanes Luis S, Barrionuevo AM, Diop DS, Perales M, Casado A, Toral S, Gutiérrez D (2024) Towards an autonomous surface vehicle prototype for artificial intelligence applications of water quality monitoring

Ding Z, Huang Y, Yuan H, Dong H (2020) Introduction to reinforcement learning. In: Deep reinforcement learning: fundamentals, research and applications, pp 47–123

Hitz G, Galceran E, Garneau MÈ, Pomerleau F, Siegwart R (2017) Adaptive continuous-space informative path planning for online environmental monitoring. J Field Robot 34(8):1427–1449

Houthooft R, Chen X, Duan Y, Schulman J, De Turck F, Abbeel P (2017) Vime: Variational information maximizing exploration

Huang Q (2020) Model-based or model-free, a review of approaches in reinforcement learning. In: 2020 international conference on computing and data science (CDS), pp 219–221

Kamarudin NA, Nordin IN, Misman D, Noh FH, Khamis N, Razif MR (2021) Development of water surface mobile garbage collector robot. Alinteri J Agric Sci 36:534–540

Kathen MJ, Samaniego FP, Flores IJ, Reina DG (2023) Aquahet-PSO: an informative path planner for a fleet of autonomous surface vehicles with heterogeneous sensing capabilities based on multi-objective PSO. IEEE Access 11:110943–110966

Knox WB, Allievi A, Banzhaf H, Schmitt F, Stone P (2023) Reward (mis)design for autonomous driving. Artif Intell 316:103829

Mnih V (2016) Asynchronous methods for deep reinforcement learning. arXiv preprint arXiv:1602.01783

Mnih V, Kavukcuoglu K, Silver D, Rusu AA, Veness J, Bellemare MG, Graves A, Riedmiller M, Fidjeland AK, Ostrovski G et al (2015) Human-level control through deep reinforcement learning. Nature 518(7540):529–533

Otto F (2021) Model-free deep reinforcement learning—algorithms and applications. In: Reinforcement learning algorithms: analysis and applications, pp 109–121

Peralta F, Yanes S, Reina DG, Toral S (2021) Monitoring water resources through a Bayesian optimization-based approach using multiple surface vehicles: the Ypacarai Lake case study. In: 2021 IEEE congress on evolutionary computation (CEC). IEEE, Piscataway, pp 1511–1518

Peralta F, Reina DG, Toral S (2023) Water quality online modeling using multi-objective and multi-agent Bayesian optimization with region partitioning. Mechatronics 91:102953

Polyak BT (1987) Introduction to optimization. Optimization Software, New York

Schaul T, Quan J, Antonoglou I, Silver D (2016) Prioritized experience replay. In: 4th international conference on learning representations (ICLR)

Schmid L, Pantic M, Khanna R, Ott L, Siegwart R, Nieto J (2020) An efficient sampling-based method for online informative path planning in unknown environments. IEEE Robot Autom Lett 5(2):1500–1507

Schulman J, Wolski F, Dhariwal P, Radford A, Klimov O (2017) Proximal policy optimization algorithms. CoRR, abs/1707.06347

Shang Y, Yu W, Zeng G, Li H, Wu Y (2024) Stereoyolo: a stereo vision-based method for maritime object recognition and localization. J Marine Sci Eng 12(1):197

Sowerby H, Zhou Z, Littman ML (2022) Designing rewards for fast learning. arXiv preprint arXiv:2205.15400

Sutton RS, Barto AG (2018) Reinforcement Learning, 2nd edn. An Introduction. MIT Press, Cambridge

Sutton RS, McAllester D, Singh S, Mansour Y (1999) Policy gradient methods for reinforcement learning with function approximation. In: Advances in neural information processing systems, vol. 12

Van Hasselt H, Guez A, Silver D (2016) Deep reinforcement learning with double q-learning. In: Proceedings of the AAAI conference on artificial intelligence, vol. 30

Wang Z, Schaul T, Hessel M, Hasselt H, Lanctot M, de Freitas N (2016) Dueling network architectures for deep reinforcement learning. In: 33rd international conference on machine learning (ICML), pp 1995–2003

World Economic Forum, Ellen MacArthur Foundation and McKinsey & Company (2016) The new plastics economy: rethinking the future of plastics

Yanes Luis S, Shutin D, Gómez JM, Reina DG, Toral SL (2024) Deep reinforcement multiagent learning framework for information gathering with local gaussian processes for water monitoring. Adv Intell Syst 6(8):2300850

MIX
Papier aus verantwortungsvollen Quellen
Paper from responsible sources
FSC® C105338

If you have any concerns about our products,
you can contact us on
ProductSafety@springernature.com

In case Publisher is established outside the EU,
the EU authorized representative is:
Springer Nature Customer Service Center GmbH
Europaplatz 3, 69115 Heidelberg, Germany

Printed by Libri Plureos GmbH
in Hamburg, Germany